Zinc Oxide Materials for Electronic and Optoelectronic Device Applications

Wiley Series in Materials for Electronic and Optoelectronic Applications

www.wiley.com/go/meoa

Series Editors

Dr Peter Capper, SELEX Galileo Infrared Ltd, Southampton, UK

Professor Safa Kasap, University of Saskatchewan, Canada

Professor Arthur Willoughby, University of Southampton, UK

Published Titles

Bulk Crystal Growth of Electronic, Optical and Optoelectronic Materials, Edited by P. Capper

Properties of Group-IV, III–V and II–VI Semiconductors, S. Adachi

Charge Transport in Disordered Solids with Applications in Electronics, Edited by S. Baranovski

Optical Properties of Condensed Matter and Applications, Edited by J. Singh

Thin Film Solar Cells: Fabrication, Characterization and Applications, Edited by J. Poortmans and V. Arkhipov

Dielectric Films for Advanced Microelectronics, Edited by M. R. Baklanov, M. Green and K. Maex

Liquid Phase Epitaxy of Electronic, Optical and Optoelectronic Materials, Edited by P. Capper and M. Mauk

Molecular Electronics: From Principles to Practice, M. Petty

Luminescent Materials and Applications, Edited by A. Kitai

CVD Diamond for Electronic Devices and Sensors, Edited by R. S. Sussmann

Properties of Semiconductor Alloys: Group-IV, III–V and II–VI Semiconductors, S. Adachi

Mercury Cadmium Telluride: Growth, Properties and Applications, Edited by P. Capper and J. Garland

Forthcoming Titles

Principles of Solar Cells, LEDs and Diodes: The Role of the PN Junction, A. Kitai

Silicon Photonics: Fundamentals and Devices, M. J. Deen and P. K. Basu

Photovoltaic Materials: From Crystalline Silicon to Third-Generation Approaches, Edited by G. J. Conibeer

Inorganic Glasses for Photonics: Fundamentals, Engineering and Applications, A. Jha, R. M. Almeida, M. C. Goncalves and P. G. Kazansky

Lead-Free Solders: Materials Reliability for Electronics, Edited by K.N. Subramanian

Zinc Oxide Materials for Electronic and Optoelectronic Device Applications

Edited by

COLE W. LITTON

Air Force Research Laboratory, Ohio, USA

DONALD C. REYNOLDS

Wright State University, Ohio, USA

THOMAS C. COLLINS

Oklahoma State University, Oklahoma, USA

WILEY

A John Wiley and Sons, Ltd, Publication

This edition first published 2011

Registered office
John Wiley & Sons Ltd, The Atrium, Southern Gate, Chichester, West Sussex, PO19 8SQ, United Kingdom

For details of our global editorial offices, for customer services and for information about how to apply for permission to reuse the copyright material in this book please see our website at www.wiley.com.

Library of Congress Cataloging-in-Publication Data

Zinc oxide materials for electronic and optoelectronic device applications /
edited by Cole W. Litton, Donald C. Reynolds, Thomas C. Collins.
p. cm. — (Wiley series in materials for electronic & optoelectronic
applications ; 35)
Includes bibliographical references and index.
ISBN 978-0-470-51971-4 (hardback)
1. Zinc oxide. 2. Electronic apparatus and appliances–Materials. 3.
Optoelectronic devices–Materials. I. Litton, Cole W. II. Reynolds, Donald C.
III. Collins, Thomas C., 1936- joint author.
TK7871.15.Z56Z56 2011
669′.52—dc22

2011000564

A catalogue record for this book is available from the British Library.

Print ISBN: 9780470519714
ePDF ISBN: 9781119991045
oBook ISBN: 9781119991038
ePub ISBN: 9781119991212

Set in 10/12pt by Times Roman, Thomson Digital, Noida, India

Cover image reprinted from F. Yun, M. A. Reshchikov, L. He, T. King, D. Huang, H. Morkoç, J. Nause, G. Cantwell, H. P. Maruska and C.W. Litton, in *Defect and Impurity Engineered Semiconductors and Devices III*, edited by S. Ashok, J. Chevallier, N.M. Johnson, B.L. Sopori and H. Okushi (Mater. Res. Soc. Symp. Proc. Volume 719, Warrendale, PA, 2002), F8.21. Reprinted with permission from MRS.

This book is dedicated to the memory of Cole W. Litton who was the driving force behind it and the lead editor. Cole passed away before its completion.

In Memoriam: Cole Litton

Cole W. Litton, the editor and compiler of *Zinc Oxide Materials for Electronic and Optoelectronic Device Applications*, died of a heart attack on Tuesday, January 26, 2010, while attending the SPIE Photonics West Conference in San Francisco.

Cole was a native of Memphis, Tennessee, born in 1930, and he attended the University of Tennessee graduating with a bachelor's degree. He served for four years as an officer in the US Air Force and then joined the Air Force Research Laboratory as a civilian scientist at Wright Patterson Air Force Base in Dayton, Ohio. There he worked on the solid-state physics team of Don Reynolds, Tom Collins, and later David Look, and was the principal designer of what became the world's highest resolution optical spectrometer. He spent 50 years with the Air Force during which time he studied at several other universities in the United States and Europe. Litton was acknowledged as a world leader in research in solid-state and semiconductor physics and crystal growth, particularly in the optical, electrical, and structural properties of compound semiconductor materials and devices. In 1971 Cole was elected a fellow of the American

Physical Society. He has been a long-time devoted member of SPIE, where he was a founder and current co-chair of the Gallium Nitride Materials and Devices Conference and also the Oxide-Based Materials and Devices Conference, two of the most successful conferences at Photonics West since their inception. In memory of Litton, the Gallium Nitride Materials and Devices Conference will now bear his name, recognizing his many contributions not only to SPIE but to advancing optics- and photonics-based research as well.

Cole Litton retired in 2006 as Senior Scientist from the Air Force Research Laboratory, but he continued to enjoy an active role in scientific workshops and symposia. At the time of his death, he had authored or co-authored about 200 scientific/technical research papers published in physics and engineering journals. He was committed to discovery and was passionate about the future of science and technology. Cole died fully engaged in the activity he most enjoyed: participating in scientific meetings. He was a unique individual with a great love of life, and he will be remembered by all who knew him.

David F. Bliss
US Air Force Laboratory
Hanscom Research Site
MA, USA

Contents

Series Preface xv
Preface xvii
List of Contributors xxi

1 Fundamental Properties of ZnO **1**
T. C. Collins and R. J. Hauenstein

1.1 Introduction 1
1.1.1 Overview 1
1.1.2 Organization of Chapter 2
1.2 Band Structure 2
1.2.1 Valence and Conduction Bands 2
1.3 Optical Properties 5
1.3.1 Free and Bound Excitons 5
1.3.2 Effects of External Magnetic Field on ZnO Excitons 6
1.3.3 Strain Field 8
1.3.4 Spatial Resonance Dispersion 9
1.4 Electrical Properties 10
1.4.1 Intrinsic Electronic Transport Properties 10
1.4.2 n-type Doping and Donor Levels 11
1.4.3 p-type Doping and Dopability 13
1.4.4 Schottky Barriers and Ohmic Contacts 17
1.5 Band Gap Engineering 19
1.5.1 Homovalent Heterostructures 20
1.5.2 Heterovalent Heterostructures 22
1.6 Spintronics 22
1.7 Summary 25
References 25

2 Optical Properties of ZnO **29**
D. C. Reynolds, C. W. Litton and T. C. Collins

2.1 Introduction 29
2.2 Free Excitons 29
2.3 Strain Splitting of the Γ_5 and Γ_6 Free Excitons in ZnO 35
2.4 Photoluminescence from the Two Polar Faces of ZnO 36
2.5 Bound-Exciton Complexes in ZnO 38
2.6 Similarities in the Photoluminescence Mechanisms of ZnO and GaN 46

2.7 The Combined Effects of Screening and Band Gap Renormalization on the Energy of Optical Transitions in ZnO and GaN 51
2.8 Closely Spaced Donor–Acceptor Pairs in ZnO 55
2.9 Summary 58
References 58

3 Electrical Transport Properties in Zinc Oxide **61**
B. Claflin and D. C. Look

3.1 Introduction 61
3.2 Hall-Effect Analysis 62
3.2.1 Single-Band Conduction 62
3.2.2 Two-Band Mixed Conduction 65
3.2.3 Conducting Surface Layers 66
3.3 Donor States and n-type Doping 66
3.3.1 Native Point Defects – Donors 68
3.3.2 Substitutional Donors 69
3.4 Hydrogen 69
3.5 Acceptor States and p-type Doping 70
3.5.1 Native Point Defects – Acceptors 71
3.5.2 Substitutional Acceptors 72
3.6 Photoconductivity 76
3.7 Summary 78
References 78

4 ZnO Surface Properties and Schottky Contacts **87**
Leonard J. Brillson

4.1 Historical Background of Schottky Contacts on ZnO 87
4.1.1 ZnO Surface Effects 88
4.1.2 Early Schottky Barrier Studies 90
4.2 Recent Schottky Barrier Studies 91
4.2.1 Surface Cleaning in Vacuum 91
4.2.2 Surface Cleaning Effects on Impurities and Defects 92
4.3 The Influence of Surface Preparation on Schottky Barriers 93
4.4 The Influence of Defects on Schottky Barriers 97
4.5 The Influence of ZnO Polarity on Schottky Barriers 102
4.6 The Influence of Chemistry 103
4.7 Charge Transport and Extended Metal–ZnO Schottky Barriers 108
4.8 Conclusion 110
Acknowledgements 110
References 110

5 Native Point Defects and Doping in ZnO **113**
Anderson Janotti and Chris G. Van de Walle

5.1 Introduction 113
5.2 Theoretical Framework 114

5.3 Native Point Defects 115
5.3.1 Oxygen Vacancies 117
5.3.2 Zinc Interstitials 119
5.3.3 Zinc Antisites 120
5.3.4 Zinc Vacancies 121
5.3.5 Defect Migration 121
5.4 Donor Impurities 125
5.4.1 Aluminum, Gallium and Indium 125
5.4.2 Fluorine 125
5.4.3 Hydrogen 125
5.5 Acceptor Impurities 129
5.5.1 Lithium 129
5.5.2 Copper 129
5.5.3 Nitrogen 129
5.5.4 Phosphorous, Arsenic and Antimony 130
5.5.5 Co-Doping 130
5.6 Isoelectronic Impurities 131
Acknowledgements 131
References 131

6 Spectral Identification of Impurities and Native Defects in ZnO 135
B.K. Meyer, D.M. Hofmann, J. Stehr and A. Hoffmann

6.1 Introduction 135
6.2 Optical Spectroscopy 136
6.2.1 Excitons Bound to Shallow Donors 136
6.2.2 Recombinations Caused by Nitrogen and Arsenic Doping 145
6.3 Magnetic Resonance Investigations 153
6.3.1 Shallow Donors 154
6.3.2 Deep Level Defects 158
6.3.3 Extrinsic Acceptors: Li, Na and N 161
6.3.4 Intrinsic Acceptors 166
References 166

7 Vapor Transport Growth of ZnO Substrates and Homoepitaxy of ZnO Device Layers 171
Gene Cantwell, Jizhi Zhang and J.J. Song

7.1 Introduction 171
7.2 Transport Theory and Comparison with Growth Data 172
7.3 Characterization 175
7.3.1 Crystallinity 175
7.3.2 Purity 176
7.3.3 Electrical 177
7.3.4 Optical 178
7.4 In-situ Doping 180
7.5 ZnO Homoepitaxy 181

7.5.1 Substrate Preparation 182
7.5.2 Homoepitaxial Films on c-plane SCVT ZnO Substrates 183
7.5.3 ZnO Homoepitaxial Films on a-plane SCVT ZnO Substrates 185
7.6 Summary 185
Acknowledgement 186
References 186

8 Growth Mechanisms and Properties of Hydrothermal ZnO 189
M. J. Callahan, Dirk Ehrentraut, M. N. Alexander and Buguo Wang

8.1 Introduction 189
8.2 Overview of Hydrothermal Solution Growth 190
8.3 Thermodynamics of Hydrothermal Growth of ZnO 190
8.3.1 Solubility of ZnO in Various Aqueous Media 190
8.3.2 ZnO Phase Stability in H_2O System 191
8.4 Hydrothermal Growth Techniques 194
8.4.1 Hydrothermal Growth of ZnO Powder 194
8.4.2 Hydrothermal Crystal Growth of ZnO Single Crystals 194
8.4.3 Industrial Growth of Large ZnO Crystals 197
8.5 Growth Kinetics of Hydrothermal ZnO 200
8.5.1 Crystallographic Structure of Hydrothermal ZnO 200
8.5.2 Growth Rates of the Crystallographic Facets of Hydrothermal ZnO 200
8.6 Properties of Bulk Hydrothermal ZnO 205
8.6.1 Extended Imperfections (Dislocations, Voids, etc.) and Surface Studies 205
8.6.2 Impurities 208
8.6.3 Electrical Properties 210
8.6.4 Optical Properties 213
8.6.5 Etching and Polishing 215
8.7 Conclusion 217
Acknowledgements 217
References 218

9 Growth and Characterization of GaN/ZnO Heteroepitaxy and ZnO-Based Hybrid Devices 221
Ryoko Shimada and Hadis Morkoç

9.1 Introduction 221
9.2 Growth of GaN/ZnO 222
9.3 Compositional Analysis 230
9.4 Structural Analysis 232
9.5 Surface Studies 235
9.6 Optical Properties 237
9.6.1 Transmission Analysis 237
9.6.2 Cathodoluminescence Analysis 239
9.6.3 Photoluminescence Analysis 242
9.7 Electrical Properties 249

9.8 GaN/ZnO Hybrid Devices 252
9.8.1 Hybrid ZnO/GaN Heterojunction LED 253
9.8.2 ZnO-based Hybrid Microcavity 259
9.9 Conclusions 261
Acknowledgements 262
References 262

10 Room Temperature Stimulated Emission and ZnO-Based Lasers 265
D.M. Bagnall

10.1 Introduction 265
10.2 Emission Mechanisms 266
10.3 Stimulated Emission 267
10.3.1 Bulk ZnO 267
10.3.2 Epitaxial Layers 267
10.3.3 Quantum wells and Superlattices 270
10.3.4 ZnMgO/ZnO Structures 270
10.3.5 ZnO/ZnCdO Structures 272
10.4 Zinc Oxide Lasers 274
10.4.1 Introduction 274
10.4.2 Microstructural Lasers 275
10.4.3 Powder Lasers 278
10.4.4 Nanowire Lasers 279
10.4.5 ZnO Laser Diodes 280
10.5 Conclusions 281
References 282

11 ZnO-Based Ultraviolet Detectors 285
Jian Zhong and Yicheng Lu

11.1 Introduction 285
11.2 Photoconductivity in ZnO 288
11.2.1 Persistent Photoconductivity 293
11.2.2 Negative Photoconductivity 295
11.3 ZnO Film-Based UV Photodetectors 297
11.3.1 Photoconductive UV Detector 297
11.3.2 Schottky Barrier UV Photodetectors 301
11.3.3 Integrated Surface Acoustic Wave and Photoconductive Wireless UV Detectors 305
11.3.4 Photodetectors Using ZnO TFT 314
11.3.5 $Mg_xZn_{1-x}O$ UV Photodetector 315
11.4 ZnO NW UV Photodetectors 318
11.4.1 Photoconductive Gain in a ZnO NW 318
11.4.2 Noise Characteristics of ZnO NW UV Photodetector 323
11.5 Conclusions 325
Acknowledgements 325
References 326

12 Room-Temperature Stimulated Emission from ZnO Multiple Quantum Wells Grown on Lattice-Matched Substrates **331**
Takayuki Makino, Yusaburo Segawa, Masashi Kawasaki and Hideomi Koinuma

12.1 Introduction 331
12.2 Experimental Details 333
12.3 Quantum Confinement Effect of Excitons in QWs 333
12.4 Exciton–Phonon Interaction in QWs 336
12.5 The Localization Mechanism of the Exciton in a QW 337
12.6 Time-Resolved Luminescence in ZnO QWs 341
12.7 Stimulated Emission in MQWs 342
12.8 Summary 346
Acknowledgements 347
References 347

Index **351**

Series Preface

WILEY SERIES IN MATERIALS FOR ELECTRONIC AND OPTOELECTRONIC APPLICATIONS

This book series is devoted to the rapidly developing class of materials used for electronic and optoelectronic applications. It is designed to provide much-needed information on the fundamental scientific principles of these materials, together with how these are employed in technological applications. The books are aimed at (postgraduate) students, researchers and technologists, engaged in research, development and the study of materials in electronics and photonics, and industrial scientists developing new materials, devices and circuits for the electronic, optoelectronic and communications industries.

The development of new electronic and optoelectronic materials depends not only on materials engineering at a practical level, but also on a clear understanding of the properties of materials, and the fundamental science behind these properties. It is the properties of a material that eventually determine its usefulness in an application. The series therefore also includes such titles as electrical conduction in solids, optical properties, thermal properties, and so on, all with applications and examples of materials in electronics and optoelectronics. The characterization of materials is also covered within the series in as much as it is impossible to develop new materials without the proper characterization of their structure and properties. Structure–property relationships have always been fundamentally and intrinsically important to materials science and engineering.

Materials science is well known for being one of the most interdisciplinary sciences. It is the interdisciplinary aspect of materials science that has led to many exciting discoveries, new materials and new applications. It is not unusual to find scientists with a chemical engineering background working on materials projects with applications in electronics. In selecting titles for the series, we have tried to maintain the interdisciplinary aspect of the field, and hence its excitement to researchers in this field.

Peter Capper
Safa Kasap
Arthur Willoughby

Preface

Zinc oxide (ZnO) powder has been widely used as a major white paint pigment and industrial processing chemical for nearly 150 years. Indeed, interest in this fascinating chemical compound dates back even to antiquity. ZnO, the first man-made zinc compound, originated many centuries ago as an impure by-product of copper smelting. The ancients discovered and put to use some of its unusual properties, which included production of the first brass metal, development of a purified ZnO for medical purposes, and the early alchemists even attempted to make gold with it. Beginning in the early 1900s, white, polycrystalline ZnO powder found extensive application in medical technology, in particular the cosmetics and pharmaceutical industries, where it is today used in facial and body powders, sun screen preparation, antibiotic lotions and salves and in dental technology for dental cements.

A modern rediscovery of ZnO and its potential applications began in the mid 1950s. At that time, science and industry alike, mostly in the US and Europe, began to realize that ZnO had many interesting novel properties that were worthy of further investigation and exploration. These novel properties included its semiconductor, piezoelectric, luminescent, ultraviolet (UV) absorption, catalytic, ferrite, photoconductive and photochemical properties. Although study of the photoluminescence and electroluminescence properties of ZnO began as early as the mid 1930s, extensive investigation of its semiconductor properties did not begin until the mid to late 1950s, once good single crystals became available, either from natural sources, or grown synthetically by vapor transport and various other techniques, for study of the optical, electrical and structural properties of semiconducting ZnO. These early ZnO single crystals, mostly needles, platelets and prisms, were, however, small and limited in size to a few millimeters. ZnO single crystals typically crystallize in the wurtzitic, hexagonal modification, are visibly transparent and have a wide, direct band gap in the near UV at 3.437 eV (at 2 K). During this same period, the late 1950s to early 1960s, it was also recognized that ZnO had very high piezoelectric coefficients which led to the development of ZnO-based piezoelectric transducers, such as sensitive strain gauges and pressure sensors, a technology which continues today. Throughout the 1960s, extensive investigations of the fundamental semiconductor properties of ZnO were made, including study of its energy band structure, band gaps, excitonic properties, electron and hole effective masses, phonon properties and the electrical transport properties of the intrinsic (undoped) material. At this time (the mid to late 1960s) it was also recognized that ZnO, like the other wide band gap II–VI materials, would be difficult to dope controllably with high concentrations of shallow donor and acceptor impurities in order to demonstrate n- and p-type conductivity and p-n junctions, which would be necessary in order to realize the full potential of ZnO in devices, such as UV diode emitters, detectors and transistors. At that time, modern epitaxial growth and doping techniques, such as molecular beam epitaxy (MBE) and metal organic chemical vapor deposition (MOCVD), had not yet been developed, and p-type doping of both epitaxial films and bulk substrates did not exist; moreover, lack of large, bulk single crystals of ZnO also hampered progress in the development of ZnO-based electronic and optoelectronic devices. Nevertheless, much progress has been made over the past four decades (1970 to present) on the development of ZnO-based transducers, varistors, white-light-emitting cathodoluminescent phosphors (in conjunction with ZnS), optically transparent electrically conducting films, optically pumped lasing, MSM-type UV detectors, based on both ZnO (near UV) and MgZnO/ZnO heterostructures (deeper UV), and surface acoustic wave devices, none of which require the use of p-type ZnO.

Demonstration of the first InGaN/GaN-based, long-lived, room-temperature, continuous wave (CW) blue light-emitting diodes (LEDs) and diode lasers in Japan in the mid 1990s, led several ZnO investigators to consider the possibility of using isomorphic, nearly lattice-matched, c-plane bulk ZnO as a substrate for GaN device epitaxy ($\sim$2% mismatch to GaN), since bulk GaN substrates did not exist, and it was clear that the large threading dislocations resulting from growth of InGaN/GaN laser device structures on lattice mismatched c-plane sapphire ($\sim$14% lattice mismatch) were degrading both the performance and lifetimes of the blue laser diodes, particularly in CW, single mode operation. Earlier a US nitride research group had already demonstrated that GaN device epitaxy could be grown by MBE on small, c-plane ZnO substrates with as much as two to three orders of magnitude reduction in threading dislocation densities within the GaN device epitaxy, in comparison with growth on highly lattice mismatched sapphire substrates. Over the past decade, this achievement led another group to successfully grow and market large (40 mm diameter), high-quality, single-crystal ZnO substrates by vapor transport techniques specifically for this purpose and more recently still another group has also developed large diameter, bulk ZnO substrates by the Pressure-Melt technique for this purpose. Work is presently underway to demonstrate the MBE growth of AlGaN/GaN-based microwave power field-effect transistor (FET) device structures, where the relatively cheap ZnO substrate will be etched away and a high thermal conductivity substrate substituted by wafer bonding techniques to improve heat dissipation from the device.

Over the past decade, a number of groups have proposed that ZnO might be a good optoelectronic device material in its own right, owing to the many similarities between the optical, electrical and structural properties of ZnO and GaN, including their band gaps (3.437 eV for ZnO and 3.50 eV for GaN at 2 K) and their lattice constants. In addition, still others have noted that ZnO has a free exciton binding energy of 60 meV, approximately twice that of GaN, which could lead to highly efficient, ZnO- and MgZnO-based, UV injection lasers (UV laser diodes and detectors) at room temperature, provided that efficient p-doping and good p-n junctions and heterojunctions can be demonstrated in these materials. p-type doping of hetero-epitaxial ZnO on sapphire has been reported by several Japanese and US groups, using N acceptor doping and several different growth techniques, with varying degrees of success, but a major breakthrough was achieved by a US group recently which reported the first MBE growth of homo-epitaxial, N-doped, p-type ZnO on high-resistivity, Li-diffused ZnO substrates. Although the temperature-dependent Hall conductivity of these p-type layers is not yet fully understood, this approach could lead rapidly to p-doping at higher hole mobilities and carrier concentration and to the formation of good p-n junctions, provided that we can achieve a better understanding of both the shallow and deep donor/acceptor compensation mechanisms in ZnO. It is important to address the questions of donor and acceptor impurity incorporation together with the likely formation of native point defect donors and acceptors in ZnO and their possible compensation mechanisms; and look into the question of possible hydrogen donor incorporation in ZnO which must be better understood if rapid progress is to be made in the p-doping of ZnO.

This book comprises some 12 chapters that are written by experts in various aspects of ZnO materials and device technology. The topics included and discussed in these chapters range from our latest understanding of the energy band structure and spintronics (Chapter 1) to our most recent understanding of the fundamental optical and electrical properties of ZnO (Chapters 2 and 3). With the generation of new devices, one has to understand and control the electronic contacts of ZnO. This is covered in Chapter 4. The latest advances in our understanding of the formation of native point defect donors and acceptors in ZnO are discussed and summarized in Chapter 5. The following chapter (Chapter 6) investigates both the intrinsic and extrinsic defects that are found in ZnO. The growth of the ZnO crystals and substrates are discussed in the next three chapters (Chapters 7, 8 and 9) along with hybrid devices, Chapter 10 reports on some recent advances in optically pumped lasing and room temperature stimulated emission from ZnO-based materials. Chapter 11 reviews the progress of UV photodetectors and points out the promise for unique applications such as single-photon detection. The final chapter (Chapter 12) presents a review of optical properties of ZnO quantum wells in which strong stimulation was observed in ZnO/ZnMgO multiple quantum wells from 5 °C to room temperature.

Cole W. Litton
Donald C. Reynolds
Thomas C. Collins

List of Contributors

M. N. Alexander, Air Force Research Laboratory, Hanscom AFB, MA, USA

D. M. Bagnall, University of Southampton, Southampton, UK

Leonard J. Brillson, The Ohio State University, Columbus, OH, USA

M. J. Callahan, Teleos Solar, Hanson, MA, USA and Air Force Research Laboratory, Hanscom AFB, MA, USA

Gene Cantwell, ZN Technology, Inc., Brea, CA, USA

B. Claflin, Wright State University, Dayton, OH, USA

T. C. Collins, Oklahoma State University, Stillwater, OK, USA

Dirk Ehrentraut, Tohoku University, Aoba-ku, Sendai, Japan

R. J. Hauenstein, Oklahoma State University, Stillwater, OK, USA

A. Hoffmann, Technical University Berlin, Berlin, Germany

D. M. Hofmann, Justus Liebig University Giessen, Giessen, Germany

Anderson Janotti, University of California, Santa Barbara, CA, USA

Masashi Kawasaki, Tohoku University, Sendai, Japan, Cross-Correlated Materials Research Group, Advanced Science Institute, RIKEN, Wako, Japan and CREST, Japan Science and Technology Agency, Tokyo, Japan

Hideomi Koinuma, The University of Tokyo, Kashiwa, Chiba, Japan

C. W. Litton, Air Force Research Laboratory, Wright-Patterson Air Force Base, OH, USA

D. C. Look, Wright State University, Dayton, OH, USA

Yicheng Lu, Rutgers University, Piscataway, NJ, USA

Takayuki Makino, Tohoku University, Sendai, Japan

B. K. Meyer, Justus Liebig University Giessen, Giessen, Germany

Hadis Morkoç, Virginia Commonwealth University, Richmond, VA, USA

D. C. Reynolds, Wright State University, Dayton, OH, USA and Air Force Research Laboratory, Wright-Patterson Air Force Base, OH, USA

Yusaburo Segawa, Advanced Science Institute, RIKEN, Wako, Japan

Ryoko Shimada, Virginia Commonwealth University, Richmond, VA, USA

J. J. Song, ZN Technology, Inc., Brea, CA, USA and University of California at San Diego, La Jolla, CA, USA

J. Stehr, Justus Liebig University Giessen, Giessen, Germany

Chris G. Van de Walle, University of California, Santa Barbara, CA, USA

Buguo Wang, Solid State Scientific Inc., Nashua, NH, USA

Jizhi Zhang, ZN Technology, Inc., Brea, CA, USA

Jian Zhong, Rutgers University, Piscataway, NJ, USA

1

Fundamental Properties of ZnO

T. C. Collins and R. J. Hauenstein
Department of Physics, Oklahoma State University, Stillwater, OK, USA

1.1 Introduction

1.1.1 Overview

Wurtzitic ZnO is a wide band gap semiconductor ($E_g = 3.437$ eV at 2 K) that has many applications, including piezoelectric transducers, varistors, phosphors, and transparent conduction films. Most of these applications only require polycrystalline materials; however, recent successes in producing large-area single crystals make possible the production of blue and UV light emitters and high temperature, high power transistors. The main advantage of ZnO as a light emitter is its large exciton binding energy ($E_b = 60$ meV). This binding energy is three times larger than that of the 20 meV exciton of GaN, which permits excitonic recombination to dominate in ZnO at room temperature (and even above). Excitonic recombination is preferable because the exciton, being an already bound system, radiatively recombines with high efficiency without requiring traps to localize carriers, as in the case in radiative recombination of electron–hole plasmas. Secondly, the deeper exciton of ZnO is more stable against field ionization due to piezoelectrically induced fields. Such piezoelectric effects are expected to increase with increasing dopant concentration for both ZnO and GaN.

For electronic applications, the attractiveness of ZnO lies in having high breakdown strength and high saturation velocity. ZnO also affords superior radiation hardness compared with other common semiconductor materials, such as Si, GaAs, CdS, and even GaN, enhancing the usefulness of ZnO for space applications. Optically pumped UV laser action in ZnO has already been demonstrated at both low and high temperatures

Zinc Oxide Materials for Electronic and Optoelectronic Device Applications, First Edition.
Edited by Cole W. Litton, Donald C. Reynolds and Thomas C. Collins.

although efficient electrically induced lasing awaits further improvements in the experimental ability to grow high quality p-type ZnO material. Nonetheless, over the past decade, researchers world-wide have made substantial theoretical and experimental progress concerning the p-type dopability of ZnO, with $10^{17}\,cm^{-3}$ range hole concentrations now plausibly achieved with material stability persisting for over 1 year, and with isolated (though often controversial) reports of hole concentrations as high as $\sim 10^{19}\,cm^{-3}$ even being reported from time to time. Finally, ZnO structures can be doped with transition metal (TM) ions to form dilute magnetic materials, denoted (Zn,TM)O, which can form a ferromagnetic state, an antiferromagnetic state as well as a general spin glass. The important point is that the Curie temperature (T_C) can be *above* room temperature. Such above-room-temperature anti- and ferromagnetic states form the basis for novel charge-based, spin-based, or even mixed spin- and charge-based devices which, collectively, are known as "spintronic" devices.

1.1.2 Organization of Chapter

The remainder of this chapter is organized as follows. In Section 1.2, a theoretical overview of the fundamental band structure of ZnO near the zone center is presented. The discussion includes the long-standing controversy over the symmetry-ordering of the valence bands at the Γ point. Next, in Section 1.3, the optical properties of intrinsic ZnO are reviewed, with particular emphasis on the excitons. Also presented in this same section are a discussion of the interaction of light, magnetic field, and strain field, as three examples of the general types of calculations done for excitons in ZnO, as well as a discussion of spatial resonance dispersion (Section 1.3.4) in which the polariton, a combined state arising out of the mixing of an exciton and light, plays a particularly important role. The electrical properties of ZnO are considered next in Section 1.4, including a discussion of intrinsic along with n-type and p-type ZnO. In particular, the important question of p-type dopability is discussed in detail in Section 1.4.3. For the implementation of optoelectronic devices, one will need Schottky barriers and ohmic contracts; recent progress in these areas is presented in Section 1.4.4. For heterojunction-based devices, band gap engineering will be required and this is considered in Section 1.5. Finally, presented in Section 1.6 is the theoretical basis for the ZnO spintronic device. The different models are based on the Heisenberg Spin Hamiltonian to describe the dilute magnetic system (Zn,TM)O. One can investigate both the interaction of the carriers with the magnetic moment of the TM as well as the TM–TM interactions. It is found that the resulting Curie temperature can be above room temperature. Spintronic devices made of (Zn,TM)O are expected to be faster and to consume less power since flipping the spin requires 10–50 times less power, and occurs roughly an order of magnitude faster, than does transporting an electron through the channel of traditional field-effect transistors (FETs).

1.2 Band Structure

1.2.1 Valence and Conduction Bands

The general electronic structure of binary III–V and II–VI compounds form semiconductors with the valence band mostly derived from the covalent bonding orbitals (s and p).

While the conduction band consists of antibonding orbitals, as one moves further outward from column IV of the periodic table, the binary compound semiconductors acquire a more ionic character. These compounds form cubic (zinc blende) and hexagonal (wurtzite) crystal structures, with ZnO crystallized in the wurtzite structure. The difference between the zinc blende and wurtzite structures is that the zinc blende is cubic while the wurtzite is a distortion of the cube in the [111] direction generally taken to be the z direction in the wurtzite.

The ionicity effect puts more electrons on the group V or group VI atoms giving the charge density more s and p characteristics of these elements in the valence band. It also causes gaps at the edge of the Brillouin zone compared with just covalent bonding materials. This translates into flatter bands across the Brillouin zone.

ZnO is a direct band gap semiconductor with valence-band maximum and conduction-band minimum occurring at the Γ point. The conduction band is s-like from Zn at Γ and is spin degenerate. The top three valence bands are p-like in character. They are split by the spin-orbit interaction in both the zinc blende and wurtzite symmetry, while wurtzite symmetry also has a crystal field splitting.

Figure 1.1 shows[1] the Quasi-cubic model[2–4] of the bottom of the con-duction band and the top of the valence band. Assuming that one has both zinc blende and wurtzite and that $\mathcal{H}_{so} = \Delta\mathbf{L} \cdot \mathbf{S}$, one can write matrices of the form:

$$\begin{pmatrix} \Delta & 0 & 0 \\ 0 & \Delta & 0 \\ 0 & 0 & -2\Delta \end{pmatrix} \tag{1.1}$$

for zinc blende, using $j = 3/2$ and $j = 1/2$ eigenstates. For wurtzite the basis is rotated as stated above, so that one has $S_+\alpha$, $S_-\beta$, $-S_-\alpha$, $S_+\beta$, $S_z\alpha$, and $S_z\beta$. This basis gives matrices (including crystal field effects δ) of the form

$$\begin{pmatrix} \Delta & 0 & 0 \\ 0 & -\Delta & -i\sqrt{2}\Delta \\ 0 & i\sqrt{2}\Delta & -\delta \end{pmatrix}. \tag{1.2}$$

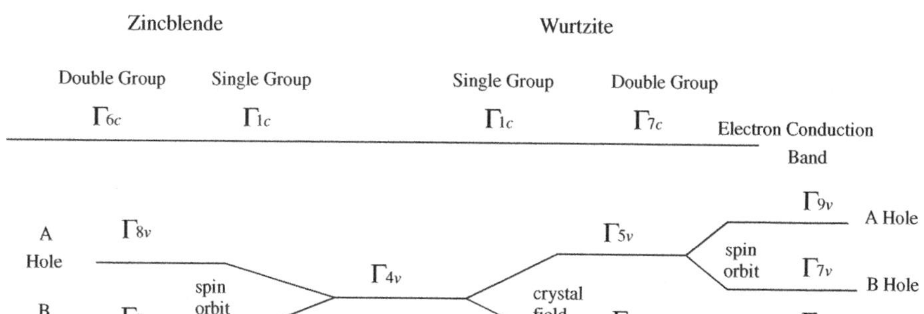

Figure 1.1 Structure and symmetries of the lowest conduction band and upmost valence bands in ZnO compounds at the Γ point

Using the values obtained by Thomas,[5] a difficulty arises. As can be seen in Equation (1.3), there are values of energy difference which can give a complex number for δ:

$$\delta = \frac{1}{2}(2E_1 + E_2) + \frac{1}{2}\sqrt{E_2^2 - 2E_1(E_1 + E_2)} \tag{1.3}$$

where E_1 is the energy difference between the Γ_9 excitons and Γ_7 exciton and E_2 is the energy difference between the two Γ_7 excitons. In fact, using Thomas[5] numbers, δ is a complex number with Γ_9 being the top valence band. This difficulty can be surmounted by assuming a negative spin-orbit splitting, which is different from all the other II–VI compounds and GaN. The physical mechanism that could produce a negative spin-orbit splitting was investigated by Cardona[6] for CuCl. In this zinc blende material, it was postulated that the valence band was formed from Cl wave functions, with a large proportion of Cu wave functions (Cu 4s3d). The inverted nature of the CuCl indicates that the Cu contributes a negative term to the spin-orbit splitting, since for this material the anion splitting is small and negative. Cardona[6] estimated the "fraction" of the metal wave function in the valence band states by writing the spin-orbit splitting of the compound as:

$$\Delta = \frac{3}{2}[\alpha\Delta_{\text{hal}} - (1-\alpha)\Delta_{\text{met}}], \tag{1.4}$$

where α is the proportion of halogen in the wave function, Δ_{hal} is the one-electron atomic spin-orbit splitting parameter of the halogen and Δ_{met} is one of the d-electrons of the metal. This gave $\alpha = 0.25$ for CuCl. It was also presented that the energy interval between the ground state of the Cu^+ ion ($3d^{10}$) and the first excited state ($3d^9$, 4s) is 2.75 eV.

Returning to ZnO, one finds the Zn d-bands below the uppermost p-like valence bands to be greater than 7 eV.[7–9] This makes it very unlikely that one has much mixing at all. Further, the d-band appeared to be relatively flat, noting again very little mixing. Also, the availability of ZnO crystals in which intrinsic exciton transitions[10] are observed in emission and their splitting in a magnetic field have led to a positive spin-orbit splitting of 16 meV. With this interpretation, the Quasi-cubic model[2–4] gives results in line with the other II–VI compounds.

In order to investigate the valence band ordering of ZnO further, Lambrecht *et al.*[11] calculated the band structure of ZnO using a linear muffin-tin potential and a Kohn–Sham local density approximation. The band gap at Γ was 1.8 eV compared with experiment of 3.4 eV and the Zn d-band was approximately 5 eV below the top of the valence band at Γ. To correct for the band gap Reynolds *et al.*[10] rigidly shifted the conduction band up to match the experimental Γ-point gap (a shift of 1.624 eV).

As is seen from above in the Quasi-cubic model,[2–4] the spin-orbit magnitude and sign are a function of the energy difference between the top of the valence and the 3d band of Zn. It is found to be greater than 7 eV.[7–9] In Reynolds *et al.*,[10] the d-band was adjusted to where the spin-orbit gave the right energy difference with the Γ_7 above the Γ_9. The d band was put at 6.25 eV below the top of the valence band. This led to a negative *g*-value and of course a negative spin-orbit parameter. This in turn matched the experimental data given in Hong *et al.*[9] but with a different interpretation of the splitting of the exciton lines. Thus, there is no agreement of the spin-orbit value for ZnO. In first principle electron structure calculations, one has an accuracy only on the order of ~100 meV whereas the splittings of the levels in the top valence band are on the order of ~10 meV!

1.3 Optical Properties

1.3.1 Free and Bound Excitons

The optical absorption (emission) of electromagnetic radiation in a ZnO crystal is dependent on the matrix element

$$\int \Psi_f^* \mathcal{H}_{\rm int} \Psi_i d\tau \tag{1.5}$$

where

$$\mathcal{H}_{\rm int} = \frac{e\hbar}{imc} \mathbf{A} \cdot \nabla. \tag{1.6}$$

Here, **A** is the vector potential of the radiation field and has the form

$$\mathbf{A} = \hat{n}|\mathbf{A}_0|e^{i(\mathbf{q}\cdot\mathbf{r}-\omega t)},$$

e is the electronic charge, m is the electron mass, c is the velocity of light, $\hat{n}$ is a unit vector in the direction of polarization, and $\mathbf{q}$ is the wave vector. Expanding the spatial part of **A** in a series gives

$$\mathcal{H}_{\rm int} \equiv \sum_{j=0}^{\infty} \mathcal{H}_{\rm int}^j \tag{1.7}$$

where

$$\mathcal{H}_{\rm int}^j = (\mathbf{q}\cdot\mathbf{r})^j \hat{n} \cdot \nabla \tag{1.8}$$

and the dipole term is then the first term ($i = 0$). The matrix element in Equation (1.5) transforms under rotation like the triple direct product

$$\Gamma_f \otimes \Gamma_r^j \otimes \Gamma_i. \tag{1.9}$$

The selection rules are then determined by which of the triple-direct-product matrix elements in question do not vanish, where Γ_r^j is the symmetry of the expansion term $\mathcal{H}_{\rm int}^j$ in Equation (1.8).

The dipole moment operator for electric dipole radiation transforms like x, y, or z dependent on the polarization. When the electric field vector **E** of the incident light is parallel to the crystal axis of ZnO, the operator corresponds to the Γ_1 representation. When it is perpendicular to the crystal axis, the operator corresponds to the Γ_5 representation. Since the crystal has a principal axis, the crystal field removes part of the degeneracy of the p-levels as seen in Figure 1.1. Including spin in the problem doubles the number of levels. Since the conduction band at $\mathbf{k} = 0$ is the Zn (4s) level[4,5] it transforms as Γ_7 while the $\mathbf{k} = 0$ at top of the valence band is made up of the O(2p) level and it splits into (p_x, p_y)Γ5 and $(p_z)\Gamma 1$. When crossed with the spin one has

$$\Gamma_5 \otimes D_{1/2} \rightarrow \Gamma_7 + \Gamma_9 \tag{1.10}$$

$$\Gamma_1 \otimes D_{1/2} \rightarrow \Gamma_7 \tag{1.11}$$

as shown in Figure 1.1.

One of the light absorption (emission) intrinsic states is an exciton, which is made up of a hole from the top of the valence band and an electron from the bottom of the conduction band. These are excitations of the *N*-particle system whereas electron structure calculations are of the $(N \pm 1)$-particle system. All solutions to the one-body calculations such as the one-particle Green's functions method do not contain the interaction of the excited "particle" with the other "particles." The more localized the excitation the more important it is to include this interaction. The more localized the excitation is, the flatter the one-electron bands, leading to heavier effective masses. This is turn leads to increased binding energy of the electron–hole pair. In ZnO, the binding energy of the ground state exciton is 60 meV. Adding in the Coulomb term of the electron–hole pair gives a hydrogenically bound pair. For the Γ_9 hole and Γ_7 electron one has

$$\Gamma_9 \otimes \Gamma_7 \rightarrow \Gamma_5 + \Gamma_6, \tag{1.12}$$

and for the Γ_7 hole and Γ_7 electron one has

$$\Gamma_7 \otimes \Gamma_7 \rightarrow \Gamma_5 + \Gamma_1 + \Gamma_2. \tag{1.13}$$

The Γ_5 and Γ_6 are doubly degenerate and the Γ_1 and Γ_2 are nondegenerate. The Hamiltonian for the exciton becomes:

$$\mathcal{H} = \mathcal{H}_e + \mathcal{H}_h + \mathcal{H}_{\text{int}}, \tag{1.14}$$

where $\mathcal{H}_e$ and $\mathcal{H}_h$ are the Hamiltonian for the electron and the hole and $\mathcal{H}_{\text{int}}$ is the interaction between the electron and hole including Coulomb, exchange and correlation. To first approximation one generally includes just the Coulomb term as noted above. Equation (1.14) gives what are referred to as the "free" excitons.

There are several extrinsic effects which modify the excitons. Most notable of these in ZnO are the bound complexes. One can have the exciton bound to an ionized donor or a neutral donor. In the case of the ionized donor, one has the molecular attraction of the exciton to the donor plus central cell corrections. For the neutral donor, one has again the molecular binding energy plus the ability of the neutral donor to be left in an excited state. Similar results are obtained with the ionized acceptor or neutral acceptor. The method of calculating these systems is to treat the system as a molecular system in the field of the crystal.

1.3.2 Effects of External Magnetic Field on ZnO Excitons

The case of an applied uniform magnetic field was developed by Wheeler and Dimmock[12] for the exciton in ZnO. It was assumed the electron bands are isotropic at least to second order in **k** with only double spin degeneracy. The exciton equation is a simple hydrogen Schrödinger equation including effective-mass and dielectric anisotropies. It is found that the mass anisotropy is small, allowing first-order perturbation calculations to be made for the energy states as well as for the magnetic-field effects.

Since the valence and conduction band extrema are at $\mathbf{k}=0$, the wave vector of the light that creates the exciton $\mathbf{k}$ will also represent the position of the exciton in $\mathbf{k}$-space. Dividing the momentum and spaces coordinates into the center-of-mass coordinates and the internal coordinates, the exciton Hamiltonian can be divided into seven terms as follows:

$$\mathcal{H}=\mathcal{H}_1+\mathcal{H}_2+\mathcal{H}_3+\mathcal{H}_4+\mathcal{H}_{k1}+\mathcal{H}_{k2}+\mathcal{H}_{k3} \tag{1.15}$$

$$\mathcal{H}_1=-\frac{\hbar^2}{2m}\left[\frac{1}{\mu_x}\left(\frac{\partial^2}{\partial x^2}+\frac{\partial^2}{\partial y^2}\right)+\frac{1}{\mu_z}\frac{\partial^2}{\partial z^2}\right]-\frac{e^2}{\epsilon\eta^{1/2}}\left(x^2+y^2+\eta^{-1}z^2\right)^{-1/2} \tag{1.16}$$

$$\mathcal{H}_2=-2i\zeta\left(\frac{A_x}{\Delta_x}\frac{\partial}{\partial x}+\frac{A_y}{\Delta_y}\frac{\partial}{\partial y}+\frac{A_z}{\Delta_z}\frac{\partial}{\partial z}\right) \tag{1.17}$$

$$\mathcal{H}_3=\frac{e}{2mc^2}\left(\frac{A_x^2}{\mu_x}+\frac{A_y^2}{\mu_y}+\frac{A_z^2}{\mu_z}\right) \tag{1.18}$$

$$\mathcal{H}_4=\frac{\zeta}{2}\sum_{i=x,y,z}(g_{e_i}S_{e_i}+g_{h_i}S_{h_i})H_i \tag{1.19}$$

$$\mathcal{H}_{k1}=\frac{-i\hbar}{2m}\left(\frac{K_x}{\Delta_x}\frac{\partial}{\partial X}+\frac{K_y}{\Delta_y}\frac{\partial}{\partial Y}+\frac{K_z}{\Delta_z}\frac{\partial}{\partial Z}\right) \tag{1.20}$$

$$\mathcal{H}_{k2}=\zeta\left(\frac{K_x}{\mu_x}A_x+\frac{K_y}{\mu_y}A_y+\frac{K_z}{\mu_z}A_z\right) \tag{1.21}$$

$$\mathcal{H}_{k3}=\frac{\hbar^2}{8m}\left(\frac{K_x^2}{\mu_x}+\frac{K_y^2}{\mu_y}+\frac{K_z^2}{\mu_z}\right), \tag{1.22}$$

where m is the free-electron mass and μ_x is the reduced effective mass of the exciton in the x direction at Γ ($\mu_x=\mu_y$). Also,

$$\zeta=\frac{e\hbar}{2mc}\qquad \eta=\frac{\epsilon_z}{\epsilon_x} \tag{1.23}$$

$$\mathbf{A}=\frac{1}{2}(\mathbf{H}\times\mathbf{r}) \tag{1.24}$$

$$\frac{1}{\Delta_y} = \left(\frac{m}{m_{ey}^*} - \frac{m}{m_{hy}^*} \right). \tag{1.25}$$

The first term is the Hamiltonian for a hydrogenic system in the absence of external fields. One could include the possibility of the mass and dielectric anisotropies. The second term is an **A**·**p** term which leads to the linear Zeeman magnetic field term. In this term, the momentum operator is $\mathbf{p}_i - 2e\,\mathbf{A}_i/c$ where $\mathbf{A}_i = (1/2)\mathbf{H} \times \mathbf{r}_i$ is the vector potential, **H** is the magnetic field, and $\mathbf{r}_i$ is the coordinate of the *i*th electron. The A^2 term is the diamagnetic field term proportional to $|H|^2$. The fourth term is the linear interaction of the magnetic field with the spin of the electron and hole. If one has small effective reduced mass for the electron and a large dielectric constant, the radii of the exciton states are much larger than the corresponding hydrogen-state radii. Hence, the spin-orbit coupling is proportional to r^{-3} and thus quite small, it is legitimate to write the magnetic field perturbations in the Paschen–Back limit as done above.

The last three terms are the **K**·**P**, **K**·**A**, and K^2 terms; **K** is the center-of mass momentum. Treating the **K**·**P** to second order and adding the K^2 term, one can obtain an energy term that appears like the center-of-mass kinetic energy. The **K**·**A** term has little effect upon the energy; however, it has very interesting properties. This term represents the quasi-electric field that an observer riding with the center-of-mass of the exciton would experience because of the magnetic field in the laboratory. This quasi-field would produce a Stark effect linear in **H**, and this would give rise to a maximum splitting interpretable as a "*g*" value.

1.3.3 Strain Field

The effective Hamiltonian formalism of Pikus[13,14] is of the form

$$\mathcal{H} = \mathcal{H}_{v-0} + \mathcal{H}_{v-p} + \mathcal{H}_c + \mathcal{H}_{\text{int}}, \tag{1.26}$$

Using the angular momentum operators $J(L = 1)$ gives

$$\mathcal{H}_{v-0} = \Delta_1 J_z^2 + \Delta_2 J_z \sigma_z + \Delta_3(\sigma_+ J_- + \sigma_- J_+), \tag{1.27}$$

where Δ_1 is the crystal field splitting of the S_x, S_y bands from the S_z band; Δ_2 is the spin-orbit splitting of the $j = 3/2$ bands from the $j = 1/2$ band, and Δ_3 is a trigonal splitting of the $|j, m_j\rangle = |\frac{3}{2}, \pm\frac{3}{2}\rangle$ bands from the $|\frac{3}{2}, \pm\frac{1}{2}\rangle$ bands.

The operator $\mathcal{H}_{v-p}$, which describes the interaction between the valence bands due to strain, can be determined from symmetry considerations by combining the strain tensor ε_{ij} with the angular momentum operators J and σ:

$$\begin{aligned}\mathcal{H}_{v-p} &= (C_1 + C_3 J_z^2)\varepsilon_{zz} + (C_2 + C_4 J_Z^2)(\varepsilon_{xx} + \varepsilon_{yy}) \\ &\quad + C_5(J_-^2 \varepsilon_+ + J_+^2 \varepsilon_-) + C_6([J_z J_+]\varepsilon_{-z} + [J_z J_-]\varepsilon_{+z})\end{aligned} \tag{1.28}$$

Here,

$$\varepsilon_\pm = \varepsilon_{xx} - \varepsilon_{yy} \pm 2i\varepsilon_{xy} \quad \varepsilon_{\pm z} = \varepsilon_{xz} \pm i\varepsilon_{yz}, \tag{1.29}$$

and $[J_iJ_j] = (J_iJ_j + J_jJ_i)/2$. These operators then describe the mixing of the p-like valence bands due to crystalline strain in the terms of material-dependent parameters (the C_i's).

The operator $\mathcal{H}_c$ represents the conduction-band energy. Because of the Γ_1 symmetry of the conduction band, the operator has the form

$$\mathcal{H}_c = E_c + d_1\varepsilon_{zz} + d_2(\varepsilon_{xx} + \varepsilon_{xy}). \tag{1.30}$$

$\mathcal{H}_{\text{int}}$ is the Coulomb interaction plus the exchange term

$$\mathcal{H}_{\text{int}} = E_b + \frac{1}{2}j\sigma_h \cdot \sigma_e \tag{1.31}$$

where in Equation (1.31)j denotes the usual exchange integral (not to be confused with the usage of j as the angular momentum quantum number in the earlier equations). The exchange term has off-diagonal matrix elements, and for example, removes the Γ_5 degeneracy. This leads to strain splittings, which are measured.

1.3.4 Spatial Resonance Dispersion

The excitons in ZnO interact with photons when the wave vector are essentially equal. The energy denominator for exciton–photon mixing is small and the mixing becomes large. These states are not to be considered as pure photon states or pure exciton states, but rather mixed states. Such a mixed state is called a polariton. When there is a dispersion of the dielectric constant, spatial dispersion has been invoked to explain certain optical effects of the crystal. This causes more than one energy transport mechanism. Spatial dispersion addresses the possibility that two different kinds of waves of the same energy and same polarization can exist in a crystal differing only in wave vector. The one with an anomalously large wave vector is an anomalous wave. In the treatment of dispersion by exciton theory, it was shown that if the normal modes of the system were allowed to depend on the wave vector, a much higher-order equation for the index of refraction would result. These new solutions occur whenever there is any curvature of the ordinary exciton band in the region of large exciton–photon coupling. These results apply to the Lorentz model as well as to quantum-mechanical models whenever there is a dependence of frequency on wave vector.

The specific dipole moment of polarization of a crystal and the electric field intensity are not in direct proportion. The two are related by a differential equation that resulted in giving Maxwell equations of higher order. This leads to the existence of several waves of the same frequency, polarization, and direction but with different indices of refraction. The index of refraction becomes

$$n^2 = \frac{k^2c^2}{\omega^2} = \epsilon + \sum_j \frac{4\pi(a_{0j} + a_{2j}K^2)\omega_{0j}^2}{\omega_{0j}^2 + (\hbar\omega_{0j}k^2/m^* - \omega^2 - i\omega\Gamma)}. \tag{1.32}$$

Here, the sum over j is to include the excitons in the frequency region of interest, and the contributions from other oscillators are included in a background dielectric constant ϵ. In Equation (1.32) the numerator and denominator have been expanded in powers of k, keeping terms to order of k^2, m^* is the sum of the effective masses of the hole and electron that comprise the exciton, and ω_{0j} is the frequency of the jth oscillator at $\mathbf{k} = 0$. Eliminating k^2 from Equation (1.32) and neglecting the linewidth of the oscillators, the equation becomes

$$n^2 = \epsilon + \sum_j \frac{4\pi(a_{0j} + a_{2j}\omega^2 n^2/c^2)\omega_{0j}^2}{\omega_{0j}^2 + (\hbar\omega_{0j}\omega^2 n^2/c^2 m^*) - \omega^2}. \tag{1.33}$$

The sum is over excitons from the top two valence bands where the "allowed" excitons have been included with $a_{0j} \neq 0$, $a_{2j} = 0$ while the "forbidden" (because excitons are included with $a_{0j} \neq 0$, $a_{2j} = 0$). Equation (1.33) reduces to a polynomial in n^2 whose roots give the wavelengths of the various "normal modes" for the transfer of energy within the crystal.

There is another structure seen in the spectra near the free exciton. This is the result of the mixing of the exciton and photon with non crossing effects of the photon dispersion. This creates a larger density of states where the mixing becomes strong enough to bind the photon curve.

1.4 Electrical Properties

Electrically, ZnO is a transparent-conducting-oxide semiconductor which at room temperature exhibits defect- or impurity-dominated n-type conductivity, even in nominally undoped materials. This defect- or impurity-dominated conductivity is a consequence of the large band gap[15,16] ($\approx$3.3 eV at room temperature) combined with the unavoidable presence of electrically active native defects and impurities with donor ionization energies typically[15] $\sim$10–100 meV at concentrations typically[15] $\sim 10^{15}$–$10^{16}\,\text{cm}^{-3}$. Furthermore, while ZnO is readily doped n-type and carrier concentrations as high as $\sim 10^{21}\,\text{cm}^{-3}$ are achievable,[17] at present p-type doping remains challenging. Furthermore, for purposes of utilizing ZnO-based materials for extending semiconductor optoelectronics further into the UV, two of the biggest technological challenges will be to develop growth and device-processing techniques to achieve control over p-type doping as well over metal/semiconductor junction formation (including both Schottky and ohmic contacts) to both n-type and p-type materials.

Most of what is known experimentally about the electronic properties of "pure" ZnO is actually based on n-type specimens. The fundamental electronic transport properties of nominally undoped and intentionally doped ZnO will be presented in Section 1.4.1. Next, in Section 1.4.2, intentional n-type doping and dopants will be considered. Then, in Section 1.4.3 will be discussed recent progress, both experimental and theoretical, concerning the technologically important question of p-type dopability in ZnO. Finally, recent progress on the fabrication of Schottky contacts to n-type, and of ohmic contacts to both n-type and p-type ZnO, will be reviewed in Section 1.4.4.

1.4.1 Intrinsic Electronic Transport Properties

To date, available data concerning fundamental electronic properties of intrinsic, accurately stoichiometric ZnO remains largely unknown.[15] In part this is because of the historical unattainability of sufficiently high-quality materials, a consequence of the difficulty (especially through bulk crystal-growth methods) in achieving adequate O incorporation into the specimen. While such a circumstance may seem merely a technological rather than a fundamental limitation, recent theoretical work[18,19] suggests that the n-type character resulting from electrically active stoichiometric native point defects such as vacancies, inter-stitials, and antisites,[18] as well as the unusually nonamphoteric and donor-like character of unintentional but ever-present hydrogen,[20] may be inherent to ZnO, at least, to materials produced through the use of near-equilibrium crystal growth methods. The question of p-type dopability shall be deferred to Section 1.4.3. For the remainder of the present section we will consider the basic electronic transport properties of nominally undoped ZnO which have been established experimentally, along with modifications due to intentional n-type doping.

In general, electrical transport properties of ZnO are directionally dependent due to the anisotropy of its wurtzite crystal structure. For the case of nominally undoped (i.e. residually n-type $\sim 10^{16}\,\mathrm{cm}^{-3}$ carrier concentration) ZnO, Figure 1.2 shows temperature-dependent Hall mobility data for current transport both parallel and perpendicular to the crystallographic c axis of bulk ZnO. As seen in Figure 1.2, the electron mobility attains a peak value of $\sim 1000\,\mathrm{cm}^2\,\mathrm{V}^{-1}\,\mathrm{s}^{-1}$ at a temperature of 50–60 K, and drops off approximately as the power law, T^p for low temperatures and T^q for high temperatures where $p \approx 3$ and $q \approx -2$ in Figure 1.2(a). The drop off at low temperature falls off faster than that expected due to ionized-impurity scattering alone whereas that at high temperatures is attributable primarily to the combination of acoustic-deformation-potential with polar-optical-phonon scattering mechanisms.[21] The slight anisotropy observed in Hall mobility data [compare Figure 1.2(a) and (b)] has been attributed exclusively to piezoelectric scattering.[15]

Also of interest is the electric-field dependence of the mobility. Figure 1.3 compares theoretically obtained plots[22] of drift velocity (v_d) vs electric field; results predicted for ZnO are contrasted with those for GaN, another direct-gap semiconductor having a closely comparable band gap ($E_g^{\mathrm{GaN}} \approx 3.4\,\mathrm{eV}$) to ZnO. A comparison of the GaN and ZnO curves at low field reveals that $\mu_{\mathrm{ZnO}} < \mu_{\mathrm{GaN}}$ while at the same time the saturated drift velocities obey the opposite relation: $v_{\mathrm{Sat}}^{\mathrm{ZnO}} > v_{\mathrm{Sat}}^{\mathrm{GaN}}$. For hot-electron devices, where fields are high and transport is nonohmic, it is v_{sat} rather than μ that can be the more important parameter to device operation.[23]

1.4.2 n-type Doping and Donor Levels

Intentional n-type doping is readily achieved through the use of column-III elements such as Al, Ga, or In, or, column-VII elements such as Cl, F, or I. All of these sit substitutionally on the appropriate cation or anion site and form reasonably shallow levels. In a 2001 review article, Look discusses the presence of three predominant donor levels in ZnO appearing at approximately 30, 60, and 340 meV.[23] Among these, the 60 meV level is that corresponding to effective mass theory (≈65 meV), whereas the 30 meV level was at the time believed[18,24] to be due to interstitial Zn ($\mathrm{Zn_i}$) and no longer believed to be due to the O vacancy ($\mathrm{V_O}$), which instead was thought to be a deep donor.[18] The identity of the 340 meV level was not established. Interestingly, also presented by Look in this same review article[23] are some of his own temperature-dependent Hall mobility data which

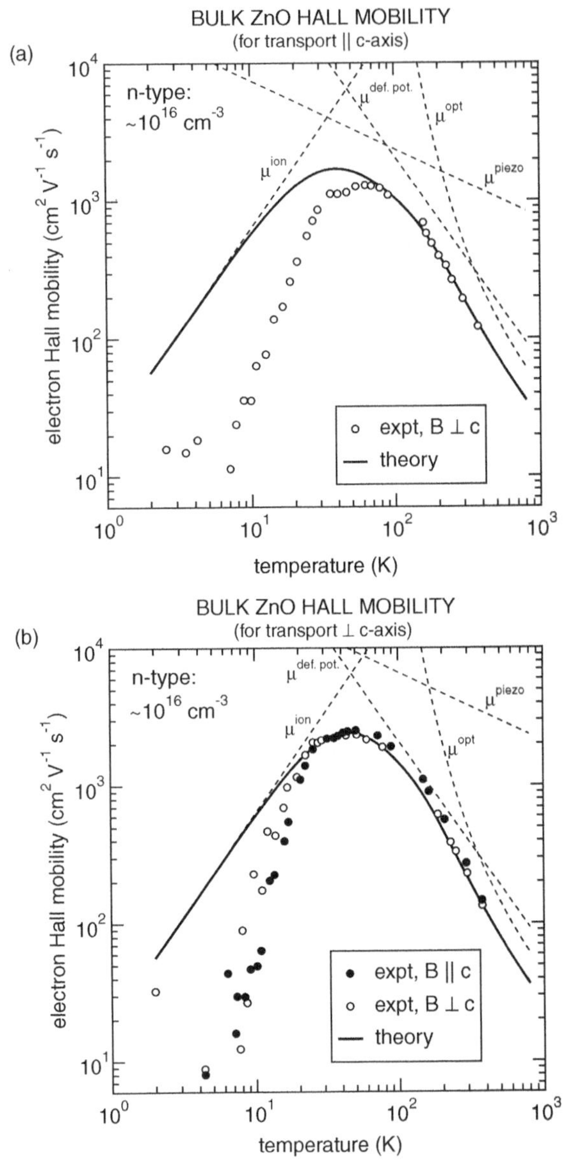

Figure 1.2 *Hall mobility vs temperature of nominally undoped ZnO. Current transport is parallel in (a) and perpendicular in (b) to the wurtzite c-axis. In each case, the magnetic field is perpendicular to the current flow. Superimposed are theoretical curves corresponding to separate scattering mechanisms. Reprinted from R. Helbig, P. Wagner, Halleffekt und anisotropie der beweglichkeit der elektronen in ZnO,. Phys. Chem. Solids 35, 327. Copyright (1974) with permission from Elsevier*

were at the time obtained on single crystals grown through the use of relatively modern bulk techniques such as the seeded vapor phase transport (SVPT) method. The SVPT-grown sample in this case was seen to exhibit a temperature-dependent mobility which compares very favorably to the best results from the older literature (shown in Figure 1.2),

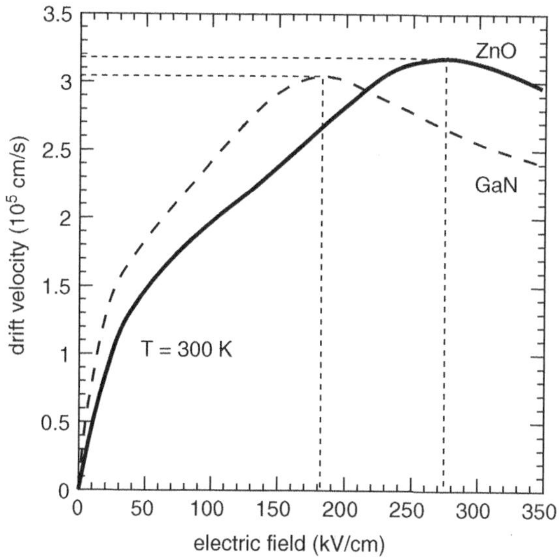

Figure 1.3 *Theoretical electron drift velocity vs electric field for ZnO (solid line) and GaN (dashed line). Though GaN exhibits higher mobility at low fields, ZnO is expected to have greater saturated drift velocity at high fields. Reprinted from J. D. Albrecht, et al., High field electron transport properties of bulk ZnO. J. Appl. Phys. 86, 6864. Copyright (1999) with permission from American Institute of Physics*

with peak mobility of $\approx$1900 $cm^2\,V^{-1}\,s^{-1}$, once again occurring near 60 K.[23] Analysis of the temperature-dependent mobility data[23] indicates for this sample the presence of the two shallow donors mentioned above (31 and 61 meV at donor concentrations of 9×10^{15} and $1 \times 10^{17}\,cm^{-3}$, respectively) along with a shallow acceptor level ($N_A = 2 \times 10^{15}\,cm^{-3}$; E_A not given).[23]

In addition to the list of above-mentioned substitutional n-type species may be added interstitial H which, while normally entering as an amphoteric species for most other semiconductors (either as H^+ or H^- as needed to effect passi-vation), for ZnO enters only in the positive charge state, and thus is able to contribute to doping rather than passivation through the formation of a O–H complex, according to a recent theoretical prediction by Van de Walle.[20] This atypical behavior of H when in ZnO is attributed to the strong O–H bond which forms and functions as an effective donor "atom" sitting on an O site.[20] As an aside, it is interesting to observe that the long-standing p-type-dopability problem in GaN was finally resolved once it was realized that, in that case, H acts as a passivating agent (through formation of complexes with the intended acceptor species). Evidently, though the roles played by H in ZnO and GaN appear to be entirely different, it would appear nonetheless important to minimize the H concentration for purposes of achieving p-type material.

1.4.3 p-type Doping and Dopability

For ZnO to be technologically successful as a UV-range optoelectronic device material, an important growth issue is the ability to reliably produce both n-type and p-type ZnO materials. As noted above, n-type doping is readily achieved (and occurs natively through Zn_i and possibly other defects in stoichiometry). However, like most wide-band-gap semiconductors, including other II–VI compounds[25] as well as III–V refractory-nitride compounds AlN (presently) and GaN (previously), ZnO exhibits an asymmetric ease of dopability one way (n-type) but not the other way (p-type). The physical origins of such dopability asymmetries in the II–VIs generally have been reviewed recently by Desnica;[25] briefly, doping-limiting mechanisms common to II–VI semiconductors include self-compensation (via native defects, lattice relaxation, amphoteric incorporation), solubility limitations, activation-efficiency limitations, in addition to other mechanisms which do not apply to higher hardness materials such as ZnO.[25]

Prior to 1999 there was only one report[26] of p-ZnO and in that case both material quality and growth reproducibility were poor. Since that time have come a number of papers, at first mostly theoretical, and more recently experimental, seeking to elucidate or obviate the difficulties of ZnO p-type doping. Among the conclusions from the aforementioned body of theoretical work[18–20,27,28] –all since 1998 and all based on *ab initio* band calculations–are the following explanations and conclusions concerning the long-standing p-type-doping problem in ZnO and how it might be solved.

According to Yamamoto and Katayama-Yoshida,[27] the doping asymmetry of ZnO is the consequence of fundamentally opposing effects of doping type on Madelung energy [n-type (p-type) doping decreases (increases) the Madelung energy]. If this is the governing mechanism, then, the proper strategy to attain p-ZnO should be to co-dope with both donors and acceptors, for example, p-ZnO:(N,Ga).

An alternative explanation was put forth by Zhang *et al.*[18] who considered the case of intrinsic ZnO and examined through *ab initio* theory the energetics of formation of native point defects as well as the donor or acceptor level(s) that these defects introduce. For each type of native stoichiometric point defect (e.g. Zn_i, Zn_O, V_O, which are all the possible native donors in ZnO, and O_i and V_{Zn} which are all the acceptors) they systematically computed, for each defect, its formation enthalpy and any donor or acceptor level(s) which it intro-duced. They concluded that native ZnO should only be able to come out n-type but never p-type, assuming growth under quasi-equilibrium conditions.[18] The argument could be stated as follows: Under Zn-rich growth conditions, Zn_i, a shallow donor, should form readily due to favorable thermodynamics, and the only possible compensators (O_i, V_{Zn}) both have unfavorable thermodynamics for formation. Thus, the prediction is that nominally undoped ZnO grown under Zn-rich conditions (again through equilibrium techniques) should come out n-type due to the existence of uncompensated Zn_i donor defects. Equally, under either O-rich or Zn-rich conditions: natively p-type ZnO should not be possible because there always should be some native donor defect (V_O, Zn_i, or Zn_O) which–shallow or deep–would be favored to form thermodynamically (which one forms would depend upon the growth conditions), and hence should be available to compensate any possible native acceptor defects (O_i, V_{Zn}) which might have formed.[18] However, it is important to note that this theory specifically presupposes equilibrium (quasi-equilibrium) crystal growth; thus, it might be theoretically possible in principle through nonequilibrium techniques to "freeze in" a metastable p-ZnO film though, evidently, the origin of the acceptor would probably have to be an impurity species rather than a native point defect of ZnO.

In another *ab initio* study, this time for extrinsic ZnO, Yan *et al.*[28] considered the role of the impurity species chemical potential in addition to those of the host (Zn and O) species considered in the earlier work by Zhang *et al.*[18] just discussed. Yan *et al.*'s work predicts that through consideration together of both dopant (N) and host (O in particular) chemical potentials during equilibrium growth that the use of certain N-containing dopant species–either NO or NO_2 in particular (as opposed to the N_2 or N_2O that had been actually attempted initially but with little success)–should enable p-type doping. It is noteworthy that both Yan *et al.*'s and Zhang *et al.*'s theories provide alternative explanations to the "co-doping" concept as proposed by Yamamoto and indeed show the latter to be incorrect. Finally, in a later theoretical overview, Zunger[29] summarized on general grounds the thermodynamics of dopant incorporation ("dopant" in this context referring either to an impurity or to a native defect often spontaneously formed) by way of the formation enthalpy $\Delta H^{(D,q)}$ for an arbitrary dopant D in a host crystal wherein the dopant is formed in charge state q ($q > 0$ for donors and $q < 0$ for acceptors):

$$\Delta H^{(D,q)}(\mu, \varepsilon_F) = q\varepsilon_F + n_D(\mu_D - \mu_H)\Delta U_b. \tag{1.34}$$

In this equation, ε_F is the Fermi energy, $\mu_{D,H}$ are the chemical potentials of dopant and host, respectively, and ΔU_b denotes any excess energy associated with local chemical bonds around the dopant, relative to the otherwise undoped crystal. The $q\varepsilon_F$ term of Equation (1.34) explains the thermodynamic origin of the spontaneous formation of so-called "killer defects". For example, if one were to attempt to dope ZnO p-type via a near-equilibrium crystal-growth technique, the lowering of ε_F would make more negative the term, $q\varepsilon_F$ (since $q > 0$ for any donor spontaneously formed), thereby enhancing the spontaneous formation through thermodynamics of compensating donor defects [by making more negative the heat of formation ΔH via Equation (1.34)], other things being equal. The terms involving $\mu_{D,H}$ show the further dependence on (quasi-equilibrium) growth conditions, with these terms affecting essentially the solubility of, e.g., a substituent such as N on the O site of a ZnO crystal (the solubility obviously being enhanced under oxygen-deficient growth conditions).

A couple of experimental results dating from around 1999 have been reported which shed additional light on the p-ZnO dopability question. First, there is the report by Joseph *et al.*,[30] who sought to implement Yamamoto's co-doping idea using N and Ga as the p-type and n-type dopants, respectively. Some of the films obtained indeed appeared to be p-type, but only under specific conditions of growth and with an odd and excessive acceptor-concentration/hole-mobility trade-off.[30] In particular, the use as dopant of N_2 proved unsuccessful and use of N_2O conditionally successful, depending also on the use of plasma excitation of the N_2 or N_2O source gas {via an electron cyclotron resonance (ECR) source in this case].[30] It should be noted from GaN MBE growth that the commercially available radio frequency (RF) sources in general have proved much more successful than the ECR sources in "cracking" the N_2 source gas down to the desired neutral-atomic N species relative to the undesired ionized dimers. This dependence on the type of plasma source (ECR vs RF) might be critical for purposes of effecting p-type doping in ZnO with plasma-excited N_2 gas.

Interestingly, another experimental paper from around the same time discusses Ga and N co-dopants in ZnO films grown using both O and N radicals produced with the aid of separate RF plasma sources on the same system.[31]

This last work, which nominally set out to effect the co-doping approach of Yamamoto and Katayama-Yoshida,[27] found instead direct experimental evidence that (i) the approach can be fraught with serious problems[31] under certain circumstances; and (ii) worse, the layer nominally intended to be "p-ZnO" in fact turned out to be a layer of the spinel compound $ZnGa_2O_4$, which formed due to metallurgical reaction during growth between the ZnO and the high concentrations of Ga needed to effect the co-doping idea[27] as proposed.

Fortunately, in the years since, and especially most recently, numerous experimental groups now have reported the successful p-type doping of ZnO.[32–41] At the time of this writing, reported hole concentrations in the $\sim 10^{17}\,cm^{-3}$ range are becoming increasingly common, with occasional reports of (ostensible) holeconcentrations as high as $\sim 10^{19}\,cm^{-3}$ though the latter are often regarded with suspicion; for example, Zhao *et al.*[33] attribute such an anomalous $10^{19}\,cm^{-3}$-range hole concentration in their putative p-ZnO ultrasonic spray pyrolysis films grown on n-Si substrates to actually be a Hall-measurement misinterpretation due to the formation of a two-dimensional hole gas (2DHG) at the heterointer face between what they believe to be actually an n-ZnO film on the n-type Si substrate: band offsets and Fermi-level pinning they argue likely give rise to a 2DHG of high sheet concentration, and it is the latter which most likely is being reflected in their Hall measurements.[33]

Many of the recent p-type doping studies have embraced the co-doping Idea,[27] often involving N on an O site of ZnO as the acceptor with a column-III element such as Ga, Al, or In as the co-donor.[42–44] In a recent study, even Zr has been employed as co-donor: Pulsed-laser-deposited ZnO doped films on Al_2O_3 with hole concentrations as high as $5.49 \times 10^{19}\,cm^{-3}$ along with low resistivity ($0.026\,\Omega\cdot cm$) and not too poor mobility ($4.38\,cm^{-2}\,V^{-1}\,s^{-1}$) have been achieved in ZnO:(Zr,N) co-doped films.[36] Besides N, other column-V elements, putatively substituting for O in ZnO, have been attempted for p-type doping. For instance, a recent study based on RF magnetron-sputtered, P-doped ZnO films has shown that, by considering the effect of post-growth anneals in O_2, N_2, and Ar ambients, it can be inferred from the activation kinetics that the identity of the P-related acceptor in this case, giving hole concentrations ranging between $\sim 10^{16}$ and $10^{19}\,cm^{-3}$, is indeed substitutional P_O, and not a $P_{Zn} + 2V_{Zn}$ complex.[34]

The principal problems which remain include a generally low acceptor solubility, large ($\sim 100\,meV$) hole binding energy (implying poor dopant activation efficiency at room temperature), a low hole mobility with accompanying excessive mobility vs concentration trade-off (though unlike the aforementioned issues this one may not be fundamental), and long-term material instability. Indeed, metastable phenomena such as persistent photoconductivity, often with accompanying type-conversion (p-type to n-type) upon illumination being observed,[37] are not uncommon. Often, due to the wide range of ZnO material fabrication techniques (many leading to decidedly nonepitaxial films) involved in the ZnO doping studies, combined with the apparent heightened significance of surface-, and grain-boundary-interfacial states in determining measured electrical properties in ZnO specimens in comparison with more traditional semiconductor materials such as Si, GaAs, or even GaN, it is challenging to sift through the vast and increasing pool of ZnO doping studies and then to reach any consensus as to a "best" method or "best" dopant species to use. At present, no such consensus exists.

Summarizing this section, theoretically it appears possible in principle to achieve at least some level of p-type doping in ZnO but such structures appear fundamentally to be: (i) probably metastable if achieved experimentally (unless current theories are wrong); (ii) evidently achievable experimentally (though by no means is there a universal consensus as to the best dopants or methods to use); and (iii) difficult to precisely reproduce and often challenging to interpret experimental results due to the wide variability of ZnO microstructural quality and character (often nonepitaxial in recent studies), as well as the apparently very important, and yet-to-be clarified, role of defect states at the typically high density of surfaces and interfaces between grain boundaries.

1.4.4 Schottky Barriers and Ohmic Contacts

For any practical implementation of ZnO-related materials as a viable optoelectronic device technology, in addition to the importance of achieving control over p-ZnO will be the control over fabrication of high quality metal/semiconductor interfaces, including both ohmic contacts as well as Schottky barriers. Progress between these two are in fact interdependent since ohmic contacts ultimately will have to be developed for both n-type and p-type material layers. However, as material control of the ZnO itself improves, so also should the control and reproducibility of the metal/semiconductor interface. Indeed, control of the material quality of any metal is trivial compared with that of the semiconductor side of the interface between the two, as has been the case in the development of every technologically significant semiconductor/metal system.

Concerning Schottky barriers, virtually all work reported to date, from the pioneering work of Mead[45] in the 1960s on cleaved ZnO surfaces to the flurry of recent work[46–53] done on the best presently available (hydrothermally or pressurized-melt-grown) bulk single-crystal materials, has been on n-type ZnO due to the historical unavailability of p-type specimens. Since ZnO is a wide-gap semiconductor with significant partially ionic bonding character, one might expect[54] Schottky barrier heights on n-type material (φ_{Bn}) to obey the simple Schottky–Mott model[55] that $\varphi_{Bn} = \Phi_M - \chi_S$, where Φ_M is the work function of the metal and χ_S is the electron affinity of the semiconductor. Instead, most φ_{Bn} values for ZnO range from 0.6 to 0.8 eV regardless of metal used, with ideality factors near unity being more the exception than the rule.[47,48] Briefly, Schottky barriers have been examined for numerous noble metals, including Ag and Au,[56–58] Pt[59–63] and Pd,[46,64–66] as well as metals such as Ni and Ir.[48] Allen *et al.*[47] have observed that only a small fraction of the available φ_{Bn} values for these metals from the literature involve "good" ideality factors (e.g. ≤ 1.1), with in general a wide variability in reported barrier height values for any given metal (see Figure 1.4). Part of the variability is due to surface polarity: for several metals, including Pt, Pd, and most recently[47,48] "silver oxide" (precise stoichiometry not given), substantially higher φ_{Bn} values have been observed on Zn-polar than on O-polar {0001} surfaces.[46,47] This is explained in terms of surface-polarity-dependent band bending due to the large spontaneous polarization of ZnO along its *c*-axis.[47] Another part of the variability has been attributed to such factors as lateral inhomogeneity.[46,64] This causes φ_{Bn} measured by *I–V* to be lower than that by *C–V* measurement, as well as not only a larger, but also a linearly temperature-dependent, ideality factor.[64] But the most important part of the variability appears to be due to the influence of surface or near-surface defect states.[48–50] For example, Allen and Durbin[48]

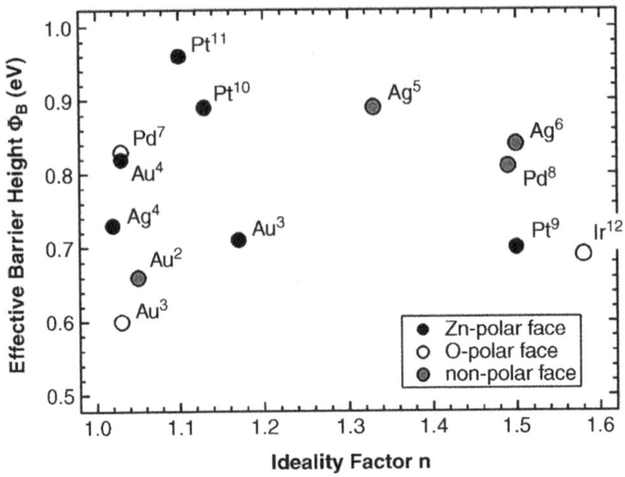

Figure 1.4 *The "best" Schottky barrier height data from the literature for several metals on n-type ZnO, from Allen et al.*[47] *Superscripts on element symbols correspond to the sources cited therein. Wide variability in reported φ_{Bn} and generally large ideality factors (≥ 1.1) are typical; a systematic dependence on surface polarity (Zn- or O-face) is also apparent. Reprinted from M. W. Allen, S. M. Durbin, and J. B. Metson, Silver Oxide Schottky contacts on n-type ZnO. Appl. Phys. Lett. 91,053512. Copyright (2007) with permission from American Institute of Physics*

have proposed that metal-induced gap states in the form of V_O, a known double donor with level approximately 0.7 ± 0.2 eV below the conduction band,[67] can form at the surface via metallurgical oxidation reaction of the metal (drawing O from the ZnO and thus forming V_O levels at enhanced concentration in a thin layer near the interface). The enhanced concentration of V_O deep donors thus formed serve to pin the Fermi level (thereby causing departure from the Schottky–Mott model of Schottky barrier formation) and might account for the previously mentioned 0.6–0.8 eV range barrier heights irrespective of metal species, plus poor ideality factor, often observed on ZnO. To support this hypothesis, the same authors investigated φ_{Bn} for silver oxide (a species already reacted with oxygen thus needing no more), and found that, unusually, the silver oxide Schottky barriers exhibited an exceptionally low (for ZnO) ideality factor of 1.06. Several other authors also have noted the beneficial effect on ideality factor of pre-treating the ZnO surface prior to metallization with an oxygen-containing chemical[59] or plasma:[50] presumably, such treatment either removes adsorbates, or reduces the near-surface V_O, or possibly does both, with the specific phenomenological details varying from metal to metal as well as depending upon inherent material quality of the ZnO specimen itself.

Concerning ohmic contacts, investigations on both n-type[68–74] and p-type[75–77] ZnO have been performed. Generally speaking, to produce the best n-type ohmic contacts the most successful approach to date has been a multilayer metallization scheme wherein the bottom layer (i.e. the layer in direct contact with ZnO) is reactive (especially if that reaction tends to remove O from ZnO) with nominally small φ_{Bn} to ZnO, and the top layer is a noble metal. For example, Kim *et al.* have investigated[70,72] Al/Pt contacts to n-type ($n \sim 10^{18}\,cm^{-3}$) ZnO, both as deposited, and, as a function of post-metallization anneal treatment. They obtain good specific contact resistance ($\sim 1 \times 10^{-5}\,\Omega\,cm^2$) even without any anneal; this they attribute to

Al/O interdiffusion resulting in the formation of a high density of near-surface V_O defects which, as we noted earlier in our discussion of Schottky barriers, constitute a double donor. This effect is enhanced upon annealing to ≈300 °C for 1 min, whereupon the specific contact resistance decreases[72] to ~$2 \times 10^{-6}\,\Omega\,cm^2$. However, higher-temperature annealing (to ≈600 °C) is seen to be deleterious: evidently, an insulating layer of aluminum oxide compound (presumably Al_2O_3) forms interfacially (hence an insulator becomes interposed between semiconductor and metal layers) when the annealing time/temperature become too high, resulting in an increase (rather than a decrease) in specific contact resistance. A somewhat analogous set of phe-nomenology is found in connection with a different[73,74] metallization scheme: Ti/Al/Pt/Au. In this case, annealing beyond even ≈200 °C is sufficient to cause degradation.[74] Finally, additional metallization schemes have been explored. Kim *et al.*, in addition to examining Ti/Au in early work[68,69] and Al/Pt subsequently,[70,72] have most recently considered a Re/Ti/Au metallization scheme,[71] wherein the latter proved to be metallurgically stable even following a high-temperature (≈700 °C, 1 min) anneal treatment as well as resulting in superior specific contact resistance of ~$2 \times 10^{-7}\,\Omega\,cm^2$ (for comparable ZnO doping levels) in comparison with Al/Pt.

Studies of ohmic contacts to p-type ZnO are fewer to date but are becoming increasingly reported as p-type ZnO material availability and quality steadily improve. In 2005, the same Korean group (Kim *et al.*) whose n-type ohmic contact studies were overviewed above also have looked at p-type material. In one study they employed a Ni/indium-tin-oxide scheme[76] and in the other a Ni/Au scheme.[75] In both, Ni from the metallization interdiffused and reacted metallurgically with the Zn from ZnO to form many intermetallic phases, ultimately yielding decent preliminary specific contact resistances (~$10^{-4}\,\Omega\,cm^2$) in each case.[75,76] Lastly, in a separate study of Au/Ni contacts by Mandalapu *et al.*[77] involving rapid thermal annealing (≈800 °C, 1 min), it was conjectured that the good specific contact resistance (again ~$10^{-4}\,\Omega\,cm^2$) was the result of Zn vacancies (due to Zn outdiffusion), coupled with activated Sb atoms, to increase the surface hole concentration which enables the formation of ohmic I–V characteristics at the metal/ZnO junction.

1.5 Band Gap Engineering

In this section we consider the prospects and survey the current status of attempts at band gap engineering within the (Zn,Mg,Cd)O heteroepitaxial system. In analogy with other semiconductors, the use of heterojunctions to confine carriers and/or light has played an important role in the performance optimization of modern commercial optoelectronic devices [e.g. separate-confinement heterostructure (SCH) lasers]. Additionally, tailoring the band gap permits a range of wavelength-tunability in the device design. Of particular significance given the present commercial and security interests in achieving UV-range optoelectronics is the possibility for pushing ZnO to even shorter wavelengths through alloying with larger-gap materials which are compatible with the crystal structure of ZnO so as to enable heteroepitaxial integration within a transparent-oxide II–VI ZnO-based system, in analogy to the (In,Al,Ga)N III–V system. Finally, in the event that adequate control for device requirements of p-type doping in ZnO proves too difficult to be practical (especially in the short term), a novel alternative approach might involve the direct injection of holes via an appropriate heterojunction combination with other suitably matched heterostructures including heterovalent combinations with III–V nitrides or with, as shall be described below, certain spinel compounds. A very brief look at some possibilities will be considered next.

1.5.1 Homovalent Heterostructures

Theoretically, the hypothetical, wurtzite-equivalent band gaps for MgO, ZnO, and CdO, all direct, are believed to correspond closely in magnitude to those of the nitrides AlN, GaN, and InN, respectively, with MgO and CdO each forming a type-I band alignment to ZnO, and with relatively smaller band offsets (as well as smaller hydrostatic deformation potentials) expected theoretically for the valence bands in comparison with the conduction band.[78] Experimentally, it is possible to enlarge the ZnO band gap through formation of wurtzite $Mg_xZn_{1-x}O$ alloys provided that x is not too large. Similarly, it is possible to create ZnO based separate-confinement structures through alloying with $Cd_yZn_{1-y}O$, again, with the restriction that y remain sufficiently small. Since only relatively recently has it become possible to fabricate ZnO itself of sufficient epitaxial quality to warrant extrapolation to ZnO-based heterostructures, presently there is not an extensive body of research to be found concerning ZnO-based heteroepitaxy.

The pioneering experimental work in this area has been accomplished by a group of researchers[79–81] based in Japan; this group was the first to significantly extend the compositional range of achievable (Zn,Mg,Cd)O alloys while retaining a reasonable degree of material quality.

Figure 1.5 summarizes the experimental state of affairs in terms of achievable ternary compositions, band gaps, and lattice constants involving $Mg_xZn_{1-x}O$ and $Cd_yZn_{1-y}O$, as of 2001.[81] Before we consider this figure in detail, it is important not to make too facile a comparison with the status of much more mature (and from the material-synthesis standpoint much less formidable) material systems such as Si–Ge, the III–V arsenides, phosphides, antimonides, or even the nitrides. In sharp contrast to any of the aforementioned systems, the material constraints imposed by thermodynamics within the (Zn,Mg, Cd)O system are fundamentally far more difficult to surmount. In particular, both CdO and MgO crystallize only in the NaCl (i.e. rocksalt) structure whereas ZnO exists only in wurtzite form. This basic structural incompatibility occurs in no other established heteroepitaxial system (including the nitrides). Not surprisingly, MgO–ZnO and CdO–ZnO pseudobinary systems have extremely limited miscibility: in thermodynamic equilibrium, the reported limits in alloy composition are $x \leq 4\%$ for the $Mg_xZn_{1-x}O$ ternary and $y \leq 2\%$ for the case of $Cd_yZn_{1-y}O$. Thus, nonequilibrium growth techniques will be essential to extend these limits; moreover, any structures so produced will be structurally metastable and subject to phase segregation beyond certain time and temperature limits of device processing.[80]

Returning now to Figure 1.5, despite the warning given, the results to date do bode very optimistically for extending the working wavelength range of prospective ZnO-based optoelectronics, in particular, in pushing it further into the UV. We see that, in metastable thin film form, $Mg_xZn_{1-x}O$ films grown by the pulsed laser deposition technique[81] have been grown out to $x_{max} \approx 33\%$, corresponding to a direct band gap close to 4.0 eV. Comparable alloy compositions have since been achieved through other nonequilibrium growth techniques such as plasma-assisted molecular beam epitaxy (P-MBE).[82] Simi-

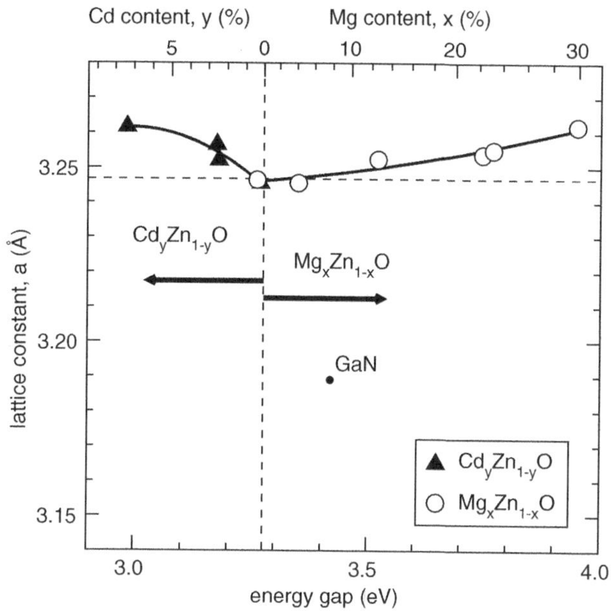

Figure 1.5 *Lattice constant vs band gap for $Mg_xZn_{1-x}O$ and $Cd_yZn_{1-y}O$ ternary films produced by pulsed laser deposition. Reprinted from T. Makino, et al., Band gap engineering based on $Mg_xZn_{1-x}O$ and $Cd_yZn_{1-y}O$ ternary alloy films. Appl. Phys. Lett. 78, 1237 (2001).*

larly, from the figure we also see that for $Cd_yZn_{1-y}O$ the miscibility limits are even more stringent than for $Mg_xZn_{1-x}O$ (metastable $y_{max} \approx 7\%$), even in epitaxial form, with the minimum Band gap (at the most Cd-rich concentration) occurring near 3.0 eV. Fortunately, the compositional range of metastable wurtzite $Cd_yZn_{1-y}O$ films has since been extended substantially, with $y_{max} \approx 69\%$, corresponding to band gaps as small as $\approx$1.8 eV, through the use of nonequilibrium growth techniques such as remote plasma-enhanced metal organic chemical vapor deposition (RPE-MOCVD) which, owing to the use of plasma excitation, enables epitaxial growth at unusually low substrate temperatures (325–400 °C) as needed for the formation of the metastable films.[83]

Recent theoretical studies tend to support the above experimental findings. An *ab initio* study of relative thermodynamic stabilities for rocksalt vs wurtzite $Mg_xZn_{1-x}O$ alloy hypothetical single crystals by Fan *et al.*[84] has confirmed the oretically, upon comparing formation enthalpies as the basis for assessing relative thermodynamic stability, that the wurtzite phase remains more stable than the rocksalt phase for sufficiently low Mg content ($x \leq 37\%$). (Of course, this assessment merely concerns the relative stability of two metastable forms: fundamentally, neither wurtzite nor rocksalt $Mg_xZn_{1-x}O$ are stable with respect to spontaneous phase segregation into a ZnO + MgO mixture, consistent with the experimentally well known 4% solubility limit for MgO in ZnO mentioned earlier.) Similarly, at high Mg concentration ($x \geq 50\%$), rocksalt theoretically is more stable than wurtzite.[84] These predictions conform nicely to the empirical observations that metastable $Mg_xZn_{1-x}O$ layers grown with $x \leq 33\%$ are found to consist of uniform wurtzite films whereas those at high concentrations (over 65%) are found to be uniform rocksalt films, with films of intermediate composition yielding a wurtzite–rocksalt mixture.[82] The same *ab initio* study[82] also considered $Cd_yZn_{1-y}O$ and reached analogous conclusions; theoretical results once are again consistent with experiment. Specifically, wurtzite should be favored over rocksalt structure for Cd content all the way up to $y \approx 70\%$.[82]

1.5.2 Heterovalent Heterostructures

An interesting way to circumvent the ZnO p-type dopability problem might be to form a heterojunction with a material which can be doped p-type readily, and then to devise a device structure which allows the ZnO layer to function as an n-type active layer in the overall structure. An early step in this direction was reported by Ohta *et al.*,[85] who fabricated and demonstrated successfully a working p-n light-emitting diode (LED) structure involving the heterojunction combination: p-$SrCu_2O_2$/n-ZnO. For this device a hard rectifying *I–V* diode characteristic was obtained with turn-on at 3 V and a prominent electroluminesence (EL) peak at 382 nm which the authors attributed to electron–hole plasma recombination in ZnO.[85] Both the electrical and spectral characteristics for this heterojunction LED appear superior to those of a homojunction ZnO LED (epitaxial p-ZnO on a n-ZnO substrate) reported by Aoki *et al.*[86] which resulted in a softer and leakier *I–V* characteristic as well as an EL spectrum which was much broader and noisier despite a strong, band-gap-dominated photoluminescence (PL) spectrum. Possibly, the different outcomes are in part due to the fact that only for the heterojunction device were all of the active layers entirely epitaxial.[85,86]

Since these earliest device attempts, much additional work on heterovalent heterostructures has been undertaken. To date, the following structures, each involving an n-type-ZnO layer in combination with a p-type non-II–VI material, have been re ported: n-ZnO/p-$SrCu_2O_2$ (already mentioned above);[85] n-ZnO/p-GaN;[87] n-ZnO/p-$Al_xGa_{1-x}N$;[88] n-ZnO/p-SiC;[89] as well as even more exotic heterostructural combinations such as n-$Mg_xZn_{1-x}O$/n-ZnO/p-$Al_xGa_{1-x}N$/p-GaN.[90] In 2007, "full-color electroluminescence" LEDs were reported, with individual RGB pixel color determined by the Cd content of a $Cd_yZn_{1-y}O$ quantum well (with $y = 7\%$, 17%, and 58% for blue, green, and red emission at room temperature, respectively).[91] In this device, the appropriate quantum well was sandwiched between a p-type-SiC layer (substrate) and an n-$Mg_{0.04}Zn_{0.96}O$ barrier layer with a capping layer of n-ZnO.[91] Overall, the device results look encouraging for electrically injected UV ZnO-based emitters but also demonstrate that achieving good electrically injected UV emission will be considerably more challenging than merely achieving good optically-pumped emission in ZnO-based materials.

1.6 Spintronics

ZnO-based dilute magnetic semiconductors can be manipulated in these low-dimensional tailored thin films for various spin-based devices to unprecedented capabilities.[92,93] In these materials, the transition metal ions are substitutional on the cation sites of ZnO, giving (Zn,TM) O. In modeling these interactions, one can use the Heisenberg Spin Hamiltonian given by

$$\mathcal{H}_{\text{eff}} = \text{const.} - 2\sum_{i<j} J_{ij}\hat{S}_i \cdot \hat{S}_j \tag{1.35}$$

where J_{ij} is the exchange term and $\hat{S}_i$ is the spin generator. There are two cases of exchange interactions: (1) the d–d superexchange between d electrons of the magnetic ions; and (2) the sp–d exchange between the d electrons and the band electrons or holes. In general this interaction is antiferromagnetic for conduction band states and ferromagnetic for valence band states. One can use the Curie–Weiss law to obtain the Curie temperature, T_C, from the magnetic measurements:

$$\frac{M}{H} = \frac{C_0 x}{T - \Theta_0 x} \tag{1.36}$$

where M is the magnetization, H is the magnetic field, T is the temperature, x is the transition metal composition, and Θ_0 and C_0 are the Curie–Weiss temperature and the Curie constant for $x = 1$, respectively. The Curie constant is expressed as[94]

$$C_0 = \frac{(g\mu_B)^2 S(S+1)_n}{3k_B} = C_n \rho / \mu \tag{1.37}$$

where g is the intrinsic Landé g factor, n is the number of cation sites per unit volume, C_n is the Curie constant for $x = 1$, ρ is the mass density, μ is the molar mass, S is the effective spin per TM ion ($S = 5/2$ for $n = 2$), μ_B is the Bohr magneton and k_B is the Boltzmann constant. Using Equations (1.35) and (1.36) one can estimate Θ_0 and C_n. The effective exchange integral between the nearest-neighbor transition ions can be estimated from

$$\frac{J_1}{k_B} = \frac{3\Theta_0}{2zS(S+1)} \tag{1.38}$$

where z equals the number of nearest neighbors; $z = 12$ for the wurtzite lattice. A negative value of J_1 indicates that the interactions between the magnetic ions are antiferromagnetic, suggesting that superexchange is likely to be the dominant mechanism.

For an external magnetic field B applied along the z-axis direction, the magnetization M_z of the dilute magnetic semiconductor (DMS) alloy containing Mn^{2+} ions is empirically written as

$$M_z = (N_0 \langle S_x \rangle) + N_0 S_{\text{sat}} B_{5/2}(5\mu_B B / k_B T_{\text{eff}}) \tag{1.39}$$

where N_0 is the density of Mn^{2+} ions and $B_{5/2}(x)$ is the Brillouin function for $S = 5/2$, S_{sat} is the saturation value for the spin of the individual Mn^{2+} ions and the rescaled temperature is $T_{\text{eff}} = T + T_0$. Along with the distribution of magnetic ions on a DMS lattice, isolated spins, pairs of spins, and triplets are distributed; hence, the magnetization is dominated by the paramagnetic response of isolated single spins, which are antiferromagnetically coupled. However, if a DMS heterostructure is considered, the ferromagnetic sp–d exchange interaction between conduction electrons and local moments results in the enhanced electromagnetic spin splitting as

$$\Delta E = g\mu_B B + f(\psi)(N_0\alpha)\langle S_z \rangle \tag{1.40}$$

where $f(\psi)$ is the wave function overlap between the combined and local moments and α is the sp–d exchange integral. The spin splitting is dominated by the second term in

Equation (1.40) because the intrinsic g factor for electrons in ZnO is small.[95] The exchange integral for the heavy-hole states is ~5 times larger than that for the conduction band and light hole states. For this reason, the spin splitting is dominated by that of the valence band states.

Progress toward understanding the (Zn,TM)O band materials and, in particular, the mechanism limiting the Curie temperature, T_C, requires careful investigation of the transport behavior; that is, it requires precise knowledge of the hole density p. The Hall resistivity in magnetic materials is given by

$$\rho_{xy} = \rho_{xy}^{0} + \rho_{xy}^{\alpha} = R_0 B + R_\alpha(\rho_{xy})M \tag{1.41}$$

where the normal contribution ρ_{xy}^{0} is proportional to the external magnetic field B, $R_0 = 1/pe$ and the anomalous contribution ρ_{xy}^{α} is proportional to the macro-scopic magnetization M. The term R_α arises from the spin-orbit interaction, which include anisotropy between scattering of spin-up and spin-down electrons. The Hall measurements should be carried out in the applied magnetic field limit where magnetization is saturated (low T and high H).

In dilute magnetic systems, ferromagnetism has been accounted for by the formation of bound magnetic polarons.[95–103] The bound magnetic polarons are formed by alignment of the spins of the transition metal ions with that of a much lower number of weakly bound carriers such as excitons within a polariton radius. The localized holes of the polarons interact with the TM impurity surrounding them, thus producing an effective magnetic field and aligning the spins. The interaction distance grows with decreasing temperature. Neighboring magnetic polarons overlap and interact via magnetic impurities forming correlated clusters of polarons. One has a ferromagnetic transition when the size of such clusters is equal to the size of the sample. This model is inherently attractive for low-carrier-density systems such as many of the electronic oxides. The polaron model is applicable to both p-type and n-type materials.[99] Even though the direct exchange interaction of the localized holes is antiferromagnetic, the interaction between bound magnetic polarons may be ferromagnetic for large concentrations of magnetic impurities. This enables ferromagnetic ordering of the Mn^{2+} ions in an otherwise insulating or semi-insulating material.

Dietl *et al.* [104] and Dietl[105] evaluated the Curie temperature for ZnO. The short bond length leads to a strong coupling between holes (on anions) and the spin of the transition metal ions on the Zn site. They showed that the holes in the extended or weakly bound states mediate the long-range interactions between the localized spins on both sides of the metal–insulator transition. This allows the use of the mean-field approximation, giving that a competition exists between the ferromagnetic and antiferromagnetic interactions. The T_C is proportional to the density of the Mn^{2+} ions and the hole density.

The theory by Dietl *et al.*[104] assumes that ferromagnetic correlations among the Mn^{2+} ions, which provide the spin, are mediated by the holes from the shallow acceptors. This model suggests that the carrier-mediated ferromagnetism in n-type materials may be observed only at low temperature with shallow donors while high temperatures for p-type materials with $T_C > 300\,K$ are predicted for ZnO.

1.7 Summary

In this chapter, we have reviewed aspects of the fundamental electronic band structure and electrical characteristics of intrinsic and doped ZnO as well as having surveyed recent important theoretical and experimental developments which, taken together, bode encouragingly for the future of epitaxial ZnO for eventual utilization as a UV-range optoelectronic device material. While ZnO in poly-crystalline form has enjoyed diverse technological use for decades, only now, through the advent of modern nonequilibrium crystal-growth techniques, combined with recent key theoretical advances (especially as concerns the long-standing p-type dopability question) does the promise, long held, of efficient UV-range epitaxial-ZnO-based optoelectronics finally appear to be reachable. As we have described, the principal inherent advantage of ZnO–in relation to other comparably wide-gap UV-range semiconductors–is the exceptionally deep exciton binding energy: at 60 meV, the binding energy of ZnO is three times that of rival GaN and fundamentally provides a highly efficient excitonic recombination mechanism which dominates at room temperature and even above. Within only the past decade, remarkable progress has been made in key areas, all essential to realizing the promise of using ZnO as the active layers of epitaxial devices. These include: quality improvement of bulk ZnO (substrate material), experimental demonstration of p-type doping, successes in the synthesis of heteroepitaxial (Mg,Cd,Zn)O metastable material combinations, and reports of significant progress in the fabrication of ohmic and Schottky contacts to ZnO. The advantage to the quality of epitaxy through having available a homoepitaxial substrate (which is lacking for GaN) cannot be underestimated. To date, numerous demonstrations of electrically injected ZnO-based LED devices have been made, with preliminary reports of electrically injected stimulated emission now beginning to appear, showing directly that, despite the long developmental road which lies ahead (which hopefully will be shortened some-what through lessons learned from the nitrides) that the future shines brightly for ZnO. Finally, over the last four years, thanks to a Curie temperature above room temperature, the potential use of ZnO in spintronics applications appears particularly promising.

References

[1] D. C. Reynolds and T. C. Collins, *Excitons, Their Properties and Uses*, Academic Press, New York, 1981, p. 29.
[2] M. Belkanski and J. des Cloizeaux, *J. Phys. Radium* **21**, 825 (1960).
[3] S. Adler and J. L. Birman, *Gen. Teleph. Electron. Corp. Res. Dev. J.* **1**, 2 (1961).
[4] J. J. Hopfield, *J. Phys. Chem. Solids* **15**, 97 (1960).
[5] D. G. Thomas, *J. Phys. Chem. Solids* **15**, 86 (1960).
[6] M. Cardona, *Phys. Rev.* **129**, 69 (1963).
[7] C. J. Vasely, R. L. Hengehold and D. W. Langer, *Phys. Rev.* **35**, 2296 (1972).
[8] R. A. Powell, W. E. Spicer and J. C. McMenamin, in *Electron Spectroscopy*, edited by D. A. Shirley, North-Holland, Amsterdam, 1972, p. 575.
[9] S. K. Hong, T. Hanada, H. Makino, H.-J. Ko, Y. Chen, T. Yao, A. Tankaka, H. Sasaki, S. Sato, D. Imai, K. Araki and M. Shinohara, *J. Vac. Sci. Technol. B* **19**, 1429 (2001).

[10] D. C. Reynolds, D. C. Look, B. Jogai, C. W. Litton, G. Cantwell and W. C. Harsch, *Phys. Rev. B* **60**, 2340 (1999).
[11] W. R. L. Lambrecht, A. V. Rodina, S. Lumpijumnong, B. Segall and B. K. Meyer, *Phys. Rev. B* **65**, 075207 (2002).
[12] R. G. Wheeler and J. O. Dimmock, *Phys. Rev.* **125**, 1805 (1962).
[13] G. E. Pikus, *Fiz. Tverd. Tela* **6**, 324 (1964) [*Sov. Phys. Solid State* **6**, 324 (1964)]
[14] D. W. Langer, R. N. Euwena, K. Era and T. Koda, *Phys. Rev. B* **2**, 4005 (1970).
[15] Landolt-Börnstein, *Physics of II-VI and I-VII Compounds, Semimagnetic Semiconductors*, edited by O. Madelung, Springer, New York, 1982, Group III, Vol. 17b
[16] H. Sheng, S. Mulhukumar, N.W. Emanetogler, and Y. Lu, *Appl. Phys. Lett.* **80**, 2132 (2002).
[17] T. Minami, H. Sato, H. Nanto and S. Takata, *Jpn. J. Appl. Phys.* **24**, L781 (1985).
[18] S. B. Zhang, S.-H. Wei and A. Zunger, *Phys. Rev. B* **63**, 75 205 (2001).
[19] E.-C. Lee, Y.-S. Kim, Y.-G. Jin and K. J. Chang, *Phys. Rev. B* **64**, 085120 (2001).
[20] C. G. Van de Walle, *Phys. Rev. Lett.* **85**, 1012 (2000).
[21] R. Helbig and P. Wagner, *J. Phys. Chem. Solids* **35**, 327 (1974).
[22] J. D. Albrecht, P. P. Ruden, S. Limpijumnong and W. R. L. Lambrecht, *J. Appl. Phys.* **86**, 6864 (1999).
[23] D. C. Look, *Mater. Sci. Eng., B* **80**, 383 (2001).
[24] D. C. Look, J. W. Hemsky, J. R. Sizelove, Phys. Rev. Lett. **82**, 2552 (1999).
[25] U. V. Desnica, *Prog. Cryst. Growth Charact.* **36**, 291 (1998).
[26] K. Minegishi, Y. Koiwai, Y. Kikuchi, K. Yano, M. Kasuga and A. Shimizu, *Jpn. J. Appl. Phys.* **36**, L1453 (1997) and references therein.
[27] T. Yamamoto and H. Katayama-Yoshida, *Jpn. J. Appl. Phys.* **38**, L166 (1999).
[28] Y. Yan, S. B. Zhang and S. T. Pantelides, *Phys. Rev. Lett.* **86**, 5723 (2001).
[29] A. Zunger, *Appl. Phys. Lett.* **83**, 57 (2003).
[30] M. Joseph, H. Tabata and T. Kawai, *Jpn. J. Appl. Phys.* **38**, L1205 (1999).
[31] K. Nakahara, H. Takasu, P. Fons, A. Yamada, K. Iwata, K. Matsubara, R. Hunter and S. Niki, *Appl. Phys. Lett.* **79**, 4139 (2001).
[32] H. B. Ye, J. F. Kong, W. Z. Shen, J. L. Zhao and X. M. Li, *Appl. Phys. Lett.* **90**, 102115 (2007).
[33] J. L. Zhao, X. M. Li, A. Krtschil, A. Krost, W. D. Yu, Y. W. Zhang, Y. F. Gu, and X. D. Gao, *Appl. Phys. Lett.* **90**, 062118 (2007).
[34] D. K. Hwang, M. S. Oh, J. H. Lim, C. G. Kang and S. J. Park, *Appl. Phys. Lett.* **90**, 021106 (2007).
[35] W. Guo, A. Allenic, Y. B. Chen, X. Q. Pan, Y. Che, Z. D. Hu and B. Liu, *Appl. Phys. Lett.* **90**, 242108 (2007).
[36] H. Kim, A. Cepler, M. S. Osofsky, R. C. Y. Auyeung and A. Pique, *Appl. Phys. Lett.* **90**, 203508 (2007).
[37] Z. Y. Xiao, Y. C. Liu, R. Mu, D. X. Zhao and J. Y. Zhang, *Appl. Phys. Lett.* **92**, 052106 (2008).
[38] X. H. Pan, J. Jiang, Y. J. Zeng, H. P. He, L. P. Zhu, Z. Z. Ye, B. H. Zhao and X. Q. Pan, *J. Appl. Phys.* **103**, 023708 (2008).
[39] S. T. Tan, X. W. Sun, Z. G. Yu, P. Wu, G. Q. Lo and D. L. Kwong, *Appl. Phys. Lett.* **91**, 072101 (2007).
[40] B. J. Kwon, H. S. Kwack, S. K. Lee, Y. H. Cho, D. K. Hwang and S. J. Park, *Appl. Phys. Lett.* **91**, 061903 (2007).
[41] Z.-Q. Fang, B. Claflin, D. C. Look, L. L. Kerr and X. Li, *J. Appl. Phys.* **102**, 023714 (2007).
[42] M. Kumar, T.-H. Kim, S.-S. Kim and B.-T. Lee, *Appl. Phys. Lett.* **89**, 112103 (2006).
[43] G. D. Yuan, Z. Z. Ye, L. P. Zhu, Q. Qian, B. H. Zhao, R. X. Fan, C. L. Perkins and S. B. Zhang, *Appl. Phys. Lett.* **86**, 202106 (2005).
[44] L. L. Chen, J. G. Lu, Z. Z. Ye, Y. M. Lin, B. H. Zhao, Y. M. Ye, J. S. Li and L. P. Zhu, *Appl. Phys. Lett.* **87**, 252106 (2005).
[45] C. A. Mead, *Solid-State Electron.* **9**, 1023 (1966).
[46] M. W. Allen, M. M. Alkaisi and S. M. Durbin, *Appl. Phys. Lett.* **89**, 103520 (2006).
[47] M. W. Allen, S. M. Durbin, and J. B. Metson, *Appl. Phys. Lett.* **91**, 053512 (2007).
[48] M. W. Allen and S. M. Durbin, *Appl. Phys. Lett.* **92**, 122110 (2008).

[49] L. J. Brillson, H. L. Mosbacker, M. J. Hetzer, Y. Strzhemechny, G. H. Jessen, D. C. Look, G. Cantwell, J. Zhang and J. J. Song, *Appl. Phys. Lett.* **90**, 102116 (2007).

[50] H. L. Mosbacker, S. El Hage, M. Gonzalez, S. A. Ringel, M. Hetzer, D. C. Look, G. Cantwell, J. Zhang, J. J. Song and L. J. Brillson, *J. Vac. Sci. Technol. B* **25**, 1405 (2007).

[51] M. Nakano, A. Tsukazaki, R. Y. Gunji, K. Ueno, A. Ohtomo, T. Fukumura and M. Kawasaki, *Appl. Phys. Lett.* **91**, 142113 (2007).

[52] M.-S. Oh, D.-K. Hwang, J.-H. Lim, Y.-S. Choi and S.-J. Park, *Appl. Phys. Lett.* **91**, 042109 (2007).

[53] R.Y. Gunji, M. Nakano, A. Tsukazaki, A. Ohtomo, T. Fukumura and M. Kawasaki, *Appl. Phys. Lett.* **93**, 012104 (2008).

[54] S. Kurtin, T.C. Mcgill, and C.A. Mead, *Phys. Rev. Lett* **22**, 1433 (1969).

[55] W. Schottky, *Naturwissenschaften* **26**, 843 (1938); N. F. Mott, *Proc. Cambridge Philos. Soc.* **34**, 568 (1938).

[56] A. Y. Polyakov, N. B. Smirnov, E. A. Knozhukhova, V. I. Vdovin, K. Ip, D. P. Norton and S. J. Pearton, *J. Vac. Sci. Technol. A* **21**, 1603 (2003).

[57] A. Y. Polyakov, N. B. Smirnov, E. A. Knozhukhova, V. I. Vdovin, K. Ip, Y. W. Heo, D. P. Norton and S. J. Pearton, *Appl. Phys. Lett.* **83**, 1575 (2003).

[58] D. C. Oh, J. J. Kim, H. Makino, T. Hanada, M. W. Cho, T. Yao and H. J. Ko, *Appl. Phys. Lett.* **86**, 042110 (2005).

[59] S.-H. Kim, H.-K. Kim and T.-Y. Seong, *Appl. Phys. Lett.* **86**, 112101 (2005).

[60] S. Kim, B. S. Kang, F. Ren, K. Ip, Y. W. Heo, D. P. Norton and S. J. Pearton, *Appl. Phys. Lett.* **84**, 1698 (2004).

[61] K. Ip, B. P. Gila, A. H. Onstine, E. S. Lambers, Y. W. Heo, K. H. Baik, D. P. Norton, S. J. Pearton, S. Kim, J. R. LaRoche and F. Ren, *Appl. Phys. Lett.* **84**, 5133 (2004).

[62] K. Ip, Y. W. Heo, K. H. Baik, D. P. Norton, S. J. Pearton, S. Kim, J. R. LaRoche and F. Ren, *Appl. Phys. Lett.* **84**, 2835 (2004).

[63] S.-H. Kim, H.-K. Kim and T.-Y. Seong, *Appl. Phys. Lett.* **86**, 022101 (2005).

[64] H. von Wenckstern, G. Biehne, R. A. Rahman, H. Hochmuth, M. Lorenz and M. Grundmann, *Appl. Phys. Lett.* **88**, 092102 (2006).

[65] U. Grossner, S. Gabrielsen, T. M. Borseth, J. Grillenberger, A. Yu. Kuznetsov and B. G. Svensson, *Appl. Phys. Lett.* **85**, 2259 (2004).

[66] R. Schifano, E. V. Monakhov, U. Grossner and B. G. Svensson, *Appl. Phys. Lett.* **91**, 193507 (2007).

[67] A. Janotti and C. G. Van de Walle, *Appl. Phys. Lett.* **87**, 122102 (2005).

[68] H.-K. Kim, S.-H. Han, T.-Y. Seong and W.-K. Choi, *Appl. Phys. Lett.* **77**, 1647 (2000).

[69] J.-M. Lee, K.-K. Kim, S.-J. Park and W.-K. Choi, *Appl. Phys. Lett.* **78**, 3842 (2001).

[70] H.-K. Kim, K.-K. Kim, S.-J. Park, T.-Y. Seong and I. Adesida, *J. Appl. Phys.* **94**, 4225 (2003).

[71] S.-H. Kim, K.-K. Kim, S.-J. Park and T.-Y. Seong, *J. Electrochem. Soc.* **152**, G169 (2005).

[72] H.-K. Kim, I. Adesida, K.-K. Kim, S.-J. Park and T.-Y. Seong, *J. Electrochem. Soc.* **151**, G223 (2004).

[73] K. Ip, Y. W. Heo, K. H. Baik, D. P. Norton and S. J. Pearton, *J. Vac. Sci. Technol. B* **22**, 171 (2004).

[74] K. Ip, Y. W. Heo, K. H. Baik, D. P. Norton and S. J. Pearton, *Appl. Phys. Lett.* **84**, 544 (2004).

[75] J.-H. Lim, K.-K. Kim, D.-K. Hwang, H.-S. Kim, J.-Y. Oh and S.-J. Park, *J. Electrochem. Soc.* **152**, G179 (2005).

[76] S.-H. Kang, D.-K. Hwang and S.-J. Park, *Appl. Phys. Lett.* **86**, 211902 (2005).

[77] L. J. Mandalapu, Z. Yang and J. L. Liu, *Appl. Phys. Lett.* **90**, 252103 (2007).

[78] A. Janotti and C. G. Van de Walle, *Phys. Rev. B* **75**, 121201 (2007).

[79] A. Ohtomo, M. Kawasaki, T. Koida, K. Masubuchi, H. Koinuma, Y. Sakurai, Y. Yoshida, T. Yasuda and Y. Segawa, *Appl. Phys. Lett.* **72**, 2466 (1998).

[80] A. Ohtomo, R. Shiroki, I. Ohkubo, H. Koinuma and M. Kawasaki, *Appl. Phys. Lett.* **75**, 4088 (1999).

[81] T. Makino, Y. Segawa, M. Kawasaki, A. Ohtomo, R. Shiroki, K. Tamura, T. Yasuda and H. Koinuma, *Appl. Phys. Lett.* **78**, 1237 (2001).

[82] Z. Vashaei, T. Minegishi, H. Suzuki, T. Hanada, M. W. Cho, T. Yao and A. Setiawan, *J. Appl. Phys.* **98**, 054911 (2005).
[83] J. Ishihara, A. Nakamura, S. Sigemori, T. Aoki and J. Temmyo, *Appl. Phys. Lett.* **89**, 091914 (2006).
[84] X. F. Fan, H. D. Sun, Z. X. Shen, J.-L. Kuo and Y. M. Lu, *J. Condens. Matter* **20**, 235221 (2008).
[85] H. Ohta, K. Kawamura, M. Orita, M. Hirano, N. Sarukura and H. Hosono, *Appl. Phys. Lett.* **77**, 475 (2000).
[86] T. Aoki, Y. Hatanaka and D. C. Look, *Appl. Phys. Lett.* **76**, 3257 (2000).
[87] R. W. Chuang, R.-X. Wu, L.-W. Lai and G.-T. Lee, *Appl. Phys. Lett.* **91**, 231113 (2007).
[88] Ya. I. Alivov, E. V. Kalinina, A. E. Cherenkov, D. C. Look, B. M. Ataev, A. K. Omaev, M. V. Chukichev and D. M. Bagnall, *Appl. Phys. Lett.* **83**, 4719 (2003).
[89] D. J. Rogers, F. Hosseini Teherani, A. Yasan, K. Minder, P. Kung and M. Razeghi, *Appl. Phys. Lett.* **88**, 141918 (2006).
[90] A. Osinsky, J. W. Dong, M. Z. Kauser, B. Hertog, A. M. Dabiran, P. P. Chow, S. J. Pearton, O. Lopatiuk and L. Chernyak, *Appl. Phys. Lett.* **85**, 4272 (2004).
[91] A. Nakamura, T. Ohashi, K. Yamamoto, J. Ishihara, T. Aoki, J. Temmyo and H. Gotoh, *Appl. Phys. Lett.* **90**, 093512 (2007).
[92] Ü. Özgür, Ya. I. Alivov, C. Liu, A. Teke, M. A. Reshchikov, S. Doğan, V. Avrutin, S.-J. Cho, and H. Morkoç, *J. Appl. Phys.* **98**, 041301 (2005).
[93] C. Weisbuch and B. Vinter, *Quantum Semiconductor Structures*, Academic Press, San Diego, 1991.
[94] T. Fukumura, Z. Jin, M. Kawasaki, T. Shono, T. Hasegawa, S. Koshihara and H. Koinuma, *Appl. Phys. Lett.* **78**, 958 (2001).
[95] K. Ando, H. Saito, Zhengwu Jin, T. Fukumura, M. Kawasaki, Y. Matsumoto and H. Koinuma, *J. Appl. Phys.* **89**, 7284 (2001).
[96] M. Berciu and R. N. Bhatt, *Phys. Rev. Lett.* **87**, 129702 (2001).
[97] A. Kaminski and S. Das Sarma, *Phys. Rev. B* **68**, 235210 (2003).
[98] T. Dietl, F. Matsukura and H. Ohno, *Phys. Rev. B* **66**, 033203 (2002).
[99] S. Das Sarma, E. H. Hwang and A. Kaminski, *Phys. Rev. B* **67**, 155201 (2003).
[100] J. Warnock and P. A. Wolff, *Phys. Rev. B* **31**, 6579 (1985).
[101] M. Sawicki, T. Dietl, J. Kossut, J. Igalson, T. Wojtowicz and W. Plesiewicz, *Phys. Rev. Lett.* **56**, 508 (1986).
[102] J. M. D. Coey, M. Venkatesan and C. B. Fitzgerald, *Nat. Mater.* **4**, 73 (2005).
[103] T. Dietl and J. Spałek, *Phys. Rev. Lett.* **48**, 355 (1982).
[104] T. Dietl, H. Ohno, F. Matsukura, J. Cibert and D. Ferrand, *Science* **287**, 1019 (2000).
[105] T. Dietl, *Nat. Mater.* **2**, 646 (2003).

2

Optical Properties of ZnO

D. C. Reynolds[1], *C. W. Litton*[2] *and T. C. Collins*[3]
[1]*Semiconductor Research Center, Wright State University, Dayton, OH, USA and Air Force Research Laboratory, Materials and Manufacturing Directorate, Wright-Patterson Air Force Base, OH, USA*
[2]*Air Force Research Laboratory, Propulsion Directorate, Turbine Engine Division, Wright-Patterson Air Force Base, OH, USA*
[3]*Oklahoma State University, Stillwater, OK, USA*

2.1 Introduction

ZnO is a wide-band-gap semiconductor material and is recognized as having potential for optoelectronic device applications. Some of the properties that support this assertion are (a) low threshold power for optical pumping at room temperature,[1–3] (b) large exciton binding energy (60 meV), which may give rise to efficient UV lasing, and (c) a tunable band gap from 2.8 to 4.0 eV.[4,5] A clear understanding of recombination mechanisms is important for achieving the potential applications. In pursuit of this goal the emission and reflection spectra of ZnO have been extensively investigated and the data have been interpreted in terms of the wurtzite crystal band structure applicable to ZnO. In this study the intrinsic exciton transitions were observed in emission from several ZnO crystals.

2.2 Free Excitons

The high quality ZnO samples were cut from a 2-in. boule grown by a seeded physical vapor transport method. Photoluminescence (PL) spectral measurements were made at 2 K with the sample immersed in liquid He.

Zinc Oxide Materials for Electronic and Optoelectronic Device Applications, First Edition.
Edited by Cole W. Litton, Donald C. Reynolds and Thomas C. Collins.

The free-exciton emission spectra were investigated, with and without a magnetic field parallel to the c-axis ($\mathbf{H} \parallel \mathbf{c}$) and with the electric-field vector either parallel ($\mathbf{E} \parallel \mathbf{c}$) or perpendicular ($\mathbf{E} \perp \mathbf{c}$) to the c-axis. The experimental data for the case ($\mathbf{E} \parallel \mathbf{c}$) and ($\mathbf{H} \parallel \mathbf{c}$) are presented in Figure 2.1.

Two weak lines are observed at $\mathbf{H} = 0$. (The weak appearance of the Γ_6 line is likely due to the finite momentum of the photon,[10–12] and that of the Γ_5 line, to unwanted collection of light with $\mathbf{E}$ not exactly parallel to $\mathbf{c}$.) Furthermore, at finite $\mathbf{H}$, we can explain the strongly split line as Γ_6, and the weakly split line as Γ_5. These assignments result from the fact that Γ_6 should split as the sum of the electron and hole g-values ($g_e + g_h$), and Γ_5, as the difference ($g_e - g_h$). Since the Γ_6 splitting gives $g = 3.09$, and since $g_e = 1.95$, from references [2,12 and 18], we derive $g_h = 1.14$. This latter number agrees well with the value $g_h = 1.2$ calculated from the Zeeman splitting of a shallow neutral-donor-bound exciton.[2,12,18] The next case investigated is $\mathbf{E} \perp \mathbf{c}$, $\mathbf{H} \parallel \mathbf{c}$ shown in Figure 2.2. Here the Γ_5 line is stronger, as expected (it is now allowed), and shows little splitting, and the Γ_6 line is still weak at $\mathbf{H} = 0$ and splits strongly (as $g_e + g_h$) at finite $\mathbf{H}$.

The reflection and emission spectra, extended to higher energies in the intrinsic region, are shown superimposed in Figure 2.3. In this case, both the ground state and excited state emission from the free excitons are observed for the orientation $\mathbf{E} \perp \mathbf{c}$. From the emission spectrum, the exciton binding energies are obtained, assuming the excited states are hydrogenic. This leads to reliable band gap energies for both the A- and B-bands, summarized in Table 2.1. Note that the emission spectrum is much more detailed than the reflection spectrum.

Emission from the C-band is not observed; however, reflection for the orientation $\mathbf{H} \parallel \mathbf{c}$ shows both the ground and first excited state of the C-band. The reflection spectra

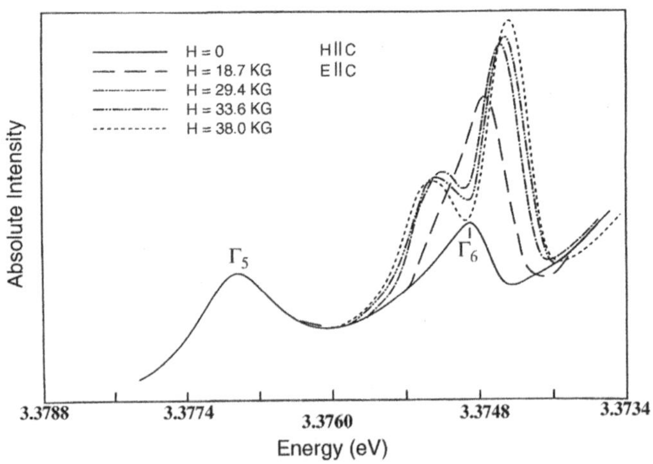

Figure 2.1 Second-order PL showing the Γ_5 and Γ_6 excitons in zero magnetic field as well as in applied fields $\mathbf{H} \parallel \mathbf{c}$, $\mathbf{E} \parallel \mathbf{c}$. Reprinted with permission from Donald Reynolds. Copyright American Institute of Physics

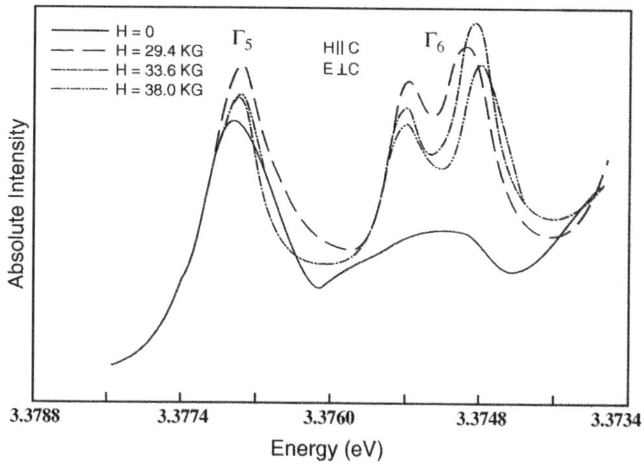

Figure 2.2 *Same as Figure 2.1 except the orientation is now* ***H*** $\|$ **c**, ***E*** $\perp$ **c**. *Reprinted with permission from Donald Reynolds. Copyright American Institute of Physics*

for the orientations **E** $\perp$ **c** and **E** $\|$ **c** are shown in Figure 2.4. The energies of the reflection minima are given for all three bands. Unfortunately, the reflection peaks do not directly give the oscillator energies associated with the three bands. The emission peak gives the energy of the oscillator, and exciton emission is observed for both the

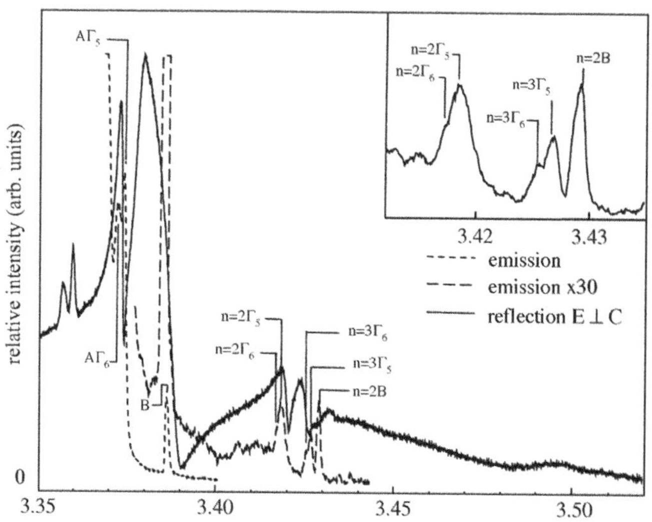

Figure 2.3 *Superposition of first-order emission and reflection spectra. An expanded view of the excited state emission transitions is shown in the inset. The energies of the emission transitions are given in Table 2.1. Reprinted with permission from Donald Reynolds. Copyright American Institute of Physics*

Table 2.1 *Parameters pertinent to ZnO*

Parameters	Values (eV)	Measured PL spectra	Derived
A-exciton Γ_5 ground state energy	3.3773	X	
$n=2$ energy	3.4221	X	
$n=3$ energy	3.4303	X	
Binding energy $n=2$	0.0597		X
Binding energy $n=3$	0.0596		X
Band gap energy	3.4370		X
A-exciton Γ_6 ground state energy	3.3756	X	
$n=2$ energy	3.4209	X	
$n=3$ energy	3.4288	X	
Binding energy $n=2$	0.060		X
Binding energy $n=3$	0.0598		X
B-exciton ground state energy	3.3895	X	
$n=2$ energy	3.4325	X	
Binding energy	0.057		X
Band gap energy	3.4465		X
E_{AB}, Γ_9-Γ_7	0.0095		X
		Measured reflection spectra	
A-exciton reflection minima	3.3776	X	
B-exciton reflection minima	3.3938	X	
C-exciton reflection minima	3.4335	X	
E_{BC}, Γ_7-Γ_7	0.0397		X
Spin-orbit parameter	0.016		X
Crystal-field parameter	0.043		X

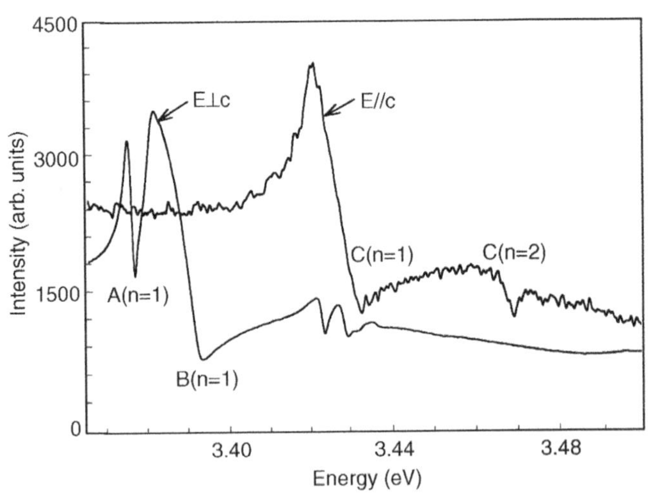

Figure 2.4 *Reflection spectra for the orientation* $\mathbf{E} \perp \mathbf{c}$ *and* $\mathbf{E} \parallel \mathbf{c}$. *The reflection minima are as follows: A, exciton 3.3776 eV; B, exciton 3.3938 eV; C, exciton 3.4335 eV; and C (n = 2), exciton 3.4700 eV. Reprinted with permission from Donald Reynolds. Copyright American Institute of Physics*

A- and B-bands. From this one can obtain the energy separation of the A- and B-bands. One can obtain an estimate of the energy separation of the B- and C-bands from the reflection minimum of each band, if one assumes that the energy of the oscillator has the same relationship to the reflection minimum in both bands. This estimate is also reported in Table 2.1.

The contributions of the spin-orbit interaction and the crystal field perturbation to the experimentally observable splittings, $E_{1,2}$ (which is the energy difference between the A and B band gaps) and $E_{2,3}$ (which is the energy difference between the B and C band gaps) have been calculated by several investigators.[7–12] For the case in which the wurtzite energy levels are treated as a perturbation of those in zinc blende, Hopfield and Thomas[12] have derived the relationships:

$$E_1 = 0$$

$$E_2 = -\frac{\delta+\Delta}{2} + \sqrt{\left[\left\{\frac{\delta+\Delta}{2}\right\}^2 - \frac{2}{3}\delta\Delta\right]}$$

$$E_3 = -\frac{\delta+\Delta}{2} - \sqrt{\left[\left\{\frac{\delta+\Delta}{2}\right\}^2 - \frac{2}{3}\delta\Delta\right]}$$

where Δ and δ represent the contributions of uniaxial field and spin-orbit interaction, respectively, to the splittings $E_{1,2}$ and $E_{2,3}$. Having observed the excited state transitions in emission from the A- and B-bands, one can make a determination of the band gap energies, the difference of which gives a reliable energy separation of those bands of 0.0095 eV. The difference between the B and C exciton transition energies is estimated from the reflection spectra since the C exciton in emission was not observed. The energies of the reflection minima were taken as the exciton transition energies, realizing that these are not the energy positions of the oscillators. Since we are concerned only with the energy difference, this allows an estimate for the value of $E_{2,3}$. Assuming that the binding energies of the B and C excitons are reasonably close, this will also be the difference between the B and C band gaps. This gives an $E_{2,3}$ value of 0.0397 eV. Substituting the $E_{1,2}$ and $E_{2,3}$ values into the quasicubic model, the spin-orbit and crystal-field parameters can be estimated. Assuming $\delta < \Delta$, the spin-orbit parameter is 16 meV, while the crystal-field parameter is 43 meV. One would expect the spin-orbit parameter to be small due to the small atomic number of oxygen. The parameters that have been determined from the exciton spectra of ZnO are compiled in Table 2.1.

Defect pair spectra have been observed in the ZnO samples being investigated.[18] This results in a number of PL lines at slightly different energies due to different pair separations. These pairs are not of the usual donor–acceptor nature, but behave simply as neutral-donor complexes; the emission then results from the collapse of excitons bound to the donors. On the high-energy side of the neutral-donor-bound exciton complex lines is a similar set of lines, which are excited states of the lower-energy complex structure. These excited states are analogous to the rotational states of the H_2 molecule. A model for the rotational states was proposed by Rorison *et al.*[19] to explain their high-magnetic-field results in InP. In this model, the donor-bound exciton D°,X is considered to be a free exciton orbiting a neutral donor; one electron was considered to be strongly correlated with the hole and the other with the donor. Some of the rotator states associated with the defect-donor-bound excitons in ZnO are shown in Figure 2.5, along with the free exciton transitions. These transitions are shown for $\mathbf{E} \perp \mathbf{c}$ in zero magnetic field, and for an applied magnetic field oriented $\mathbf{H} \perp \mathbf{c}$ and $\mathbf{H} \parallel \mathbf{c}$. In zero magnetic field, the Γ_5 free exciton as well as the Γ_5 exciton associated with D°,X rotator states are observed. When $\mathbf{H} \perp \mathbf{c}$ is turned on, the Γ_6 free exciton as well as the Γ_6 exciton associated with the rotator state appear. The 3.3702 eV Γ_6 rotator state and the 3.3714 eV Γ_5 rotator state are associated with the 3.3594 eV D°,X transition, not shown here but reported in Reynolds *et al.*[18] When $\mathbf{H} \parallel \mathbf{c}$ is turned on, both the Γ_6 free exciton and the Γ_6 rotator state show a magnetic field splitting. Assuming the unallowed Γ_6 excitons were Γ_2 excitons they would not be expected to split in a magnetic field since the Γ_2 is a singlet exciton. This is consistent with a $\Gamma_7 \times \Gamma_9$ transition rather than a $\Gamma_7 \times \Gamma_7$ transition. It is seen in Figure 2.5 that the Γ_6 free exciton and the Γ_6 rotator exciton show essentially the same magnetic field splitting. Additional splittings are observed on the high energy side of the 3.3702 eV Γ_6 rotator state. These result from Γ_6 excitons associated with other donor-bound excitons, where the Γ_6 rotators are masked by Γ_5 rotators associated with different D°,X lines in zero field. The defect pair spectra will be further discussed in Section 3.5.

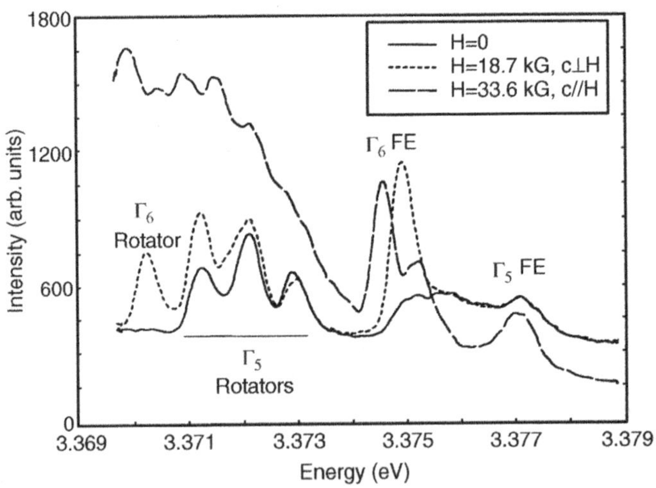

Figure 2.5 *The Γ_5 and Γ_6 free exciton transitions along with several donor-bound exciton rotator state transitions. These transitions are shown in zero field and with applied fields in the orientations $\mathbf{H} \perp \mathbf{c}$ and $\mathbf{H} \parallel \mathbf{c}$. In zero magnetic field the lowest energy Γ_5 rotator state occurs at 3.3714 eV. In an applied magnetic field ($\mathbf{H} \perp \mathbf{c}$), the lowest energy Γ_6 rotator state occurs at 3.3702 eV. In an applied magnetic field of 33.6 kG ($\mathbf{H} \parallel \mathbf{c}$), the split components of the Γ_6 free exciton occur at 3.3752 eV and 3.3746 eV, and the split components of the Γ_6 rotator state occur at 3.3704 eV and 3.3698 eV. Reprinted with permission from Donald Reynolds. Copyright American Institute of Physics*

2.3 Strain Splitting of the Γ_5 and Γ_6 Free Excitons in ZnO

A strain splitting of the Γ_5 and Γ_6 excitons in ZnO has also been observed. The splitting of exciton lines in wurtzite crystals when exposed to an applied stress was first reported by Koda and Langer.[20] In the case of wurtzite crystals all of the orbital degeneracies of the valence band are lifted by the trigonal crystal field and spin-orbit interactions. The above phenomena could not be explained by the one-electron band scheme and deformation potential theory. A theoretical interpretation was provided by Akimoto and Hasegawa.[21] They found that the combined effects of stress and the electron hole exchange interaction in a quasicubic model were able to predict the splitting and polarization pattern of the free exciton. These investigations were extended to several materials by Langer *et al.*[22] Their investigations primarily involved reflection measurements; as a result the individual excitons that make up the band were not resolved. In the analysis of Akimoto and Hasegawa[21] only the splitting of the Γ_5 exciton was treated. The Γ_6 and Γ_2 excitons were not considered since they are forbidden. In the current experiment the excitons are being observed in emission and both the Γ_5 and a forbidden exciton are resolved. In the absence of a magnetic field the forbidden exciton is only observed in samples containing in-grown strain. It would be expected that strain would relax selection rules since it changes the symmetry of the sample. Not only is the forbidden exciton observed in the presence of strain but it also splits. When viewing the Γ_5 and Γ_6 excitons in grating second order, a well-resolved splitting of the Γ_6 exciton is observed (solid line Figure 2.6). The emission line from the Γ_5 exciton is broadened, suggesting a splitting which is not resolved. In Akimoto and Hasegawa[21] it was pointed out that it is the combined effect of strain and exchange coupling which causes line splitting. If either one is zero, the line splitting is also zero. This allows the identification

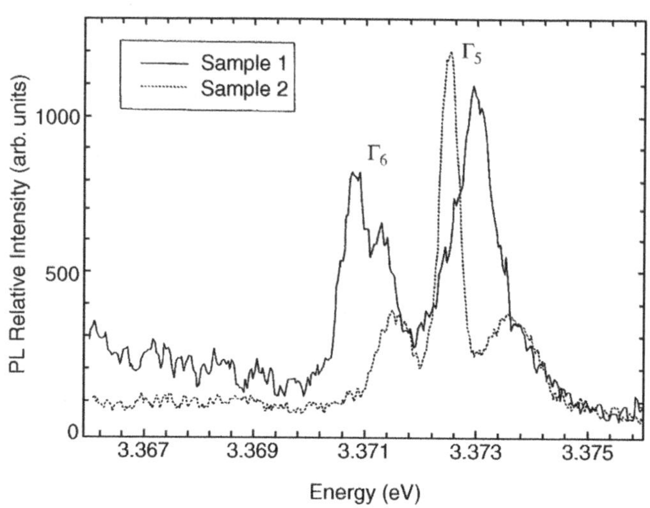

Figure 2.6 *Second-order emission spectra in Γ_5, Γ_6 free-exciton region of ZnO. Reprinted with permission from Donald Reynolds. Copyright American Institute of Physics*

of the unallowed exciton as a Γ_6 exciton rather than a Γ_2 exciton since the latter is a singlet and would not split. These data are therefore consistent with the top valence band in ZnO having Γ_9 symmetry.

2.4 Photoluminescence from the Two Polar Faces of ZnO

The crystal structure of ZnO is wurtzite and the stacking sequence of atomic layers along the "*c*"-axis is not symmetric. As a result, a ZnO crystal surface that is normal to the "*c*"-axis exposes one of two distinct polar faces, with $(000\bar{1})$ being considered the O face and (0001) the Zn face. PL measurements on the two faces reveal a striking difference. Two transitions are observed in PL from the O face that are *not* in PL from the Zn face. These lines are identified as phonon replicas of a particular D°,X transition using energy separations, excitation dependence, and time-resolved PL measurements. In addition, PL emission from free excitons and from excited states of donor-bound excitons is found to be more intense from the O face than from the Zn face.

PL spectra from the O face of the ZnO sample are shown in Figure 2.7(a) for various excitation intensities. The strong set of lines near 3.36 eV result from neutral-donor-bound exciton complexes (D°,X) associated with defect pairs.[18] The peak at 3.3735 eV is emission from the excited states of the donor-bound excitons, while the peak at 3.3779 eV is the free exciton emission. Note the strong excitation dependence of the 3.3622 eV line compared with that of other D°,X lines–at the lowest exciting intensity, it is not even visible. PL spectra from the Zn face of the same sample are shown in Figure 2.7(b). Except for a small shift in energy, the D°,X lines behave similarly on both faces. In contrast, the free exciton and D°,X excited state lines observed in the O-face PL are *essentially absent* in the Zn-face PL. A more striking contrast in PL from the two faces is evident in Figure 2.8, which displays spectra for energies below 3.35 eV. The spectra from the O face in Figure 2.8(a) display lines at 3.2367 eV and 3.3115 eV that are not evident in the spectra from the Zn face in Figure 2.8(b). These lines also depend more strongly on excitation intensity than do the surrounding spectral features, reminiscent of the 3.3622 line in Figure 2.7(a).

Since the 3.3622 eV line results from direct recombination of a particular D°,X complex, it is suggested that the 3.3115 eV line results from collapse of the same complex along with emission of an E_1-TO phonon, and the 3.2367 eV line results from collapse of the complex with emission of both an E_1-TO phonon and an E_1-LO phonon. The observed energy separations imply energies of 50.7 meV for the E_1-TO phonon and 74.8 meV for the E_1-LO phonon, in close agreement with 50.5 meV and 72.3 meV obtained from Raman measurements.[23] Curiously, no phonon replica is observed near 3.2874 implying that the D°,X complex does not collapse and excite only a single E_1-LO phonon. This may suggest that the coupling between the E_1-LO phonon and the transverse exciton is weak.

Both reduced free-exciton emission and the absence of phonon replicas suggest that the local crystal environment near the Zn face is different from that near the O face. Perhaps the Zn face is more subject to atmospheric contamination than the O face. The resulting surface state defects could lead to band tailing that would absorb the near-band-edge free-

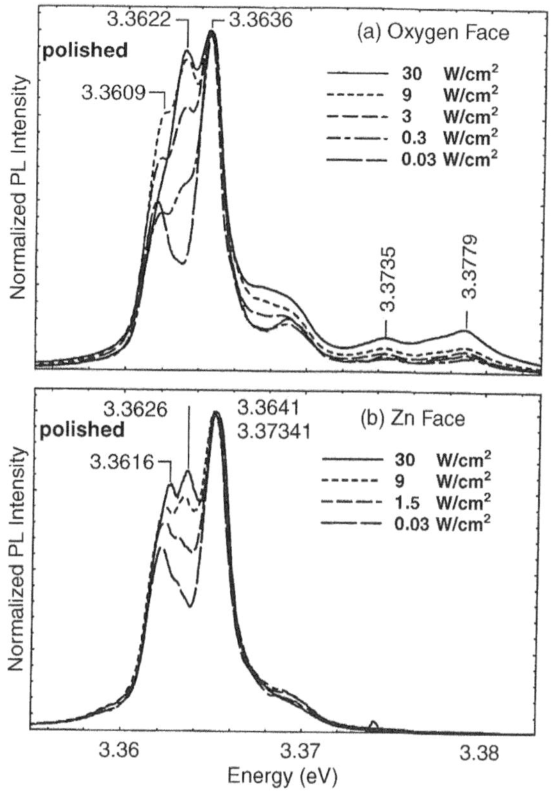

Figure 2.7 *Normalized ZnO PL spectra (above 3.35 eV) from (a) the O face and (b) the Zn face. Excitation intensities are given in the key. The O-face spectra at the two highest excitation intensities exhibited small shifts in emission energy, possibly from band-renormalization effects. These shifts were removed from the figure for improved clarity: the 30 W cm^{-2} spectrum was shifted by +0.59 meV, and the 9 W cm^{-2} spectrum was shifted by +0.25 meV. Reprinted with permission from Donald Reynolds. Copyright American Institute of Physics*

exciton and excited donor-bound-exciton emissions. The fact that PL from the Zn face clearly exhibits the parent D°,X transition [Figure 2.7(b)] but *not* the phonon replicas [Figure 2.8(b)] indicates that either the D°,X transition cannot couple to phonons near the Zn face, or the requisite phonons are not created near the Zn face. Either case might result if the Zn face is more susceptible to polishing damage than the O face, with the resulting sub-surface damage affecting the phonon spectrum.

To further verify the identification of the phonon replicas, time-resolved PL was used to measure the decay lifetimes for the 3.3622 eV, 3.3115 eV and 3.2367 eV lines. Results are shown in Figure 2.9. The data display single exponential decays for all of the transitions over several times the radiative recombination lifetime. Decay times were obtained from a least-squares fit of the data to a single exponential for intensities between 0.85 and 0.05. The decay time for the D°,X transition at 3.3622 eV was 440 ps while the decay times for the phonon replicas at 3.3115 eV and 3.2367 eV were 490 ps and 480 ps, respectively.

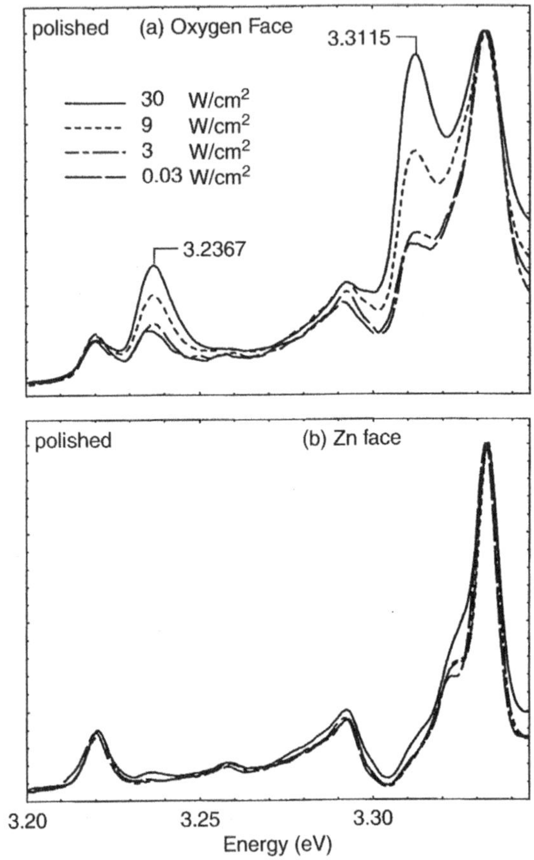

Figure 2.8 Normalized ZnO PL spectra (below 3.35 eV) from (a) the O face and (b) the Zn face. The excitation intensities in the key apply to both (a) and (b). Reprinted with permission from Donald Reynolds. Copyright American Institute of Physics

Since the phonon replicas presumably result from low-probability decay branches for the D°,X state, one would expect their intensities to track the decaying D°,X state population. This would make the replica decay lifetimes match the parent transition lifetime, as is observed.

2.5 Bound-Exciton Complexes in ZnO

Defect pair spectra have been extensively studied in GaAs.[24–34] A large number of PL lines were observed in ZnO resulting from excitons being bound to the defect pairs (the term "defect" can include both foreign impurities as well as native defects). Several of the lines were observed to be strongly polarized.[30] The polarization of the lines suggests that during growth, defect complexes are preferentially incorporated in certain crystallographic orientations. The crystal will be strained in the vicinity of the defect pairs with the strain

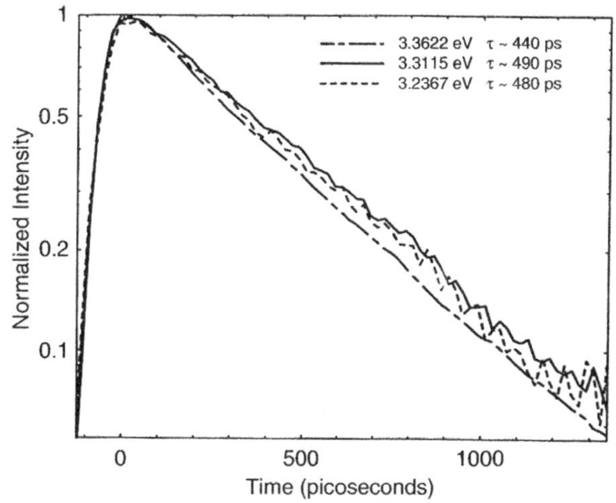

Figure 2.9 *Time-resolved PL intensity decays for the D°,X line at 3.3622 eV and its phonon replicas at 3.3115 eV and 3.2367 eV. Reprinted with permission from Donald Reynolds. Copyright American Institute of Physics*

being oriented in the direction of the pair. This results in the electric vector being parallel to the strain direction. Similar polarization properties were reported by Langer *et al.*[35] from uniaxial pressure measurements on ZnO crystals.

The defect pairs have the properties of neutral donors and the emission lines result from the collapse of excitons bound to the neutral donor complexes. The chemical make-up of the defect pair is not known, but from the PL analysis, it must simulate a neutral donor. One would speculate that in the growth process the first component of the pair would be incorporated, perhaps at a lattice site. This would then be conducive to the incorporation of the second component which would occupy a nearest neighbor lattice, or interstitial, or more distant site. The pair would then be crystallographically oriented. Annealing studies show that as the annealing temperature is increased, the higher energy PL lines disappear and, at an annealing temperature of 800 °C, essentially all of the emission intensity goes into the lowest energy emission line, which is believed to be the near neighbor alignment. It would appear that annealing results in defect diffusion which ultimately produces nearest neighbor defect pairs. The PL spectrum of this final pair also shows polarization properties which are consistent with defect pair structure.

On the high energy side of the neutral donor-bound-exciton complex lines is a similar set of lines which are believed to be excited states of the lower energy complex structure. Similar emission lines have been observed in many other materials, such as ZnSe,[36] CdTe,[37] GaAs,[37,39] and CdS.[40] These transitions were first interpreted as excited states of the neutral donor-bound-exciton D°,X but with very little detail as to their nature. Later, Guillaume and Lavallard[41] proposed a rigid rotation model to explain these excited states in CdTe. In this model the hole is excited to rotate around the fixed donor, analogous to rotation of diatomic molecules. This model had difficulty in predicting the observed energies for the excited-state transitions. A nonrigid-rotator model was subsequently

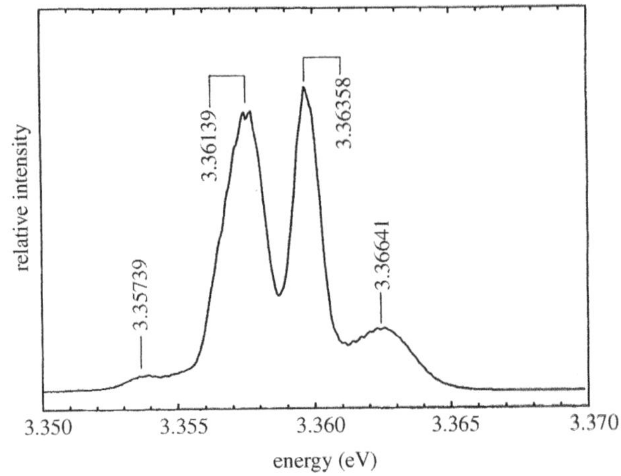

Figure 2.10 *Neutral donor defect pair spectra in ZnO, first-order spectrum. Reprinted with permission from Donald Reynolds. Copyright American Institute of Physics*

proposed by Ruhle and Klingenstein,[42] which was successful in predicting the excited state energies in InP and GaAs. A more sophisticated model was applied to the D°,X ground and excited states by Herbert;[43] this model predicts the energy ordering of the excited states.

A final model was proposed by Rorison *et al.*[19] to explain their high-magnetic-field results in InP. In this model D°,X is considered to be a free exciton orbiting a neutral donor; one electron was considered to be strongly correlated with the hole and the other with the donor. This model was capable of explaining the relative intensities of the PL transition in the ground- and excited-state regions of InP.

The neutral donor defect pair spectra are shown in Figure 2.10. The spectra are recorded in first order and result from excitons bound to the neutral donor defect pairs. The same spectra recorded in second order are shown in Figure 2.11. In these spectra more lines are resolved. Defect pair spectra would be expected to show polarization effects. Local strains oriented in the direction of the pair will result, and the electric vector will orient in the direction of the strain. Polarization effects are shown in Figure 2.12. The solid line shows the spectra with the electric vector perpendicular to the "*c*"-axis of the crystal, $\mathbf{E} \perp \mathbf{c}$. The dashed line shows the spectra with the electric vector parallel to the "*c*"-axis of the crystal, $\mathbf{E} \parallel \mathbf{c}$. Some of the lines show strong polarization, indicating the direction of the pairs. This evidence supports the contention that these emission lines are associated with defect pair complexes. The emission results from the collapse of excitons bound to the defect pair complexes which simulate neutral donors. These complexes are shown to have the electronic character of neutral donors. The magnetic field splitting of the lines is consistent with neutral-donor-bound-excitons in the wurtzite structure. Plotted in Figure 2.13 is the splitting as a function of magnetic field, with the crystalline "*c*"-axis oriented perpendicular to the magnetic field ($\mathbf{c} \perp \mathbf{H}$). The magnetic field data are for the line at 3.36012 eV. One would expect a doublet splitting (with $\mathbf{c} \perp \mathbf{H}$) to arise from an exciton bound to a neutral donor or

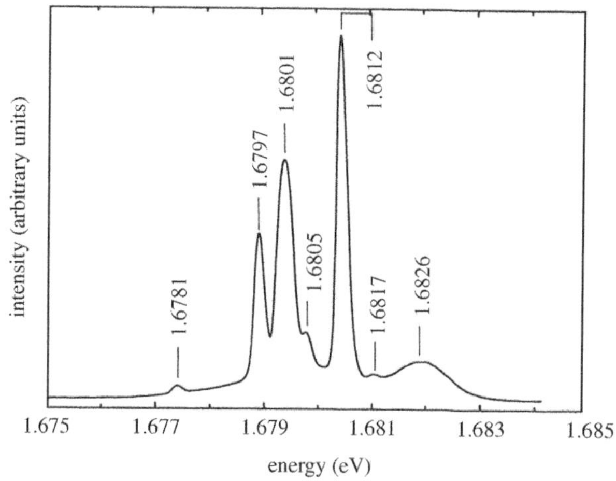

Figure 2.11 *Second-order spectrum of defect pair spectra in Figure 2.10. Reprinted with permission from Donald Reynolds. Copyright American Institute of Physics*

acceptor in the wurtzite symmetry. In this orientation the hole g-value in the upper state goes to zero ($g_h = g_{h||} \cos\theta$), so the magnetic field splitting results from the electron spin splitting in the final state. From these data, the electron g-value is measured, giving a value $g_e = 1.85$ in good agreement with the previously measured $g_e = 1.95$.[47] A magnetic field splitting for the orientation $\mathbf{c} \parallel \mathbf{H}$ was not observed. In this orientation, a contribution from both the electron and hole spins is expected. The spin-up, spin-down transitions leading to a sum of the

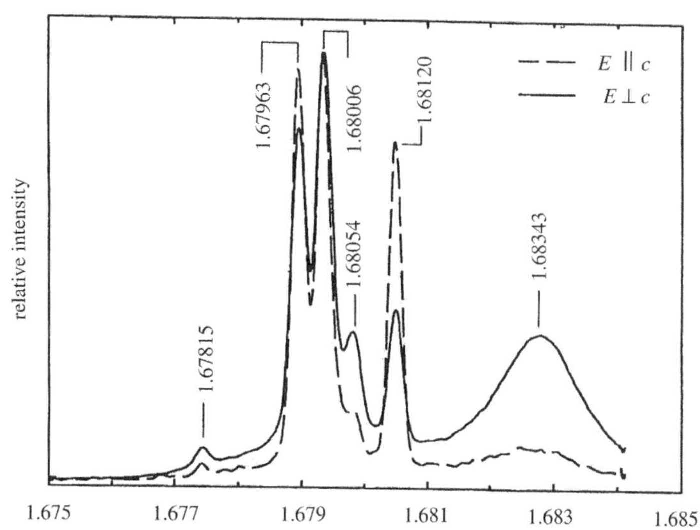

Figure 2.12 *Polarized spectra of Figure 2.11. Reprinted with permission from Donald Reynolds. Copyright American Institute of Physics*

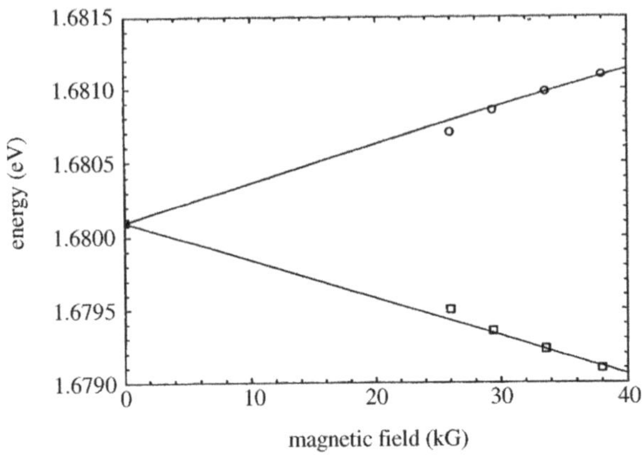

Figure 2.13 *Magnetic field splitting of the 3.36012 eV defect donor-bound-exciton line in the orientation* **c**$\perp$***H***. *Reprinted with permission from Donald Reynolds. Copyright American Institute of Physics*

g-values ($g_e + g_h$), are not allowed. The transitions leading to a difference of the g-values ($g_e - g_h$) are the spin conserving transitions. These transitions are allowed but the resulting g-value is small so that the splitting is not resolved. The other D°,X transitions showed similar magnetic field splitting.

Another characteristic of neutral-donor-bound-exciton transitions is the two-electron transitions.[45] For this case the exciton collapses and the neutral donor returns to the ground state, or it may pick up energy from the exciton, leaving the electron on the donor in an excited state, in the final state. The energy of the transition is:

$$E_T = E_{FX} - E_b - \Delta E \quad (2.1)$$

where E_T is the transition energy, E_{FX} is the free exciton energy, E_b is the energy with which the exciton is bound to the donor and ΔE is the energy necessary to put the donor into an excited state. Transitions of this type are shown in Figure 2.14. The solid curve shows the donor-bound-exciton transitions (D°,X) at 3.3636 eV and 3.3614 eV and the respective two electron transitions at 3.3220 eV and 3.3189 eV. From these energies, the donor binding energies can be calculated, assuming the excited states are hydrogenic. The donor at 3.3636 has a binding energy of 55.5 meV; the donor at 3.3614 eV has a binding energy of 56.7 meV. The dashed curve shows the sample after annealing at 800 °C. The D°,X emission essentially goes into the line at 3.3570 eV, for which the two-electron transitions are $n = 2$, $E = 3.3137$; and $n = 3$, $E = 3.3058$ eV. The $n = 2$ state gives a donor binding energy of 57.7 meV and the $n = 3$ state gives a donor binding energy of 57.6 meV. It is believed that the defect pairs at lower energies are moving closer together. It also appears that the donor binding energy is increasing as the pairs move closer together.

As alluded to above, sample annealing moves all of the D°,X emission into the lowest energy line. It appears that the more distant pairs are the first to break up and move to

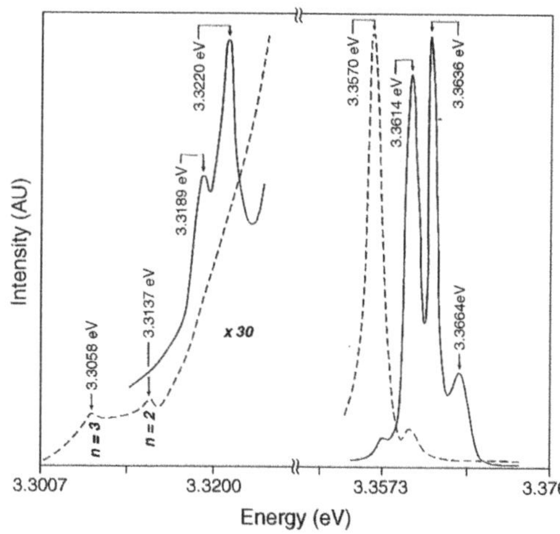

Figure 2.14 *Two electron transitions associated with the ground state defect donor-bound excitons. Reprinted with permission from Donald Reynolds. Copyright American Institute of Physics*

closer spacing. There is a near conservation of the total emission intensity suggesting that the pairs are not eliminated, but simply reconfigure. The total integrated intensity of all of the lines as a function of annealing temperature is shown in the inset of Figure 2.15. It is noted that the total intensity of all of the D°,X lines is conserved within less than a factor of two. The shift of the emission intensity to the lowest energy line with annealing temperature occurs rather dramatically between 700 °C and 800 °C. It is also noted that

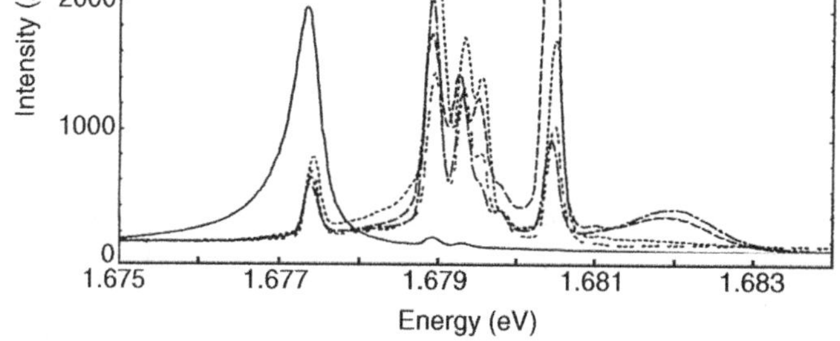

Figure 2.15 *Integrated intensity of the defect donor-bound exciton lines as a function of annealing temperature. Reprinted with permission from Donald Reynolds. Copyright American Institute of Physics*

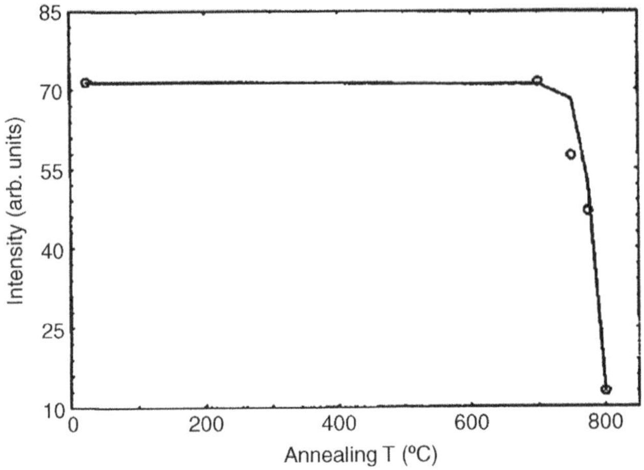

Figure 2.16 First-order annealing curve for the defect donor-bound excitons. Reprinted with permission from Donald Reynolds. Copyright American Institute of Physics

the lowest energy line broadens dramatically as the total intensity is culminated in that line. This may be a strain broadening as all of the pairs move to near neighbor distances. From the annealing temperatures, an activation energy can be obtained. Using the expression for first-order annealing:

$$PL_{i+1} + PL_i e^{-\nu t e^{-E/kT_i}} \qquad (2.2)$$

where the prefactor $\nu = 1.744 \times 10^{13}$ (optical phonon frequency), t is the annealing time, E is the activation energy, and T_i is the annealing temperature; the curve in Figure 2.16 is obtained. This gives a value $E = 3.6$ eV for the activation energy. The activation energies for the diffusion of Zn in ZnO have been previously determined.[47–49] These activation energies fall within the range 3.0–3.3 eV. Thus an activation energy of 3.6 eV would appear to be a reasonable value for promoting the motion of the defect pairs.

Excited states associated with the D°,X ground state transitions are observed. These are observed at high resolution in second order, on the high energy side of the ground state transitions and are analogous to the excited state transitions described above. The transitions 3.3662 eV (Γ_6) and 3.3670 eV (Γ_5) in Figure 2.17 are excited states analogous to rotational states of the H_2 molecule. These states are rotational states associated with the 3.3564 eV ground state, and are not electronic excited states. This is the first time these transitions have been observed when the neutral donor itself is a complex center. As observed from Figure 2.17, these transitions are on the low energy side of the 3.3772 eV (Γ_5) and 3.3750 eV (Γ_6) free-exciton (FE) transitions. The solid curve in the figure represents spectra with an applied magnetic field of 18 kG. The Γ_6 exciton is an unallowed transition that becomes allowed in the presence of an applied magnetic field. The dashed curve shows the same transition in zero magnetic field. Note that the rotator state associated with the Γ_6 exciton is observed. The two lowest energy rotation states are

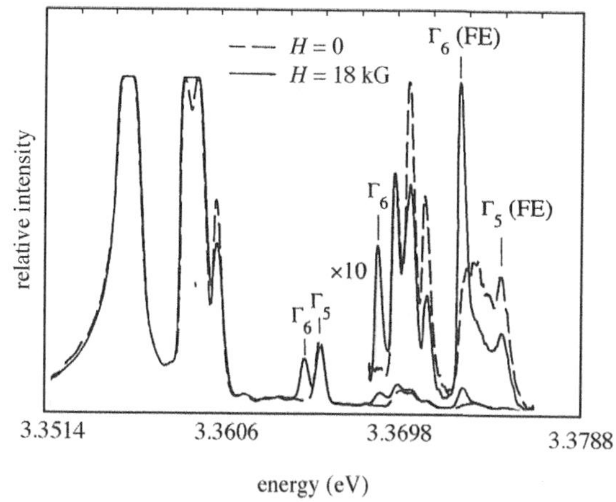

Figure 2.17 *Excited rotational states associated with the defect donor-bound excitons. Note that Γ_6 rotational excitons are observed. These are second-order spectra. Reprinted with permission from Donald Reynolds. Copyright American Institute of Physics*

associated with the lowest energy 3.3564 eV D°,X transition. The next two lowest energy rotator states, 3.3714 eV (Γ_5) and 3.3702 eV (Γ_6), are associated with the next lowest energy 3.3594 eV D°,X transition. It is noted that again one of the rotator states is associated with the Γ_6 exciton. Other rotator states associated with the Γ_6 exciton are most likely not resolved since they would come in the energy region where they would not be resolved from other Γ_5 rotator states. This is the first observation of rotator states associated with the Γ_6 unallowed exciton, and lend support to the model that the exciton itself rather than the hole is rotating.[19]

Following the arguments of Guillaume and Lavallard,[41] using the Hellmann–Feynman theorem, one derives the energy difference:

$$\Delta E \approx \mathrm{J}(\mathrm{J}+1)\sigma E_{\mathrm{D}}/r^2 \tag{2.3}$$

for the rotator states. J is the rotational quantum number, E_{D} is the binding energy of the donor, $\sigma = m_e/m_n$ and r is the radius of the excitonic molecule. According to Akimoto and Hanamura,[50] r is between 1.44 and 3.47 times the Bohr radius of the free exciton. From the observed data, one obtains an average value of $E_{\mathrm{D}} = 56.9$ meV. Assuming r to be twice the Bohr radius one obtains $\Delta E \approx 6$ meV. This agrees satisfactorily with the experimental value of 10.6 meV. Taking the experimental value of 10.6 meV, and inserting it into Equation (2.3), a value of 1.5 is obtained for the Bohr radius, which is in the same range 1.44–3.47 given in Linder.[46] One would expect the Bohr radius to be reduced in ZnO due to the greater binding energy.

The energy level diagram of the transitions shown in Figures 2.14 and 2.17 is shown in Figure 2.18, for the as-grown sample and for the sample after an 800 °C anneal. The fourteen transitions in the as-grown sample reduce to five after annealing. The energies of the transitions are shown. It is noted that after annealing, all of the higher energy D°,

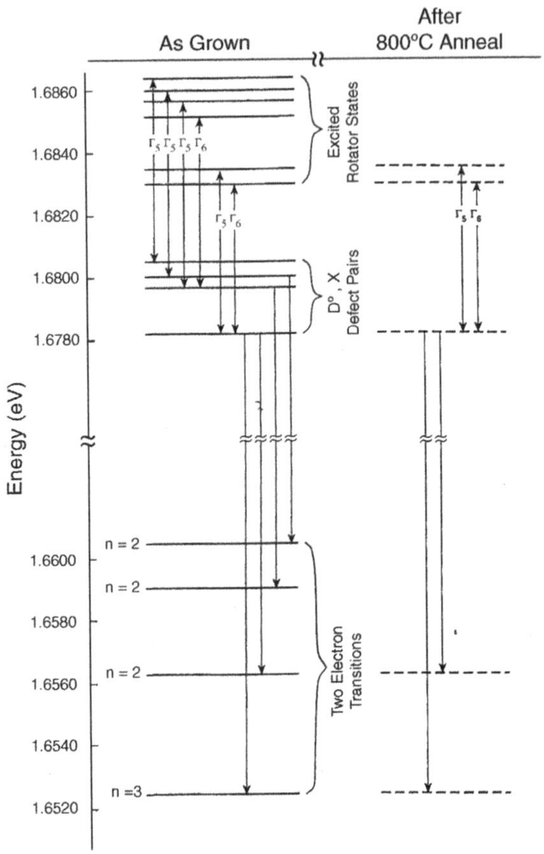

Figure 2.18 *Energy level diagram showing the transitions in Figures 2.14 and 2.17. The energies of the transitions as well as their identities are given for the as-grown sample and for the sample after an 800 °C anneal. Reprinted with permission from Donald Reynolds. Copyright American Institute of Physics*

X transitions disappear and only the lowest energy D°,X transition remains. If one assumes that the higher energy excited rotator transitions result from the rotation of the exciton, the Γ_5 and Γ_6 excitons are labeled. For the two lowest energy D°,X states, the Γ_5 and Γ_6 rotator states are clear, but for the two highest energy D°,X states the Γ_6 rotator states will not be resolved from the Γ_5 rotator states. After annealing, only two excited rotator states remain and they are clearly identified with the application of a magnetic field.

2.6 Similarities in the Photoluminescence Mechanisms of ZnO and GaN

The III–V nitrides have recently attracted much attention because of their large band gaps, high thermal conductivities, and high melting points. Many of the problems encountered by researchers working with these materials are the lack of consistent high quality material. The highest quality GaN grown to date has been grown by hydride vapor phase epitaxy (HVPE), molecular beam epitaxy (MBE), and by metal organic chemical vapor deposition (MOCVD) on a variety of substrates. The lattice mismatch, as well as the stacking order mismatch between many of the currently used substrate materials and GaN, is substantial. The substrate mismatch leads to strain-induced defects and dislocations. Efforts are now being made to obtain closely lattice-matched substrates that will minimize strain-induced defects. One of the promising materials that provides a close lattice match to GaN is ZnO, which also has the wurtzite stacking order. In addition to having the same crystal structure and close lattice match, ZnO also exhibits optical properties similar to those observed in GaN. It is found that models used to explain a specific property in one material may be applicable in explaining a similar property in the other material. In particular the well known broad yellow band in GaN and a similar broad band, historically referred to as the green band in ZnO, appear to have similar origins.

A number of studies have been devoted to the origin of the yellow band in GaN. Controversy still exists, however, concerning the position if the electronic levels giving rise to the transition that produces the yellow band as well as the complex nature of the deep level. Ogino and Aoki[51] as well as Hofmann *et al.*[52] propose a model in which the transition between a shallow donor and a deep level gives rise to the yellow band. Alternatively Glaser *et al.*[53] propose a model in which the transition proceeds from a deep level to an effective mass acceptor. Here it is shown that the transition responsible for the green band in ZnO supports the model of Ogino and Aoki[51] and Hofmann *et al.*[52] and can also account for the very large width of the band. This same explanation may also apply to the yellow band in GaN.

Representative PL spectra from GaN exhibit free- and bound-exciton transitions near the absorption edge. At somewhat lower energies a series of donor–acceptor pair lines are observed (zero phonon line at 3.2880 eV) where the zero phonon line is followed at lower energies by phonon-assisted transitions. These transitions are shown in Figure 2.19 (dashed curve). The transition at 3.2880 eV results from the recombination of a shallow donor and a Si acceptor. It is recognized that Si is an amphoteric impurity in GaN and normally produces donor states. The transitions at 3.1973 eV and 3.1050 eV are phonon-assisted transitions. The transition at 3.2167 eV results from the recombination of the shallow donor with the Mg acceptor. This transition is close to the first phonon transition associated with the Si acceptor. The assignment of the Si and Mg acceptor transitions is based on the published binding energies of these acceptors as well as the knowledge that the MBE reactor probably contained residual Mg, resulting from previous Mg doping.

The transitions observed in ZnO, shown in Figure 2.19 (solid curve), are remarkably similar to the established donor–acceptor pair transitions in GaN. The donor–acceptor pair bands have been previously identified in ZnO.[54,55] The first donor–acceptor transition occurs at 3.3219 eV with phonon replicas at 3.2167 eV and 3.1449 eV. The phonon replicas in the two spectra are separated by the longitudinal optical phonon energies associated with the respective materials. While the identity of the donors and acceptors in ZnO is unknown, the similarity of the band gaps of ZnO and GaN together with the appearance of the pair transitions in the same energy region would

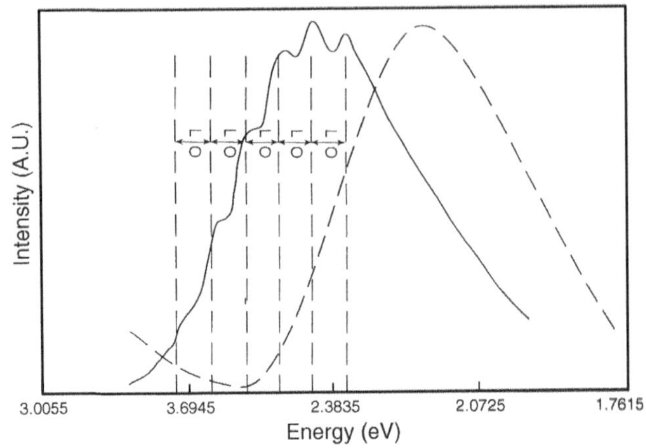

Figure 2.19 *Donor–acceptor pair transitions in GaN (dashed curve) and in ZnO (solid curve). The zero phonon transitions along with their phonon replicas are identified in the text. Reprinted with permission from Donald Reynolds. Copyright American Institute of Physics*

indicate that the donor and acceptor binding energies are not greatly different in these two materials.

At somewhat lower energies than the donor–acceptor pair transitions (2.2–2.3 eV) the well-known yellow band is observed in GaN as shown in Figure 2.20 (dashed curve). This band is seen in high conductivity MBE material (N deficient) and disappears in semi-insulating material (N rich). A similar band, historically called the green band, has been observed in ZnO and is also shown in Figure 2.20 (solid curve). It has been shown in ZnO

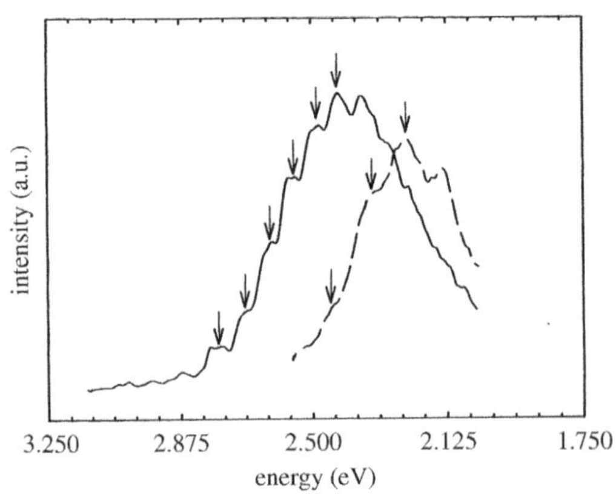

Figure 2.20 *The yellow emission band (dashed curve) observed in GaN and the green emission band (solid curve) observed in ZnO. The structured peaks occurring on the high energy side of the green band are separated in energy by that of the longitudinal optical phonon in ZnO. Reprinted with permission from Donald Reynolds. Copyright American Institute of Physics*

that the relative intensities of the phonon-assisted peaks and the green band depend markedly on the electrical conductivity of the samples, the intensities of the two peaks becoming nearly equal at high conductivities.[54,55] It has also been shown that the conductivity of ZnO is affected strongly by excess Zn, with the conductivity increasing rapidly with increasing Zn concentration.[56] Similar studies show that the conductivity of GaN decreases with increasing N concentration.[56] These similarities suggest that the green band in ZnO and the yellow band in GaN may be associated with a related defect mechanism.

The stoichiometry considerations discussed above suggest certain point defects as viable candidates for the yellow band in GaN and the green band in ZnO. That is, in Ga-rich GaN, we would expect high quantities of the N vacancy V_N, Ga interstitial Ga_i and/or the Ga antisite Ga_N. In Zn-rich ZnO, the analogous defects would be V_O, Zn_i and Zn_O. Recent theoretical calculations[57] in GaN predict shallow donor states for V_N and Ga_i but mid-gap unfilled levels for Ga_N. The latter defect could presumably act as a double donor or double acceptor since the neutral state contains two electrons in a four fold degenerate level. Thus, from this point of view, it is possible that the yellow band in GaN involves Ga_N, and the green band in ZnO involves Zn_O.

Experiments by Ogino and Aoki[51] and by Hofmann *et al.*[52] suggest that the yellow band in GaN results from the recombination between a shallow donor and a deep level. The complex nature of the deep level is unclear. Ogino and Aoki[51] propose a complex consisting of a Ga vacancy and a C on a N site (C_N). Hofmann *et al.*[52] propose that the deep level may be a double donor though an acceptor cannot be ruled out. Suski *et al.*[58] suggest that the deep level is an antisite (N_{Ga}). Neugebauer and Van de Walle[59] report that their calculations show that the complex consisting of a Ga vacancy and C_N as well as the N_{Ga} are thermodynamically unstable.

They propose that the deep level is a Ga vacancy or related complex. It is clear that a consensus has not been reached on the make-up of the deep level.

A modulated structure is observed on the high energy side of both the green band in ZnO and the yellow band in GaN as shown in Figure 2.20. The modulated structure can be explained from the model shown in Figure 2.21. The PL emission results from the recombination between the shallow donor level and the deep level. Hot electrons in the conduction band are pumped up by the HeCd excitation source. Peaks in the PL emission band occur whenever the energy of the PL peak coincides with the sum of the energies of the donor level plus an integral multiple of a principal optical phonon energy. At adjacent energy values an equilibrium number of electrons will arrive at the donor level and thus take part in the recombination with the deep level. The deep level will also have accompanying excited states due to interaction with local vibrational modes as well as lattice modes. It would be expected that the dominant transition would occur between the shallow donor and the ground state of the deep level. Transitions will also occur between the shallow donor and the excited states of the deep level, with reduced oscillator strengths. This model agrees with the model proposed by Ogino and Aoki[51] and by Hofmann *et al.*[52] for GaN and has the added advantage that it can explain the width of the emission band. The width of the yellow band in GaN and the green band in ZnO is extremely broad and would not be explained by the width of the impurity levels. The energy separation between the modulated peaks in ZnO and GaN corresponds to the longitudinal optical phonon

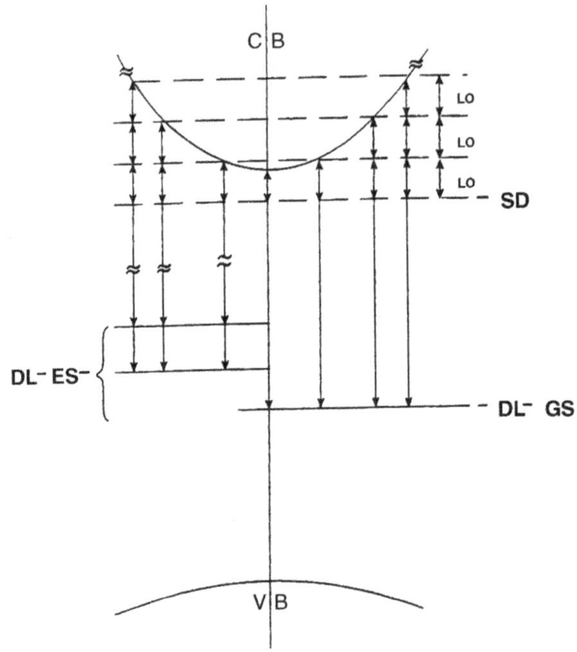

Figure 2.21 *Model to explain the green-band emission in ZnO. Band model in K-space shows the conduction band, the shallow donor level (SD), the deep level ground state (DL-GS), and the deep level excited states (DL-ES). The longitudinal optical phonons are designated as (LO). The energies of the above states are all related to the valence band. Reprinted with permission from Donald Reynolds. Copyright American Institute of Physics*

energies in their respective materials: 0.072 eV in ZnO and 0.092 eV in GaN. It is noted in Figure 2.20 that the modulated structure does not occur on the low energy side of the bands. This would be expected since the phonons that are involved in cascading hot electrons from the conduction band to the donor level are not involved with the low energy emission. This emission is accounted for by recombination of donor electrons with excited states of the deep level. The microscopic nature of the deep level is uncertain; it may be a complex center whose excited states consist of both local vibrational modes and lattice modes. These excited states are so distributed that they do not produce a resolvable modulated structure on the low energy side of the green band.

If one now assumes that there is a relationship between the green band in ZnO and the yellow band in GaN, then a similar model would apply to the latter. This would support the model of Ogino and Aoki[51] and Hofmann *et al.*[52] and would also explain the very large width of the yellow band. The modulated structure observed for the green band in ZnO could not be accounted for by the alternative model, proposed by Glaser *et al.*,[53] to explain the yellow band in GaN. In Glaser *et al.*'s model the electrons transfer from a shallow donor to the deep state in a nonradiative process followed by radiative decay from the deep state to a shallow acceptor.

2.7 The Combined Effects of Screening and Band Gap Renormalization on the Energy of Optical Transitions in ZnO and GaN

Free-carrier screening will modify a spherical Coulomb potential, resulting in a reduced exciton binding energy. This will result in a blue shift of the optical transitions associated with both free excitons and bound excitons. The effects of free-carrier screening have been reported in both bulk materials such as Ge[60–62] and GaSe[63] and quantum-well structures.[64–66] In this paper we report on the position of the optical transitions in both GaN and ZnO crystals when they are excited simultaneously with two lasers. This experiment involves a HeCd laser at 3250 Å which excites free electrons and holes, and an Ar^+ ion laser at 5145 Å, which excites electrons from the valence band to an intermediate level in the band and then from the intermediate level to the conduction band. The increased number of free electrons excited by the Ar^+ ion laser will effectively screen the free exciton transitions as well as the bound exciton transitions. Screening results in a decrease in the binding energy of both types of transitions. In ZnO, both the free- and bound-exciton transitions show a red shift as the exciting intensity of the Ar^+ ion laser is increased. The same transitions in GaN show essentially no energy shift with increased exciting intensity of the Ar^+ ion laser. The increased number of free electrons excited by the Ar^+ ion laser also produce many-body effects in intrinsic semiconductors, which lead to a renormalization of the band gap. The renormalization results in a red shift of the optical transitions due to reduction of the band gap energy. The energies of the optical transitions are then the resultant of the blue shift due to screening and the red shift due to renormalization.

The steady-state concentration of additional electrons cannot be calculated since the cross-sections for the various transitions are not known. An estimate of the number of electrons participating in the screening process can be made if one assumes that substantial screening will occur when the Debye length and the Bohr radius are comparable. Free carrier screening modifies the Coulomb potential $\varphi \sim e/4\pi \in r$ by adding a multiplicative factor, $\exp(-r/\lambda)$, where r is the distance from the center of the charge, and λ, the Debye length, given by:[67]

$$\lambda(\in kT/e^2 n_{\mathrm{eff}})^{\frac{1}{2}} \tag{2.4}$$

The major part of the charge cloud will be at the Bohr radius $\alpha_B = 0.529\in/m^*$. In GaN $m^* \simeq 0.22$ and $\in \simeq 9.5$[68] giving a Bohr radius of $\simeq 22$ Å. The Debye length will be comparable when $n_{\mathrm{eff}} \sim 1 \times 10^{16}\,\mathrm{cm}^{-3}$. In ZnO the dielectric constant is smaller, 8.12 vs 9.5, and the mass is larger, 0.318 vs 0.22, giving a smaller Debye length as well as a smaller Bohr radius. Therefore, roughly the same number of free electrons will produce substantial screening in ZnO as well.

Having an estimate of the number of free electrons required for the screening process, one can estimate the screened donor binding energy from the common empirical relationship used to describe donor screening

$$E_D = E_{D0} - \alpha N_D^{\frac{1}{2}} \tag{2.5}$$

where E_D is the screened donor binding energy, E_{D0} is the unscreened donor binding energy, and the suggested value of α for n-type GaN is $\alpha \simeq 2.1 \times 10^{-5}$ meV cm.[69] For ZnO we can approximate α as $[(\epsilon_{ZnO}/\epsilon_{GaN})(m^*_{GaN}/m^*_{ZnO})]\alpha_{GaN}$ giving a value $\alpha \simeq 1.2 \times 10^{-5}$ meV cm. One sees that the screening in ZnO is somewhat weaker than that in GaN. Assuming $1\times10^{16}\,cm^{-3}$ free electrons, one calculates from Equation (2.5) a reduction in the donor binding energy in ZnO of $\simeq$2.7 meV due to screening. From screening alone, this should result in a blue shift of the donor-bound-exciton (D°,X) transition.

The transition energy of D°,X is the free exciton energy minus the energy with which the exciton is bound to the donor. By Haynes's rule[70] the exciton is bound to the donor by some fraction of the donor binding energy. It would be expected that the free exciton binding energy will be screened very similar to that of the donor binding energy. Therefore, in the screened case, the D°,X transition energy will be that of the screened free exciton energy minus the energy with which the exciton is bound to the screened donor. A reasonable estimate of the blue energy shift of D°,X with screening would then be obtained from Equation (2.5). The D°,X PL spectra as a function of Ar^+ ion laser intensity are shown in Figure 2.22. The D°,X intensity decreases as the Ar^+ ion intensity increases, as expected. As the excitons are screened from the donors, the D°,X decay route is eliminated as one of the processes by which excitons are dissipated. Therefore, as more excitons are screened from the donors, the D°,X intensity will decrease and the free exciton intensity will increase. This is verified in Figure 2.23 where the free exciton intensity is plotted as a function of the Ar^+ ion laser intensity. Here the Γ_5 and Γ_6 free excitons are clearly identified. One other feature is observed in both the D°,X spectra and the free exciton spectra and that is the energy red shift.

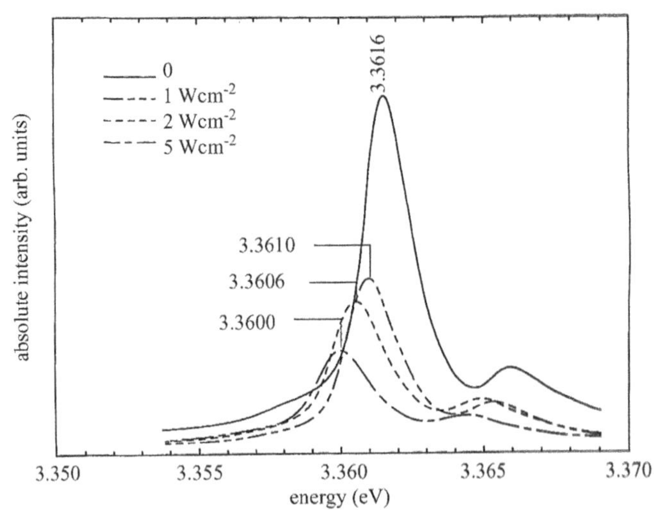

Figure 2.22 *Intensity and energy position of the D,°X transitions in ZnO as a function of the Ar^+ ion laser intensity. Reprinted with permission from Donald Reynolds. Copyright American Institute of Physics*

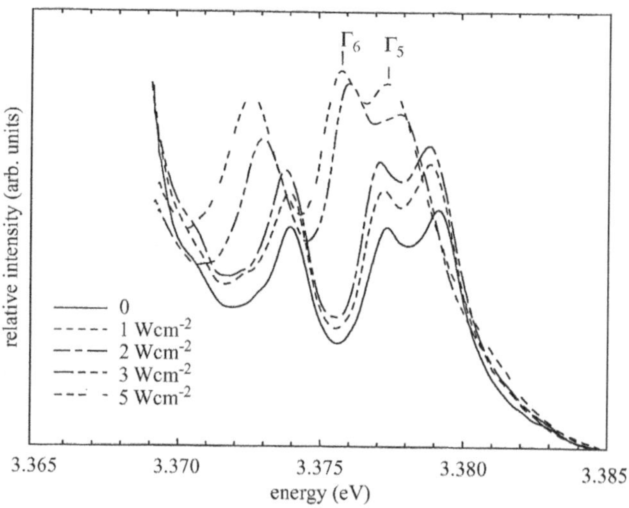

Figure 2.23 *Intensity and energy of the free-exciton transitions in ZnO as a function of the Ar^+ ion laser intensity. Reprinted with permission from Donald Reynolds. Copyright American Institute of Physics*

The GaN sample was excited in exactly the same way as the ZnO sample. Making the assumption of $1 \times 10^{16}\,cm^{-3}$ free electrons for GaN, then from Equation (2.5) a reduction in the donor binding energy of ~4.5 meV due to screening is obtained. The PL emission spectra from GaN as a function of Ar^+ ion laser intensity are shown in Figure 2.24.

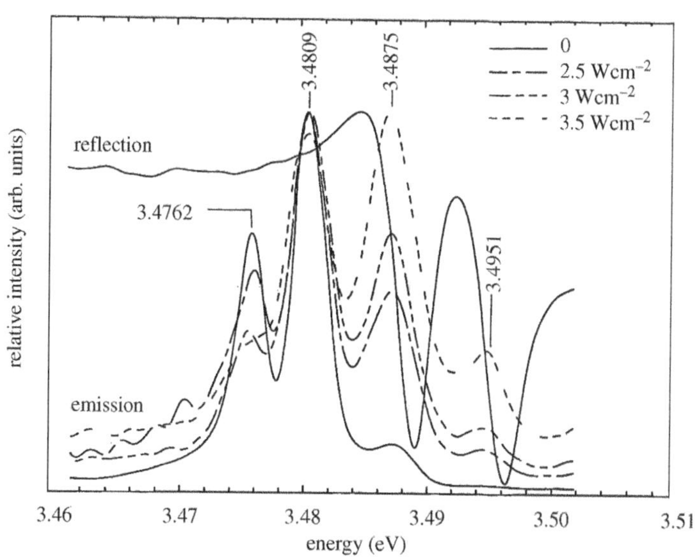

Figure 2.24 *The PL emission spectra from GaN as a function of the Ar^+ ion laser intensity. Reprinted with permission from Donald Reynolds. Copyright American Institute of Physics*

The D°,X emission peak (3.4809 eV) as well as the ionized-donor-bound-exciton (D^+,X) peak (3.4762 eV) are dominant when the Ar^+ ion laser intensity is zero. As the Ar^+ intensity is increased, more free electrons are created which neutralize the ionized donors, and at the highest Ar^+ intensity the ionized donor-bound excitons are essentially all neutralized. The D°,X centers experience an increase in numbers as the ionized donors are neutralized, but they also experience a decrease due to screening. At the highest intensity of the Ar^+ ion laser the free exciton associated with the A valence band (3.4875 eV) is the dominant transition and the free exciton associated with the B valence band (3.4951 eV) is clearly evident. In GaN, as in ZnO, the bound excitons decrease in intensity while the free excitons increase in intensity with an increase in the Ar^+ ion laser intensity. The top solid line in Figure 2.24 shows the reflection minima, which coincide with the A- and B-band free excitons. In the case of GaN, the bound and free excitons do not show an obvious blue or red shift as the Ar^+ ion laser intensity is increased.

The additional electrons excited by the Ar^+ ion laser may also result in many-body effects in an intrinsic semiconductor, which will lead to a renormalization of the fundamental energy gap. The renormalization reduces the energy gap as shown in Figure 2.25. In this figure, the energy shift is estimated from density functional theory within the local density approximation using the Hedin– Lundquist parameterization.[71] Other parameterized forms exist and yield similar results.[72] The reduction of the renormalized energy gap, as a function of electron density, is shown for both GaN and ZnO in Figure 2.25. Here it is seen that renormalization reduces the ZnO band gap more than the GaN band gap for the same electron density. The PL energy positions are determined by a combination of renormalization and screening, the former giving a red

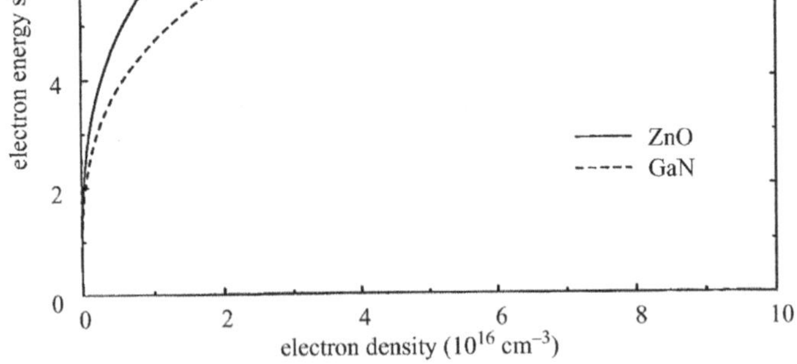

Figure 2.25 *The reduction of the renormalized energy gap as a function of electron density for both ZnO and GaN. Reprinted with permission from Donald Reynolds. Copyright American Institute of Physics*

Table 2.2 *The blue and red energy shifts of the optical transitions in both ZnO and GaN as the electron density is increased*

Sample	Electron density (cm^{-3})	Renormalization red shift (meV)	Screening blue shift (meV)
ZnO	5×10^{15}	4.5	2.1
GaN	5×10^{15}	3.5	3.6
ZnO	1×10^{16}	5.7	2.7
GaN	1×10^{16}	4.5	4.5

shift and the latter an offsetting blue shift. For a given electron density, the blue shift can be obtained from Equation (2.5) and the red shift from Figure 2.25. Taking electron densities within an order of magnitude of those calculated to give substantial screening, the blue and red shifts for both ZnO and GaN are given in Table 2.2. From the table, one would expect to observe an increasing red shift of the exciton transitions from ZnO as the Ar^+ ion laser intensity is increased. In the case of GaN, essentially no shift in the energy positions of the exciton transitions would be expected. These predictions are verified in Figures 2.22–2.24

These results show that the energy positions of optical transitions are modified when additional free electrons are introduced into the system. The energy positions are determined by the combined effects of screening and band gap renormalization resulting from the additional free electrons. This study establishes trends for the optical transition energies, but is not intended to give the exact energy positions.

2.8 Closely Spaced Donor–Acceptor Pairs in ZnO

Closely spaced donor–acceptor pair spectra have been observed in the PL radiative recombination of selected donor bound exciton complexes in ZnO single crystals, that have been ion implantation-doped with H and Zn atoms and subsequently annealed in a nitrogen (N_2) atmosphere at temperatures in the range of 700–800 °C.[73] Observations of the donor–acceptor pair spectra are explained as the bound states of Hopfield's bound exciton complex model,[74] in terms of the effective mass approximation of the wurtzitic ZnO energy band structure.

The closely spaced donor–acceptor pair spectra for a H-implanted ZnO sample are shown in Figure 2.26; sample A (solid line), was implanted into the O face, while sample B (dashed line), was implanted into the Zn face. Both samples were then annealed in N_2 vapor at 700 °C for 1 h. The sample implanted into its Zn face shows an emission line at 3.2888 eV, and an LO phonon replica at 3.2155 eV, the same energies as observed for the implanted sample prior to annealing. The sample implanted into the O face shows an emission line at 3.2845 eV, and an LO phonon replica at 3.2118 eV. It is noted that the higher energy emission lines at 3.3223 and 3.3356 eV do not show an energy shift with N_2 annealing. It was reported by Hopfield,[74] that when donor–acceptor pairs are close enough together, they can bind neither an

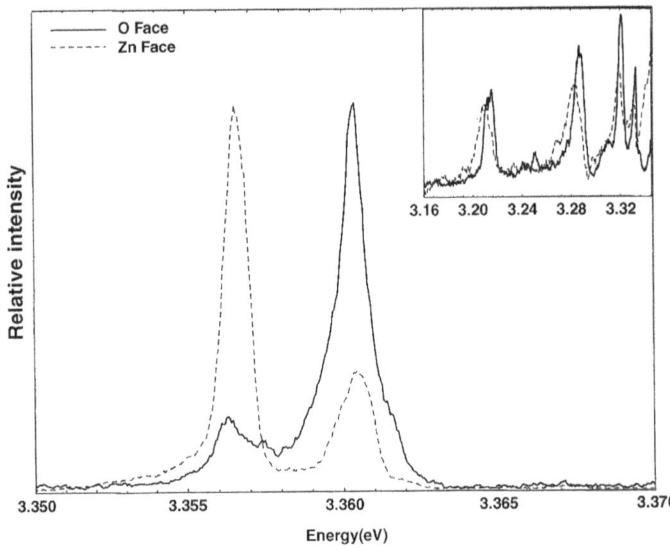

Figure 2.26 *Comparison of ZnO faces, H-implanted. Reprinted with permission from Donald Reynolds. Copyright American Institute of Physics*

electron nor a hole but they can bind an exciton. The exciton has an interaction energy with the donor and acceptor ions and this interaction energy is negative. This will move the transition to lower energy, as is observed. It is believed that the emission lines at 3.2888 and 3.2155 eV are associated with more distant spaced pairs that bind an electron and then an exciton. These lines are observed from the O face. The emission lines at 3.2845 and 3.2118 eV, from the Zn face, result from more closely spaced pairs that bind only the exciton. The energies are shifted due to the total interaction energy of the exciton with the donor and acceptor ions.

In the same samples, emission lines are observed at higher energies of 3.3606 and 3.3567 eV, as shown in Figure 2.26 (inset). Here again, from the O face (solid line), the higher energy line is much more intense; however, when the Zn face (dashed line) is examined, the line intensities are reversed, showing a shift to lower energy with annealing in N_2. As was observed with the lower energy lines, it is believed that the line at 3.3606 eV results from more distant spaced pairs that bind an electron and then the exciton, whereas the line at 3.3567 eV results from closely spaced pairs that bind only the exciton. The energy shift is comparable with that of the lower energy lines at 3.2888 and 3.2845 eV, suggesting that we are observing the interaction energy of the exciton with the closely spaced donor–acceptor pair ions.

Similar spectra from Zn-implanted ZnO samples are shown in Figure 2.27. Here again, sample A (solid line) is implanted into the O face and sample B (dashed line) into the Zn face. The samples were then annealed, in this case at 800 °C, for 1 h in N_2 vapor. Here we note that there are lines at 3.2903 and 3.2173 eV that are observed from the Zn face (dashed curve), whereas there are lower energy lines at 3.2836 and 3.2123 eV that are observed from the O face (solid curve). The energy shift to a lower energy is observed on the opposite face from that which was observed for the H-implanted samples.

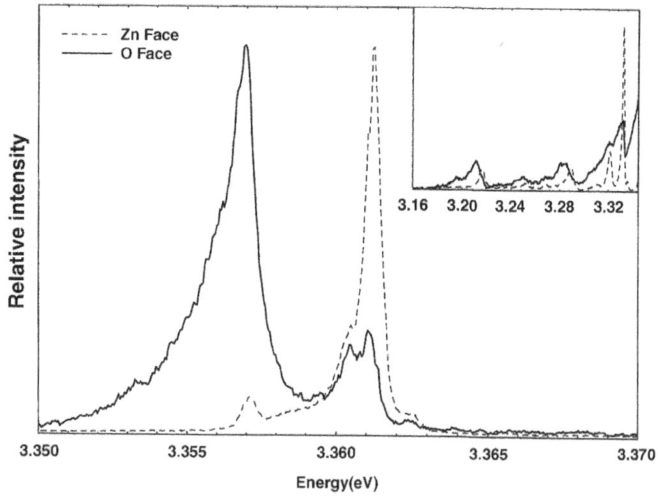

Figure 2.27 *Comparison of ZnO faces, Zn-implanted. Reprinted with permission from Donald Reynolds. Copyright American Institute of Physics*

Again, as in the H-implanted samples, higher energy lines are observed at 3.3618 and 3.3571 eV, as shown in Figure 2.27 (inset). In the Zn-implanted samples, the line at 3.3618 eV does not shift to a lower energy after annealing in N_2, when viewed from the Zn face (dashed line), while the more intense line at 3.3618 eV does shift to a lower energy at 3.3571 eV in the O face sample (solid line). Here again the shift to a lower energy is believed to be caused by an exciton bound to a closely spaced pair. This shift was observed from the O face, and it was also observed from the lower energy line at 3.2903 eV. The energy shift from the two faces of the Zn-implanted samples is reversed from those of the two faces of the H-implanted samples.

The Zn implant will not penetrate as deeply into a sample as will the H implant. When a sample is annealed in N_2, N ions will in-diffuse to occupy O sites near the surface, and pair with donor-created sites near the surface. The more closely spaced pairs will form near the surface, which in our case will be the O face. This accounts for the observation of the shift to lower energy when observing the O face in Zn-implanted samples. In the H-implanted samples, the H ion will penetrate deeper into the sample. In this case, when the sample is annealed in N_2 vapor, the N ion must diffuse deeper into the sample to create acceptors closer to the donors. It is believed that the acceptor results from N ions replacing O sites and that these dominant acceptor sites are near the crystal surface, when viewing the O face. Therefore, in the H-implanted samples, the closely spaced, donor–acceptor pairs would not be created near the O surface. The N ions will diffuse deeper to occupy O sites nearer the donors which may be H interstitials. This would result in the energy shift occurring when observing the Zn face, as is shown, and would account for the reversal of the faces in Zn-implanted and H-implanted samples. It is believed that the acceptor in the donor–acceptor pair results from a N ion occupying an O site. The source of the donor is not established and will require more study.

2.9 Summary

We have observed the Γ_5 and Γ_6 free excitons associated with the top valence band in ZnO, as well as these same excitons associated with rotator states. The unallowed, Γ_6 excitons, split in an applied magnetic field in both states, giving a g-value of 3.09, which is the sum of the electron and hole g-values. This is the predicted splitting for the Γ_6 exciton from effective mass and group theory, which is consistent with the top valence band having Γ_9 symmetry. We have also observed the splitting of the Γ_6 exciton due to combined strain and electron–hole spin exchange, which also reflects the Γ_9 symmetry of the top valence band in ZnO.

Many bound-exciton transitions are observed in ZnO, and these include excitons bound to neutral and ionized centers, as well as complex centers, like defect pairs. In this case, many lines are observed which are related to the donor–acceptor spacing in the defect pair complex. Free-to-bound and bound-to-bound transitions are observed which are not associated with excitonic transitions. The well-known green band is associated with pair spectra where the transition energy is associated with the donor–acceptor separation. Similarities in some of the optical transitions in ZnO and GaN have also been observed. This is to be expected since these materials have the same crystal structure and their band gaps are similar.

References

[1] M. Kawasaki, A. Ohtomo, I. Ohkubo, H. Koinuma, Z. K. Tang, P. Yu, G. K. L. Wong, B. P. Zhang and Y. Segawa, *Mater. Sci. Eng. B* **56**, 239 (1998).

[2] D. M. Bagnall, Y. F. Chen, Z. Zhu, T. Yao, S. Koyama, M. Y. Shen and T. Goto, *Appl. Phys. Lett.* **70**, 2230 (1997).

[3] Z. K. Yang, P. Yu, G. K. L. Wong, M. Kawasaki, A. Ohtomo, H. Koinuma and Y. Segawa, *Solid State Commun.* **103**, 459 (1997).

[4] R. D. Vispute, V. Talyansky, S. Choopun, R. P. Sharma, T. Ven Katesan, M. He, X. Tang, J. B. Halpern, M. G. Spencer, Y. X. Li, L. G. Salamanca-Biba, A. A. Lliadis, and K. A. Jones, *Appl. Phys. Lett.* **73**, 348 (1998).

[5] A. Ohtomo, M. Kawasaki, I. Ohkubo, H. Koinuma, T. Yasuada,and Y. Segawa, *Appl. Phys. Lett.* **75**, 980 (1999).

[6] E. Mollwo and Z. Aganew, *Physik* **6**, 257 (1954).

[7] D. G. Thomas, *J. Phys. Chem. Solids* **15**, 86 (1960).

[8] E. Tomzig and R. Helbig, *J. Lumin.* **14**, 403 (1976).

[9] K. Hümmer, *Phys. Status Solidi B* **56**, 249 (1973).

[10] J. J. Hopfield, *J. Phys. Chem. Solids* **15**, 97 (1960).

[11] D. C. Reynolds and T. C. Collins, *Phys. Rev.* **185**, 1099 (1969).

[12] J. J. Hopfield and D. G. Thomas, *Phys. Rev.* **122**, 35 (1961).

[13] K. Kawabe, R. H. Tredgold and Y. Inuishi, *Electron. Eng. Jpn.* **87**, 62 (1967).

[14] G. A. Slack, *J. Phys. Chem. Solids* **34**, 321 (1973).

[15] G. A. Slack and T. F. McNelly, *J. Cryst. Growth* **34**, 263 (1976).

[16] S. Adler and J. Birman,GTE Laboratory Report (unpublished).

[17] M. Balkanski and J. Des Cloizeaux, *J. Phys. Radium* **21**, 825 (1960).

[18] D. C. Reynolds, D. C. Look, B. Jogai, C. W. Litton, T. C. Collins, W. Harsch and G. Cantwell, *Phys. Rev.* **57**, 12151 (1998).

[19] J. Rorison, D. C. Herbert, P. J. Dean and M. S. Skolnick, *J. Phys. C* **7**, 6453 (1984).

[20] T. Koda and D. W. Langer, *Phys. Rev. Lett.* **20**, 50 (1968).

[21] O. Akimoto and H. Hasegawa, *Phys. Rev. Lett.* **20**, 916 (1968).
[22] D. W. Langer, R. N. Euwema, K. Era and T. Koda, *Phys. Rev.* **2**, 4005 (1970).
[23] T. C. Damen, S. P. S. Porto and B. Tell, *Phys. Rev.* **142**, 570 (1966).
[24] H. Kunzel and K. Ploog, *Appl. Phys. Lett.* **37**, 416 (1980).
[25] P. J. Dobson, G. B. Scott, J. H. Neave and B. A. Joyce, *Solid State Commun.* **43**, 917 (1982).
[26] F. Briones and D. M. Collins, *J. Electron. Mater.* **11**, 847 (1982).
[27] P. K. Bhattacharya, H. J. Buhlmann and M. Hegems, *J. Appl. Phys.* **53**, 6391 (1983).
[28] A. P. Roth, R. G. Goodchild, S. Charbonneau and D. F. Williams, *J. Appl. Phys.* **54**, 3427 (1983).
[29] D. C. Reynolds, K. K. Bajaj, C. W. Litton, E. B. Smith, P. W. Yu, W. T. Masselink, F. Fischer and H. Morkoc, *Solid State Commun.* **52**, 685 (1984).
[30] M. S. Skolnick, T. D. Harris, C. W. Tu, T. M. Brennan and M. D. Sturge, *Appl. Phys. Lett.* **46**, 427 (1985).
[31] A. C. Beye and G. Neu, *J. Appl. Phys.* **58**, 3549 (1985).
[32] A. C. Beye, B. Gil, G. Neu and C. Verie, *Phys. Rev. B* **37**, 4514 (1988).
[33] S. Charbonneau, W. G. McMullan, M. O. Henry and M. L. W. Thewalt, *Mater. Res. Soc. Symp.* **104**, 549 (1988).
[34] S. Charbonneau and M. L. W. Thewalt, *Phys. Rev. B* **41**, 8221 (1990).
[35] D. W. Langer, R. N. Euwema, K. Era and T. Koda, *Phys. Rev. B* **2**, 4005 (1970).
[36] J. L. Merz, H. Kukimoto, K. Nassau and J. W. Shiever, *Phys. Rev. B* **6**, 545 (1972).
[37] P. Hiesinger, S. Suga, F. Willmann and W. Dreybrot, *Phys. Status Solidi B* **67**, 64 (1975).
[38] A. M. White, P. J. Dean, L. L. Taylor, R. C. Clarke, P. J. Ashen and J. B. Mullin, *J. Phys. C* **5**, 1727 (1972).
[39] A. M. White, P. J. Dean and B. Day, *J. Phys. C* **7**, 1400 (1974).
[40] C. H. Henry and K. Nassau, *Phys. Rev. B* **2**, 977 (1970).
[41] C. Benoit a la Guillaume and P. Lavallard, *Phys. Status Solidi B* **70**, K143 (1975).
[42] W. Ruhle and W. Klingenstein, *Phys. Rev. B* **18**, 7011 (1977).
[43] J. C. Herbert, *J. Phys. C* **10**, 3327 (1977).
[44] D. C. Reynolds and T. C. Collins, *Phys. Rev.* **185**, 1099 (1969).
[45] P. J. Dean, J. D. Cuthbert, D. G. Thomas and R. T. Lynch, *Phys. Rev. Lett.* **18**, 122 (1967).
[46] R. Linder, *Acta Chem. Scand.* **6**, 457 (1952).
[47] E. A. Secco, *Discuss. Faraday Soc.* **28**, 94 (1959).
[48] E. A. Secco, *Reactivity of Solids,* John Wiley & Sons, Ltd, New York, 1969, p. 523.
[49] J. R. Roberts and C. Wheeler, *Trans. Faraday Soc.* **56**, 570 (1960).
[50] O. Akimoto and E. Hanamura, *J. Phys. Soc. Jpn.* **33**, 1537 (1972).
[51] T. Ogino and M. Aoki, *Jpn. J. Appl. Phys.* **19**, 2395 (1980).
[52] D. M. Hofmann, D. Kovalev, G. Steude, B. K. Meyer, A. Hoffmann, L. Eckey, R. Heitz, T. Detchprom, H. Amano and I. Akasaki, *Phys. Rev. B* **52**, 16702 (1995).
[53] E. R. Glaser, T. A. Kennedy, K. Doverspike, L. B. Rowland, D. K. Gaskill, J. A. Freitas Jr, J. Asif Khan, D. T. Olson, J. M. Kuznia and D. K. Wickenden, *Phys. Rev.* **51**, 1326 (1995).
[54] G. Heiland, E. Mollno and F. Stockmann,in *Solid State Physics*, edited by F. Seitz and D. Turnbull, Academic Press, Inc., New York, 1959, Vol. **8**
[55] B. Andress, *Z. Physik* **170**, 1 (1962).
[56] D. C. Look, D. C. Reynolds, W. Kim, O. Oktas, A. Botchkarev, A. Salvador and H. Morkoc, *Appl. Phys. Lett.* (to be published).
[57] P. Boguslawski, E. L. Briggs and J. Bernholc, *Phys. Rev. B* **51**, 17255 (1995).
[58] T. Suski, P. Berlin, H. Teisseyre, M. Leszcynsky, I. Grzegory, J. Jun, M. Bockowski, S. Porowski and T. D. Moustakas, *Appl. Phys. Lett.* **67**, 2188 (1995).
[59] J. Neugebauer and C. G. Van de Walle, *Appl. Phys. Lett.* **69**, 503 (1996).
[60] H. Schweitzer, A. Forchel, H. Hangleiter, S. Schmitt-Rink, J. P. Lowenau and H. Huag, *Phys Rev. Lett.* **51**, 698 (1983).
[61] I. Balslev, *Phys. Rev. B* **30**, 3203 (1984).
[62] S. K. Chun, *J. Appl. Phys.* **80**, 4773 (1996).
[63] A. M. Bakiev, Yu V. Vandyshev, G. S. Volkov, V. S. Dneprovskii, Z. D. Kovalyuk, A. R. Lesiv, S. V. Savinov and A. I. Furtichev, *Sov. Phys. Solid State* **28**, 579 (1986).

[64] C. Guillemot, *Phys. Rev. B* **31**, 1428 (1985).
[65] J. A. Brum, G. Bastard and C. Guillemot, *Phys. Rev. B* **30**, **905** (1984).
[66] D. C. Reynolds, B. Jogai, P. W. Yu, K. R. Evans and C. E. Stutz, *Phys. Rev. B* **46**, 15274 (1992).
[67] D. C. Look, *Electrical Characterization of GaAs Materials and Devices*, John Wiley & Sons, Ltd, New York, 1989.
[68] P. Perlin, E. Litwin-Staszewska, B. Suchanek, W. Knap, J. Camassel, T. Suski, R. Piotrkowski, I. Grzegory, S. Porowski, E. Kaminska and J. C. Chervin, *Appl. Phys. Lett.* **68**, 1114 (1996).
[69] B. K. Meyer, D. Volm, A. Graber, H. C. Alt, T. Detchprohm, A. Amano and I. Akasaki, *Solid State Commun.* **95**, 597 (1995).
[70] J. R. Haynes, *Phys. Rev. Lett.* **4**, 361 (1960).
[71] L. Hedin and B. I. Lundquist, *J. Phys. C* **4**, 2064 (1971).
[72] O. Gnnnavson and B. I. Lundquist, *Phys. Rev. B* **13**, 4274 (1976).
[73] D. C. Reynolds, C. W. Litton, T. C. Collins, J. E. Hoelscher and J. Nause, *Appl. Phys. Lett.* **88**, 141919 (2006).
[74] J. J. Hopfield, in *Proceedings of the 7th International Conference on the Physics of Semiconductors*, Academic Press, New York, 1965, p. 725.

3

Electrical Transport Properties in Zinc Oxide

B. Claflin and D. C. Look
Semiconductor Research Center, Wright State University, Dayton, OH, USA

3.1 Introduction

Zinc oxide (ZnO) is a truly remarkable and versatile material. It has long been used for a wide variety of commercial applications ranging from phosphors, ceramics, abrasives, and as a pigment for paints, to serving as the active (UV-absorbing) ingredient in sunscreen, a key component in cosmetic and medical ointments, a source of Zn for mineral fortification in foods, and as an additive for plastic and rubber processing.[1,2] It also possesses useful semiconducting properties which were identified in the naturally occurring mineral form, zincite, for crystal radio receivers,[3] and led to early electrical and optical characterization efforts.[4–6] However, as-grown, unintentionally doped ZnO is almost always found to be n-type because of high concentrations of background donors, so this early work focused primarily on surface effects[7–11] for catalysis and gas sensing applications[12] and on the photoresponse[13] of the material. The inability to identify a suitable p-type dopant for ZnO limited progress in the development of electronic devices since p-n homojunctions could not be formed. This doping asymmetry is fairly common in wide band gap semiconductors;[14–19] similar problems have limited the development of diamond films[20] and initially slowed the progress of GaN-based devices.[21] However, demonstrations of GaN-based blue light-emitting diodes (LEDs) in the early 1990s and diode lasers[22–25] in the mid-1990s generated a renewed interest in the electrical and optical properties of ZnO. Reports of successful p-type ZnO growth[26–33] a few years later, while viewed with

Zinc Oxide Materials for Electronic and Optoelectronic Device Applications, First Edition.
Edited by Cole W. Litton, Donald C. Reynolds and Thomas C. Collins.

considerable skepticism, sparked an ongoing debate about doping and fueled interest in the material as well. Intensive research efforts were initiated aimed at a number of electronic and optoelectronic device applications such as blue/UV light emitters and detectors, energy efficient white lighting, transparent thin film transistors (TTFTs) as well as transparent conductors for large area displays to provide a more cost effective alternative to tin-doped indium oxide (ITO). LEDs and laser diodes (LDs) aim to take advantage of the large band gap (E_g ~3.35 eV at room temperature) and large free exciton binding energy (~60 meV)[34] in ZnO, and have been demonstrated[35,36] to produce blue/near-UV emission. Combining such an efficient UV emitter with appropriate phosphors for solid-state white lighting could save as much as \$130B yr^{-1} in the USA by the year 2025 according to some estimates.[37,38] For electronic applications, ZnO TTFTs operating at frequencies in the GHz range have recently been demonstrated.[39] However, to take full advantage of the promise of these devices, it is necessary to understand the bulk electrical characteristics of ZnO as well as the influence of impurities, defects and surface layer effects on transport properties. In this chapter, these fundamental aspects of the material are reviewed and some of the major challenges that remain to be solved are discussed. One of the most common and powerful probes of the electrical properties in semiconductors is the temperature-dependent Hall effect since it provides a wealth of information about charge transport in these materials including details about carrier scattering and about the donor and acceptor levels which control n- and p-type doping. We will begin with a brief review of the basic theory of the Hall effect for bulk conduction in a single band. Extensions of these concepts will then be presented which allow for quantitative analysis of conduction by surfaces as well as mixed conduction effects which can play an important role in the performance of p-type samples.

3.2 Hall-Effect Analysis

3.2.1 Single-Band Conduction

Carrier transport can be accurately described by the Boltzmann transport equation which can be generalized to include photo-excited carriers:[40]

$$\frac{\partial f}{\partial t} + \dot{r} \cdot \nabla_r f + \dot{k} \cdot \nabla_k f - \left(\frac{\partial f}{\partial t}\right)_{\text{coll}} + G - R = 0 \tag{3.1}$$

Here f is the occupation distribution function in phase space, G is the generation rate and R is the recombination rate of photo-excited carriers. Using the relaxation time (τ) approximation (RTA) to the Boltzmann transport equation, the current, conductivity, and Hall coefficient for n-type material are given by:[41]

$$j_x = \frac{ne^2\langle\tau\rangle}{m^*} E_x = -ne\mu_c E_x \tag{3.2}$$

$$R_H = \frac{E_y}{j_x B_z} = \frac{-1}{ne}\frac{\langle\tau^2\rangle}{\langle\tau\rangle^2} = \frac{-r}{ne} \tag{3.3}$$

$$\mu_H = R_H\sigma = \left(-\frac{1}{ne}\frac{<\tau^2>}{<\tau>^2}\right)\left(\frac{ne^2<\tau>}{m^*}\right) = \frac{e}{m^*}\frac{<\tau^2>}{<\tau>} \tag{3.4}$$

where

$$\langle\tau^n(E)\rangle = \frac{\int_0^\infty \tau^n(E)E^{3/2}\,e^{-E/kT}dE}{\int_0^\infty E^{3/2}\,e^{-E/kT}dE} \tag{3.5}$$

for nondegenerate electrons (Boltzmann statistics) where electric and magnetic fields are applied in the x- and z-directions, respectively. Equations (3.1)–(3.5) can also be used to describe nondegenerate hole transport by making the substitutions n → p and e → -e. The relaxation time arises from carrier scattering by a variety of mechanisms:

$$\tau^{-1} = \tau_{ii}^{-1} + \tau_{ac}^{-1} + \tau_{po}^{-1} + \tau_{pe}^{-1} + \tau_{dis}^{-1} \tag{3.6}$$

where the most common scattering sources include ionized impurities (τ_{ii}), acoustic phonons (τ_{ac}), polar-optical phonons (τ_{po}), the piezoelectric potential (τ_{pe}) and charged dislocation defects (τ_{dis}), each with its own scattering rate (inverse relaxation time) which simply add to produce the total scattering rate as shown. Standard approximations for these scattering mechanisms have been developed[40–46] and provide the connection between experimental Hall-effect measurements and intrinsic materials parameters of interest as well as donor, acceptor, and dislocation concentrations. For completeness, expressions for these scattering modes are listed below (constants and materials-related parameters are listed in Table 3.1):

$$\tau_{ii}(E) = \frac{2^{9/2}\pi\varepsilon_o^2(m^*)^{1/2}E^{3/2}}{e^4(2N_A+n)[\ln(1+y)-y/(1+y)]} \tag{3.7}$$

Table 3.1 *Physical constants and ZnO specific materials parameters for Hall-effect analysis*

$\hbar$	1.055×10^{-34} J s
e	1.602×10^{-19} C
k_B	1.380×10^{-23} J K^{-1}
m^*	$0.3^*m_e = 2.73\times10^{-31}$ kg
ε_o(300 K)	$8.12^*\varepsilon = 7.19\times10^{-11}$ F m^{-1}
ε_∞(300 K)	$3.72^*\varepsilon = 3.29\times10^{-11}$ F m^{-1}
T_{po}	837 K
E_1	15 eV
ρ_d	5.675×10^3 kg m^{-3}
s	6.006×10^3 m s^{-1}
P_{pe}	0.21
c_{latt}	5.2069×10^{-10} m

for scattering from ionized impurities in nondegenerate, n-type material where ε_0 is the static dielectric constant, $y = 8\varepsilon_0 m^* k_B T E/(\hbar^2 e^2 n)$, k_B is the Boltzmann constant, and $\hbar$ is Planck's constant divided by 2π. This is the well-known Brooks–Herring formula which is valid[46] when $ka_o/2 \gg 1$, where a_o is the Bohr radius. Equation (3.7) can be applied to p-type samples by replacing $(2N_A + n)$ with $(2N_D + p)$. The relaxation time for acoustic mode deformation potential scattering is:

$$\tau_{ac}(E) = \frac{\pi\hbar^4 \rho_d s^2 E^{-1/2}}{2^{1/2} e^2 E_1^2 (m^*)^{3/2} k_B T} \tag{3.8}$$

where ρ_d is the density, s is the speed of sound in the material and E_1 is the deformation potential. Since scattering from polar optical phonons is inelastic, we can only get an approximate expression for the relaxation time:[41,47]

$$\tau_{po}(E) = \frac{2^{3/2}\pi\hbar^2 (e^{T_{po}/T} - 1)[0.5446E^{1/2} + 0.5888(k_B T_{po})^{1/2} - 0.1683(k_B T_{po})^{-1/2} E]}{e^2 k_B T_{po} (m^*)^{1/2} (\varepsilon_\infty^{-1} - \varepsilon_o^{-1})} \tag{3.9}$$

where T_{po} is the Debye temperature and ε_∞ is the high-frequency dielectric constant. Polar-optical mode scattering can be significant at room temperature and above, but this scattering weakens considerably in ZnO for $T < \sim 150$ K and the RTA analysis provides quite accurate results. The relaxation time for acoustic mode piezoelectric-potential scattering can be written:

$$\tau_{pe}(E) = \frac{2^{3/2}\pi\hbar^2 \varepsilon_o E^{1/2}}{e^2 P^2 (m^*)^{1/2} k_B T} \tag{3.10}$$

where $P = (h_{pz}{}^2/\rho s^2 \varepsilon_0)^{1/2}$ is the coefficient of piezoelectric coupling. Finally,

$$\tau_{dis}(E) = \frac{\hbar^3 \varepsilon_o^2 c_{latt}^2 \left(1 + \frac{8m^* \lambda^2 \varepsilon}{\hbar^2}\right)^{3/2}}{N_{dis} m^* e^4 \lambda^4} \tag{3.11}$$

for scattering from dislocations, where $\lambda = (\varepsilon_0 k_B T/e^2 n)^{1/2}$ is the screening parameter. Since dislocation densities of $10^{10}\,\mathrm{cm}^{-2}$ and higher are quite common for ZnO samples grown on non-native substrates, this mechanism can be very important and may reduce the room temperature Hall mobility to values $\mu_H < 100\,\mathrm{cm^2\,V^{-1}\,s^{-1}}$ for epitaxial samples.

We can use these analytic expressions for the dominant carrier scattering mechanisms to fit experimental Hall mobility data as a function of temperature, $\mu_H(T)$. Of particular interest is the scattering from ionized impurities, which usually dominates at low temperatures. Using Equation (3.7), we can determine the concentration of acceptors, N_A, in n-type material or the donor concentration, N_D, for p-type material. In practice, it is necessary to simultaneously fit *both* $\mu_H(T)$ and $n_H(T)$ since the carrier concentration appears in Equation (3.7). The fit of $n_H(T)$ provides shallow donor concentrations and donor energies through the charge-balance equation (CBE):

$$n+N_{\mathrm{A}}=\sum_{i}\frac{N_{D_i}}{1+n/\phi_{D_i}} \tag{3.12}$$

where the sum extends over separate donor levels, i, and

$$\phi_{D_i}=\frac{g_{0_i}}{g_{1_i}}e^{\frac{\alpha_{D_i}}{k_{\mathrm{B}}}}N_{\mathrm{C}}'T^{3/2}e^{-\frac{E_{D0_i}}{k_{\mathrm{B}}T}} \tag{3.13}$$

For each donor, g_0/g_1 is a degeneracy factor, $N_{\mathrm{C}}'=2(2\pi m_{\mathrm{n}}^{*}k_{\mathrm{B}})^{3/2}/h^3$ is the effective density of states at 1 K, E_{D_i} is the energy of the ith donor and $E_{\mathrm{D}_i}=E_{\mathrm{D}0i}-\alpha_{\mathrm{D}i}T$. For effective-mass-like donor levels, the analysis can be generalized to account for hydrogenic-type excited states:

$$\phi_{\mathrm{D}_i}=\frac{g_{0_i}}{g_{1_i}}e^{\frac{\alpha_{\mathrm{D}_i}}{k_{\mathrm{B}}}}N_{\mathrm{C}}'T^{3/2}\frac{e^{-\frac{E_{\mathrm{D}0_i}}{k_{\mathrm{B}}T}}}{1+\sum_2^m j^2 e^{-\left(1-\frac{1}{j^2}\right)\frac{E_{\mathrm{D}0_i}}{k_{\mathrm{B}}T}}} \tag{3.14}$$

where it is assumed that $g_j/g_0=j^2$ (similar to the H atom) and α_{D_i} is assumed to be the same for each excited state as in the ground state (i.e. independent of j).

Similar expressions can be derived for p-type samples, where the CBE becomes:

$$p+N_{\mathrm{D}}=\sum_{i}\frac{N_{\mathrm{A}_i}}{1+n/\phi_{\mathrm{A}_i}} \tag{3.15}$$

with

$$\phi_{\mathrm{A}_i}=\frac{g_{1_i}}{g_{0_i}}e^{\frac{\alpha_{\mathrm{A}_i}}{k_{\mathrm{B}}}}N_{\mathrm{V}}'T^{3/2}e^{-\frac{E_{\mathrm{A}0_i}}{k_{\mathrm{B}}T}} \tag{3.16}$$

where $N_{\mathrm{V}}'=2(2\pi m_{\mathrm{p}}^{*}k_{\mathrm{B}})^{3/2}/h^3$ and $E_{\mathrm{A}_i}=E_{\mathrm{A}0_i}-\alpha_{\mathrm{A}_i}T$. Note that the degeneracy factor, g_1/g_0, is inverted from the analogous term in Equation (3.13) because the degeneracies refer to electrons rather than holes.[41]

3.2.2 Two-Band Mixed Conduction

One important special case of two-layer conduction in ZnO involves a thin conducting n-type layer on p-type bulk material. In this case, for transport by electrons in the conduction band and holes in the valence band, the current, conductivity, and Hall coefficient take the form:[41]

$$j_x=j_{\mathrm{p}x}+j_{\mathrm{n}x} \tag{3.17}$$

$$\sigma=\sigma_{\mathrm{p}}+\sigma_{\mathrm{n}}=e(p\mu_{\mathrm{p}}+n\mu_{\mathrm{n}}) \tag{3.18}$$

$$R_{\mathrm{H}}=\frac{R_{\mathrm{p}}\sigma_{\mathrm{p}}^2+R_{\mathrm{n}}\sigma_{\mathrm{n}}^2}{(\sigma_{\mathrm{p}}+\sigma_{\mathrm{n}})^2}=\frac{p\mu_{\mathrm{p}}^2-n\mu_{\mathrm{n}}^2}{e(p\mu_{\mathrm{p}}+n\mu_{\mathrm{n}})^2} \tag{3.19}$$

The sign of R_H is used to assign the sample 'type' from a given Hall-effect measurement; a sample will be classified as n-type if $p\mu_p^2 < n\mu_n^2$ or as p-type when $p\mu_p^2 > n\mu_n^2$. Notice that when $p\mu_p^2 = n\mu_n^2$ the Hall coefficient $R_H = 0$. This feature results in a singularity in the apparent carrier concentration [Equation (3.3)] when a simple single-band analysis is used and a change of sign for R_H across this transition. Likewise, since the Hall mobility is derived from the Hall coefficient and conductivity [Equation (3.4)], it will go to zero when $p\mu_p^2 = n\mu_n^2$. These behaviors provide clear signatures of the presence of mixed conduction effects.

3.2.3 Conducting Surface Layers

As was pointed out in the introduction, ZnO surfaces are highly reactive and surface conducting or depletion layers can dramatically change the electrical characteristics of the material. For such a case, where the carrier concentration and mobility are depth dependent, the analysis provided above can be generalized:

$$\sigma_{sq} = \int_0^d \sigma(z)\mathrm{d}z = e\int_0^d n(z)\mu(z)\mathrm{d}z \tag{3.20}$$

$$R_{Hsq}\sigma_{sq}^2 = \int_0^d n(z)\mu^2(z)\mathrm{d}z \tag{3.21}$$

where σ_{sq} and $R_{H_{sq}}$ denote sheet (areal) quantities (cm^{-2}) rather than volume quantities (cm^{-3}). For the simple case of a two-layer system, with a thin surface conducting layer on a thick, bulk sample, the effective mobility and carrier concentration can be determined explicitly:

$$\mu = \frac{n_b\mu_b^2 + n_s\mu_s^2}{n_b\mu_b + n_s\mu_s} \tag{3.22}$$

$$n = \frac{(n_b\mu_b + n_s\mu_s)^2}{n_b\mu_b^2 + n_s{\mu_s}^2} \tag{3.23}$$

The impact of such conductive surface layers on Hall-effect measurements can be substantial, as seen in Figures 3.1 and 3.2. Recent work[47–49] has extended this analysis to provide more quantitative information about the characteristics of the surface. Examples of surface effects will be discussed in more detail later.

3.3 Donor States and n-type Doping

As-grown, unintentionally doped ZnO generally exhibits n-type conduction, with typical electron concentrations on the order of $n \sim 10^{15}$–10^{17} cm^{-3}.[50] For such samples,

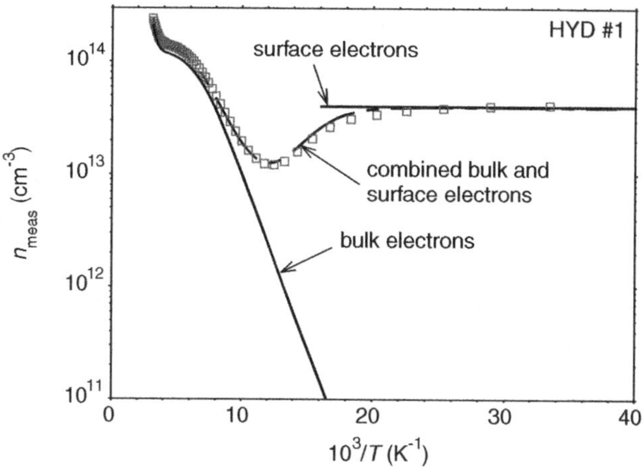

Figure 3.1 *Representative temperature-dependent Hall-effect data and two-layer model fits showing the impact of a surface conducting layer on carrier concentration*

temperature-dependent Hall-effect measurements usually show one or more shallow donors with binding energies in the range of E_d ~30–75 meV.[51–53] Surprisingly, a definitive assignment for the source of these background carriers has not been clearly established and this question continues to be actively studied and debated.[54] Measured values of the room temperature electron mobility are typically μ_e ~200 cm^2 V^{-1} s^{-1}[55] for high quality samples and peak values μ_{peak} >2000 cm^2 V^{-1} s^{-1} have been observed at low temperature. Material prepared by vapor-phase-growth typically exhibits the highest mobility although melt-grown material can sometimes have peak values that are almost as high. Hydrothermal ZnO shows lower values of mobility, sometimes much lower, partly because of much higher background acceptor impurity concentrations.

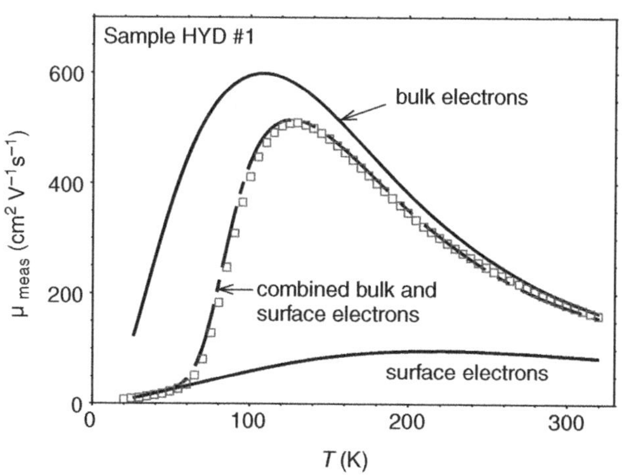

Figure 3.2 *Representative temperature-dependent Hall-effect data and two-layer model fits showing the impact of a surface conducting layer on carrier mobility*

3.3.1 Native Point Defects – Donors

Basic bonding chemistry tells us that native point defects, such as oxygen vacancies (V_O) or zinc interstitials (Zn_i), should be donors, and for many years it was widely believed that these defects were responsible for the intrinsic n-type behavior observed[6,56] in unintentionally doped material, since ZnO growth is often performed under Zn-rich conditions. This explanation continues to be cited frequently in the current literature even though there is no direct evidence to support the assignment of these defects as the dominant donors.[16,57,58] While variations in stoichiometry do appear to play an important role in determining the surface conductivity of ZnO,[7] this native defect model was challenged[59] for bulk transport by theoretical calculations that suggested that V_O and Zn_i do not have shallow levels, but are in fact deep donors, and have high formation energies in n-type material.[54,60–67] However, it was quickly realized that Zn_i is actually a shallow donor, but the formation energy is still high. Electron paramagnetic resonance (EPR) and positron annihilation spectroscopy (PAS) measurements[68–74] have reported that the V_O donor level is deep, and that it is not observed in as-grown material. Electron irradiation experiments, which have been used to produce point defects in ZnO,[71–77] show that V_O is stable up to 400 °C.[78] These irradiation experiments have also demonstrated[75,76] that Zn_i is actually a shallow donor ($E_d \sim E_c - 30$ meV), but for n-type growth conditions, the Zn_i concentration should still be very low because of its high formation energy. In addition, the Zn_i defect diffuses rapidly at 500 °C[79–81] so it is unlikely to be stable at room temperature.[74] Ion bombardment experiments indicate point defects are unstable above ~300 °C.[82,83]

There has been some debate whether the Zn anti-site defect (Zn occupying an O sublattice site, Zn_O) is a shallow or deep donor[59–61] but recent work[62] suggests that it induces a lattice distortion resulting in a shallow level. However, Zn_O has an even higher formation energy than either V_O or Zn_i defects, so it is not expected to be present in any significant quantity, especially in n-type material. No evidence of electrical or optical activity for the Zn_O level has been observed experimentally.

Recent theoretical analyses[60–62] conclude that isolated native defects are not responsible for the high residual electron concentration found in unintentionally doped material. There is, however, some experimental evidence that complexes[84] involving Zn_i and acceptors such as N or Li substituting for O (N_O, Li_O) may exist as shallow donors in n-type material and have low enough formation energies to be produced in significant concentrations. Density functional calculations indicate that such defect complexes can form through Coulombic interactions[85] or as a result of quantum mechanical hybridization effects.[86] It has even been suggested recently that an *attractive* interaction between V_O and Zn_i may exist and produce a donor defect complex (V_O-Zn_i) with a low formation energy.[86] A complex between a deep V_O donor and an unidentified shallow donor has been observed using optically detected EPR.[78]

For intentionally doped material, the formation energies of V_O, Zn_i, and Zn_O defects are large enough under n-type growth conditions that even the shallow levels should have low enough concentrations that they will not influence the electrical transport properties. However, in p-type material, since the Fermi level (E_F) is close to the valence band maximum, these defects will have lower formation energies, especially under Zn-rich growth conditions, and may limit p-type doping through self-compensation.

3.3.2 Substitutional Donors

The group III elements, Al, Ga, and In, are all well-known donors[6,87–91] when substituting in the Zn sublattice, and recent work[92–94] indicates that B is also a donor, as expected. One or more of these elements is almost always present in ZnO, as seen by photoluminescence (PL) measurements,[53,95,96] and they are the most common dopants used to produce n-type material, with electron concentrations in the mid-$10^{20}\,cm^{-3}$ range[87–91] achieved easily. The donor levels[53] for Al ($E_d \sim E_c - 52\,meV$), Ga ($E_d \sim E_c - 55\,meV$), and In ($E_d \sim E_c - 63\,meV$) are all in the range of values observed in undoped material, so residual concentrations of one or more of these impurities are likely to contribute to the background carriers responsible for the bulk n-type conductivity in these films.[88]

There have also been a few reports of n-type doping with the group VII elements, F or Cl, replacing O.[97–103] High electron concentrations have been achieved using the alkali halides, with measured mobilities comparable to those seen for doping with group III metals. However, F and Cl are not commonly seen as residual impurities in unintentionally doped material, and they are unlikely to be utilized for intentional doping given the safety and environmental issues associated with them. The ease of doping ZnO with group III metals makes them preferable.

A number of transition metals have also been investigated as potential dopants, such as Fe, Cr, Co, Ni, Mn, Mo and V.[6,83,104–113] However, these elements do not provide shallow levels and their impact on the electrical properties of ZnO is quite limited. The main interest in these impurities is for potential magnetic/spintronic applications.[113,114]

Hydrogen has also been suggested as a promising shallow donor candidate;[115–119] given the potentially important role played by H in the electronic properties of ZnO, it will be discussed separately, and in more detail in the next section.

3.4 Hydrogen

Subjecting ZnO to annealing in a H ambient or exposure to H plasma has long been known to increase its conductivity.[6,120–122] Hydrogen diffuses easily in ZnO, with early measurements indicating an activation energy for diffusion, $E_a = 0.91\,eV$,[6,120,121] although more recent studies have determined a much lower value, $E_a = 0.17\,eV$,[123] and indicate that all of the H can be removed from a sample by annealing at $T > 500\,°C$.[123–130] In most semiconductors, H is an amphoteric impurity,[131] incorporating as an H^- acceptor in n-type material and as an H^+ donor in p-type material. However, this is not the case in ZnO, where calculations using density functional theory show interstitial H incorporates only in the H^+ donor configuration for all Fermi level positions[115] because of the strong O-H bond. Experiments using EPR and muon spin rotation confirmed this doping behavior and determined a donor energy level of $E_d \sim E_c - 30\,meV$.[116,118] Similar values for the H donor level, $E_d \sim E_c - 25–70\,meV$,[76,118,119,125,127,132–134] are determined by electrical and optical measurements. However, infrared (IR) spectroscopy measurements looking at O-H and O-D (deuterium) vibrations suggest that H occupies the anti-bonding configuration but this molecular complex is unstable at room temperature.[135,136] More recent theoretical work suggests that H can form complexes with Zn vacancies

(V_{Zn})[137] or substitute in the O sublattice by forming multi-center bonds,[117,128] although there is some disagreement about the latter.[138] Of course, H^- can also serve to passivate acceptors in ZnO, as has been observed in several reports.[27,76,84,133,135,139–149] Passivation of acceptors by H can significantly impact the electrical properties if the acceptor concentration is large,[142,150] as shown in Figure 3.3 for a hydrothermal ZnO sample. If a sample with a high concentration of acceptors is exposed to H at sufficient temperature for diffusion to occur, passivation of these acceptors will increase the Fermi level by repopulating existing donors that were ionized as a result of compensation, as can be seen in Table 3.2.

3.5 Acceptor States and p-type Doping

The development of reproducible and stable p-type doping in ZnO has proven to be a very difficult problem.[151] For most of the twentieth century, the inability to identify any shallow acceptors resulted in a widespread belief that it was impossible to dope ZnO p-type. Not surprisingly, early reports of successful p-type doping[26,27,30–33] were largely dismissed, but as a result of a large amount of work over the past decade, a general consensus has begun to emerge that p-type material does, in fact, exist. Reports are even beginning to appear[152–155] which claim to have produced stable p-type films. Considerable skepticism still remains[156] but a concerted effort is now underway to produce better p-type material and to develop a detailed understanding of the physics of acceptor states in ZnO. In general, films require thermal annealing treatments, often in air or an O_2 ambient,

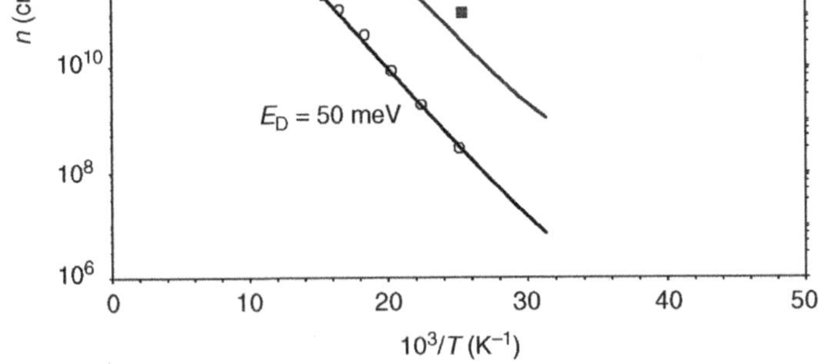

Figure 3.3 Increase in electron density produced by forming gas anneal and best-fit two-layer Hall-effect model. No new donor levels introduced and no change in N_D for the two dominant donors, but significant reduction in N_A most likely caused by passivation with H. Reprinted with permission from B. Claflin and D. C. Look, J. Vac. Sci. Tech. B 27, 1722 (2009). Copyright 2009, American Vacuum Society

Table 3.2 Best-fit two-layer Hall-effect model parameters for hydrothermal ZnO sample in different states. Donor concentrations unaffected by annealing in forming gas but acceptor concentration decreases most likely as a result of passivation. Reprinted with permission from B. Claflin and D. C. Look, J. Vac. Sci. Tech. B 27, 1722 (2009). Copyright 2009, American Vacuum Society

Sample	N_D (50 meV) 10^{16} cm^{-3}	N_D (400 meV) 10^{16} cm^{-3}	N_A 10^{16} cm^{-3}
As-received	1.228	1.5	1.215
After photo-Hall	2.0	2.0	1.98
Forming gas anneal	2.0	2.0	1.2

to exhibit p-type characteristics.[30,85,152,157–162] One of the challenging aspects of developing p-type ZnO is the low value of the hole mobility – typically around $\mu \sim 1\ cm^2\,V^{-1}\,s^{-1}$ at room temperature. Similar limitations are also seen in p-type GaN. Such low hole mobilities make Hall-effect measurements difficult because the Hall voltages involved are small, and other effects, such as persistent photoconductivity in ZnO,[163,164] and sample homogeneity,[165–168] can significantly impact the results.

3.5.1 Native Point Defects – Acceptors

Theoretical calculations indicate that the Zn vacancy (V_{Zn}), O interstitial (O_i), and O antisite (O_{Zn}) are acceptors[59–62] with V_{Zn} having the lowest formation energy of all the acceptor-type native point defects. For n-type material grown under O-rich conditions, V_{Zn} is expected to be the dominant acceptor, and this has been confirmed experimentally.[72,169] The V_{Zn} defect has been well-characterized by PAS measurements since in n-type material the ionized vacancy is negatively charged and traps positrons easily. Temperature-dependent Hall-effect measurements on unintentionally doped ZnO deposited by seeded-vapor growth[169] show small concentrations of V_{Zn} (low 10^{15} cm^{-3}), which account for all of the background acceptors observed in this sample. However, these measurements do not provide any information about the V_{Zn} acceptor energy level because these defects function as compensation centers in n-type material and thus do not require thermal activation to be ionized. The V_{Zn} acceptor concentration is determined by fitting temperature-dependent mobility data, involving Equations (3.4–3.11), and in particular, by analysis of ionized impurity scattering at low temperature using Equation (3.7). In p-type material, the high formation energy of V_{Zn} precludes its generation in significant concentrations.[62]

The O_i defect is predicted to be unstable at the tetrahedral lattice site,[62] inducing a lattice relaxation to form a split interstitial defect, O_i(split), which is effectively an O_2 molecule embedded in the ZnO lattice.[62] The O_i(split) defect is expected to be electrically neutral for all positions of the Fermi level. However, interstitial O can also occupy the octahedral lattice site, O_i(oct) as a deep acceptor. The theoretical formation energy for O_i(oct) is very high and it is not expected to be present in any significant concentration. Similarly, O_{Zn} is predicted to have the highest formation energy of all the native acceptor point defects and is expected to have negligible concentrations. There have been no experimental observations of either O_i(oct) or O_{Zn} defects, even after high-energy electron irradiation.

3.5.2 Substitutional Acceptors

The most obvious acceptor dopant for ZnO is substitutional N on the O site (N_O),[50,170–172] since these two elements have similar ionic radii, which allows for a high N solubility. Successful N_O doping has been confirmed using EPR and X-ray photoelectron spectroscopy (XPS)[159,173,174] and concentrations as high as [N] $\sim 10^{19}$–10^{21} cm^{-3} have been demonstrated.[26,175–180] Recent reports indicate a surface polarity dependence to N incorporation, with higher levels for Zn-face growth at low temperatures.[181] However, only weak p-type electrical behavior has been observed (see Figure 3.4) in N-doped ZnO samples,[26,163,164,175] while in several reports, n-type conduction is seen[28,182,183] despite the p-type dopants. Furthermore, in a few cases, the p-type conduction that has been observed in N-doped samples is unstable over time,[151,184] although the mechanism responsible for this degradation has not been clearly established. Measured values of the N acceptor energy range from $E_a \sim E_v + 90$–200 meV [26,47,53,163,164,185–190] with a value quoted frequently of $E_a \sim E_v + 165$ meV.[53] Several theories have been proposed to explain the difficulty of producing p-type ZnO by N-doping:[191] (i) limited N solubility;[192] (ii) self-compensation by the formation of Zn_i or V_O defects for low concentrations of N, and Zn_O defect complexes at high N concentrations;[193,194] (iii) formation of N-N complexes that produce donor levels which compensate the N_O acceptor levels;[158,159,173,174,195] and (iv) compensation by donors as a result of unintentional impurity doping, particularly H.[115,117] As discussed above, high levels of N incorporation have been demonstrated in ZnO, and the formation energy of Zn_O is too large, even under Zn-rich growth conditions, to produce significant concentrations of anti-site defects, so these mechanisms are not likely to limit p-type doping with N. Compensation by unintentional donors, N-N complexes, or V_O defects, because of their low formation energy in p-type material, are more reasonable explanations for the weak p-type behavior observed[170] but more effort is needed to clarify this issue.

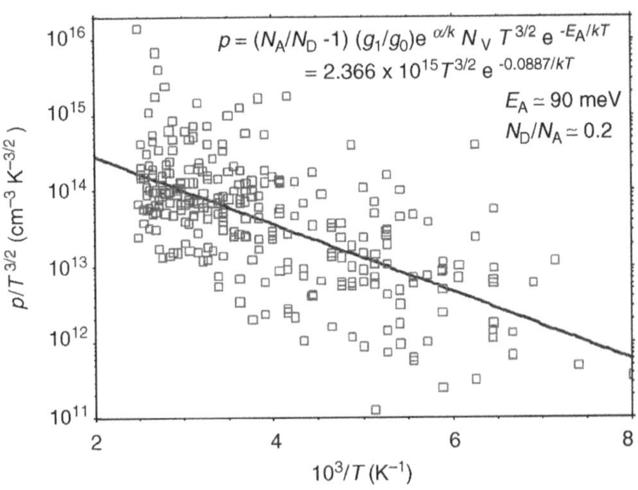

Figure 3.4 Weak p-type behavior observed in N-doped ZnO sample grown by MBE. D. C. Look, B. Claflin, Ya. I. Alivov, and S. J. Park. The future of ZnO light emitters, Phys. Stat. Sol. A. 2004. 201. 2203. Copyright Wiley-VCH Verlag GmbH & Co. KGaA. Reproduced with permission

Other group V elements, P, As, and Sb, also have been investigated as p-type dopants, with surprising success,[30,113,152–154,160–162,185,196–220] since these ions are significantly larger than O[199] and are expected to have low solubility for incorporation into the O sublattice.[196,199] In addition, theoretical calculations predict deep acceptor levels for these dopants.[85,170,171,221] However, temperature-dependent photo-Hall-effect measurements have unequivocally demonstrated p-type doping[163,164] in P-doped ZnO via observation of a mixed conduction transition from p- to n-type behavior, according to Equations (3.19) and (3.20), under exposure to above band gap light, as shown in Figure 3.5. The n-type conduction persists, even after the light source is removed, but

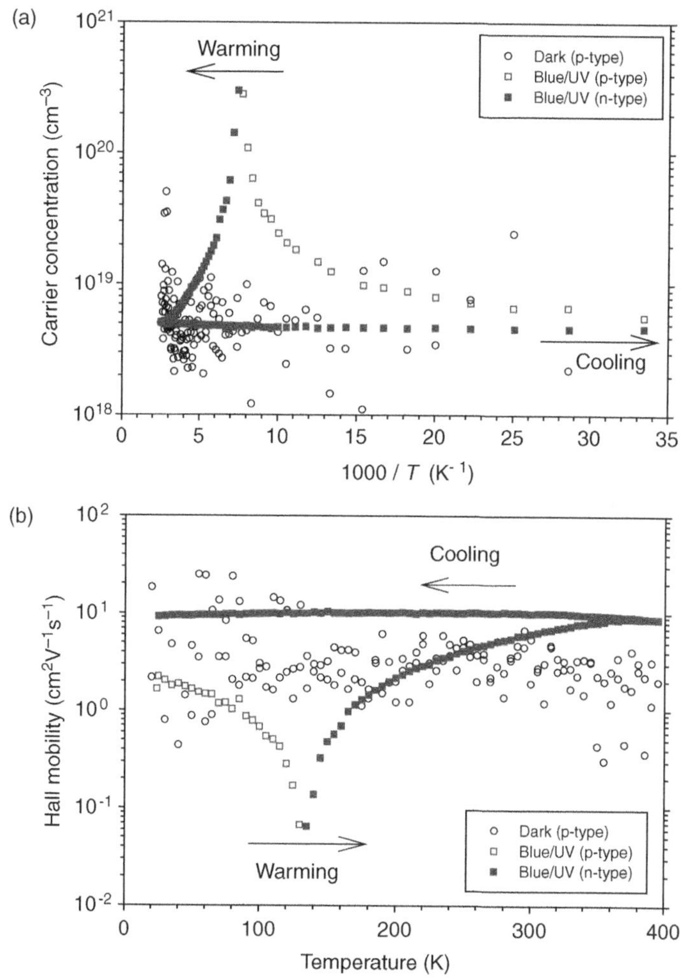

Figure 3.5 *Mixed conduction transition from p- to n-type conduction for P-doped ZnO sample under UV illumination in vacuum. The sample exhibits n-type PPC when the light source is removed which slowly decays if the sample is annealed at 400 K or exposed to air. Reprinted from J. Cryst. Gr., 287, 16, B. Claflin, D.C. Look, S.J. Park, G. Cantwell, Persistent n-type photoconductivity in p-type ZnO, 16–22, Copyright (2006) with permission from Elsevier*

the original p-type state can be restored by modest annealing in vacuum in the dark.[164] Further evidence of successful p-type doping is provided by reports of light emission from ZnO homojunction LEDs[179,180,222–230] using group V dopants for the p-layer. Hole concentrations as high as $p = 1.7 \times 10^{19}\,\text{cm}^{-3}$[196] and mobilities as high as $\mu_h = 39.3\,\text{cm}^2\,\text{V}^{-1}\,\text{s}^{-1}$ at room temperature[161] have been reported for P-doped samples, although mobilities closer to $\mu_h \sim 1\,\text{cm}^2\,\text{V}^{-1}\,\text{s}^{-1}$ are more common. Since simple substitution of P, As, or Sb for O is unable to account for the shallow acceptor levels observed, defect complex models have been proposed[85,231,232] which involve the group V dopants incorporating in the Zn sublattice (anti-site substitutions) and spontaneously generating two neighboring V_{Zn} defects. These models predict acceptor levels of $E_a \sim E_v + 180$ meV for P[231] and $E_a \sim E_v + 150$ meV for As.[85] These values are in agreement with experimentally determined values, $E_a \sim E_v + 120$–150 meV for P,[153,205–207] $E_a \sim E_v + 100$–180 meV for As,[30,209,233] and $E_a \sim E_v + 115$–200 meV for Sb.[216,218] However, much higher acceptor energies, $E_a \sim E_v + 336$–400 meV for P, have been reported recently,[234] and there is some theoretical disagreement about the stability of the group V anti-site – $2V_{Zn}$ defect complex.[62] In addition, XPS measurements suggest P substitutes on the O site,[200,201,208] while evidence for As_{Zn} anti-site incorporation has also been reported,[235] so this question is not settled. At present, no other models have been proposed which can account for the observed p-type conduction produced by P, As, and Sb doping, or the low acceptor energies measured experimentally.

Efforts have also been made to produce p-type ZnO through substitution of group Ia (Li, Na, K) or Ib (Cu, Ag, Au) dopants in the Zn sublattice,[6,146,236–251] instead of replacing O with a group V element, to introduce shallow acceptor levels. First-principles calculations predict acceptor energies of $E_a \sim E_v + 40$–90 meV for Li, $E_a \sim E_v + 110$–170 meV for N, and $E_a \sim E_v + 320$ meV for K.[146,170] Lithium has been shown to induce ferroelectric behavior[68,240,244–248] and has long been known to diffuse easily in ZnO at moderate temperatures. A historical review of group Ia doping in ZnO,[236] including recent results, indicates that Li and Na produce similar shallow acceptor levels ($E_a \sim E_v + 300$ meV) which do not induce significant lattice relaxations, as well as deep acceptor levels, $E_a \sim E_v + 800$ meV for Li_{Zn} and $E_a \sim E_v + 600$ meV for Na_{Zn}, with large lattice relaxations.[237] To achieve useful p-type conduction, even with these shallow group Ia acceptor energies, requires a doping concentration of greater than $10^{20}\,\text{cm}^{-3}$,[236] which has not been realized to date. The hole concentration goes through a maximum of $p = 6.0 \times 10^{17}\,\text{cm}^{-3}$ with increased Li doping, and for high concentrations ([Li] > 1.2%), [238] Li interstitials (Li_i), which act like donors, compensate the Li_{Zn} acceptors and pin the Fermi level in the gap. As a result of this amphoteric self-compensation, ZnO samples doped with group Ia elements generally exhibit semi-insulating behavior. Similarly, bulk, unintentionally doped ZnO crystals grown using the hydrothermal method, exhibit high resistivity or semi-insulating characteristics because of Li incorporation from the mineralizer. The effectiveness of p-type doping using group Ib elements faces similar limitations as the group Ia elements. Theoretical investigations suggest deep acceptor levels for Cu ($E_a \sim E_v + 700$ meV), Ag ($E_a \sim E_v + 400$ meV), and Au ($E_a \sim E_v + 500$ meV),[252] although experimentally determined values are shallower, $E_a \sim E_v + 380$ meV for Cu, $E_a \sim E_v + 200$ meV for Ag, and $E_a \sim E_v + 450$ meV for Au.[253] However, even for these shallower acceptor energies, high doping concentrations are needed, without the formation of self-compensating donor centers, to produce the hole densities required for practical applications.

Finally, attempts have been made to produce p-type ZnO by intentionally introducing two dopants simultaneously, also known as co-doping (not to be confused with doping using the transition metal Co). In fact, a patent has even been awarded in the United States for p-type ZnO growth using the co-doping approach.[254] Still, these efforts have been viewed skeptically[113,163,198,255] and generally fall into two categories: (i) doping with two acceptor-type elements (groups Ia, Ib, and V);[168,189,190,256–258] and (ii) incorporation of one acceptor-type and one donor-type impurity.[147,192,224,259–266] The goals of the added process complexity associated with co-doping are:[184,192,267,268] (i) to increase the solubility of the acceptors; (ii) to reduce the acceptor ionization energy; and (iii) to decrease the formation of self-compensating donor-type defects. Hole concentrations as high as $p \sim 10^{18}\,cm^{-3}$ and mobilities, $\mu_h \sim 140–167\,cm^2\,V^{-1}\,s^{-1}$, have been reported for co-doped samples[148,224,260] but these values are surprisingly high and should be viewed cautiously.[149,163] As pointed out above, N and P are reasonably shallow acceptors and can be incorporated in high concentrations into the lattice without the addition of another dopant. A recent report utilizing co-doping of Li and N[258] determined a N acceptor level similar to that seen for N doping without Li, $E_a \sim E_v + 138\,meV$. Likewise, it is not clear if the intentional addition of compensating donor centers, which are predicted to form acceptor-donor-acceptor complexes,[192] will be beneficial in any practical application since it is the excess acceptor concentration, $N_A - N_D$, that ultimately controls the resulting electrical properties. Another immediate consequence of intentionally adding donor impurities will be an increase in ionized impurity scattering from the charged compensation centers which may result in a reduced hole mobility.[26,263] Since the hole mobility in p-type ZnO is already quite low, $\mu_h \sim 1\,cm^2\,V^{-1}\,s^{-1}$, any process that will potentially decrease it further will not be attractive. However, one possible exception to this limitation is co-doping with an acceptor-type impurity and H as the donor, combined with a post-growth anneal to remove the H and activate the acceptor.[27,146–148] This approach has been used successfully with Mg and H co-doping to produce p-type GaN.[255,269] The addition of H during p-type ZnO growth will shift the Fermi level farther into the gap, away from the valence band. As a result of this shift in E_F, the formation energy for acceptor incorporation will be reduced and the formation energy for the creation of native donor defects increases, suppressing the formation of compensation centers. However, first principles pseudopotential calculations indicate an energy barrier of at least 1.25 eV for dissociation of H–N bonds.[141] Another problem is that if substitutional H_O[117,128] is removed by post-growth annealing, it will leave behind a V_O defect, so a compensating donor still remains, limiting the effectiveness of this approach. Co-doping with H has not been systematically investigated in a controlled manner to evaluate its effectiveness. However, since H is often present during film deposition, especially for vapor-phase transport, hydrothermal, and metal organic chemical vapor deposition (MOCVD) growth, and p-type conduction is only observed following post-growth annealing, this mechanism may already be in widespread use serendipitously, although poorly controlled. This lack of process control over H during growth may contribute to some of the variation in film characteristics reported in the literature. A careful study of the growth parameter space, including H, would be helpful to address this question.

3.6 Photoconductivity

Early work on ZnO[6] showed that exposure to above band gap light usually results in an increase in conductivity, i.e. photoconductivity (PC). Fast and slow photoconductive processes were identified; the fast process produces a small effect and is reversible in vacuum when the light source is removed. This fast process can also be observed using longer wavelength light, below the absorption edge, and is most likely related to excitations from electronic levels in the forbidden gap. In contrast, the slow process produces a much larger effect, increasing the conductivity in some cases by several orders of magnitude before saturating, and is irreversible in vacuum.[6,142,164,270] This latter effect is also known as persistent photoconductivity (PPC), and, like the PC, is usually observed to be n-type in ZnO. These PPC effects slowly decay as a result of modest thermal annealing treatments[6,164] or by exposure to dry O_2 or air.[6,142,270,271] Temperature-dependent photo-Hall-effect measurements show that the PC and PPC can have a significant impact on surface conduction. Figure 3.6 shows temperature-dependent Hall-effect and photo-Hall-effect data[142] for an as-received, hydrothermal ZnO sample, which has a surface layer that dominates the conduction at low temperature ($T < 60$ K) and bulk conduction controlled by two donor levels, $E_{d1} \sim E_c - 50$ meV and $E_{d2} \sim E_c - 400$ meV, for $T > 100$ K. In the surface layer, the electron concentration is nearly independent of temperature (i.e. the carriers are degenerate), and they have a very low mobility, $\mu_{e\ surf} < 30\,cm^2\,V^{-1}\,s^{-1}$, compared with the experimentally measured bulk value of $\mu_{e\ bulk} \sim 300\,cm^2\,V^{-1}\,s^{-1}$ at 60 K. It is important to note that the electron concentration shown in Figure 3.6(a) is presented as a volume concentration, which assumes uniform bulk conduction throughout the sample, because Hall-effect measurements only determine a sheet carrier concentration. At high temperature, this assumption is usually valid and the analysis provides accurate results. However, at low temperature, where conduction proceeds primarily through the surface layer, this analysis underestimates the electron density in the surface layer by a factor d_{bulk}/d_{surf}.[48] Unfortunately, it is very difficult to determine the thickness of the surface layer, although a minimum thickness for this layer can be estimated.[48] Under vacuum illumination with blue/UV light, produced by a 150 W halogen lamp and passing through a CS 7-59 notch filter with a peak transmission at 3.35 eV, the surface conduction dominates the electrical transport over almost the entire measurement range [see Figure 3.6(a)]; conduction through the bulk of the sample is only observed at the highest temperatures investigated. It is not surprising that the PC has a significant surface component since the absorption coefficient at the band edge, α is $\sim 2 \times 10^5\,cm^{-1}$ at room temperature[272] so the absorption depth of the light, d is ~ 0.1 μm. The resulting PPC, when the sample is returned to the dark, exhibits an increase in surface conduction compared with the as-received state, now dominating the transport characteristics for $T < 120$ K. When the PPC is subsequently allowed to fully decay, following exposure to air, surface conduction is no longer observed at any temperature,[142] and the 50 meV donor level controls the electron concentration over five orders of magnitude, down to $n_{bulk} \sim 3.0 \times 10^8\,cm^{-3}$. This volume concentration determines an upper limit for any possible surface sheet charge in the sample, $n_{sh\ surf} \sim 1.67 \times 10^7\,cm^{-2}$. These results, in conjunction with previous observations, suggest that irradiation with band gap light in vacuum modifies the surface, most likely through desorption of O-containing species. An alternate model of the PPC in ZnO based upon metastable charge state

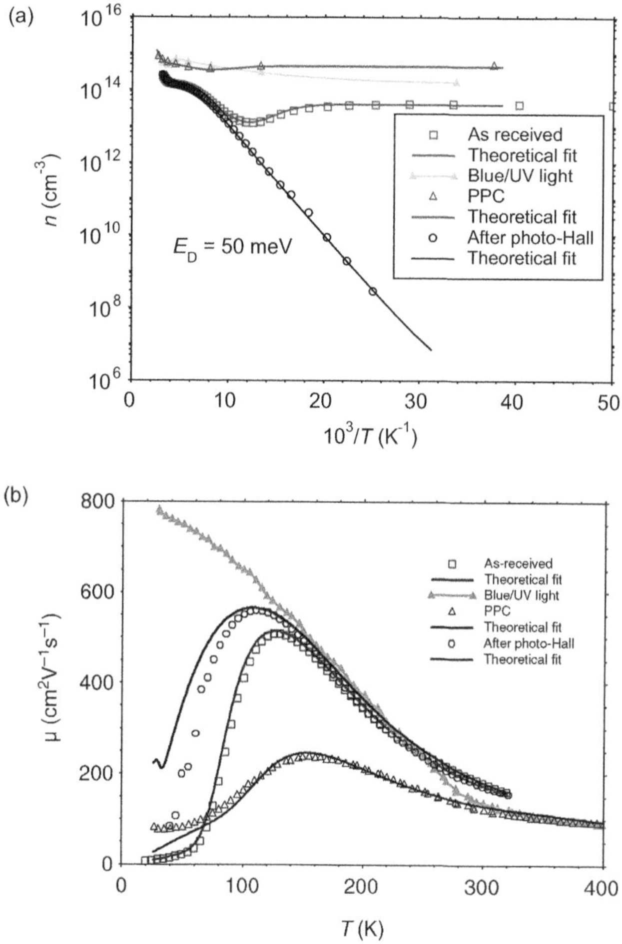

Figure 3.6 *Temperature-dependent Hall-effect and photo-Hall-effect measurements on hydrothermal ZnO showing dramatic changes in surface conducting layer resulting from UV light illumination in vacuum and subsequent exposure to air. Reprinted with permission from B. Claflin and D. C. Look, J. Vac. Sci. Tech. B, 27, 1722 (2009). Copyright 2009, American Vacuum Society*

configurations of V_O has been suggested[67] but in this case it is not clear that it can explain the absence of a surface conducting layer following vacuum illumination or the sensitivity to gas ambient, particularly O.

The n-type PPC can have a particularly significant impact on p-type material, in some cases producing a mixed conduction transition from p- to n-type, and ultimately producing solid n-type Hall-effect measurements, as shown in Figure 3.5. Even weak light exposure can produce important PPC effects in p-type material, in large part as a result of the substantial difference in electron and hole mobilities,[164] as seen in Figure 3.7. Since the Hall coefficient, R_H, varies as $p\mu_p^2 - n\mu_n^2$ [Equation (3.19)], for reasonable room temperature values of $\mu_n \sim 200\,cm^2\,V^{-1}\,s^{-1}$ and $\mu_p \sim 1\,cm^2\,V^{-1}\,s^{-1}$, an electron concentration as low as $n \sim 2.5 \times 10^{12}\,cm^{-3}$ can substantially distort a Hall-effect measurement for a sample with a hole

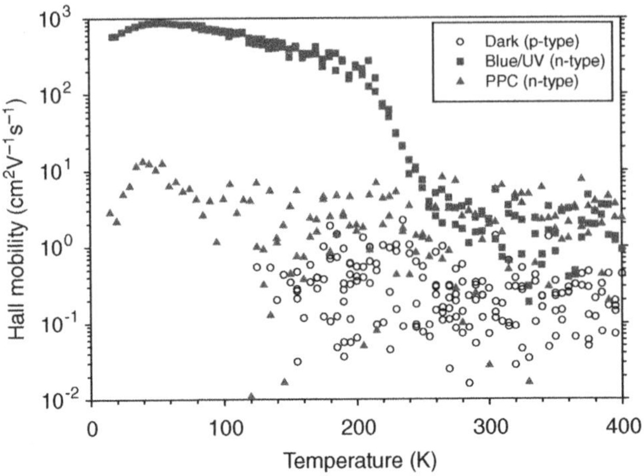

Figure 3.7 *Large difference in measured electron and hole mobilities in the same N-doped ZnO sample. Hole mobilities measured in the dark and electron mobilities measured under UV illumination. Reprinted from J. Cryst. Gr., 287, 16, B. Claflin, D.C. Look, S.J. Park, G. Cantwell, Persistent n-type photoconductivity in p-type ZnO, 16–22, Copyright (2006) with permission from Elsevier*

concentration of $p \sim 1 \times 10^{17}\,\text{cm}^{-3}$. This effect could be even more pronounced at low temperatures since peak electron mobilities can be higher than $\mu_n \sim 2000\,\text{cm}^2\,\text{V}^{-1}\,\text{s}^{-1}$. Such mixed conduction effects may help to explain some of the unusually large room temperature hole concentrations that have been reported for p-type material.

3.7 Summary

As a result of recent, worldwide research, considerable progress has been made toward understanding and controlling the electrical properties of ZnO, particularly the development of stable and reproducible p-type material. There is no doubt that p-type ZnO exists, and intensive efforts are now underway to develop higher quality material which will support commercialization of electronic devices based on ZnO p-n homojunctions. Significant challenges remain to be solved, in particular the development of improved growth processes to reduce native defects and background impurity levels. Hydrogen impacts the electrical properties of ZnO in several different ways and understanding the details of these interactions and learning how to control them will play a key role in future progress. Likewise, recent advances have demonstrated the importance of surface conducting layers but more work is needed to either suppress or enhance them depending on the requirements of specific applications.

References

[1] H. E. Brown, *ZnO Rediscovered*, The New Jersey Zinc Company, New York, 1957.
[2] W. H. Hirschwald, *Acc. Chem. Res.* **18**, 228 (1985).

[3] United States patent #912,726 (1909).
[4] W. Jander and W. Stamm, *Z. Anorg. Allg. Chem.* **199**, 165 (1931).
[5] H. H. von Baumback and C. Z. Wagner, *Z. Phys. Chem., B* **22**, 199 (1933).
[6] G. Heiland, E. Mollwo and F. Stockmann, in *Solid State Physics*, edited by F. Seitz and D. Turnbull, Academic Press, New York, 1959, Vol. 8, p. 191.
[7] J. J. Lander, *J. Phys. Chem. Solids* **14**, 137 (1960) and references therein.
[8] D. Kohl, M. Henzler and G. Heiland, *Surf. Sci.* **41**, 403 (1974).
[9] M. Grunze, W. Hirschwald and D. Hofmann, *J. Cryst. Growth* **52**, 241 (1981).
[10] D. Eger, Y. Goldstein and A. Many, *RCA Review* **36**, 508 (1975).
[11] H. Harreis and G. Heiland, *Surf. Sci.* **24**, 643 (1971).
[12] G. Heiland, *Sens. Actuators* **2**, 343 (1982) and references therein.
[13] G. Heiland, *J. Phys. Chem. Solids* **22**, 227 (1961) and references therein.
[14] S. Lany, J. Osorio-Guillén and A. Zunger, *Phys. Rev. B* **75**, 241203(R) (2007).
[15] Y. Yan, J. Li, S.-H. Wei and M. M. Al-Jassim, *Phys. Rev. Lett.* **98**, 135506 (2007).
[16] C. G. Van de Walle, *J. Phys.: Condens. Matter* **20**, 064230 (2008).
[17] Y. Yan and S.-H. Wei, *Phys. Status Solidi B* **245**, 641 (2008).
[18] D. B. Laks, C. G. Van de Walle, G. F. Neumark and S. T. Pantelides, *Appl. Phys. Lett.* **63**, 1375 (1993).
[19] T. V. Butkhuzi, B. E. Tsekvava, N. P. Kekelidze, E. G. Chikoidze, T. G. Khulordava and M. M. Sharvashidze, *J. Phys. D: Appl. Phys.* **32**, 2683 (1999).
[20] J. P. Goss, R. J. Eyre and P. R. Briddon, *Phys. Status Solidi B* **245**, 1679 (2008).
[21] H. Amano, M. Kito, K. Hiramatsu and I. Akasaki, *Jpn. J. Appl. Phys.* **28**, L2112 (1989).
[22] I. Akasaki, H. Amano, M. Kito and K. Hiramatsu, *J. Lumin.* **48/49**, 666 (1991).
[23] S. Nakamura, M. Senoh and T. Mukai, *Jpn. J. Appl. Phys.* **30**, L1708 (1991).
[24] S. Nakamura, T. Mukai and M. Senoh, *Jpn. J. Appl. Phys.* **30**, L1998 (1991).
[25] S. Nakamura, M. Senoh, S. Nagahama, N. Iwasa, T. Yamada, T. Matsushita, H. Kiyoku and T. Sugimoto, *Jpn. J. Appl. Phys.* **35**, L74 (1996).
[26] D. C. Look, D. C. Reynolds, C. W. Litton, R. L. Jones, D. B. Eason and G. Cantwell, *Appl. Phys. Lett.* **81**, 1830 (2002).
[27] K. Minegishi, Y. Koiwai, Y. Kikuchi, K. Yano, M. Kasuga and A. Shimizu, *Jpn. J. Appl. Phys.* **36**, L1453 (1997).
[28] Y. Sato and S. Sato, *Thin Solid Films* **281–282** 445 (1996).
[29] T. V. Butkhuzi, A. V. Bureyev, A. N. Georgobiani, N. P. Kekelidze and T. G. Khulordava, *J. Cryst. Growth* **117**, 366 (1992).
[30] Y. R. Ryu, S. Zhu, D. C. Look, J. M. Wrobel, H. M. Joeng and H. W. White, *J. Cryst. Growth* **216**, 330 (2000).
[31] M. Joseph, H. Tabata, H. Saeki, K. Ueda and T. Kawai, *Physica B* **302–303** 140 (2001).
[32] X. -L. Guo, H. Tabata and T. Kawai, *J. Cryst. Growth* **223**, 135 (2001).
[33] T. V. Butkhuzi, M. M. Sharvashidze, N. M. Gamkrelidze, Kh. V. Gelovani, T. G. Khulordava, N. P. Kekelidze and E. E. Kekelidze, *Semicond. Sci. Technol.* **16**, 575 (2001).
[34] D. G. Thomas, *J. Phys. Chem. Solids* **15**, 86 (1960).
[35] Y. R. Ryu, T. S. Lee, J. A. Lubguban, H. W. White, B. J. Kim, Y. S. Park and C. J. Youn, *Appl. Phys. Lett.* **88**, 241108 (2006).
[36] E. S. P. Leong, S. F. Yu and S. P. Lau, *Appl. Phys. Lett.* **89**, 221109 (2006).
[37] Navigant Consulting, Inc., *Energy Savings Potential of Solid State Lighting in General Illumination Applications*, U S Department of Energy, (2003), p. 34.
[38] T. Whitaker, *Compound Semiconductor* **10**, 20 (2004).
[39] B. Bayraktaroglu, K. Leedy and R. Neidhard, *IEEE Elect. Dev. Lett.* **29**, 1024 (2008).
[40] B. R. Nag, *Electron Transport in Compound Semiconductors*, Springer-Verlag, Berlin, 1980, p. 133.
[41] D. C. Look, *Electrical Characterization of GaAs Materials and Devices*, John Wiley & Sons, Ltd, New York, 1989, Ch. 1.
[42] A. C. Beer, *Galvanomagnetic Effects in Semiconductors*, Academic Press, New York, 1963.
[43] E. H. Putley, *The Hall Effect and Semiconductor Physics*, Dover, New York, 1960.

[44] D. L. Rode, in *Semiconductors and Semimetals*, edited by R. K. Willardson and A. C. Beer, Academic Press, New York, 1975, Vol. 10.
[45] J. D. Wiley, in *Semiconductors and Semimetals*, edited by R. K. Willardson and A. C. Beer, Academic Press, New York, 1975, Vol. 10.
[46] J. R. Meyer and F. J. Bartoli, *Phys. Rev. B* **23**, 5413 (1981).
[47] D. C. Look, *J. Appl. Phys.* **104**, 063718 (2008).
[48] D. C. Look, *Surf. Sci.* **601**, 5315 (2007).
[49] D. C. Look, B. Claflin and H. E. Smith, *Appl. Phys. Lett.* **92**, 122108 (2008).
[50] A. Zeuner, H. Alves, D. M. Hofmann, B. K. Meyer, A. Hoffmann, U. Haboeck, M. Strassburg and M. Dworzak, *Phys. Status Solidi B* **234**, R7 (2002).
[51] P. Wagner and R. Helbig, *J. Phys. Chem. Solids* **35**, 327 (1974).
[52] D. C. Look, D. C. Reynolds, J. R. Sizelove, R. L. Jones, C. W. Litton, G. Cantwell and W. C. Harsch, *Solid State Commun.* **105**, 399 (1998).
[53] B. K. Meyer, H. Alves, D. M. Hofmann, W. Kriegseis, D. Forster, F. Bertram, J. Christen, A. Hoffmann, M. Straßburg, M. Dworzak, U. Haboeck and A. V. Rodina, *Phys. Status Solidi B* **241**, 231 (2004).
[54] S. Lany and A. Zunger, *Phys. Rev. Lett.* **98**, 045501 (2007).
[55] A. R. Hutson, *Phys. Rev.* **108**, 222 (1957).
[56] F. A. Kröger, *The Chemistry of Imperfect Crystals*, North-Holland, Amsterdam, 1974.
[57] C. G. Van de Walle, *Phys. Status Solidi B* **229**, 221 (2002).
[58] C. G. Van de Walle, *Physica B* **308–310**, 899 (2001).
[59] A. F. Kohan, G. Ceder, D. Margan and C. G. Van de Walle, *Phys. Rev. B* **61**, 15019 (2000).
[60] S. B. Zhang, S.-H. Wei and A. Zunger, *Phys. Rev. B* **63**, 075205 (2001).
[61] F. Oba, S. R. Nishitani, S. Isotani, H. Adachi and I. Tanaka, *J. Appl. Phys.* **90**, 824 (2001).
[62] A. Janotti and C. G. Van de Walle, *Phys. Rev. B* **76**, 165202 (2007).
[63] P. Erhart, K. Albe and A. Klein, *Phys. Rev. B* **73**, 205203 (2006).
[64] F. Oba, A. Togo, I. Tanaka, J. Paier and G. Kresse, *Phys. Rev. B* **77**, 245202 (2008).
[65] A. Janotti and C. G. Van de Walle, *Appl. Phys. Lett.* **87**, 122102 (2005).
[66] T. R. Paudel and W. R. L. Lambrecht, *Phys. Rev. B* **77**, 205202 (2008).
[67] S. Lany and A. Zunger, *Phys. Rev. B* **72**, 035215 (2005).
[68] P. Kasai, *Phys. Rev.* **130**, 989 (1963).
[69] F. A. Selim, M. H. Weber, D. Solovdonikov and K. G. Lynn, *Phys. Rev. Lett.* **99**, 085502 (2007).
[70] R. Laiho, L. S. Vlasenko and M. P. Vlasenko, *J. Appl. Phys.* **103**, 123709 (2008).
[71] F. Tuomisto, K. Saarinen and D. C. Look, *Phys. Status Solidi A* **201**, 2219 (2004).
[72] F. Tuomisto, K. Saarinen, D. C. Look and G. C. Farlow, *Phys. Rev. B* **72**, 085206 (2005).
[73] L. S. Vlasenko and G. D. Watkins, *Phys. Rev. B* **72**, 035203 (2005).
[74] L. S. Vlasenko and G. D. Watkings, *Physica B* **376**, 677 (2006).
[75] D. C. Look, J. W. Hemsky and J. R. Sizelove, *Phys. Rev. Lett.* **82**, 2552 (1999).
[76] D. C. Look, *J. Electron. Mater.* **35**, 1299 (2006).
[77] D. C. Look, D. C. Reynolds, J. W. Hemsky, R. L. Jones and J. R. Sizelove, *Appl. Phys. Lett.* **75**, 811 (1999).
[78] L. S. Vlasenko and G. D. Watkins, *Phys. Rev. B* **71**, 125210 (2005).
[79] G. Neumann, in *Current Topics in Materials Science*, edited by E. Kaldis, North-Holland, Amsterdam, 1981, Vol. 7, p. 279.
[80] P. Erhart and K. Albe, *Appl. Phys. Lett.* **88**, 201918 (2006).
[81] P. Erhart and K. Albe, *Phys. Rev. B* **73**, 115207 (2006).
[82] S. O. Kucheyev, P. N. L. Deenapanray, C. Jagadish, J. S. Williams, M. Yano, K. Kioke, S. Sasa, M. Inoue and K. Ogata, *Appl. Phys. Lett.* **81**, 3350 (2002).
[83] S. O. Kucheyev, C. Jagadish, J. S. Williams, P. N. K. Deenapanray, M. Yano, K. Koike, S. Sasa, M. Inoue and K. Ogata, *J. Appl. Phys.* **93**, 2972 (2003).
[84] D. C. Look, G. C. Farlow, P. Reunchan, S. Limpijumnong, S. B. Zhang and K. Nordlund, *Phys. Rev. Lett.* **95**, 225502 (2005).
[85] S. Limpijumnong, S. B. Zhang, S.-H. Wei and C. H. Park, *Phys. Rev. Lett.* **92**, 155504 (2004).
[86] Y.-S. Kim and C. H. Park, *Phys. Rev. Lett.* **102**, 086403 (2009).

[87] T. Makino, Y. Segawa, S. Yoshida, A. Tsukazaki, A. Ohtomo and M. Kawasaki, *Appl. Phys. Lett.* **85**, 759 (2004).
[88] D. M. Hofmann, D. Pfisterer, J. Sann, B. K. Meyer, R. Tena-Zaera, V. Munoz-Sanjose, T. Frank and G. Pensl, *Appl. Phys. A* **88**, 147 (2007).
[89] K. -K. Kim, S. Niki, J.-Y. Oh, J.-O. Song, T.-Y. Seong, S.-J. Park, S. Fujita and S.-W. Kim, *J. Appl. Phys.* **97**, 066103 (2005).
[90] T. Yamada, T. Morizane, T. Arimitsu, A. Miyake, H. Makino, N. Yamamoto and T. Yamamoto, *Thin Solid Films* **517** 1027 (2008).
[91] M. N. Jung, J. E. Koo, S. J. Oh, B. W. Lee, W. J. Lee, S. H. Ha, Y. R. Cho and J. H. Chang, *Appl. Phys. Lett.* **94**, 041906 (2009).
[92] X. L. Chen, B. H. Xu, J. M. Xue, Y. Zhao, C. C. Wei, J. Sun, Y. Wang, X. D. Zhang and X. H. Geng, *Thin Solid Films* **515**, 3753 (2007).
[93] J. Sun, D. A. Mourey, D. Garg and T. N. Jackson, *Electrochem. Solid State Lett.* **11**, D47 (2008).
[94] B. J. Lokhande, P. S. Patil and M. D. Uplane, *Physica B* **302–303**, 59 (2001).
[95] K. Johnston, M. O. Henry, D. McCabe, E. McGlynn, M. Dietrich, E. Alves and M. Xia, *Phys. Rev. B* **73**, 7 (2006).
[96] S. Müller, D. Stichtenoth, M. Uhrmacher, H. Hofsäss and C. Ronning, *Appl. Phys. Lett.* **90**, 012107 (2007).
[97] A. Guillén-Santiago, M. de la Olvera, A. Maldonado, R. Asomoza and D. R. Acosta, *Phys. Status Solidi A* **201**, 952 (2004).
[98] P. M. Ratheesh Kumar, C. Sudha Kartha, K. P. Vijayakumar, F. Singh and D. K. Avasthi, *Mater. Sci. Eng. B* **117**, 307 (2005).
[99] S. Ilican, Y. Caglar, M. Caglar and F. Yakuphanoglu, *Appl. Surf. Sci.* **255**, 2353 (2008).
[100] H.S. Yoon, K.S. Lee, T.S. Lee, B. Cheong, D.K. Choi, D.H. Kim and W.M. Kim, *Sol. Energy Mater. Sol. Cell* **92**, 1366 (2008).
[101] E. Chikoidze, M. Nolan, M. Modreanu, V. Sallet and P. Galtier, *Thin Solid Films* **516**, 8146 (2008).
[102] E. Chikoidze, M. Modreanu, V. Sallet, O. Gorochov and P. Galtier, *Phys. Status Solidi A* **205**, 1575 (2008).
[103] J. Rousset, E. Saucedo and D. Lincot, *Chem. Mater.* **21**, 534 (2009).
[104] R. Janisch, P. Gopal and N. A Spaldin, *J. Phys.: Condens. Matter* **17**, R657 (2005).
[105] S.J. Pearton, D.P. Norton, M.P. Ivill, A.F. Hebard, J.M. Zavada, W.M. Chen and I.A. Buyanova, *J. Electron. Mater.* **36**, 462 (2007).
[106] W. E. Fenwick, M. H. Kane, Z. Fang, T. Zaidi, N. Li, V. Rengarajan, J. Nause and I. T. Ferguson, *Mater. Res. Soc. Symp. Proc.* **957**, 0957-K04-10 (2007).
[107] S. J. Pearton, D. P. Norton, M. P. Ivill, A. F. Hebard, W. M. Chen, I. A. Buyanova and J. M. Zavada, *Mater. Res. Soc. Symp. Proc.* **957**, 0957-K04-10 (2007).
[108] N. H. Hong, J. Sakai and A. Hassini, *J. Phys.: Condens. Matter* **17**, 199 (2005).
[109] D. H. Hill, D. A. Arena, R. A. Bartynski, P. Wu, G. Saraf, Y. Lu, L. Wielunski, R. Gateau, J. Dvorak, A. Moodenbaugh and Y. K. Yeo, *Phys. Status Solidi A* **203**, 3836 (2006).
[110] P. Wu, G. Saraf, Y. Lu, D. H. Hill, R. Gateau, L. Wielunski, R. A. Bartynski, D. A. Arena, J. Dvorak, A. Moodenbaugh, T. Siegrist, J. A. Raley and Y. K. Yeo, *Appl. Phys. Lett.* **89**, 012508 (2006).
[111] K. Masuko, A. Ashida, T. Yoshimura and N. Fujimura, J. *Vac. Sci. Technol. B* **27**, 1760 (2009).
[112] J. N. Duenowa, T. A. Gessert, D. M. Wood, T. M. Barnes, M. Young, B. To and T. J. Coutts, *J. Vac. Sci. Technol. A* **25**, 955 (2007).
[113] S. J. Pearton, D. P. Norton, K. Ip, Y. W. Heo and T. Steiner, *J. Vac. Sci. Technol. B* **22**, 932 (2004).
[114] T. Dietl, H. Ohno, F. Matsukura, J. Cibert and D. Ferrand, *Science* **287** 1019 (2000).
[115] C. G. Van de Walle, *Phys. Rev. Lett.* **85**, 1012 (2000).
[116] S. F. J. Cox, E. A. Davis, S. P. Cottrell, P. J. C. King, J. S. Lord, J. M. Gil, H.V. Alberto, R. C. Vilão, J. P. Duarte, N. Ayres de Campos, A. Weidinger, R. L. Lichti and S. J. C. Irvine, *Phys. Rev. Lett.* **86**, 2601 (2001).
[117] A. Janotti and C. G. Van de Walle, *Nat. Mater.* **6**, 44 (2007).

[118] D. M. Hofmann, A. Hofstaetter, F. Leiter, H. Zhou, F. Henecker, B. K. Meyer, S. B. Orlinskii, J. Schmidt and P. G. Baranov, *Phys. Rev. Lett.* **88**, 045504 (2001).
[119] K. Shimomura, K. Nishiyama and R. Kadono, *Phys. Rev. Lett.* **89**, 255505 (2002).
[120] E. Mollwo, *Z. Phys.* **138**, 478 (1954).
[121] D. G. Thomas and J. J. Lander, *J. Chem. Phys.* **25**, 1136 (1956).
[122] J. J. Lander, *J. Phys. Chem. Solids* **3**, 87 (1957).
[123] K. Ip, M. E. Overberg, Y. W. Heo, D. P. Norton, S. J. Pearton, C. E. Stutz, B. Luo, F. Ren, D. C. Look and J. M. Zavada, *Appl. Phys. Lett.* **82**, 385 (2003).
[124] Y. M. Strzhemechny, H. L. Mosbacker, D. C. Look, D. C. Reynolds, C. W. Litton, N. Y. Garces, N. C. Giles, L. E. Halliburton, S. Niki and L. J. Brillson, *Appl. Phys. Lett.* **84**, 2545 (2004).
[125] D. C. Look, R. L. Jones, J. R. Sizelove, N. Y. Garces, N. C. Giles and L. E. Halliburton, *Phys. Status Solidi A* **195**, 171 (2003).
[126] N. H. Nickel and K. Fleischer, *Phys. Rev. Lett.* **90**, 197402 (2003).
[127] Y. M. Strzhemechny, J. Nemergut, P. E. Smith, J. Bae, D. C. Look and L. J. Brillson, *J. Appl. Phys.* **94**, 4256 (2003).
[128] J. Bang and K. J. Chang, *Appl. Phys. Lett.* **92**, 132109 (2008).
[129] E. V. Monakhov, A. Yu. Kuznetsov, J. S. Christensen, K. Maknys and B. G. Svensson, *Superlatt. Microstruct.* **38**, 472 (2005).
[130] Y. J. Li, T. C. Kaspar, T. C. Droubay, Z. Zhu, V. Shutthanandan, P. Nachimuthu and S. A. Chambers, *Appl. Phys. Lett.* **92**, 152105 (2008).
[131] J. I. Pankove and N. M. Johnson (Eds), *Hydrogen in Semiconductors, Semiconductors and Semimetals*, Academic Press, Boston, 1991, Vol. 34.
[132] H. Qiu, B. Meyer, Y. Wang and C. Wöll, *Phys. Rev. Lett.* **101**, 236401 (2008).
[133] C. H. Seager and S. M. Myers, *Appl. Phys. Lett.* **94**, 2888 (2003).
[134] E. V. Lavrov, F. Herklotz and J. Weber, *Phys. Rev. B* **79**, 165210 (2009).
[135] M. D. McCluskey and S. J. Jokela, in *Proceedings of the NATO Advanced Workshop on Zinc Oxide*, NATO Science Series II, edited by N.H. Nickel and E. Terukov, Springer, Dordrecht, 2005, Vol. **194**, p. 125.
[136] M. D. McCluskey, S. J. Jokela and W.M. Hlaing Oo, *Physica B* **376–377** 690 (2006).
[137] S. Zh. Karazhanov, E. S. Marstein and A. Holt, *J. Appl. Phys.* **105**, 033712 (2009).
[138] H. Takenaka and D. J. Singh, *Phys. Rev. B* **75**, 241102(R) (2007).
[139] X. Li, B. Keyes, S. Asher, S. B. Zhang, S.-H. Wei, T. J. Coutts, S. Limpijumnong and C. G. Van de Walle, *Appl. Phys. Lett.* **86**, 122107 (2005).
[140] S. Limpijumnong, X. Li, S.-H. Wei and S. B. Zhang, *Physica B* **376–377**, 686 (2006).
[141] J. Hu, H. Y. He and B. C. Pan, *J. Appl. Phys.* **103**, 113706 (2008).
[142] B. Claflin and D. C. Look, *J. Vac. Sci. Technol., B* **27**, 1722 (2009).
[143] N. Ohashi, T. Ishigaki, N. Okada, T. Sekiguchi, I. Sakaguchi and H. Haneda, *Appl. Phys. Lett.* **80**, 2869 (2002).
[144] N. Ohashi, T. Ishigaki, N. Okada, H. Taguchi, I. Sakaguchi, S. Hishita, I. Sakaguchi and H. Haneda, *J. Appl. Phys.* **93**, 6386 (2003).
[145] N. Ohashi, Y.-G. Wang, T. Ishigaki, Y. Wada, H. Taguchi, I. Sakaguchi, T. Ohgaki, Y. Adachi and H. Haneda, *J. Cryst. Growth* **306**, 316 (2007).
[146] E.-C. Lee and K. J. Chang, *Physica B* **376–377**, 707 (2006).
[147] J. G. Lu, S. Fujita, T. Kawaharamura and H. Nishinaka, *Chem. Phys. Lett.* **441**, 68 (2007).
[148] J. M. Bian, X. M. Li, X. D. Gao, W. D. Yu and L. D. Chen, *Appl. Phys. Lett.* **84**, 541 (2004).
[149] D. C. Look and B. Claflin, *Phys. Status Solidi B* **241**, 624 (2004).
[150] D. C. Look, *Superlatt. Microstruct.* **42**, 284 (2007).
[151] T. M. Barnes, K. Olson and C. A. Wolden, *Appl. Phys. Lett.* **86**, 112112 (2005).
[152] K.-K. Kim, H.-S. Kim, D.-K. Hwang, J.-H. Lim and S.-J. Park, *Appl. Phys. Lett.* **83**, 63 (2003).
[153] A. Allenic, W. Guo, Y. Chen, M. B. Katz, G. Zhao, Y. Che, Z. Hu, B. Liu, S. B. Zhang, and X. Pan, *Adv. Mater.* **19**, 3333 (2007).
[154] Z. Ye, J. Wang, Y. Wu, X. Zhou, F. Chen, W. Xu, Y. Miao, J. Huang, J. Lu, L. Zhu and B. Zhao, *Front. Optoelectron. China* **1**, 147 (2008).

[155] M. Lorenz, B. Cao, G. Zimmermann, G. Biehne, C. Czekalla, H. Frenzel, M. Brandt, H. von Wenckstern and M. Grundmann, *J. Vac. Sci. Technol., B* **27**, 1693 (2009).
[156] N. Izyumskaya, V. Avrutin, Ü. Özgür, Y. I. Alivov and H. Morkoç, *Phys. Status Solidi B* **244**, 1439 (2007).
[157] L. Dunlop, A. Kursumovic and J. L. MacManus-Driscoll, *Appl. Phys. Lett.* **93**, 172111 (2008).
[158] L. Li, C. X. Shan, B. H. Li, B. Yao, J. Y. Zhang, D. X. Zhao, Z. Z. Zhang, D. Z. Shen, X. W. Fan and Y. M. Lu, *J. Phys. D: Appl. Phys.* **41**, 245402 (2008).
[159] N. Y. Garces, N. C. Giles, L. E. Halliburton, G. Cantwell D. B. Eason, D. C. Reynolds and D. C. Look, *Appl. Phys. Lett.* **80**, 1334 (2002).
[160] Y. J. Li, Y. W. Heo, Y. Kwon, K. Ip, S. J. Pearton and D. P. Norton, *Appl. Phys. Lett.* **87**, 072101 (2005).
[161] V. Vaithianathan, B.-T. Lee, and S. S. Kim, *J. Appl. Phys.* **98**, 043519 (2005).
[162] V. Vaithianathan, Y. H. Lee, B.-T. Lee, S. Hishita and S. S. Kim, *J. Cryst. Growth* **287**, 85 (2006).
[163] D. C. Look, B. Claflin, Ya. I. Alivov and S. J. Park, *Phys. Status Solidi A* **201**, 2203 (2004).
[164] B. Claflin, D. C. Look, S. J. Park and G. Cantwell, *J. Cryst. Growth* **287**, 16 (2006).
[165] T. Ohgaki, N. Ohashi, S. Sugimura, H. Ryoken, I. Sakaguchi, Y. Adachi and H. Haneda, *J. Mater. Res.* **23**, 2293 (2008).
[166] O. Bierwagen, T. Ive, C. G. Van de Walle and J. S. Speck, *Appl. Phys. Lett.* **93**, 242108 (2008).
[167] A. Krtschil, D.C. Look, Z.-Q. Fang, A. Dadgar, A. Diez and A. Krost, *Physica B* **376–377**, 703 (2006).
[168] A. Krtschil, A. Dadgar, N. Oleynik, J. Bläsing, A. Diez and A. Krost, *Appl. Phys. Lett.* **87**, 262105 (2005).
[169] F. Tuomisto, K. Saarinen and D. C. Look, *Phys. Rev. Lett.* **91**, 205502 (2003).
[170] C. H. Park, S. B. Zhang and S.-H. Wei, *Phys. Rev. B* **66**, 073202 (2002).
[171] T. Yamamoto and H. Katayama-Yoshida, *Physica B* **302–303**, 155 (2001).
[172] A. Kobayashi, O. F. Sankey and J. D. Dow, *Phys. Rev. B* **28**, 946 (1983).
[173] N. Y. Garces, Lijun Wang, N. C. Giles, L. E. Halliburton, G. Cantwell and D. B. Eason, *J. Appl. Phys.* **94**, 519 (2003).
[174] C. L. Perkins, S.-H. Lee, X. Li, S. E. Asher and T. J. Coutts, *J. Appl. Phys.* **97**, 034907 (2005).
[175] X. Li, Y. Yan, T. A. Gessert, C. L. Perkins, D. Young, C. DeHart, M. Young and T. J. Coutts, *J. Vac. Sci. Technol., A* **21**, 1342 (2003).
[176] X. Li, Y. Yan, T. A. Gessert, C. Dehart, C. L. Perkins, D. Young and T. J. Coutts, *Electrochem. Solid State Lett.* **6**, C56 (2003).
[177] B. S. Li, Y. C. Liu, Z. Z. Zhi, D. Z. Shen, Y. M. Lu, J. Y. Zhang, X. W. Fan, R. X. Mu and D. O. Henderson, *J. Mater. Res.* **18**, 8 (2003).
[178] T. E. Murphy, D. Y. Chen, E. Cagin and J. D. Phillips, *J. Vac. Sci. Technol., B* **23**, 1277 (2005).
[179] S. J. Jiao, Z. Z. Zhang, Y. M. Lu, D. Z. Shen, B. Yao, J. Y. Zhang, B. H. Li, D. X. Zhao, X. W. Fan and Z. K. Tang, *Appl. Phys. Lett.* **88**, 031911 (2006).
[180] B.S. Li, Y.C. Liu, Z.Z. Zhi, D.Z. Shen, Y.M. Lu, J.Y. Zhang, X.W. Fan, R.X. Mu and D. O. Henderson, *J. Mater. Res.* **18**, 8 (2003).
[181] S. Lautenschlaeger, S. Eisermann, B. K. Meyer, G. Callsen, M. R. Wagner and A. Hoffmann, *Phys. Status Solidi. – Rapid Res. Lett.* **3**, 16 (2009).
[182] K. Iwata, P. Fons, A. Yamada, K. Matsubara and S. Niki, *J. Cryst. Growth* **209**, 526 (2000).
[183] H. Matsui, H. Saeki, T. Kawai, H. Tabata and B. Mizobuchi, *J. Appl. Phys.* **95**, 5882 (2004).
[184] L. G. Wang and A. Zunger, *Phys. Rev. Lett.* **90**, 256401 (2003).
[185] M. Brandt, H. von Wenckstern, H. Schmidt, A. Rahm, G. Biehne, G. Benndorf, H. Hochmuth, M. Lorenz, C. Meinecke, T. Butz and M. Grundmann, *J. Appl. Phys.* **104**, 013708 (2008).
[186] T. Makino, A. Tsukazaki, A. Ohtomo, M. Kawasaki and H. Koinuma, *J. Phys. Soc. Jpn.* **75**, 073701 (2006).
[187] H. von Wenckstern, R. Pickenhain, H. Schmidt, M. Brandt, G. Biehne, M. Lorenz, M. Grundmann and G. Brauer, *Superlatt. Microstruct.* **42**, 14 (2007).
[188] H. von Wenckstern, R. Pickenhain, H. Schmidt, M. Brandt, G. Biehne, M. Lorenz, M. Grundmann and G. Brauer, *Appl. Phys. Lett.* **89**, 092122 (2006).

[189] J. G. Lu, Y. Z. Zhang, Z. Z. Ye, L. P. Zhu, L. Wang, B. H. Zhao and Q. L. Liang, *Appl. Phys. Lett.* **88**, 222114 (2006).
[190] X.H. Wang, B. Yao, D.Z. Shen, Z.Z. Zhang, B.H. Li, Z.P. Wei, Y.M. Lu, D.X. Zhao, J.Y. Zhang, X.W. Fan, L.X. Guan and C.X. Cong, *Solid State Commun.* **141**, 600 (2007).
[191] Y. Marfaing and A. Lusson, *Superlatt. Microstruct.* **38**, 385 (2005).
[192] T. Yamamoto and H. Katayama-Yoshida, *Jpn. J. Appl. Phys., Part* 2 **38**, L166 (1999).
[193] E. C. Lee, Y.-S. Kim, Y.-G. Jin and K. J. Chang, *Phys. Rev. B* **64**, 085120 (2001).
[194] K. G. Saw, K. Ibrahim, Y. T. Lim and M. K. Chai, *Thin Solid Films* **515**, 2879 (2007).
[195] Y. Yan, S. B. Zhang and S. T. Pantelides, *Phys. Rev. Lett.* **86**, 5723 (2001).
[196] Y. W. Heo, S. J. Park, K. Ip, S. J. Pearton and D. P. Norton, *Appl. Phys. Lett.* **83**, 1128 (2003).
[197] Y. W. Heo, K. Ip, S. J. Park, S. J. Pearton and D. P. Norton, *Appl. Phys. A* **78**, 53 (2004).
[198] S.B. Zhang, S.-H. Wei and Y. Yan, *Physica B* **302–303**, 135 (2001).
[199] Y. W. Heo, K. Ip, S. J. Pearton and D. P. Norton, *Phys. Status Solidi A* **201**, 1500 (2004).
[200] F. G. Chen, Z. Z. Ye, W. Z. Xu, B. H. Zhao, L. P. Zhu and J. G. Lv, *J. Cryst. Growth* **281**, 458 (2005).
[201] G. Hu, H. Gong, E. F. Chor and P. Wu, *Appl. Phys. Lett.* **89**, 251102 (2006).
[202] D.-K. Hwang, M.-S. Oh, J.-H. Lim, C.-G. Kang and S.-J. Park, *Appl. Phys. Lett.* **90**, 021106 (2007).
[203] B. J. Kwon, H. S. Kwack, S. K. Lee, Y. H. Cho, D. K. Hwang and S. J. Park, *Appl. Phys. Lett.* **91**, 061903 (2007).
[204] H. von Wenckstern, G. Benndorf, S. Heitsch, J. Sann, M. Brandt, H. Schmidt, J. Lenzner, M. Lorenz, A.Y. Kuznetsov, B.K. Meyer and M. Grundmann, *Appl. Phys. A* **88**, 125 (2007).
[205] A. Allenic, W. Guo, Y. B. Chen, Y. Che, Z. D. Hu, B. Liu and X. Q. Pan, *J. Phys. D: Appl. Phys.* **41**, 025103 (2008).
[206] G. Hu and H. Gong, *Acta Mater.* **56**, 5066 (2008).
[207] D.-K. Hwang, H.-S. Kim, J.-H. Lim, J.-Y. Oh, J.-H. Yang, S.-J. Park, K.-K. Kim, D. C. Look and Y. S. Park, *Appl. Phys. Lett.* **86**, 151917 (2005).
[208] V. Vaithianathan, K. Asokan, J. Y. Park and S. S. Kim, *Appl. Phys. A* **94**, 995 (2008).
[209] Y. R. Ryu, T. S. Lee and H. W. White, *Appl. Phys. Lett.* **83**, 87 (2003).
[210] D. C. Look, G. M. Renlund, R. H. BurgenerII and J. R. Sizelove, *Appl. Phys. Lett.* **85**, 5269 (2004).
[211] Veeramuthu Vaithianathan, Byung-Teak Lee and Sang Sub Kim, *Appl. Phys. Lett.* **86**, 062101 (2005).
[212] C. Yuen, S. F. Yu, E. S. P. Leong, S. P. Lau, K. Pita, H. Y. Yang and T. P. Chen, *J. Appl. Phys.* **101**, 094905 (2007).
[213] E. Przeździecka, E. Kamińska, K. P. Korona, E. Dynowska, W. Dobrowolski, R. Jakieła, Ł. Kłopotowski and J. Kossut, *Semicond. Sci. Technol.* **22**, 10 (2007).
[214] M. Kumar, S.-Y. Choi, *Appl. Surf. Sci.* **255**, 2173 (2008).
[215] T. Aoki, Y. Shimizu, A. Miyake, A. Nakamura, Y. Nakanishi and Y. Hatanaka, *Phys. Status Solidi B* **229**, 911 (2002).
[216] F. X. Xiu, Z. Yang, L. J. Mandalapu, D. T. Zhao, J. L. Liu and W. P. Beyermann, *Appl. Phys. Lett.* **87**, 152101 (2005).
[217] O. Lopatiuk-Tirpak, L. Chernyak, F. X. Xiu, J. L. Liu, S. Jang, F. Ren, S. J. Pearton, K. Gartsman, Y. Feldman, A. Osinsky and P. Chow, *J. Appl. Phys.* **100**, 086101 (2006).
[218] W. Guo, A. Allenic, Y. B. Chen, X. Q. Pan, Y. Che, Z. D. Hu and B. Liu, *Appl. Phys. Lett.* **90**, 242108 (2007).
[219] L. J. Mandalapu, F. X. Xiu, Z. Yang and J. L. Liu *J. Appl. Phys.* **102**, 023716 (2007).
[220] U. Grossner, J. S. Christensen, B. G. Svensson and A. Y. Kuznetsov, *Superlatt. Microstruct.* **42**, 294 (2007).
[221] Z. G. Yu, H. Gong and P. Wu, *Chem. Mater.* **17**, 852 (2005).
[222] W.F. Liu, J.M. Bian, L.Z. Hu, H.W. Liang, H.Q. Zang, J.C. Sun, Z.W. Zhao, A.M. Liu and G.T. Du, *Solid State Commun.* **142**, 655 (2007).
[223] Z. Z. Zhang, Z. P. Wei, Y. M. Lu, D. Z. Shen, B. Yao, B. H. Li, D. X. Zhao, J. Y. Zhang, X. W. Fan and Z. K. Tang, *J. Cryst. Growth* **301–302**, 362 (2007).

[224] F. Zhuge, L. P. Zhu, Z. Z. Ye, D. W. Ma, J. G. Lu, J. Y. Huang, F. Z. Wang, Z. G. Ji and S. B. Zhang, *Appl. Phys. Lett.* **87**, 092103 (2005).
[225] T. Aoki, Y. Hatanaka and D. C. Look, *Appl. Phys. Lett.* **76**, 3257 (2000).
[226] Y. R. Ryu, T. S. Lee, J. H. Leem and H. W. White, *Appl. Phys. Lett.* **83**, 4032 (2003).
[227] W. F. Liu, J. M. Bian, L. Z. Hu, H. W. Liang, H. Q. Zang, J. C. Sun, Z. W. Zhao, A. M. Liu and G. T. Du, *Solid State Commun.* **142**, 655 (2007).
[228] A. Tsukazaki, M. Kubota, A. Ohtomo, T. Onuma, K. Ohtani, H. Ohno, S. F. Chichibu and M. Kawasaki, *Jpn. J. Appl. Phys.* **44**, L643 (2005).
[229] A. Tsukazaki, A. Ohtomo, T. Onuma, M. Ohtani, T. Makino, M. Sumiya, K. Ohtani, S. F. Chichibu, S. Fuke, Y. Segawa, H. Ohno, H. Koinuma and M. Kawasaki, *Nat. Mater.* **4**, 42 (2005).
[230] J. M. Bian, X. M. Li, C. Y. Zhang, W. D. Yu and X. D. Gao, *Appl. Phys. Lett.* **85**, 4070 (2004).
[231] W.-J. Lee, J. Kang and K.J. Chang, *Physica B* **376–377**, 699 (2006).
[232] W.-J. Lee, J. Kang and K. J. Chang, *Phys. Rev. B* **73**, 024117 (2006).
[233] C. Morhain, M. Teisseire, S. Vézian, F. Vigué, F. Raymond, P. Lorenzini, J. Guion, G. Neu and J.-P. Faurie, *Phys. Status Solidi B* **229**, 881 (2002).
[234] A. Allenic, X. Q. Pan, Y. Che, Z. D. Hu and B. Liu, *Appl. Phys. Lett.* **92**, 022107 (2008).
[235] U. Wahl, E. Rita, J. G. Correia, A. C. Marques, E. Alves and J. C. Soares, *Phys. Rev. Lett.* **95**, 215503 (2005).
[236] B. K. Meyer, J. Stehr, A. Hofstaetter, N. Volbers, A. Zeuner and J. Sann, *Appl. Phys. A* **88**, 119 (2007) and references therein.
[237] B. K. Meyer, J. Sann and A. Zeuner, *Superlatt. Microstruct.* **38**, 344 (2005).
[238] J. G. Lu, Y. Z. Zhang, Z. Z. Ye, Y. J. Zeng, H. P. He, L. P. Zhu, J. Y. Huang, L. Wang, J. Yuan, B. H. Zhao and X. H. Li, *Appl. Phys. Lett.* **89**, 112113 (2006).
[239] O. F. Schirmer, *J. Phys. Chem. Solids* **29**, 1407 (1968).
[240] A. Valentini, F. Quaranta, M. Rossi and G. Battaglin, *J. Vac. Sci. Technol., A* **9**, 286 (1991).
[241] Y. Kanai, *Jpn. J. Appl. Phys.* **30**, 703 (1991).
[242] Y. Kanai, *Jpn. J. Appl. Phys.* **30**, 2021 (1991).
[243] X. Pan, Z. Ye, J. Li, Y. Zeng, X. Gu, L. Zhu, B. Zhao and Y. Che, *Appl. Surf. Sci.* **253**, 6060 (2007).
[244] J. A. Savage and E. M. Dodson, *J. Mater. Sci.* **4**, 809 (1969).
[245] A. Onedera, N. Tamaki, K. Jin and H. Yamashita, *Jpn. J. Appl. Phys.* **36**, 6008 (1997).
[246] G. Weise, E. M. Fechner, G. Owsian and D. Kraut, *Thin Solid Films* **32**, 87 (1976).
[247] H. Wolk, S. Deubler, D. Forkel, H. Foettinger, M. Iwatschenko-Borho, F. Meyer, M. Renn, W. Witthuhn and R. Helbig, *Mater. Sci. Forum* **10–12**, 863 (1986).
[248] T. Nagata, T. Shimura, Y. Nakano, A. Ashida, N. Fujimura and T. Ito, *Jpn. J. Appl. Phys.* **40**, 5615 (2001).
[249] W. Jun and Y. Yintang, *Mater. Lett.* **62**, 1899 (2008).
[250] B. D. Ahn, H. S. Kang, J. H. Kim, G. H. Kim, H. W. Chang and S. Y. Lee, *J. Appl. Phys.* **100**, 093701 (2006).
[251] S.S. Lin, J.G. Lu, Z.Z. Ye, H.P. He, X.Q. Gu, L.X. Chen, J.Y. Huang and B.H. Zhao, *Solid State Commun.* **148**, 25 (2008).
[252] Y. Yan, M. M. Al-Jassim and S.-H. Wei, *Appl. Phys. Lett.* **89**, 181912 (2006).
[253] A. N. Gruzintsev, V. T. Volkov, I. I. Khodos, T. V. Nikiforova and M. N. Koval'chuk, *Russ. Microelectron.* **31**, 200 (2002).
[254] United States patent #652, 7858 (2003).
[255] C. G. Van de Walle and J. Neugebauer, *J. Appl. Phys.* **95**, 3851 (2004).
[256] A. Dadgar, A. Krtschil, F. Bertram, S. Giemsch, T. Hempel, P. Veit, A. Diez, N. Oleynik, R. Clos, J. Christen and A. Krost, *Superlatt. Microstruct.* **38**, 245 (2005).
[257] T. H. Vlasenflin and M. Tanaka, *Solid State Commun.* **142**, 292 (2007).
[258] Y.-Z. Zhang, J.-G. Lu, Z.-Z. Ye, H.-P. He, L.-L. Chen and B.-H. Zhao, *Chin. Phys. Lett.* **26**, 046103 (2009).
[259] M. Joseph, H. Tabata and T. Kawai, *Jpn. J. Appl. Phys.* **38**, L1205 (1999).
[260] L. L. Chen, J. G. Lu, Z. Z. Ye, Y. M. Lin, B. H. Zhao, Y. M. Ye, J. S. Li and L. P. Zhu, *Appl. Phys. Lett.* **87**, 252106 (2005).

[261] H. B. Ye, J. F. Kong, W. Z. Shen, J. L. Zhao and X. M. Li, *J. Phys. D: Appl. Phys.* **40**, 5588 (2007).
[262] D.-S. Liu, C.-S. Sheu and C.-T. Lee, *J. Appl. Phys.* **102**, 033516 (2007).
[263] H. B. Ye, J. F. Kong, W. Z. Shen, J. L. Zhao and X. M. Li, *Appl. Phys. Lett.* **90**, 102115 (2007).
[264] J. F. Kong, H. Chen, H. B. Ye, W. Z. Shen, J. L. Zhao and X. M. Li, *Appl. Phys. Lett.* **90**, 041907 (2007).
[265] Y. R. Sui, B. Yao, Z. Hua, G. Z. Xing, X. M. Huang, T. Yang, L. L. Gao, T. T. Zhao, H. L. Pan, H. Zhu, W. W. Liu and T. Wu, *J. Phys. D: Appl. Phys.* **42**, 065101 (2009).
[266] J. G. Lu, Z. Z. Ye, F. Zhuge, Y. J. Zeng, B. H. Zhao and L. P. Zhu, *Appl. Phys. Lett.* **85**, 3134 (2004).
[267] T. Yamamoto, *Jpn. J. Appl. Phys.* **42**, 514 (2003).
[268] T. Yamamoto, *Phys. Status Solidi A* **193**, 423 (2002).
[269] S. Nakamura, N. Iwasa, M. Senoh and T. Mukai, *Jpn. J. Appl. Phys.* **31**, 1258 (1992).
[270] B. Claflin, D. C. Look and D. R. Norton, *J. Electron. Mater.* **36**, 442 (2007).
[271] O. Schmidt, P. Kiesel, C. G. Van de Walle, N. M. Johnson, J. Nause and G. H. Döhler, *Jpn. J. Appl. Phys.* **44**, 7271 (2005).
[272] J. F. Muth, R. M. Kolbas, A. K. Sharma, S. Oktyabrsky and J. Narayan, *J. Appl. Phys.* **85**, 7884 (1999).

4

ZnO Surface Properties and Schottky Contacts

Leonard J. Brillson
The Ohio State University, Columbus, OH, USA

4.1 Historical Background of Schottky Contacts on ZnO

With the development of exciting new optoelectronic and microelectronic device applications for ZnO, there is now renewed interest in the understanding and control of its electrical contact properties. Transparent thin film transistors, blue/UV light-emitting diodes (LEDs) and lasers, high electron mobility transistors, electronic nanostructures, and spintronics all require metal contacts and an understanding of metal–ZnO interfaces and processing techniques. Historically, there has been considerable research devoted to the surface physics and chemistry of ZnO, in large part due to the strong effect that surface and polarity have on ZnO interface charge transfer. This chapter will examine the results of surface and interface research produced over the past fifty years in the context of Schottky barriers.

The challenge of ZnO surfaces and interfaces is evident from the wide and variable range of Schottky barriers measured for the same metal on the ZnO surface. In the case of Au/ZnO diodes, Schottky barrier heights Φ_{SB} can range from 0 to 1.2 eV, depending on the experimental conditions. Similarly wide n-type $\Phi_{SB}{}^{n}$ energy ranges are observed for Pt/ZnO and Ta/ZnO diodes. Furthermore, high work function metals exhibit lower than expected Φ^{n}_{SB}. Since the ZnO electron affinity $\chi_{ZnO} = \sim 4.2$ eV, $\Phi_{SB}{}^{n}$ should be $\Phi_M - \chi_{ZnO} = 5.65 - 4.2 = 1.45$ eV for Pt, whereas measured Φ_{SB} are limited to 0.96 eV for air-exposed surfaces and 0.75 eV for high vacuum-cleaved surfaces. Wide barrier variations are evident for

Zinc Oxide Materials for Electronic and Optoelectronic Device Applications, First Edition.
Edited by Cole W. Litton, Donald C. Reynolds and Thomas C. Collins.

other high work function metals such as Pd and Au with various air-exposed treatments. This strong dependence on surface preparation indicates that extrinsic factors such as crystal quality and surface treatment have a large effect on ZnO barrier heights. Understanding these variations is the aim of this chapter.

4.1.1 ZnO Surface Effects

Investigations of ZnO surfaces and interfaces under controlled ambient conditions, i.e. in an ultrahigh vacuum (UHV) chamber, began in the 1960s with the pioneering work of G. Heiland on ZnO surface conductivity. Analogous to Schottky barrier formation, gas adsorption on ZnO involves charge transfer at the semiconductor interface that alters the carrier concentration within the surface space charge region. Band bending that induces charge accumulation near the surface changes the surface conductivity in a manner similar to transistor action. For adsorbate-induced band bending, however, charge transfer between the surface and the adsorbate replaces the gate bias action of the transistor. This effect is especially pronounced for wide band gap semiconductors such as GaN, ZnO, and other metal oxides whose intrinsic carrier concentrations are very low.

In the case of ZnO, surface conductivity is strongly dependent on gas adsorption. Figure 4.1 illustrates the effect of ZnO exposure to oxygen or atomic hydrogen on the semiconductor's charge distribution, band bending, and electron concentration.[1]

Oxygen adsorption on clean ZnO surfaces results in charge transfer of electrons to the O atoms, producing a negatively charged surface and positively charged donors within the surface space charge region. This n-type (upward) band bending results in a surface band bending qV_B and a carrier concentration $n(z)$ that decreases toward the surface. Conversely, adsorption of atomic hydrogen on ZnO results in charge transfer from the adsorbate to the semiconductor. This produces a positively charged surface and negatively charged acceptors within the surface space charge region. This p-type (downward) band bending moves the conduction band edge below the Fermi level, inducing an accumulation of electrons near the surface. In both cases, the carrier concentration $n(z)$ near the surface differs by orders-of-magnitude from the bulk carrier concentration n_B. These charge transfer processes are the basis for a variety of ZnO thin film and nanostructure gas sensors.

Elemental surface composition and bonding can significantly affect ZnO interface charge transfer. As an example, conductivity changes of ZnO surfaces with gas exposure exhibit an orientation dependence. Hydrogen exposure increases conductivity σ faster on oxygen-terminated surfaces while oxygen exposure increases σ faster on Zn-terminated surfaces. Figure 4.2 illustrates the change in sheet conductance for atomic hydrogen on Zn- vs O-terminated ZnO surfaces.

For mobility μ and carrier density n, the sheet conductance $g_\square$ is defined as:

$$g_\square = q_0 \int_0^l \mu \cdot n \, dz \tag{4.1}$$

integrated over a layer thickness l. With initial H exposure, $g_\square$ increases to values over an order of magnitude higher on the oxygen face vs the zinc face. Ultimately, $g_\square$ values

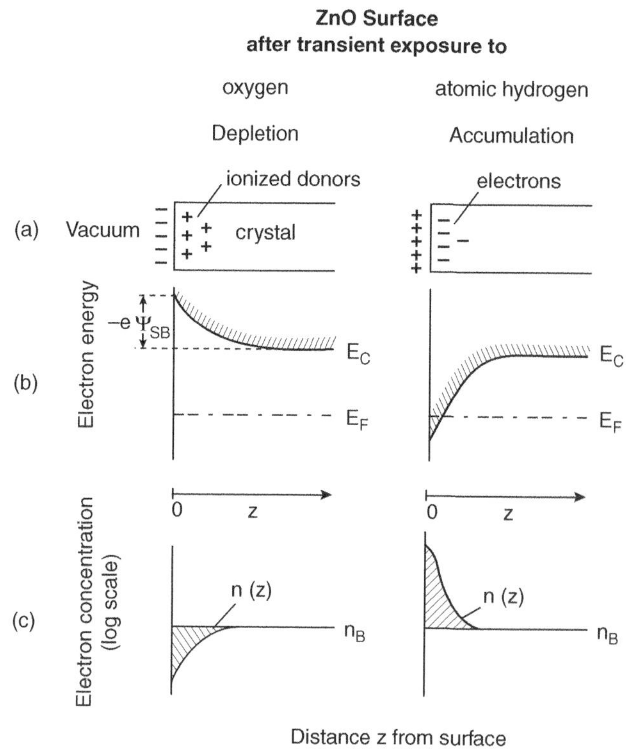

Figure 4.1 *ZnO surface (a) charge distribution, (b) band bending and (c) electron concentration before and after transient exposure to oxygen and atomic hydrogen. Oxygen depletes the surface of electrons, while atomic hydrogen induces electron accumulation. Reprinted from G. Heiland, Polar surfaces of zinc oxide crystals, Surf. Sci. 13, Issue 1, 72–84. Copyright (1969) with permission from Elsevier*

for the two surfaces converge since the final band bending and sheet conductance depend on the energy and density of the states, regardless of the rate of charge transfer. Nevertheless, Figure 4.2 shows that the chemical composition and bonding of the outer surface layer affects the rate of charge transfer at adsorbate–semiconductor interfaces. Since hydrogen donates electrons to the ZnO for both polar surfaces, one expects the same band bending and charge transfer once the adsorbate–semiconductor bonding forms the same localized states. However, the surface atom termination affects the rate at which chemical bonding and impurity centers form and at which charge transfer takes place. Such temporal effects play an important role in the sensitivity and response of ZnO devices. Furthermore, hydrogen and, to a lesser extent, oxygen can diffuse into metals such as Pd, Pt, and Au, thereby affecting the metal–ZnO Schottky barrier. The sensitivity of ZnO to hydrogen and oxygen indicates that specific surface chemical treatments can have pronounced electronic effects that can dominate Schottky barrier formation.

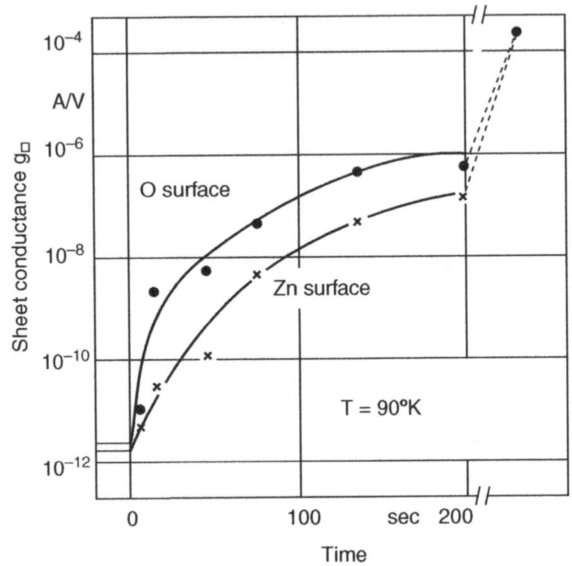

Figure 4.2 Sheet conductance vs time for atomic hydrogen exposure to Zn- terminated vs O-terminated ZnO surfaces. The O-terminated surface exhibits over an order of magnitude larger response. Reprinted from G. Heiland, Polar surfaces of zinc oxide crystals, Surf. Sci. 13, Issue 1, 72–84. Copyright (1969) with permission from Elsevier

4.1.2 Early Schottky Barrier Studies

Schottky barrier studies of metals on ZnO began in the mid 1960s as part of a larger effort to understand semiconductor surface states and Fermi level pinning. Mead prepared clean metal–ZnO contacts by cleaving ZnO crystals in a stream of evaporating metal within a high vacuum chamber.[2] He observed a relatively wide $\Phi_{SB}{}^{n}$ range, in contrast to more covalent compound semiconductors such as GaAs and Si. Kurtin *et al.* interpreted this covalent–ionic contrast in terms of surface states more likely to form due to higher bond disruption with covalent lattice termination.[3] Nevertheless the $\Phi_{SB}{}^{n}$ range was still smaller than that expected with different Φ_M. Indeed, Brillson showed that ZnO $\Phi_{SB}{}^{n}$ appeared to saturate for high work function metals whether plotted vs Φ_M or the interface heat of reaction ΔH_R, a measure of the metal reactivity with a semiconductor.[4] Such saturation is evident for both ionic and covalent semiconductor–metal junctions in general; it indicates states in the semiconductor band gap that limit Fermi level movement.

Mead's early work also highlighted the different barrier heights measured by current–voltage (*I–V*), capacitance–voltage (*C–V*), and internal photoemission spectroscopy (IPS). These differences underscore the need to use more than one measurement technique in determining Schottky barriers because of, e.g., image force lowering, lateral barrier inhomogeneities, nonuniform doping, recombination, and parasitic capacitances. In addition, diodes formed on ZnO prepared by chemical methods in air can introduce interfacial dielectric layers that alter Schottky barriers significantly. Until the past few years, most Schottky barrier studies have not involved surfaces prepared under clean conditions, leading to a wide array of barrier heights and a lack of predictability.

4.2 Recent Schottky Barrier Studies

With the increased interest in ZnO for device applications, researchers have readdressed ZnO Schottky barriers under more systematic conditions. Central to such efforts have been barrier studies on well-characterized ZnO single crystal surfaces prepared in high vacuum or UHV. In addition, researchers have investigated the role of impurities and defects in limiting efforts to achieve p-type ZnO, critical for ZnO LEDs and lasers. In the course of examining impurities and defects in general, researchers have now found considerable evidence that defects play an important role in Schottky barrier formation as well. Thus ZnO crystal quality has become a significant factor in Φ_{SB} measurements.

4.2.1 Surface Cleaning in Vacuum

The preparation of clean metal–ZnO surfaces is challenging for all but UHV-cleaved surfaces. Cleaved surfaces require bulk crystals, whereas commercial ZnO is typically available only in thin film or platelet form. Surface cleaning by ion bombardment results in damaged surfaces with high concentrations of defects. High temperature annealing in vacuum or an oxygen ambient also damages the ZnO surface (see below). On the other hand, a remote oxygen plasma provides a viable method of cleaning ZnO surfaces.

Coppa *et al.* measured Schottky barriers on ZnO Zn-polar (0001) and O-polar ($000\bar{1}$) surfaces prepared by oxygen (20% O_2/80% He) plasma treatment in a UHV chamber. Their surface characterization of untreated ZnO polar surfaces showed significant hydroxide and carbon contamination.[5,6] Annealing in a remote oxygen plasma at 525 °C for 30 min removed all detectable hydrocarbons and left less than half a monolayer of OH as measured by X-ray photoelectron spectroscopy (XPS). Their studies also showed the effect of different temperatures, treatment times, and conditions of subsequent cooling on surface chemical composition, microstructure, and electronic structure.[6] Significantly, they also found that annealing at higher temperatures, e.g. 600–700 °C, in just a pure oxygen ambient could eliminate all surface adsorbates but resulted in thermal decomposition of the ZnO surface that degraded the surface microstructure. The latter result is significant since thermal annealing in oxygen has sometimes been used by researchers to prepare ZnO for surface electronic studies. Overall, metal diodes to ZnO cleaned with a remote oxygen plasma at elevated temperatures display large differences in *I–V* characteristics as a function of surface preparation.

Strzhemechny *et al.*,[7] Mosbacker *et al.*[8] and Dong *et al.*[9] have prepared clean ZnO (0001) and ($000\bar{1}$) surfaces using a room temperature remote oxygen (20% O_2/80% He) plasma (ROP) in a UHV chamber. XPS measurements showed complete hydrocarbon and nearly complete OH removal along with a clear low energy electron diffraction pattern.[8] Atomic force microscopy (AFM) showed no significant surface disruption by the remote plasma treatment. However, this ROP treatment produced a dramatic change in Φ_{SB} for Au diodes from ohmic for as-received surfaces to rectifying with *I–V* barriers of 0.5–0.6 eV.[8] Ohmic Au/ZnO contacts are frequently observed for air-exposed ZnO surfaces and have been attributed to an accumulation layer induced by OH adsorbates. OH removal appears to eliminate this accumulation layer.

4.2.2 Surface Cleaning Effects on Impurities and Defects

Besides removing surface hydroxides, the ROP treatment also removes H from within the ZnO surface.[7] This is evidenced by a strong reduction in the photoluminescence (PL) intensity of the I_4 peak attributed to H in ZnO.[10] The effect of such H is to introduce new shallow donors that increase n-type carrier concentration.

Another important effect of ROP treatment is to remove subsurface defects. The presence of electronically active defects and their increase with proximity to the ZnO surface was first shown by researchers using depth-resolved cathodoluminescence spectroscopy (DRCLS). See, for example, Brillson[11] for a review of this technique. Essentially, an incident electron beam penetrates the solid surface, producing a cascade of secondaries that lose energy by creating first plasmons and then by impact ionization and formation of electron–hole pairs. These electron–hole pairs recombine via transitions involving band-to-band transitions, band-to-defect transitions or transitions due to formation of new dielectric phases. The rate of electron–hole pair creation reaches a maximum at depths that increase with increasing electron beam energy E_B. For energies in the range of 0.1–5 keV, these depths are only a few tens of nanometers or less. For minority carrier diffusion lengths on this scale, it is possible to resolve variations in electronic features as a function of depth on a nanometer scale. With decreasing E_B, the ratio of 2.5 eV "green" deep level emission intensity I(DL) due to deep level native point defects relative to the 3.3 eV near band edge (NBE) luminescence intensity I(NBE) increases.

Figure 4.3(a) illustrates cathodoluminescence spectra for a ZnO(000$\bar{1}$) surface measured with low energy incident electron beams. Besides the NBE peak at 3.3 eV, there is pronounced mid-gap emission at ~2.5 eV, commonly termed "green" emission. The ratio of this deep level emission intensity I(DL) vs the near band edge emission intensity I(NBE) increases with decreasing E_B and proximity to the free ZnO surface.[8]

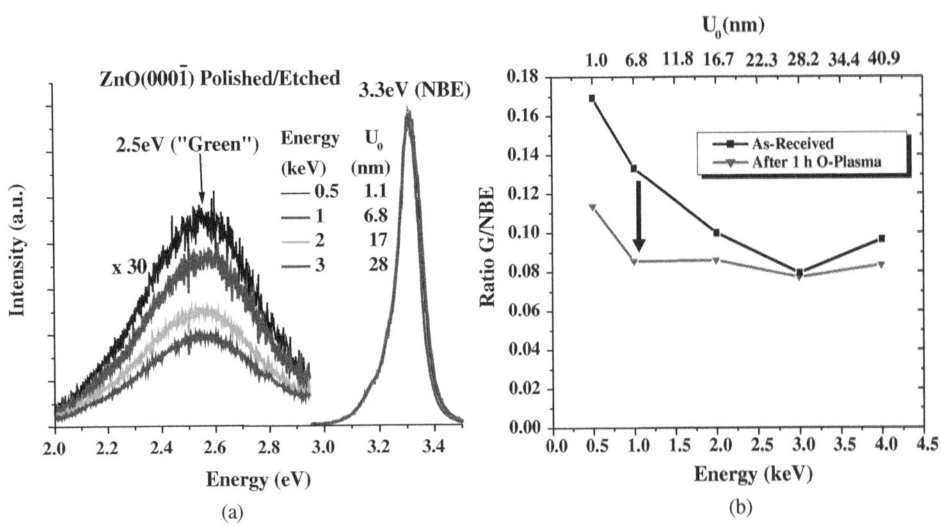

Figure 4.3 (a) DRCL spectra of ZnO(000$\bar{1}$) showing increase of deep level "green" luminescence with decreasing E_B. (b) I(DL)/I(NBE) vs incident beam energy and its reduction with ROP processing

This increase occurs within the first 50–100 nm and indicates a segregation of native point defects toward the free ZnO surface. Also shown in Figure 4.3(b) is the reduction of *I*(DL)/*I*(NBE) with ROP processing, indicative of the sub-surface reduction in native point defects. As Section 4.4 will show, the presence of such defects can have major effects on the measurement of Schottky barrier heights.

4.3 The Influence of Surface Preparation on Schottky Barriers

A wide range of Schottky barriers are observed for a given metal on ZnO, depending on surface preparation prior to metal deposition. Over the past decade, researchers have explored the effect of different chemical treatments in order to develop reliable rectifying contacts. A summary of these results appears in Table 4.1.

Table 4.1 lists Φ_{SB}^{n} for various metals on ZnO surfaces prepared by different methods.[2,9,12–36] Entries in Table 4.1 are described not only in terms of surface treatment and measurement technique used but also by crystal quality. The importance of crystal quality will become apparent in the next section. Table 4.1 shows that Φ_{SB}^{n} ranges from 1.2 eV down to ohmic, depending on the metal and on surface treatment. Reliably p-type ZnO is still under development. A wide Φ_{SB}^{n} energy range is observed even for the same metal, e.g. Pt, Au, and Ta. This strong dependence on surface preparation indicates that extrinsic factors such as crystal quality and surface treatment have a large effect on ZnO barrier heights. In general, $\Phi_{SB}(C\text{–}V) \geq \Phi_{SB}(I\text{–}V)$, with the exception of Ag oxide. Similarly, with the exception of Ag, low work function metals such as In, Al, and Ti yield low Φ_{SB}s. (Ag oxidizes easily, producing high barrier heights that depend on the degree of oxidation.) These results indicate that tunneling lowers $\Phi_{SB}(I\text{–}V)$ except where interfacial oxide layers form.

The influence of ZnO surface preparation and ZnO bulk crystal quality is evident from transport measurements. *I–V* measurements of current transport across the metal–semiconductor interface follow the thermionic emission relationship:

$$J = A^{*}T^{2}\exp(-q\Phi_{SB}/k_{B}T)\{\exp[(qV - JR_{S})/nk_{B}T] - 1\} \tag{4.2}$$

where J is the current density, T the temperature, V the applied voltage, R_S the series resistance and n the ideality factor.

From Table 4.1, the highest Φ_{SB}^{n} values obtained from *I–V* experiments can be plotted vs ideality factor, as shown in Figure 4.4. Figure 4.4 indicates n values considerably larger than unity, due to image force lowering, thermionic field emission, and lateral contact inhomogeneity.[38] Higher Φ_{SB}^{n} values for Zn-polar vs O-polar surfaces are also evident from this plot, which features mostly air-exposed ZnO. UHV clean metal–ZnO contacts also display higher Φ_{SB}^{n} values for Zn-polar vs O-polar surfaces as well.[9]

Note the high Φ_{SB} values of Ag–ZnO even though the Ag work function of 4.26 eV is much less than that of, e.g., 5.65 eV for Pt. This is attributed to the oxidation of Ag, which increases the Ag work function substantially.[39]

Allen *et al.* also investigated the effect of ZnO growth methods on the *I–V* characteristics.[35] Figure 4.5 illustrates *I–V* plots for hydrothermal vs melt-grown ZnO as well as their different polar or nonpolar surfaces.

Table 4.1 *$\Phi_{SB}{}^{n}$ measurements by various techniques vs orientation and crystal quality for metals on ZnO surfaces prepared by different cleaning methods. The metal column includes the corresponding work function. Reprinted from H.B. Michaelson, The work function of the elements and its periodicity, J. Appl. Phys. 48, 4729. Copyright (1977) with permission from American Institute of Physics*

Metal	$q\Phi_{SB}{}^{n}$ (eV)	Ideality factor	Surface treatment	Measurement technique	Ref.
Pt (5.65)	0.75 nonpolar	—	Vac-cleave	IPS	[2]
Pt	0.42 (000$\bar{1}$)	3.45	HD, ROP1A	*I–V*	[16]
Pt	Ohmic (000$\bar{1}$)	NA	LD, ATMA	*I–V*	[16]
Pt	0.39 (000$\bar{1}$)	1.00	LD, ROP1A	*I–V*	[16]
Pt	0.61 (0001)	1.70	PLD, AG	*I–V*	[21]
Pt	0.85, 0.73 (0001)	1.77	ATMA, LA	*C–V, I–V*	[22]
Pt	0.96 (0001)	1.1	OCA	*I–V*	[23]
Pt	0.6 (0001)	3.1	OCA	*I–V*	[23]
Pt	0.70 (0001)	1.5	UVOA	*I–V*	[28]
Pt	0.93, 0.89 (0001)	1.15	HPA	*C–V, I–V*	[32]
Pt	0.55 (0001)	2.0	CCA	*I–V*	[33]
Pt	0.72, 0.68 (000$\bar{1}$)	1.2	CCA	*C–V, I–V*	[33]
Ir (5.27)	0.65 (000$\bar{1}$)	2.62	HD, ROP1A	*I–V*	[16]
Ir	0.69 (000$\bar{1}$)	1.58	HD, ROP1A	*I–V*	[14]
Ir	0.54 (000$\bar{1}$)	1.66	LD, ATMA	*I–V*	[16]
Ir	0.64(000$\bar{1}$)	1.36	LD, ROP1A	*I-V*	[14]
Pd (5.12)	0.68 nonpolar	—	Vac-cleave	IPS	[2]
Pd	0.59,0.61,0.60	—	Acid etch	IPS, *C–V, I–V*	[12]
Pd	0.73, 0.53 (0001)	1.3	ROP2	*C–V, I–V*	[9]
Pd	0.68, 0.61 (000$\bar{1}$)	1.2	ROP2	*C–V, I–V*	[9]
Pd	1.14, 0.81 (0001) PLD	1.49	AG	*C–V, I–V*	[24]
Pd	0.74 (000±1)	2.0	Acetone Acid etch	*I–V*	[25]
	0.60 (000±1)	1.4		*I–V*	
Pd	0.83 (000$\bar{1}$)	1.03	HPA	*I-V*	[26]
Pd	1–1.2 (000$\bar{1}$)	1.8	HPA	*C–V, I–V*	[27]
Pd	0.55 (0001)	2.0	CCA	*I–V*	[33]
Pd	0.59, 0.59 (000$\bar{1}$)	1.2	CCA	*I–V, C–V*	[33]
Au (5.1)	0.65 nonpolar	—	Vac-cleave	IPS	[2]
Au	0.645,0.67,0.66 –	—	CCA	IPS, *C–V, I–V*	[12]
Au	0.67,0.60 (000$\bar{1}$)	1.86, 1.03	ROPHT, ROPRT	*I–V*	[13]
Au	0.71 (0001),	1.17,	ROPHT,	*I–V*	[31]
	0.60 (000$\bar{1}$)	1.03	ROPRT		[31]
Au	Ohmic (0001)	NA	LD, ATMA	*I–V*	[2]
Au	1.2, 0.81 (0001)	1.2	LD, ROP2	*C–V, I–V*	[9]
Au	1.07, 0.77 (000$\bar{1}$)	1.3	LD, ROP2	*C–V, I–V*	[9]
Au	0.48 (000$\bar{1}$)	1.30	LD, ROP1A	*I–V*	[14]
Au	Ohmic (000$\bar{1}$)	NA	LD, ROP1, 650 °C	*I–V*	[15]
Au	0.43 (000$\bar{1}$)	3.57	HD, ROP1A	*I–V*	[14]
Au	0.46 (000$\bar{1}$)	1.56	LD, ATMA	*I–V*	[14]
Au	0.48 (000$\bar{1}$)	1.30	LD, ROP1A	*I–V*	[16]
Au	Ohmic (0001)	NA	ATMA	*I–V*	[17]
Au	0.63 (0001)	1.15	HPA	*I–V*	[17]
Au	0.65 (0001)	—	ATMA	*C–V*	[18]

Table 4.1 (Continued)

Metal	$q\Phi_{SB}{}^{n}$ (eV)	Ideality factor	Surface treatment	Measurement technique	Ref.
Au	0.66 (0001) ZnO:N	1.8	OCA	*I–V*	[20]
Au	0.37 (000$\bar{1}$) ZnO:N	3.5	OCA	*I–V*	[20]
Au	0.71, 0.70 (0001)	1.4	CCA	*I–V, C–V*	[33]
Au	0.70, 0.69 (000$\bar{1}$)	1.1	CCA	*C–V, I–V*	[33]
PEDOT:PSS (5.0)	0.9, 0.7 (0001)	1.2	HT, CCA	*I–V, C–V*	[36]
Ti (4.33)	<0.3 nonpolar	—	Vac-cleave	*I–V*	[2]
Ti	Ohmic (0001)	NA	OCA	*I–V*	[29]
Cu (4.65)	0.45 nonpolar	—	Vac-cleave	*I–V*	[2]
Al (4.28)	0.0 nonpolar	—	Vac-cleave	*I–V*	[2]
Ag (4.26)	0.68 nonpolar	—	Vac-cleave	IPS	[2]
Ag	0.92,0.89(11$\underline{2}$0)	1.33	AG	*C–V, I–V*	[19]
Ag	0.69 (0001)	—	ATMA	*C–V*	[18]
Ag	0.84 (11$\underline{2}$0)	1.5	O_2 plasma	*I–V*	[30]
Ag	0.80, 0.78 (0001)	1.2	CCA	*C–V, I–V*	[33]
Ag	1.11, 1.08 (0001)	1.08	CCA highest	*I–V, C–V*	[34]
Ag	0.80, 0.77 (000$\bar{1}$)	1.1	CCA	*C–V, I–V*	[33]
Ag	0.99, 0.97 (000$\bar{1}$)	1.06	CCA highest	*I–V, C–V*	[34]
Ag oxide (~5.0)	1.20, 0.93 (0001)	1.03	HT, OCA, OPS	*I–V, C–V*	[35]
Ag oxide	0.99, 0.79 (000$\bar{1}$)	1.04	HT, OCA, OPS	*I–V, C–V*	[35]
Ag oxide	1.03 (0001)	1.14	MG, OCA,OPS	*I–V*	[35]
Ag oxide	0.98, 0.89 (000$\bar{1}$)	1.10	MG, OCA,OPS	*I–V, C–V*	[35]
Ag oxide	1.02 M-plane	1.10	MG, OCA,OPS	*I–V*	[35]
Ta (4.25)	Ohmic (000$\bar{1}$)	NA	ROP1A	*I–V*	[15]
Ta	Blocking (000$\bar{1}$)	NA	LD, ROP1A, 350 °C	*I–V*	[15]
Ta	Blocking (000$\bar{1}$)	NA	LD, ROP1A, 550 °C	*I–V*	[15]
Ta	Ohmic (000$\bar{1}$)	NA	HD, ROP1A	*I–V*	[15]
	Leaky (000$\bar{1}$)	NA	HD, ROP1A, 550 °C	*I–V*	[15]
In (4.12))	<0.3 nonpolar	—	Vac-cleave	*I–V*	[2]

Crystal quality: Low defect (LD); high defect (HD); pulsed laser deposited (PLD); hydrothermal (HT); melt-grown (MG).
Surface treatment: Vacuum cleaved in stream of evaporating metal (Vac-cleave); H_3PO_4, HCl, deionized water (Acid etch); chemically cleaned and air-exposed (CCA); laser annealed (LA); remote oxygen plasma cleaned at high temperature (ROPHT), plus room temperature re-exposure (ROPRT); room temperature remote oxygen plasma for 1 h and air exposure (ROP1A) or for 2 h without breaking vacuum (ROP2); hydrogen peroxide and air exposed (HPA); acetone, trichloroethylene, methanol and air-exposed (ATMA); organic clean and air exposed (OCA); UV ozone and air-exposed (UVOA); oxygen-plasma sputtered (OPS); as-grown (AG).
Measurement technique: Internal photoemission spectroscopy (IPS); current–voltage (*I–V*); capacitance–voltage (*C–V*).

Here Φ_{SB} values decrease with increasing n, due to inhomogeneity that produce a distribution of barrier heights as well as tunneling contributions to the transport. Barriers with n values approaching unity are considerably higher than those for conventional nonideal contacts. Also Ag oxide contacts exhibit higher Φ_{SB} for Zn-polar vs O-polar surfaces similar to those reported for elemental metals.

Barrier inhomogeneity can also account for the difference between barriers measured by *I–V* vs *C–V*. Von Wenckstern *et al.* have compared effective barrier heights $\Phi_{SB,eff}$ for

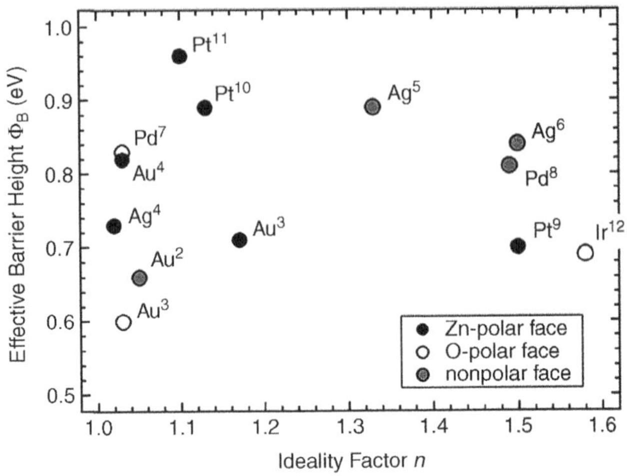

Figure 4.4 *N-type barrier height Φ_{SB}^n vs ideality factor n for the highest reported barriers on ZnO. Reprinted from M.W. Allen, S.M. Durbin, and J.B. Metson, Silver oxide Schottky contacts on n-type ZnO, Appl. Phys. Lett. 91, 053512. Copyright (2007) with permission from American Institute of Physics*

Pd on ZnO measured by *I–V* with barriers measured by *C–V*, $\Phi_{SB,CV}$.[40] Figure 4.6(a) illustrates $\Phi_{SB,eff}$, $\Phi_{SB,CV}$ and their difference (inset) vs temperature.

Lateral fluctuations in Φ_{SB} can be modeled by a Gaussian distribution of barrier heights with a standard deviation σ around a mean value $\Phi_{SB0,m}$ (for zero bias) according to:[41]

$$\Phi_{SB,eff}(T) = \Phi_{SB0,m}(T) - q\sigma^2/2k_BT \tag{4.3}$$

The Figure 4.6(a) inset line yields σ. Figure 4.6(b) shows the variation of $\Phi_{SB,eff}$ with $1/T$, which should be linear if σ in Equation (4.3) is temperature independent. Both $\Phi_{SB0,m}$ and σ can vary with applied voltage V_A with proportionality factors ρ_2 and ρ_3, respectively, according to:

$$\Phi_{SB,m} = \Phi_{SB0,m} + \rho_2 V_A \tag{4.4a}$$

and

$$\sigma^2 = \sigma_0^2 + \rho_3 V_A \tag{4.4b}$$

such that the ideality factor is given (for temperature-independent ρ_2 and ρ_3) as:

$$n = 1/[1 - \rho_2 + e\rho_3/(2k_BT)] \tag{4.5}$$

The slope and straight line fits in Figure 4.6(b) yield ρ_2 and ρ_3. In this Pd study, $\sigma = (125 \pm 15)$ meV.[24] This approach provides a method for quantifying the homogeneity of Schottky barriers.

Pearton *et al.* have reviewed Schottky barriers to chemically treated ZnO and noted that barrier heights do not seem to follow the difference in metal work functions.[42] They identified Au and Ag contacts on epiready ZnO(0001) surfaces cleaned with just

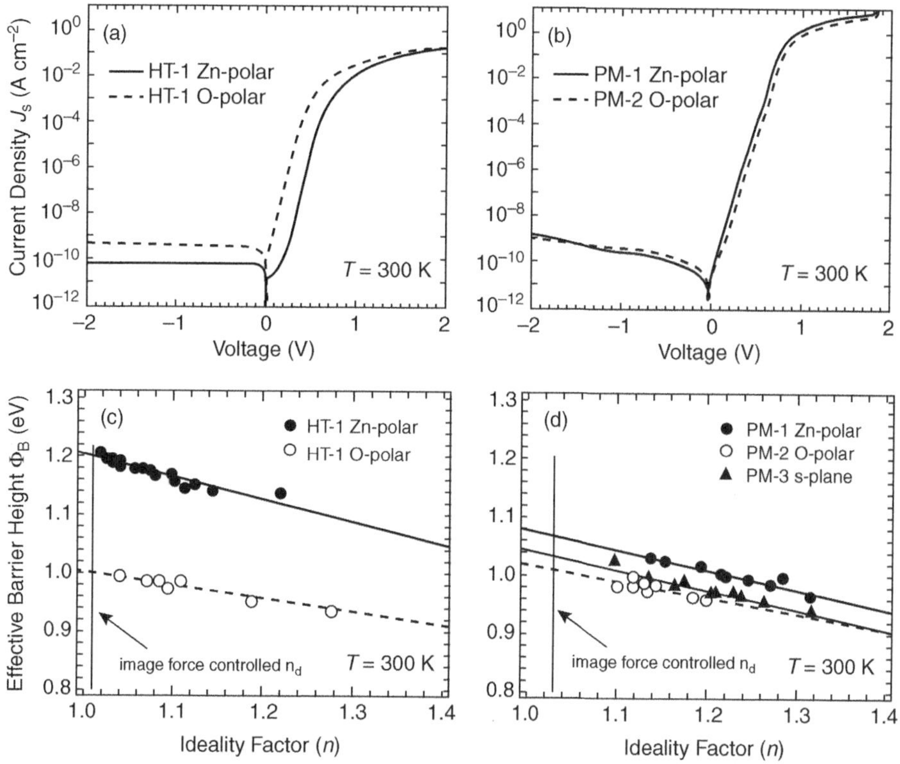

Figure 4.5 *Room temperature I–V characteristics (a, b) and effective Φ_{SB} values vs ideality factors (c, d) for silver oxide diodes on hydrothermal and melt bulk ZnO and for different surface orientations. Reprinted from M.W. Allen, S.M. Durbin, and J.B. Metson, Silver oxide Schottky contacts on n-type ZnO, Appl. Phys. Lett. 91, 053512. Copyright (2007) with permission from American Institute of Physics*

organic solvents as Schottky contacts with the lowest reverse current values. However, the low thermal stability of these contacts to ZnO limits their use in device applications.

As already mentioned, high work function metals exhibit lower than expected $\Phi_{SB}{}^{n}$ based on a ZnO electron affinity $\chi_{ZnO} = \sim 4.2$ eV. Highest Au $\Phi_{SB}{}^{n}$ were measured for ROP-cleaned surfaces of low defect (LD) density ZnO. Highest Pd Φ_{SB} were measured for hydrogen peroxide oxidized surfaces. The increased barrier with oxidized interfacial layers indicates a dipole contribution to Φ_{SB}. Similarly, the blocking contact formed at the Ta–ZnO interface indicates the formation of a Ta oxide layer.[14,15] Indeed, cathodoluminescence (CL) spectra of the Ta–ZnO interface reveal the presence of a Ta_2O_3-like layer. On the other hand, H diffusion can form a dipole at the Pd–ZnO interface that lowers Φ_{SB}.[9]

4.4 The Influence of Defects on Schottky Barriers

While many barrier variations can be ascribed to interface preparation, other extrinsic factors appear to play a role. Only recently have factors such as crystalline

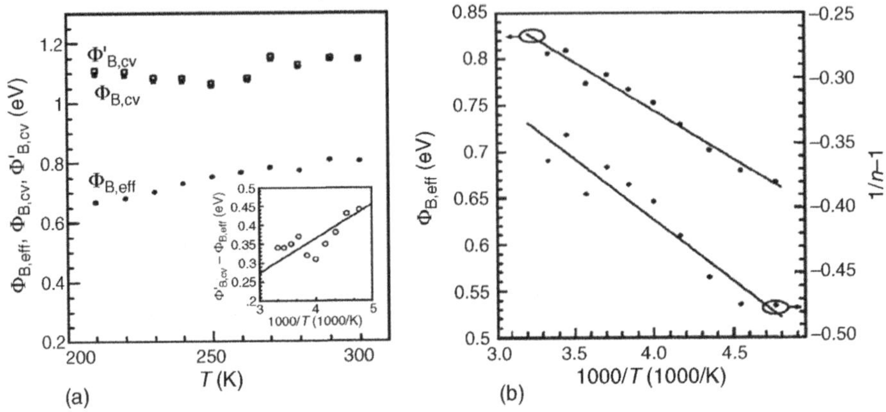

Figure 4.6 (a) Effective barrier height $\Phi_{SB,\ eff}$, $\Phi_{SB,\ CV}$ and $\Phi'_{SB,\ CV}$ vs temperature for Pd/ZnO (0001). The inset shows the difference between $\Phi_{SB,\ eff}$ and $\Phi'_{SB,\ CV}$ with the line through the origin yielding the standard deviation σ. (b) $\Phi_{SB,\ eff}$ and 1/n-1 vs 1/T. Reprinted from H. von Wenckstern, Mean barrier height of Pd Schottky contacts on ZnO thin films, Appl. Phys. Lett. 88, 092102. Copyright (2006) with permission from American Institute of Physics

quality, impurity content, and native point defect concentration become serious considerations.

In this regard, Schottky barrier measurements should take into account the fact that the quality of ZnO crystals can vary dramatically, depending on the growth method, annealing conditions, and subsequent polishing or etching. Figure 4.7 displays four examples of DRCLS spectra taken from ZnO crystals grown from the melt or by hydrothermal methods.[43] Each low temperature spectrum contains near band edge features including phonon replicas as well as broad deep level features at ~2.1 and 2.4 eV.

The depth U_0 of peak electron–hole pair creation rate varies with incident beam energy E_B such that $U_0 = 55$, 85, 300 and 990 nm for $E_B = 2$, 5, 10 and 20 keV, respectively, for the bare ZnO surface. Figure 4.7(a) and (b) are from the same crystal grower yet near band edge (NBE) emission in (a) is an order of magnitude higher but has a much less uniform distribution of deep level defects at 2.1 and 2.4 eV than in (b). Specifically, deep level emission in Figure 4.7(a) increases by nearly two orders of magnitude between the near-surface and the bulk. NBE emission in Figure 4.7(c) is even higher than in (a) plus there is no measurable deep level emission. In contrast, Figure 4.7(d) shows DRCL spectra from a hydrothermal crystal grown by a different vendor from (a) and (b) that exhibits orders of magnitude lower NBE, high deep level emissions, and significant lower uniformity with depth. This figure shows that defect concentrations can vary by orders of magnitude, not only from different growers but also with depth on a nanometer scale.

To illustrate the effect of native point defects on I–V features, Figure 4.8 shows how contacts to remote oxygen plasma (ROP)-cleaned ZnO(000$\bar{1}$) surfaces can change from ohmic to Schottky-like as surface impurities and sub-surface native point defects are removed. The contributions of surface contamination, sub-surface H, and native point defects can be separated at different stages of this conversion process. As in Figure 4.3, DRCLS of a ZnO surface shows that the ~2.5 eV increases proportionally as excitation

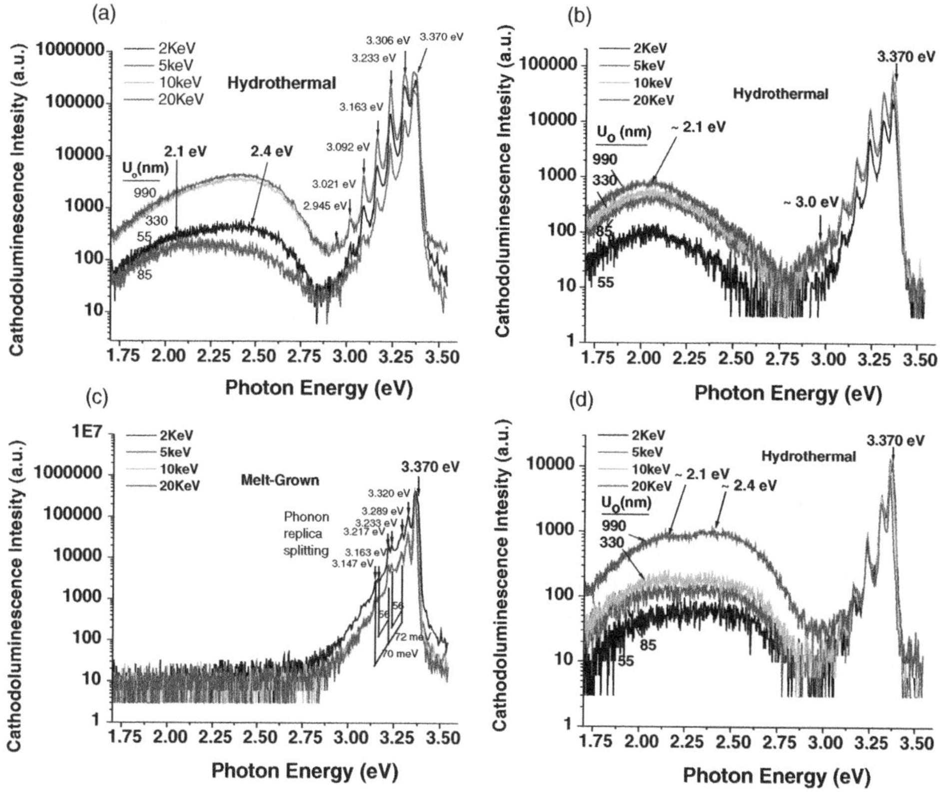

Figure 4.7 *10 K DRCL spectra of ZnO crystals from the same supplier (a) and (b) grown hydrothermally, (c) melt growth, and (d) grown hydrothermally by a second vendor. All spectra recorded at constant power and optical resolution*

occurs closer to the surface. Figure 4.8(a) illustrates this increase in the $I(2.5\,\text{eV})/I(3.36\,\text{eV})$ ratio vs depth of peak excitation rate.[15] It also shows the strong decrease of this ratio with ROP treatment within the outermost few tens of nanometers.

Furthermore, this defect ratio continues to decrease even after 30 min ROP processing, which XPS and low-energy electron diffraction measurements show cleans the surface completely. Figure 4.8(b) shows the effect of these treatments. Au diodes on the as-received surface are ohmic. With 30 min ROP surface cleaning, the I–V characteristic becomes asymmetric with $\Phi_{SB}{}^{n} = 0.44\,\text{eV}$ and ideality $n > 5$. Additional ROP processing further increases rectification: $\Phi_{SB}{}^{n}$ increases to 0.51 eV while n decreases to 1.51. Such studies show that sub-surface defects rather than surface contaminants determine the Schottky barrier changes.

Hydrogen within ZnO can act as donors, as predicted theoretically[44] and observed experimentally,[45,46] which in turn can lead to heavy n-type doping that narrows the surface space charge region and promotes tunneling. This is evident from the ohmic contact (straight line) formed on ZnO exposed to a remote H plasma in Figure 4.8(b).

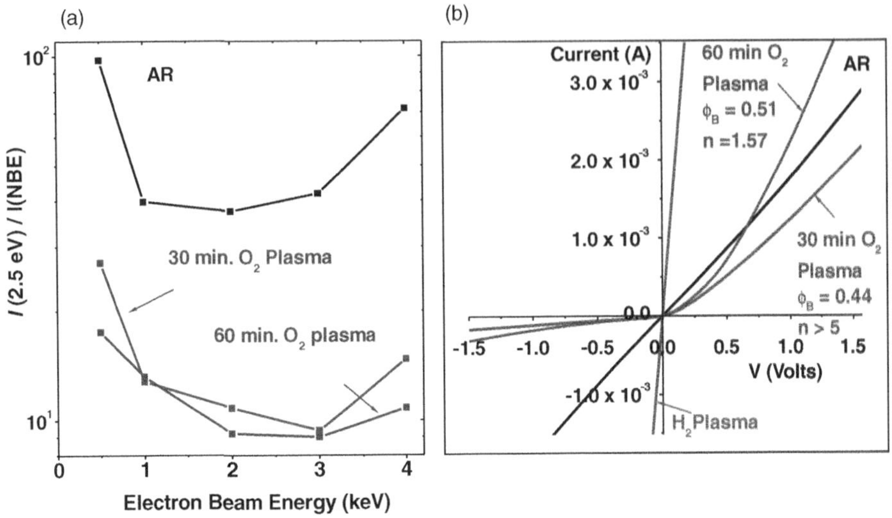

Figure 4.8 (a) DRCLS ratio I(2.5 eV)/I(3.36 eV) vs depth of peak excitation vs ROP treatment time. (b) I–V characteristics of Au/ZnO(000$\bar{1}$) diodes on these surfaces change: (i) as-received and chemically etched (AR); (ii) with 30 min ROP cleaning; (iii) 60 min ROP cleaning; and (iv) subsequent H plasma exposure. Reprinted from L.J. Brillson, et al., Dominant effect of near-interface native point defects on ZnO Schottky barriers, Appl. Phys. Lett. 90, 102116. Copyright (2007) with permission from American Institute of Physics

Figure 4.9 further illustrates how the density of native point defects within the ZnO crystal before metallization strongly affects Schottky barriers. Figure 4.9(a) shows *I*(2.5 eV)/*I*(3.36 eV) DRCLS intensity ratios for ROP-cleaned ZnO crystals with "high", "medium" and "low" defect concentrations. Figure 4.9(a) shows that the defect ratios can vary by over two orders of magnitude for different ZnO crystals.[14]

Figure 4.9(b) shows that the reverse currents vary by two orders of magnitude for Au contacts to these surfaces. "High" defect ZnO shows nearly ohmic behavior, "medium" defect ZnO shows only weak rectification, while the strongest rectification appears for the "low" defect ZnO. In general, $\Phi_{SB}{}^n$ are higher for "low" vs "high" defect ZnO crystals with correspondingly lower ideality factors.[14] The direct correlation between reverse current leakage and defect density underscores the importance of using low defect crystals in Schottky barrier research.

The presence of defects and other electrically active sites are strongly affected by surface preparation as well. Besides the defect variations with depth into the surface space charge regions shown in Figure 4.7, sizeable variations in ZnO surface potential are evident that depend sensitively on surface cleaning and polishing. In Figure 4.10(a), DRCLS spectra show relatively low defect emissions within 40 nm of the free surface but orders-of-magnitude higher defects deeper into the crystal.[47] This reduction is due to a chemomechanical polishing that achieves near-monolayer surface roughness as pictured in the AFM map in Figure 4.10(b). Notwithstanding this apparently smooth surface, a KFPM map obtained simultaneously with the AFM map shows surface

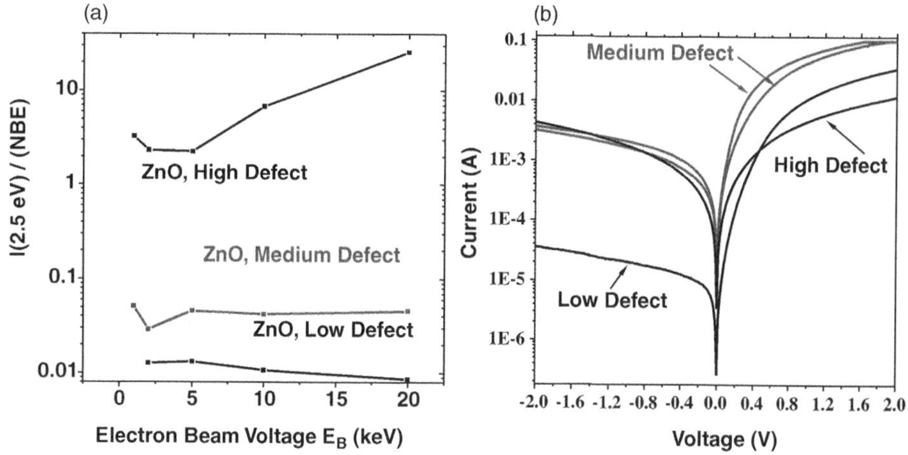

Figure 4.9 (a) Defect-to-NBE intensity ratio vs depth for "high", "medium" and "low" defect density ZnO. (b) Corresponding I–V characteristics. Reprinted from L.J. Brillson, et al., Dominant effect of near-interface native point defects on ZnO Schottky barriers, Appl. Phys. Lett. 90, 102116. Copyright (2007) with permission from American Institute of Physics

potentials that vary by well over 100 meV on the same surface on a scale of hundreds of nanometers. In general, one finds larger variations in potential across otherwise smooth surfaces in crystals with relatively high DRCLS defect emissions below the surface. Such potential variations are even larger for surfaces with larger rms roughness and surface asperities.[47] Similar potential variations across hundreds of nanometers in other crystals can easily account for the distribution of Schottky barrier heights discussed in Figure 4.6.

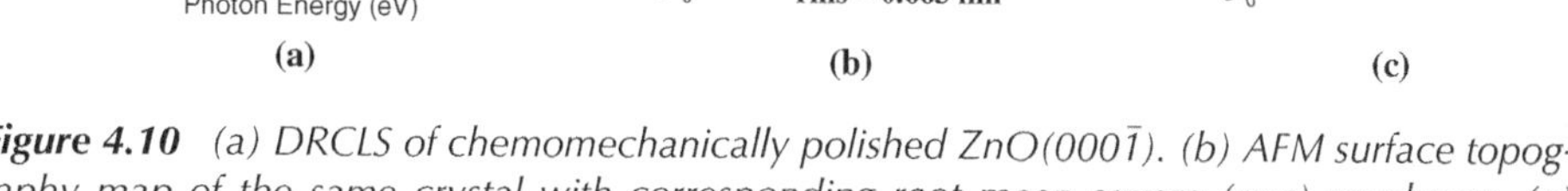

Figure 4.10 (a) DRCLS of chemomechanically polished ZnO(000$\bar{1}$). (b) AFM surface topography map of the same crystal with corresponding root mean square (rms) roughness. (c) Simultaneously acquired Kelvin force probe microscope (KFPM) map. Reprinted from D.R. Doutt, et al., Impact of near-surface native point defects, chemical reactions, and surface morphology on ZnO interfaces. J. Vac. Sci. Technol. B. 26, 1477. Copyright (2008) with permission from American Vacuum Society

4.5 The Influence of ZnO Polarity on Schottky Barriers

Several studies indicate that, as with air-exposed ZnO contacts, metals on Zn-polar surfaces yield significantly higher Φ_{SB} than on O-polar surfaces. This is evident from Table 4.1, and Figures 4.4 and 4.5 for several metals. It can be attributed to higher sub-surface defect concentrations for O-polar crystals.[9] DRCLS, C^{-2}–V, and deep level transient spectroscopy (DLTS) studies show that Zn-face diodes have higher barrier heights, lower sub-surface deep level defects, and lower sub-surface free carrier concentrations than O-face diodes of the same ZnO crystal. Figure 4.11(a) illustrates the ratio I_{DEF}/I_{NBE} of 2.5 eV deep level defect to near-band-edge emission for Zn and O faces of the same ZnO crystal, both before and after chemical and ROP cleaning and as a function of depth below the ZnO surface. Defect emissions are always higher for O-face surfaces.[9] Figure 4.11(b) shows the effect of these differences on Au and Pd Schottky barriers on the

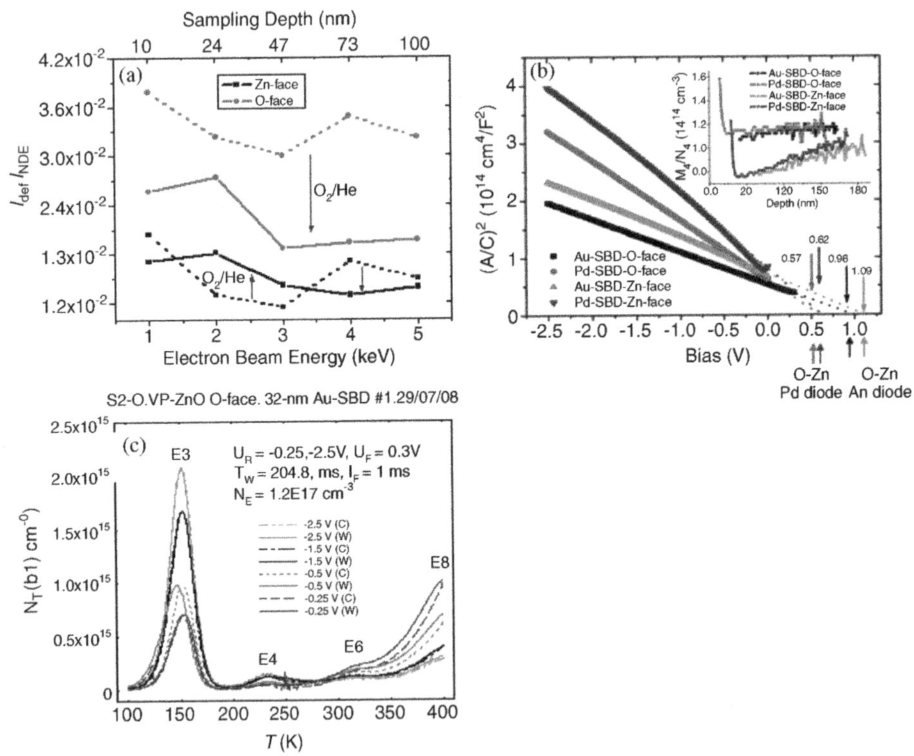

Figure 4.11 (a) Ratio I_{DEF}/I_{NBE} of 2.5 eV deep level defect to near-band-edge emission for Zn and O faces of the same ZnO crystal, both before and after chemical and ROP cleaning. Zn-terminated surfaces have lower deep level defect densities.[9] (b) C^{-2} –V plots for Au and Pd diodes on Zn-polar and O-polar surfaces of ROP-cleaned ZnO.[9] Inset shows higher carrier densities for the O-polar surfaces. (c) DLTS trap emissions vs depth for 32 nm Au on ZnO(000$\bar{1}$) showing an increase in trap E8 with decreasing depth. Reprinted from Z.-Q. Fang, et al., Surface traps in vapor-phase-grown bulk ZnO studied by deep level transient spectroscopy, J. Appl. Phys.104, 063707. Copyright (2008) with permission from American Institute of Physics

same ROP-cleaned surfaces. Zn-polar surfaces yield higher Schottky barriers than O-polar surfaces for both Au and Pd UHV-deposited contacts formed immediately after ROP cleaning.

The inset in Fig. 4.11(b) illustrates the change in carrier concentration vs depth obtained from C^{-2}–V measurements for all four diodes. The O-face ZnO exhibits substantially higher carrier densities vs the Zn-terminated surface, apparently from additional acceptors near the Zn-polar surface.[48] Furthermore, strong increases in carrier density are apparent for both surfaces in the 70–100 nm depth range. (C–V measurements at shallower depths may be subject to forward bias effects.)

Associated with these defect and doping effects, DLTS reveals a strong increase in trap states approaching the surface. Figure 4.11(c) illustrates DLTS spectra taken from one of the Au- ZnO(000$\bar{1}$) diodes in (b), showing an increase in trap feature E8 located 0.9 eV below the conduction band.[48] This feature increases by over an order of magnitude relative to the bulk E3 trap signal as depth decreases from ~180 to ~60 nm. Thus both doping and trap densities are changing dramatically with distance to the free surface. It is also significant that the trap depth of 0.9 eV corresponds closely to an energy level ~2.5 eV above the valence band, consistent with the 2.5 eV defect emissions in Figures 4.3 and 4.7. Overall, Figure 4.11 emphasizes the strong coupling between polarity, native point defects and doping at or near the ZnO surface.

4.6 The Influence of Chemistry

Chemistry also plays a role at metal–ZnO interfaces. There is a qualitative difference between reactive vs nonreactive metal–ZnO interfaces. Metals such as Al, Ta, Ti, and Ir that react with ZnO to form oxides produce defects associated with oxygen vacancies, whereas metals such as Au, Pd, and Pt that form eutectics with Zn produce defects associated with zinc vacancies. Depending on the density of defects and the extent of reaction in the sub-surface region, reactive metals can form either ohmic or blocking contacts.[14,15]

Finally, native point defects created by chemical reaction and diffusion at the metal–ZnO interface produce large changes in barrier height. Figure 4.12 illustrates the creation of new native point defects at the Au– and Al–ZnO interfaces.[14] The ability to excite CL emission from interfaces is particularly useful for studying localized electronic states at metal–semiconductor interfaces. In turn, these states can be related to the Schottky barriers that form with charge transfer across the interfaces. Figure 4.12 illustrates CL spectra for ZnO(000$\bar{1}$) surfaces with and without a metal overlayer. The Figure 4.12(a) insert illustrates a metal overlayer on clean, ordered ZnO along with the incident electron beams and exciting photons. For the 30 nm thick metal overlayers shown, the probe depth of 2 keV into the bare ZnO is equivalent to the depth excited by 5 keV through the metal.

In the case of Au on ZnO(000$\bar{1}$) both the metallized and bare surface ZnO show almost identical spectra with defect emission almost three orders of magnitude below the NBE emission. This Au/ZnO diode exhibits rectifying I–V characteristics. In the case of Al on ZnO(000$\bar{1}$), Figure 4.12(b), the metallized surface reveals an increase in the deep level

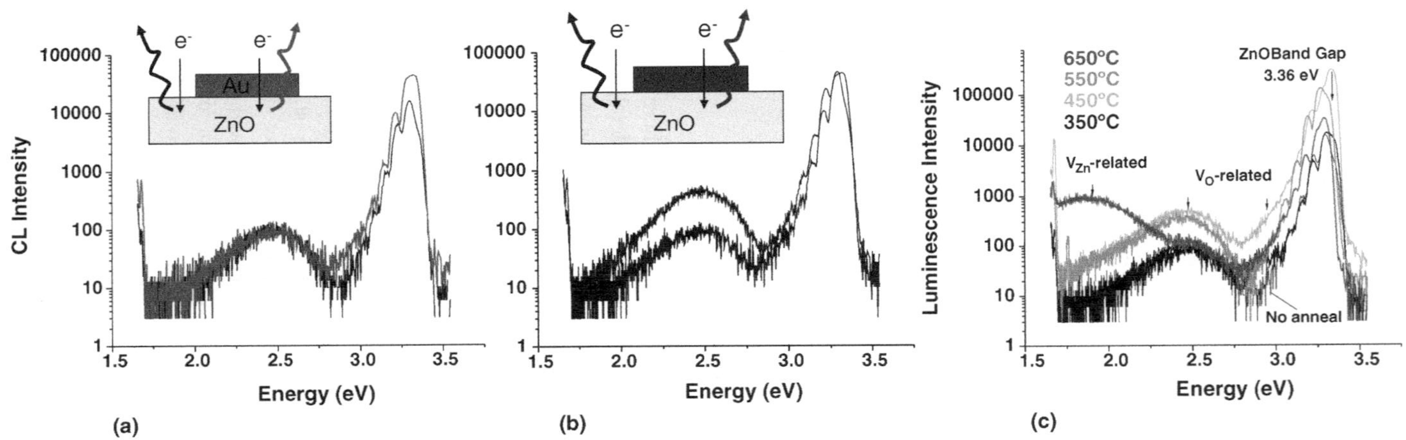

Figure 4.12 *CL spectra of bare ZnO(000$\bar{1}$) vs with and without a Au (a), Al (b) or annealed Au (c) diode overlayer on the semiconductor. Reprinted from L.J. Brillson, et al., Dominant effect of near-interface native point defects on ZnO Schottky barriers, Appl. Phys. Lett. 90, 102116. Copyright (2007) with permission from American Institute of Physics*

emission centered at 2.5 eV by nearly an order of magnitude. This increase signifies the creation of new defect states within a depth of only a few nanometers from the metal interface. Significantly, these new states occur at the same energy of native point defects that are intrinsic to the ZnO crystal. This Al/ZnO diode displays ohmic *I–V* characteristics.

In the case of the same Au/ZnO(000$\bar{1}$) diode at different annealing temperatures, Figure 4.12(c), little or no change takes place in the spectra for temperatures up to 550 °C. With a 650 °C anneal, however, a new emission appears centered at ~2.0 eV. This peak corresponds to emission frequently seen in bulk ZnO and has been attributed to Zn vacancies. Recent experiments on implanted ZnO in fact show a strong correlation as a function of depth between the density of Zn vacancies or vacancy clusters obtained from positron annihilation spectroscopy (PAS) with the intensity of the ~2 eV ZnO deep level emission measured by DRCLS.[49] Likewise, the 2.5 eV feature is commonly associated with oxygen vacancies.[14]

Consistent with the effect of interface metal–ZnO chemistry on Schottky barriers, the defect emissions and the Schottky barrier characteristics show corresponding changes as a function of annealing temperature. The CL features shown in Figure 4.12 match the changes in *I–V* characteristics as a function of temperature. Figure 4.13 illustrates *I–V* characteristics for Al, Au, and Ta vs 1-h anneal at the temperatures indicated. For Al, Figure 4.13(a) shows that the diode retains its ohmic character with increasing anneal temperature. This is expected since Al continues to react at higher temperatures with O in ZnO to create more O vacancies. Figure 4.13(b) shows the dramatic change from rectifying to ohmic as the annealing temperature nears the Zn–Au eutectic temperature and Zn vacancy-related defects are created.[15] Figure 4.13(c) displays another chemically driven change in diode properties. Here Ta reacts strongly with ZnO to produce Ta oxide and forms a blocking contact in addition to new O vacancy-related defects. The result is that the reverse current characteristic decreases by nine orders of magnitude.[15] Interestingly, even the zero bias point of the *I–V* curve is displaced due to a built-in dipole at the Ta–ZnO interface. For ZnO crystals with higher initial defect densities, this blocking feature also appears at intermediate temperatures. However, reverse current leakage increases with higher temperature annealing as the additional defects provide hopping conduction pathways through the barrier layer.

Figures 4.12 and 4.13 lead to several conclusions: (1) metals on semiconductors can induce the formation of localized interface states; (2) these new states appear at energies corresponding to point defects native to the bulk semiconductor; and (3) the nature of the chemical interaction between metal and ZnO determines what states form and at what temperature. The appearance of new states in Figure 4.12(c) occurs at a temperature just above the Au–Zn eutectic temperature (625 °C), consistent with the extraction of Zn from the ZnO. Conversely, the formation of new states at Al–ZnO interfaces is consistent with the formation of a metal oxide that extracts oxygen from the ZnO. Indeed, the behavior exhibited by these Au and Al diodes on ZnO are characteristic of many other metals. Thus relatively unreactive metals such as Au, Pd, and Pt that form eutectics with Zn display rectifying *I–V* characteristics and increases in ~2.1 eV defect emission, particularly at elevated temperatures. Conversely, more reactive metals such as Al, Ta, and Ir react with the crystal's oxygen to form oxides and display ohmic behavior. This can explain the $\Phi_{SB}{}^{n}$ vs ΔH_R plots for metals on ZnO that show characteristic "S-shape" plateaus in barrier height for reactive vs unreactive metals.[4]

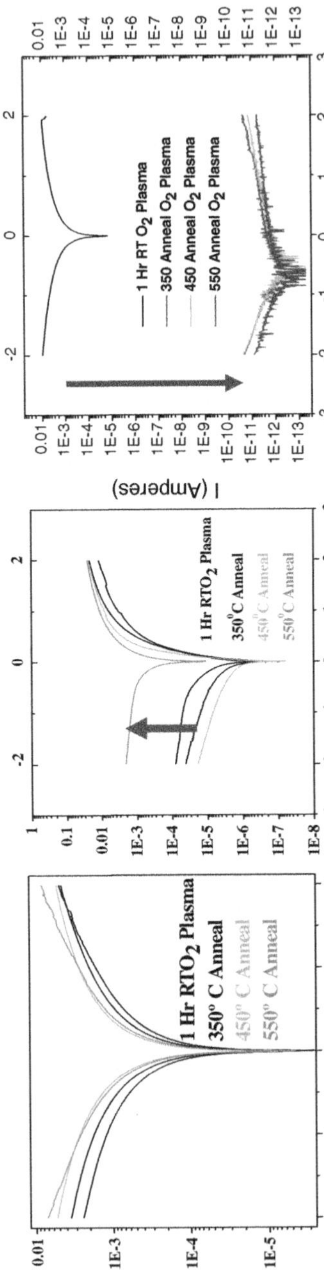

Figure 4.13 *I–V characteristics for (a) Al, (b) Au, and (c) Ta diodes on clean ZnO(000$\bar{1}$)*

Recently, Allen and Durbin recognized the role of metal–ZnO chemical reaction in forming oxygen vacancies and gap states that "pin" the Fermi level, pointing to the increase in Φ_{SB}^{n} with free energy of formation of its metal oxide[50] [see Figure 4.14(a)]. Robertson *et al.* find a similar trend for an even wider set of metal oxides[51] (see Figure 4.14(b)]. Allen and Durbin noted that their Ag oxide contacts were unlikely to form such oxides with ZnO and thus enabled less E_F pinning. The role of native point defects in E_F stabilization was first noted in studies of metal–InP interfaces, where outdiffusion of anions or cations promoted by the interface chemistry leads to two distinct Schottky barrier levels, depending on the anion/cation nonstoichiometry.[52,53] Metals that react

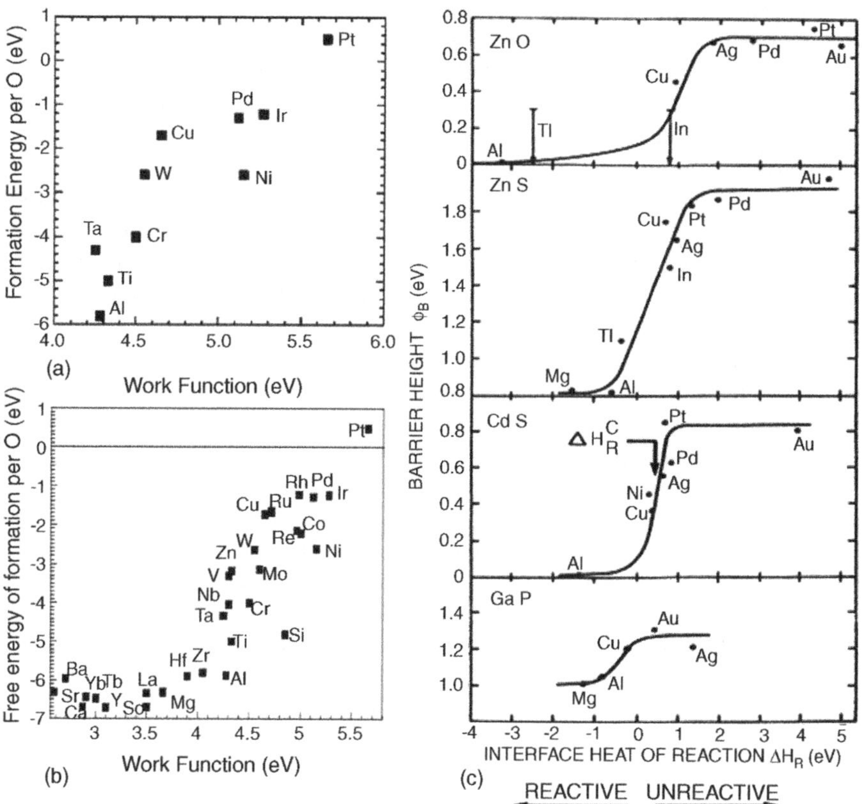

Figure 4.14 *(a) Free energy of metal oxide formation per O atom vs metal work function for common metal contacts to ZnO. Reprinted from M.W. Allen and S.M. Durbin, Influence of oxygen vacancies on Schottky contacts to ZnO, Appl. Phys. Lett. 92, 122110. Copyright (2008) with permission from American Institute of Physics (b) Experimental values of MO_X per O atom vs work function for a wider range of metals. Reprinted from J. Robertson, O. Sharia, and A. A. Demkov, Fermi level pinning by defects in HfO2-metal gate stacks, Appl. Phys. Lett. 91, 132912. Copyright (2007) with permission from American Institute of Physics (c) Transition in Schottky barrier with interface chemical reactivity for ZnO and other compound semiconductors. Reprinted from L.J. Brillson, Transition in Schottky Barrier Formation with Chemical Reactivity, Phys. Rev. Lett. 40, 260. Copyright (1978) with permission from American Physical Society*

strongly with semiconductor anions form low or zero barriers while relatively unreactive metals form high barriers. Figure 4.14(c) illustrates the transition in $\Phi_{SB}{}^{n}$ for ZnO with interface heat of reaction,

$$\Delta H_{R} = (1/x)[H_{F}(CA) - H_{F}(M_{x}A)] \tag{4.6}$$

for the reaction $M + (1/x)CA \Rightarrow (1/x)[M_{x}A + C]$ for metal M, anion A, and cation C, where M_xA is the most stable metal–anion reaction product.[54] In this case, C = Zn and A = O. Metals that form oxides form low barriers while relatively unreactive metals that only form eutectics with Zn form barriers in the ~0.7 eV range. This transition in Schottky barrier between two limiting values is common to most if not all compound semiconductors, regardless of ionicity. Together with the observed formation of native point defects at the metal–semiconductor interface with electronic states in the band gap, Figure 4.14(c) demonstrates that the same chemically induced defect principle applies to ZnO as well. Overall, these interface-specific states indicate that native point defects can play an important role in Schottky barrier formation.

Along with the increase in defect emission measured with DRCLS at the intimate metal–ZnO contact, the corresponding *I–V* characteristics show increased leakage or current blockage, depending on whether or not the metal forms an interfacial oxide. These ZnO results demonstrate that native point defects resident in the bulk crystal as well as similar defects produced by chemical interaction at the ZnO interface have a strong effect on Schottky barrier formation. This finding has broad significance for Schottky barrier studies given Figure 4.7 and the orders-of-magnitude difference in defect densities and density variations with depth between ZnO crystals grown by different growth methods – hydrothermal, melt, vapor phase transport, and molecular beam epitaxy, the same growth method but different sources, and even crystals grown by the same method from the same source! Low defect ROP-cleaned ZnO displays higher Schottky barriers and lower idealities than the corresponding metal diode on ROP-cleaned high defect material.[14] Similarly, low defect ZnO shows much lower increases in defect creation and depth redistribution than high defect ZnO. Thus considerable evidence exists to suggest that Schottky barrier features depend heavily on native point defect densities resident in the crystal prior to diode formation. Therefore, a knowledge of these defect concentrations and their depth distribution is a prerequisite for Schottky barrier studies.

4.7 Charge Transport and Extended Metal–ZnO Schottky Barriers

The effects described thus far can be represented schematically in terms of a competition of several interface charge transport mechanisms as well as in terms of chemically active interface material structures. Figure 4.15(a) represents these charge transport mechanisms schematically. Here thermionic emission of carriers over the Schottky barrier is augmented by tunneling through the barrier and hopping transport through states within the band gap. Impurities or defects that increase carrier density inside the surface space charge region will decrease the barrier width, thereby increasing the contribution of tunneling to the overall current. Similarly, increasing defect concentrations near the semiconductor surface enables charge movement through the semiconductor depletion region, especially at defect

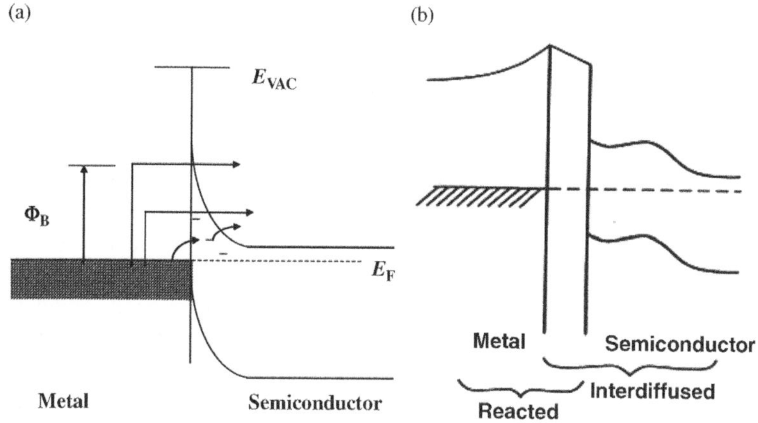

Figure 4.15 *(a) Competing charge transport mechanisms in ZnO including thermionic emission, tunneling, and hopping transport through defect levels in the band gap. (b) Schematic band diagram of extended metal–semiconductor interface including reacted and interdiffused regions with associated changes in band bending and interface dielectric phases*

densities at which the gap state wave functions begin to overlap. Indeed, if enough defects are introduced, e.g., by near-surface segregation or surface mechanical damage, the otherwise Schottky barriers become ohmic.

Beyond the multiple channel charge transport pictured in Figure 4.15(a), Figure 4.15(b) illustrates how chemical interactions at the microscopic metal–ZnO interface alter the classical Schottky barrier picture.[55,56] Metal–ZnO reactions can introduce new dielectric layers that block transport or change the effective work function at the intimate junction. The chemical interactions evident in Figure 4.12 demonstrate that such interactions are common in ZnO and other oxides. Figure 4.15(b) also shows changes in the band bending region of the semiconductor due to semiconductor anion and/or cation outdiffusion as well as metal or other impurity indiffusion that change the effective carrier concentration at different depths. Such chemical changes thereby contribute to the tunneling and/or hopping transport pictured in Figure 4.15(a).

These charge transport mechanisms are in fact more general than ZnO alone, extending to compound semiconductors overall. They highlight the importance of initial semiconductor crystal quality, surface preparation, contact formation, subsequent thermal or other processing, and the resultant interface chemical reaction and/or and diffusion leading to the formation of electronically active states within the semiconductor band gap.

Based on these factors, controllable ZnO Schottky barriers require at minimum: (1) high quality single crystals whose defect concentrations are several orders of magnitude below unintentionally doped crystal concentrations; (2) surface preparation that removes chemical contamination such as hydroxides and hydrocarbons without creating surface asperities beyond monolayer roughness levels; (3) contact formation by metal deposition that does not introduce additional near-interface contamination, reaction, or other semiconductor disruption; and (4) any subsequent thermal processing kept below temperatures at which that particular metal–semiconductor reaction can occur.

For high barriers, Table 4.1 shows that high work function metals such as Au, Pd, and Pt are good candidates although ambient gas diffusion through Au and Pd introduces interface dipoles that can alter the barriers. Ag provides relatively high barriers once oxidized and if the final reaction product is thermally stabilized. Ta reacts to form Ta oxide interlayers that block transport effectively. However, excessive annealing coupled with high background defect densities can degrade the blocking layer. In this regard, Ir is found to yield relatively high barriers due to its high work function, weak reactivity with ZnO, and stability in air. A new strategy to avoid interface chemical interaction is to use high work function and chemically inert conducting polymers to form high barriers.[36] For ohmic contacts, low work function metals such as Al, Ti and In are good candidates in low defect ZnO as long as they do not react to form blocking oxides. Reactive metals such as Al and Ti will also form interface defects that promote hopping transport.

4.8 Conclusion

Schottky barriers to ZnO show a wide range of phenomena and electronic behavior. These can all be directly tied to the chemical and structural changes that researchers are now beginning to probe systematically on an atomic scale. Work on ZnO contact has only recently been carried out on high quality material, nearly free from complicating factors such as impurities, morphological and native point defects. These advances will help accelerate our understanding and control of ZnO electrical contacts.

Acknowledgements

The author gratefully acknowledges support for this work by National Science Foundation Grant No. DMR-0513968, DMR-0803276 (Verne Hess). He acknowledges the technical contributions of Drs H. L. Mosbacker, Y. Strzhemechny, Y. Dong and M. Hetzer as well as PhD student D. Doutt.

References

[1] G. Heiland, *Surf. Sci.* **13**, 72 (1969).
[2] C.A. Mead, *Solid State Electron.* **9**, 1023 (1966).
[3] S. Kurtin, T.C. McGill and C.A. Mead, *Phys. Rev. Lett.* **22**, 1433 (1969).
[4] L.J. Brillson, *Phys. Rev. Lett.* **40**, 260 (1978).
[5] B.J. Coppa, R.F. Davis and R.J. Nemanich, *Appl. Phys. Lett.* **82**, 400 (2003).
[6] B.J. Coppa, C.C. Fulton, P.J. Hartlieb, R.F. Davis, B.J. Rodriquez, B.J. Shields and R.J. Nemanich, *J. Appl. Phys.* **95**, 5856 (2004).
[7] Y.M. Strzhemechny, H.L. Mosbacker, D.C. Look, D.C. Reynolds, C.W. Litton, N.Y. Garces, NC. Giles, L.E. Halliburton, S. Niki and L.J. Brillson, *Appl. Phys. Lett.* **84**, 2545 (2004).
[8] H.L. Mosbacker, Y.M. Strzhemechny, B.D. White, P.E. Smith, D.C. Look, D.C. Reynolds, C.W. Litton and L.J. Brillson, *Appl. Phys. Lett.* **87**, 012102 (2005).

[9] Y. Dong, Z-Q. Fang, D.C. Look, G. Cantwell, J. Zhang, J.J. Song and L.J. Brillson, *Appl. Phys. Lett.* **93**, 072111 (2008).
[10] B. Meyer, H. Alves, D.M. Hofmann, W. Kriegsels, D. Forster, F. Bertram, J. Christen, A. Hoffmann, M. Straβburg, M. Dworzak, U. Haboeck and A.V. Rodina, *Phys. Status Solidi B* **241**, 231 (2004).
[11] L.J. Brillson, *J. Vac. Sci. Technol., B* **19**, 1762 (2001).
[12] R.C. Neville and C.A. Mead, *J. Appl. Phys.* **41**, 3795 (1970).
[13] B.J. Coppa, R.F. David and R.J. Nemanich, *Appl. Phys. Lett.* **82**, 400 (2003).
[14] L.J. Brillson, H.L. Mosbacker, M.J. Hetzer, Y. Strzhemechny, G.H. Jessen, D.C. Look, G. Cantwell, J. Zhang and J.J. Song, *Appl. Phys. Lett.* **90**, 102116 (2007).
[15] H.L. Mosbacker, C. Zgrabik, M.J. Hetzer, A. Swain, D.C. Look, G. Cantwell, J. Zhang, J.J. Song and L.J. Brillson, *Appl. Phys. Lett.* **91**, 072102 (2007).
[16] H.L. Mosbacker, S. El Hage, M. Gonzalez, S.A. Ringel, M. Hetzer, D.C. Look, G. Cantwell, J. Zhang, J.J. Song and L.J. Brillson, *J. Vac. Sci. Technol., B* **25**, 1405 (2007).
[17] Q.L. Gu, C.K. Cheung, C.C. Ling, A.M.C. Ng, A.B. Djurišić, L.W. Lu, X.D. Chen, S. Fung, C.D. Beling and H.C. Ong, *J. Appl. Phys.* **103**, 093706 (2008).
[18] A.R. Polyakov, N.B. Smirnov, E.A. Kozhukhova, V.I. Vdovin, K.Ip, D.P. Norton,and S.J. Pearton, *J. Vac. Sci. Technol., A* **21**, 1603 (2003).
[19] H. Sheng, S. Muthukumar, N.W. Emanetoglu and Y. Lu, *Appl. Phys. Lett.* **80**, 2132 (2002).
[20] D.C. Oh, J.J. Kim, H. Makino, T. Handa, M.W. Cho, T. Yao and H.J. Ko, *Appl. Phys. Lett.* **86**, 042110 (2005).
[21] K. Ip, Y.W. Heo, K.H. Baik, D.P. Norton, S.J. Pearton, S. Kim, J.R. LaRoche and F. Ren, *Appl. Phys. Lett.* **84**, 2835 (2004).
[22] M.-S. Oh, D.-K. Hwang, J.-H. Lim, Y.-S. Choi and S.-J. Park, *Appl. Phys. Lett.* **91**, 042109 (2007).
[23] H. Endo, M. Sugibuchi, K. Takahashi, S. Goto, S. Sugimura, K. Hane and Y. Kashiwaba, *Appl. Phys. Lett.* **90**, 121906 (2007).
[24] H.v. Wenckstern, G. Biehne, R.A. Rahman, H. Hochmuth, M. Lorenz and M. Grundmann, *Appl. Phys. Lett.* **88**, 092102 (2006).
[25] H.v. Wenckstern, E.M. Kaidashev, M. Lorenz, H. Hochmuth, G. Biehne, J. Lenzner, V. Gottschalch, R. Pickenhain and M. Grundmann, *Appl. Phys. Lett.* **84**, 79 (2004).
[26] U. Grossner, S. Gabrielsen, T.M. Børseth, J. Grillenberger, A.Y. Kuznetsov and B.G. Svensson, *Appl. Phys. Lett.* **85**, 2259 (2004).
[27] R. Schifano, E.V. Monakhov, U. Grossner and B.G. Svensson, *Appl. Phys. Lett.* **91**, 193507 (2007).
[28] K. Ip, B.P. Gila, A.H. Onstine, E.S. Lambers, Y.W. Heo, K.H. Baik, D.P. Norton, S.J. Pearton, S. Kim, J.R. LaRoche and F. Ren, *Appl. Phys. Lett.* **84**, 5133 (2004).
[29] H.S. Yang, D.P. Norton, S.J. Pearton and F. Ren, *Appl. Phys. Lett.* **87**, 212106 (2005).
[30] S. Liang, H. Sheng, Y. Liu, Z. Huo, Y. Lu and H. Shen, *J. Cryst. Growth* **225**, 110 (2001).
[31] B. Coppa, C.C. Fulton, S.M. Kiesel, R.F. Davis, C. Pandarinath, J.E. Burnette, R.J. Nemanich and D.J. Smith, *J. Appl. Phys.* **97**, 103517 (2005).
[32] S.-H. Kim, H.-K. Kim and T.-Y. Seong, *Appl. Phys. Lett.* **86**, 112101 (2005).
[33] M.W. Allen, M.M. Alkaisi and S.M. Durbin, *Appl. Phys. Lett.* **89**, 103520 (2006).
[34] M.W. Allen, P. Miller, R.J. Reeves and S.M. Durbin, *Appl. Phys. Lett.* **90**, 062104 (2007).
[35] M.W. Allen, S.M. Durbin and J.B. Metson, *Appl. Phys. Lett.* **91**, 053512 (2007).
[36] M. Nakano, A. Tsukazaki, R.Y., Gunji, K. Ueno, A. Ohtomo, T. Fukumura and M. Kawasaki, *Appl. Phys. Lett.* **91**, 142113 (2007).
[37] H.B. Michaelson, *J. Appl. Phys.* **48**, 4729 (1977).
[38] R.T. Tung, *Mater. Sci. Eng. R.* **35**, 1 (2001).
[39] K. Hong and J.-L. Lee, *Electrochem. Solid State Lett.* **11**, H29 (2007).
[40] H. von Wenckstern, G. Biehne, R.A. Rahman, H. Hochmuth, M. Lorenz and M. Grundmann, *Appl. Phys. Lett.* **88**, 092102 (2006).
[41] J.H. Werner and H.H. Güttler, *J. Appl. Phys.* **69**, 1522 (1991).
[42] S.J. Pearton, D.P. Norton, K. Ip, Y.W. Heo and T. Steiner, *Superlatt. Microstruct.* **34**, 3 (2003).

[43] L.J. Brillson, H.L. Mosbacker, D. Doutt, M. Kramer, Z.L. Fang, D.C. Look, G. Cantwell, J. Zhang and J.J. Song, *Superlatt. Microstruct.* **45**, 206–213 (2009).
[44] C.G. Van de Walle, *Phys. Rev. Lett.* **85**, 1012 (2000).
[45] Y.M. Strzhemechny, H.L. Mosbacker, D.C. Look, D.C. Reynolds, C.W. Litton, N.Y. Garces, NC. Giles, L.E. Halliburton, S. Niki and L.J. Brillson, *Appl. Phys. Lett.* **84**, 2545 (2004).
[46] D.M. Hoffman, A. Hofstaetter, F. Leiter, H. Zhou, F. Henecker and B.K. Meyer, *Phys. Rev. Lett.* **88**, 045505 (2002).
[47] D.R. Doutt, C. Zgrabik, H.L. Mosbacker and L.J. Brillson, *J. Vac. Sci. Technol.* **26**, 1477 (2008).
[48] Z.-Q. Fang, B. Claflin, D.C. Look, Y.F. Dong, H.L. Mosbacker and L.J. Brillson, *J. Appl. Phys.* **104**, 063707 (2008).
[49] Y. Dong, A. Kuznetsov, F. Tuomisto and L.J. Brillson, *Phys. Rev. B*, **81**, 081201(R) (2010).
[50] M.W. Allen and S.M. Durbin, *Appl. Phys. Lett.* **92**, 122110 (2008).
[51] J. Robertson, O. Sharia and A. A. Demkov, *Appl. Phys. Lett.* **91**, 132912 (2007).
[52] L.J. Brillson, L.J. Brillson, C. F. Brucker, A. D. Katnani, N.G. Stoffel and G. Margaritondo, *Appl. Phys. Lett.* **38**, 784 (1981).
[53] L.J. Brillson,in *The Chemical Physics of Solid Surfaces and Heterogeneous Catalysis*, edited by D. A. Kingand D. P. Woodruff, Elsevier, Amsterdam, 1988, Vol. **5**, Ch. 4, p. 119.
[54] L.J. Brillson, *Phys. Rev. Lett.* **40**, 260 (1978).
[55] L.J. Brillson, C.F. Brucker, A.D. Katnani, N.G. Stoffel, R. Daniels and G. Margaritondo, *J. Vac. Sci. Technol.* **19**, 661 (1981).
[56] L.J. Brillson, C.F. Brucker, A.D. Katnani, N.G. Stoffel, R. Daniels and G. Margaritondo, *Surf. Sci.* **132**, 212 (1983).

5

Native Point Defects and Doping in ZnO

Anderson Janotti and Chris G. Van de Walle
Materials Department, University of California, Santa Barbara, CA, USA

5.1 Introduction

Controlled introduction of impurities ("doping") forms the basis of much of semiconductor technology; indeed, p-type (acceptor-doped) and n-type (donor-doped) layers and the junctions between them control carrier confinement, carrier flow, and ultimately the device characteristics. Achieving both n-type and p-type conductivity has traditionally proved to be a challenge in wide-band-gap semiconductors.

The doping can be affected by the presence of native point defects such as vacancies (V_{Zn} and V_O), self-interstitials (Zn_i and O_i) and antisites (Zn_O and O_{Zn}). Such defects may cause *self-compensation*: e.g. when one tries to dope the material p-type, certain native defects that act as donors may spontaneously form and compensate the deliberately introduced acceptors. In ZnO, specific native defects have long been believed to play an even more important role. As-grown ZnO frequently exhibits high levels of unintentional n-type conductivity, and native point defects have often been invoked to explain this behavior. In particular, the oxygen vacancy and the Zn interstitial, which act as donors, have most often been mentioned as sources of n-type conductivity in ZnO. We will see there is no convincing evidence that this is actually the case.

A lot of progress has been made in the last few years in our understanding of impurities and native defects in ZnO: the availability of higher-quality bulk crystals and epitaxial layers has allowed for better experiments, and first-principles calculations have provided a sound theoretical foundation.

Zinc Oxide Materials for Electronic and Optoelectronic Device Applications, First Edition.
Edited by Cole W. Litton, Donald C. Reynolds and Thomas C. Collins.

In Section 5.2 we discuss the theoretical framework for describing incorporation of defects and impurities, reviewing the results of different approaches to overcome the band-gap problem in density functional theory (DFT). Section 5.3 focuses on results for native point defects, paying particular attention to the oxygen vacancy, which has frequently (and incorrectly) been invoked as the source of n-type conductivity in ZnO. Section 5.4 discusses donor impurities, and Section 5.5 acceptor impurities. The chapter ends with a brief discussion of isoelectronic impurities in Section 5.6.

5.2 Theoretical Framework

The equilibrium concentration of an impurity or native defect is given by:

$$c = N_{\text{sites}} \exp^{(-E^{\text{f}}/k_{\text{B}}T)} \tag{5.1}$$

where E^{f} is the *formation energy*, N_{sites} is the number of sites the defect or impurity can be incorporated on, k_{B} is the Boltzmann constant, and T is the temperature. Equation (5.1) shows that defects with a *high* formation energy will occur in *low* concentrations.

The formation energy is not a constant but depends on the growth conditions. For example, the formation energy of an oxygen vacancy is determined by the relative abundance of Zn and O atoms, as expressed by the chemical potentials μ_{Zn} and μ_{O}, respectively. If the vacancy is charged the formation energy depends further on the Fermi level (E_{F}), which acts as a reservoir for electrons. Forming an oxygen vacancy requires the removal of one oxygen atom; the formation energy is therefore:[1,2]

$$E^{\text{f}}(V_{\text{O}}^{q}) = E_{\text{tot}}(V_{\text{O}}^{q}) - E_{\text{tot}}(\text{ZnO}) + \frac{1}{2}E_{\text{tot}}(\text{O}_2) + \mu_{\text{O}} + qE_{\text{F}}. \tag{5.2}$$

$E_{\text{tot}}(V_{\text{O}}^{q})$, $E_{\text{tot}}(\text{ZnO})$ and $E_{\text{tot}}(\text{O}_2)$ are the total energies of ZnO with an oxygen vacancy in charge state q, ZnO perfect crystal, and an O_2 molecule, respectively, and are usually calculated using a supercell that is periodically repeated in the three dimensions. The oxygen chemical potential μ_{O} is the energy of the O reservoir, and E_{F} is the Fermi level. Similar expressions apply to the other native defects and to the various impurities.

The chemical potentials μ_{O} and μ_{Zn} depend on the experimental growth conditions, which can be either Zn-rich or O-rich (or anything in between). They should therefore be explicitly regarded as variable in the formalism. However, in thermodynamic equilibrium it is possible to place *bounds* on the chemical potentials. The O chemical potential μ_{O} is subject to an upper bound given by the energy of O in an O_2 molecule, corresponding to extreme O-rich conditions ($\mu_{\text{O}}^{\text{max}} = 0$); similarly, the Zn chemical potential μ_{Zn} is upper bounded by the energy of Zn in zinc metal, corresponding to extreme Zn-rich conditions ($\mu_{\text{Zn}}^{\text{max}} = 0$). It should be kept in mind that these chemical potentials, which are free energies, are temperature and pressure dependent.

The upper bounds in μ_{O} and μ_{Zn} defined above lead to lower bounds given by the stability of ZnO:

$$\mu_{\text{Zn}} + \mu_{\text{O}} = \Delta H_{\text{f}}(\text{ZnO}) \tag{5.3}$$

where ΔH_f(ZnO) is the formation enthalpy of ZnO. The upper limit on the O chemical potential ($\mu_O^{max} = 0$) then results in a lower limit on the Zn chemical potential:

$$\mu_{Zn}^{min} = \Delta H_f(\text{ZnO}) \tag{5.4}$$

Similarly, the upper limit on the Zn chemical potential ($\mu_{Zn}^{max} = 0$) results in a lower limit on the O chemical potential:

$$\mu_O^{min} = \Delta H_f(\text{ZnO}) \tag{5.5}$$

By combining Equation (5.4) or Equation (5.5) with Equation (5.3), we observe that the host chemical potentials thus vary over a range corresponding to the magnitude of the enthalpy of formation of ZnO. The calculated values of ΔH_f(ZnO) are usually in good agreement with the experimental value of –3.61 eV.[3]

The Fermi level E_F in Equation (5.2) is usually referenced to the valence-band maximum (VBM). It is not an independent parameter, but is determined by the condition of charge neutrality. In principle equations such as (5.2) can be formulated for every native defect and impurity in the material; the complete problem (including free-carrier concentrations in valence and conduction bands) can then be solved self-consistently, imposing charge neutrality. However, it is instructive to examine formation energies as a function of E_F in order to examine the behavior of defects and impurities when the doping level changes.

Defects and impurities often introduce levels in the band gap; these *transition levels* can be obtained based on the calculated formation energies. The transition level $\varepsilon(q+1/q)$ is defined as the Fermi-level position for which the formation energies of charge states $q+1$ and q are equal. The experimental significance of this level is that for Fermi-level positions below $\varepsilon(q+1/q)$, charge state $q+1$ is stable, while for Fermi-level positions above $\varepsilon(q+1/q)$, charge state q is stable.

Calculations for native point defects in ZnO have been performed with a variety of methods, but in this Chapter we will limit the discussion to calculations based on state-of-the-art first principles techniques. Such calculations, used to derive E_{tot} in Equation (5.2), do not require any adjustable parameters or any input from experiment. They are based on DFT,[4] usually using the local density approximation (LDA) or generalized gradient approximation (GGA), using a supercell geometry and *ab initio* pseudopotentials. Details of the computational approach can be found in the literature.[2,5,6]

5.3 Native Point Defects

Density functional calculations for native point defects in ZnO have been reported by several groups.[6–20] However, the fact that the band gap of ZnO is severely underestimated by the commonly used LDA or GGA functionals makes the interpretation of the calculations very difficult. Defects often induce occupied states in the band gap. These states have a certain ratio of valence- vs conduction-band character and, therefore, their positions with respect to the VBM can be underestimated. This uncertainty affects the prediction of transition levels and formation energies, leading to potentially significant errors, especially in the case of wide-band-gap semiconductors such as ZnO.

Different approaches to overcome the DFT-LDA or -GGA deficiencies in predicting band gaps have been employed in the investigation of point defects in ZnO. These include self-interaction corrections, the LDA + U, and the B3LYP and HSE hybrid functionals.[6,8,14–20] Although uncertainties still exist in the numerical values of formation energies, important qualitative conclusions can be extracted.

Most of the calculations agree that oxygen vacancies and zinc vacancies are the lowest energy defects, followed by the Zn interstitial and the ZnO antisite. Oxygen interstitials and O_{Zn} antisites were found to be high in energy. The defects that are favored under Zn-rich conditions (V_O, Zn_i, and Zn_O) all act as donors, while those that are favored under O-rich conditions (V_{Zn}, O_i, and O_{Zn}) all act as acceptors. In Figure 5.1 we show the calculated defect formation energies as a function of Fermi level from Janotti and Van de Walle.[6] These results were obtained using an extrapolated scheme based on LDA and LDA + U calculations, as discussed in the literature.[6,14,15]

Despite the qualitative similarities, it is important to discuss the differences between the results given by the various approaches employed to calculate transition levels and formation energies of defects in ZnO. Calculations that are based purely on LDA or GGA functionals carry a large uncertainty in the transition levels and formation energies due to the severe underestimation of the band gap of ZnO. In these cases, transition levels related to defects that induce (single-particle) states in the band gap can be significantly underestimated. When these single-particle states are occupied with electrons, the formation energies for the relevant charge states will be underestimated as well.

In an attempt to overcome this issue, Zhang *et al.* included empirical corrections to the bare DFT-LDA results.[8] As a main result, they have found that V_O has a high formation

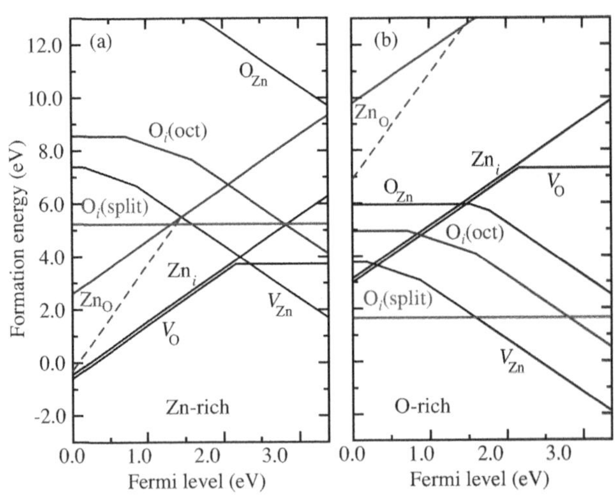

Figure 5.1 *Formation energies as a function of Fermi-level position for native point defects in ZnO for Zn-rich (a) and O-rich (b) conditions. The zero of Fermi level corresponds to the valence-band maximum. Only segments corresponding to the lowest-energy charge states are shown. The slope of these segments indicates the charge state. Kinks in the curves indicate transitions between different charge states. Reprinted from A. Janotti and C. G. Van de Walle, Native point defects in ZnO, Phys. Rev. B 76, 165202. Copyright (2007) with permission from The American Physical Society*

energy in n-type ZnO, with the $\varepsilon(2+/0)$ transition level located in the upper part of the band gap. Lany and Zunger[17] performed LDA + U calculations, which partially correct the band gap, and used these results to correct the position of the VBM in ZnO. Otherwise, the results were based on LDA and a rigid shift of the conduction-band minimum (CBM) to correct the band gap, while leaving the positions of deep levels unchanged. Lany and Zunger obtained the V_O $\varepsilon(2+/0)$ transition level at ~1.6 eV above the VBM.

Using LDA + U, Paudel and Lambrecht concluded that the V_O $\varepsilon(2+/0)$ transition level is located near the VBM.[19] Their scheme includes an application of U to the Zn s states that dominate the character of the conduction band, in addition to applying U to the Zn d states. This seems to go against the nature of the LDA + U correction, which is intended to correct the energies of localized states that are underbound in LDA. While the semicore Zn d states are indeed quite localized, the Zn s states that make up the conduction band are clearly delocalized extended states. Since the V_O related state in the gap has a large contribution from Zn s states, the application of U to Zn s states will also affect the position of the V_O related state in an unphysical way.

Janotti *et al.* recognized that LDA + U affects both valence and conduction bands of ZnO[21] and that the single-partice defect states are corrected according to their valence- vs conduction-band character. Because LDA + U only partially corrects the band gap, an extrapolation scheme based on the LDA and LDA + U calculations was then employed to obtain gap-corrected transition levels and formation energies that can be quantitatively compared with experimental results.[6,14,15]

Using the B3LYP hybrid functional, Patterson carried out calculations of V_O in ZnO.[16] The B3LYP results for the electronic structure of V_O in ZnO are consistent with those obtained by Janotti and Van de Walle.[6,14,15] However, Patterson's interpretation of the transition levels based on the results for the single-particle states is not correct. The position of the transition levels cannot be directly extracted from the position of the single-particle states. Transition levels must be calculated based on differences in formation energies (as explained in Section 5.2). This is particularly important for defects which exhibit very different lattice relaxations in different charge states, which as we will see is the case of V_O in ZnO.

Oba *et al.* recently performed calculations for point defects in ZnO using the HSE hydrid Functional.[20] The calculated single-particle band structure for oxygen vacancy using the HSE is shown in Figure 5.2.[20] The position of the transition levels in the HSE are in good agreement with the results of Janotti and Van de Walle.[6] However, the formation energy of V_O is relatively low, indicating that these defects would be present in significant concentrations in n-type ZnO. This is inconsistent with the results of recent experiments on ZnO single crystals, in which electron paramagnetic resonance (EPR) signals identifying oxygen vacancies are not present in as-received crystals.[22] Oxygen vacancies are observed only after irradiating the samples with high energy electrons.

5.3.1 Oxygen Vacancies

The oxygen vacancy has frequently been invoked as the source of unintentional n-type conductivity in ZnO. However, all density functional calculations indicate that this assignment cannot be correct. According to DFT calculations[6,8,9,11,12,14,16,17,19,20] V_O is a *deep* rather than a *shallow* donor and, consequently, cannot contribute to n-type

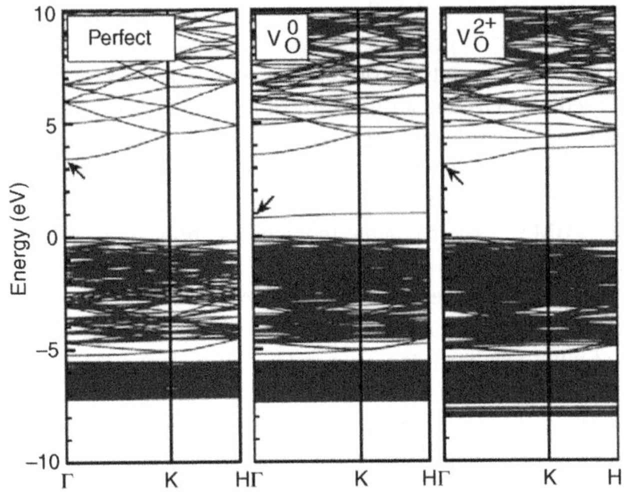

Figure 5.2 *Band structure for the ZnO perfect crystal, and oxygen vacancy (V_O) in the neutral and +2 charge states obtained using the HSE hybrid functional. The zero in energy corresponds to the valence-band maximum in the perfect crystal. Reprinted from F. Oba, A. Togo, I. Tanaka, J. Paier, and G. Kresse, Defect energetics in ZnO: A hybrid Hartree-Fock density functional study, Phys. Rev. B 77, 245202. Copyright (2008) with permission from The American Physical Society*

conductivity. Although the calculations reported in the literature differ on the values for transition levels and formation energies due to the different approaches to correct the band gap, they unanimously agree that V_O is a deep donor. According to Figure 5.1, the $\varepsilon(2+/0)$ transition level is located at ~1 eV below the CBM, i.e. V_O is stable in the neutral charge state in n-type ZnO. The oxygen vacancy is a "negative-U" center, implying that $\varepsilon(2+/+)$ lies above $\varepsilon(+/0)$ in the band gap.[6,14] As the Fermi level moves upward, the charge-state transition is thus directly from the +2 to the 0 charge state.

It should be noted that, while V_O cannot contribute to the n-type conductivity in ZnO because it assumes the neutral charge state when the Fermi level is near the CBM, it can be a source of compensation in p-type ZnO. In this case, V_O assumes the +2 charge state when the Fermi level is near the VBM and has relatively low formation energies as shown in Figure 5.1.

An alternative way to express the negative-U behavior of V_O is to say that V_O^+ is always higher in energy than either V_O^{2+} or V_O^0, for any Fermi-level position. The positive charge state is thus never thermodynamically stable. This is an important finding, because it is the positive charge state, with its unpaired electron, that is detectable by magnetic resonance techniques. An EPR signal associated with V_O should thus not be observed under thermodynamically stable conditions. It is, of course, possible to create oxygen vacancies in the +1 charge state in a metastable manner, for instance by excitation with light. Once generated, V_O^+ does not immediately decay into the +2 or neutral charge state because of energetic barriers. These barriers are associated with the large lattice relaxations that occur around the oxygen vacancy;[14] these relaxations are very different for the different charge states as shown in Figure 5.3. Therefore, it is possible to detect EPR signals due to V_O^+ upon photoexcitation at low enough temperatures, but if the excitation is removed and the temperature is raised, these signals decay.

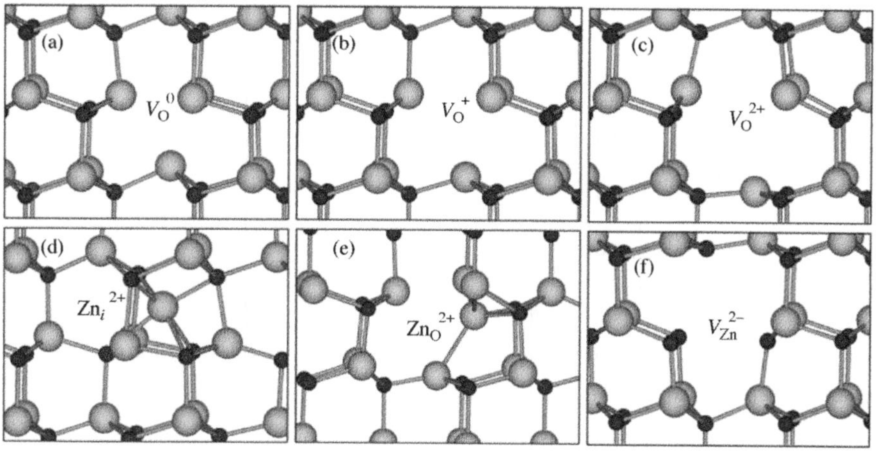

Figure 5.3 *Ball and stick model of the local atomic relaxations around the oxygen vacancy in the (a) neutral, (b) +1 charge state and (c) +2 charge states. In the neutral charge state, the four Zn nearest neighbors are displaced inward by 12% of the equilibrium Zn-O bond length. In the +1 charge state, the four Zn nearest neighbors are displaced outward by 3%, and for the +2 charge states the displacements are outward by 23%. Local structure of the zinc interstitial (d), zinc antisite (e), and zinc vacancy (f).*

Vlasenko and Watkins have carried out optically detected electron paramagnetic ressonance (ODEPR) experiments using high quality ZnO single crystals.[22] They report very interesting results related to oxygen vacancies that are in good agreement with the first principles results shown in Figure 5.1. First, the EPR signals related to V_O could be detected only after irradiating the ZnO samples with high energy electrons. This indicates that oxygen vacancies are not present in the as-grown (or as-received) ZnO single crystals, supporting the results shown in Figure 5.1, in which V_O has a high formation energy in n-type ZnO. Recent experiments by Evans *et al.* also confirm that the EPR signals related to oxygen vacancies are not detectable in as-grown ZnO single crystals, but only after irradiation.[23]

Secondly, Vlasenko and Watkins have reported that V_O^+ can be observed only upon excitation with photon energies above ~2 eV.[22] Evans *et al.* report an threshold excitation energy of 2.1 eV.[23] These results clearly confirm that V_O^+ is not thermodynamically stable as discussed in the literature.[6,14] Moreover, the excitation energy is in good agreement with the optical transitions extracted from the calculated configuration coordinate diagram from Janotti and Van de Walle,[14] reproduced in Figure 5.4.

5.3.2 Zinc Interstitials

Zinc interstitials are stable at the octahedral interstitial site as shown in Figure 5.3(d). The $\varepsilon(2+/+)$ and $\varepsilon(+/0)$ levels of Zn_i lie above the CBM,[6,8,9] implying that the zinc interstitial will always donate electrons to the conduction band, thus acting as a shallow donor. Note that electrons can be trapped in an extended (hydrogenic) state around Zn_i at low temperatures.

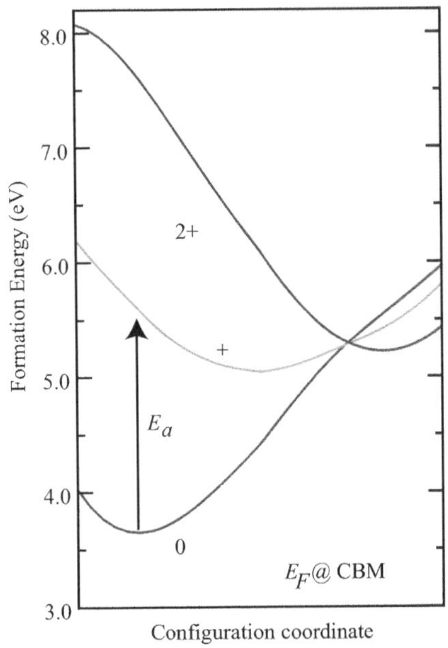

Figure 5.4 *Calculated configuration coordinate diagram for V_O^0, V_O^+ and V_O^{2+} for Fermi energy (E_F) at the conduction-band minimum (CBM). The arrow represents an optical transition from V_O^0 to V_O^+ in n-type ZnO. Reprinted from A. Janotti and C. G. Van de Walle, Oxygen vacancies in ZnO, Appl. Phys. Lett. 87, 122102. Copyright (2005) with permission from American Institute of Physics*

Under n-type conditions (Fermi level high in the band gap), Zn_i has quite a high formation energy, even under extreme Zn-rich conditions,[6,8,15] as shown in Figure 5.1. Zinc interstitials are thus unlikely to be responsible for unintentional n-type conductivity. Note that the formation energy of Zn_i^{+2} decreases rapidly when the Fermi level moves lower in the band gap; the zinc interstitial is thus also a potential source of compensation in p-type ZnO.

The fact that Zn interstitials act as shallow donors was experimentally established by heating ZnO crystals in Zn vapor followed by a rapid quench;[24] Hall measurements indicated the appearance of a shallow donor with ionization energy of 51 meV.[25]

Evidence for Zn interstitials as shallow donors also resulted from high-energy electron irradiation experiments by Look *et al.*[26] The donor was identified as a Zn-sublattice defect based on a much higher production rate for the Zn (0001) face than for the O $(000\bar{1})$ face, and was found to have an ionization energy of 30 meV. Alternatively, it was suggested that a Zn-interstitial related complex could be responsible for the observations. Look *et al.* suggested that Zn_i-N_O would form under N ambient conditions and contribute to the unintentional n-type conductivity.[27]

5.3.3 Zinc Antisites

First-principles calculations[6,8] have found Zn_O antisites to be higher in energy than the other donor-like native defects (V_O and Zn_i). The calculations of Oba *et al.*[9] found Zn_O to be comparable in energy with V_O under Zn-rich conditions. Recent calculations revealed a large off-site displacement of the Zn atom,[6] as shown in Figure 5.3(e). All calculations agree that zinc antisites behave as shallow donors, however they have high formation energies in n-type ZnO. Zinc antisites could potentially play a role under nonequilibrium conditions, for instance under irradiation.[26] When created, they would contribute to n-type conductivity.

5.3.4 Zinc Vacancies

Zinc vacancies have high formation energies in p-type ZnO, but their formation energies are low enough in n-type ZnO (Figure 5.1) to occur in modest concentrations.[6,15] They would introduce a small amount of compensation. The relaxations around V_{Zn} in the –2 charge state is shown in Figure 5.3(f).

Several EPR observations of zinc vacancies have been reported. EPR signals with *g* factors in the range 2.0018–2.056 were observed in irradiated single crystals by Taylor *et al.*[28] It was proposed that one subset of these lines would be due to Zn vacancies. Galland and Herve[29] observed lines with *g* factors between 2.0024 and 2.0165 in irradiated single crystals, also attributing them to Zn vacancies. Recent first-principles calculations[6] indicate that the zinc vacancy is indeed a deep acceptor with low formation energy in n-type ZnO. Zinc vacancies are also a likely cause of green luminescence in ZnO,[6,7] as discussed in Section 5.3.5.2.

5.3.5 Defect Migration

When studying point defects in ZnO, it is also important to know their mobility in the ZnO lattice. Point defects mediate self-diffusion and impurity diffusion, and often play an important role in impurity incorporation during growth and processing. Neumann summarized experimental results up to 1981 for diffusion of host atoms as well as impurities in ZnO.[30] Values for activation energies of Zn self-diffusion range from 1.9 to 3.3 eV, while activation energies for oxygen self-diffusion vary in a much wider range, from 1.5 to 7.5 eV.

We note that interpreting these results or using them in a predictive manner is not straighforward. It is important to keep in mind that the activation energy for self-diffusion is the sum of the formation energy of the defect responsible for self-diffusion and the migration energy of the defect. The latter is a well defined quantity, which can be obtained from density functional calculations as the energy difference between the saddle point along the migration path and the initial or stable state. The first term in the activation energy, however, namely the formation energy of the defect, strongly depends on the experimental conditions, as should be clear from Equation (5.2): details of Fermi-level position and, in particular, chemical potentials can cause large changes in the formation energy. It is usually not straightforward to assess how the environmental conditions will affect the formation energy and hence the activation energy for self-diffusion. This explains the wide spread in the reported values, and makes it difficult to offer predictive values.

Tomlins *et al.*[31] reported an activation energy of 3.86 eV for self-diffusion of Zn in ZnO, and suggested that Zn self-diffusion is controlled by a vacancy mechanism. The activation energy is the sum of the formation energy of the defect responsible for self-diffusion (either the Zn vacancy or the Zn interstitial, in this case) and the migration energy of the defect. Based on experiments involving heating crystals in Zn vapor followed by rapid quenching, Thomas reported a migration barrier of 0.55 eV for Zn interstitials in ZnO.[24] Recent density functional calculations revealed that Zn interstitial migrates through a kick-out mechanism with a very low migration barrier of 0.57 eV.[6] However, because the formation energy of the Zn interstitial is very high in n-type samples, even under Zn-rich conditions, it is unlikely that Zn interstitial mediates Zn self-diffusion in the samples that have been studied to date, which are invariably n-type. Instead, it is more likely that Zn self-diffusion is mediated by Zn vacancies which require a higher energy barrier of 1.40 eV, but have a much lower formation energy in n-type ZnO.[6]

Note that a migration barrier of 0.57 eV implies that Zn interstitials are mobile at room temperature; zinc interstitials are thus unlikely to occur as isolated interstitials, but will have a high tendency to either diffuse out of the sample or to bind with other defects or impurities. This renders it less likely that Zn interstitials can contribute to unintentional n-type conductivity in ZnO.

For diffusion of oxygen in ZnO, Tomlins and Routbort[32] reported an activation energy between 3.6 eV and 4.2 eV. Here the point defect responsible for diffusion would be the oxygen vacancy or the oxygen interstitial. Oxygen interstitials can be stable as deep acceptors at the octahedral interstitial site, or as a split interstitial, where the extra oxygen atom shares a regular lattice site with a host oxygen atom.[6,15] Oxygen interstitials at the octahedral site [O_i(oct)] have a low migration barrier, but in order to contribute to self-diffusion they would ultimately need to become substitutional again, by exchanging positions with oxygen atoms at the regular lattice sites. Calculations indicate that this process involves a high energy barrier.[6] Oxygen split interstitials have high formation energies and cannot explain the results of Tomlins and Routbort either.[32] Since Tomlins and Routbort stated that their experiments were performed in semi-insulating crystals, first-principles calculations thus indicate that oxygen self-diffusion is mediated by doubly ionized oxygen vacancy (V_O^2+) with a calculated activation energy of 4.5 eV.[6]

5.3.5.1 EPR Observations of Point Defects in ZnO

There are numerous reports of EPR measurements of oxygen vacancies in the literature; they are summarized in Table 5.1.[6] The observations fall into two broad categories: those with $g \approx 1.96$, and those with $g \approx 1.99$. The $g \approx 1.99$ line has been consistently assigned to oxygen vacancies, and we support that assignment. Indeed, this signal has only been observed after irradiation of the samples, clearly indicating it is related to a point defect. Also, it has been found that illumination is necessary to observe the center, consistent with the theoretical result that some type of excitation is required in order to generate the paramagnetic +1 charge state.

Smith and Vehse[48] were the first to provide a conclusive assignment of the $g \approx 1.99$ EPR line to the oxygen vacancy in ZnO. Using a ZnO crystal that had been irradiated by high-energy electrons, they observed that the $g \approx 1.99$ center is light sensitive (light being essential to create the V_O^+ charge state). They also observed hyperfine interaction lines associated with the ^{67}Zn neighbors of the vacancy. Similar hyperfine structure was observed by Gonzalez *et al.*[49] Soriano and Galland[50] have also shown that illumination is necessary to detect the $g \approx 1.99$ line, and measured its decay after illumination is turned off. The light sensitivity and metastability is consistent with the current model of oxygen vacancies in ZnO.[8,10,14]

Table 5.1 *Overview of electron paramagnetic resonance observations of donors in ZnO*

g value	Sample	Treatment	Assignment	Reference
$g_{\parallel}=1.956$, $g_{\perp}=1.955$	single crystal	1150 °C, 7 h	els. bound at donors	[33]
$g_{\parallel}=1.957$, $g_{\perp}=1.956$	powder	900 °C, 2 h	V_O	[34]
$g_{\parallel}=1.956$, $g_{\perp}=1.955$	single crystal	–	els. in CB or donor band	[35]
$g \sim 1.96$	powder	975 °C, 1-20 h		[36]
$g=1.9539$	powder	–	V_O	[37]
$g=1.9564, 1.9600$	powder	575 K, vac + O_2		[38]
$g_{\parallel}=1.9576$, $g_{\perp}=1.9557$	single crystal	–	V_O	[39]
$g=1.9557$	powder	–	V_O	[40]
$g=1.956$	powder	–	V_O	[41]
$g = 1.955$	powder	930 °C, H_2 then O_2	V_O	[42]
$g_{\parallel}=1.9573$, $g_{\perp}=1.9557$	powder	–	V_O	[43,44]
$g=1.9564$, $g=1.9596$	powder	700-900 °C, N_2,H_2	V_O	[45]
$g_{\parallel}=1.9570$, $g_{\perp}=1.9551$	single crystal	–	delocalized els.	[46]
$g_{\parallel}=1.957$, $g_{\perp}=1.956$	single crystal	–	shallow donors	[47]
$g_{\parallel}=1.9948$, $g_{\perp}=1.9963$	single crystal	irradiation, illumination	V_O	[48]
$g_{\parallel}=1.9948$, $g_{\perp}=1.9961$	single crystal	irradiation	V_O	[28]
$g_{\parallel}=1.9945$	single crystal	irradiation, illumination	V_O	[49]
$g_{\parallel}=1.9945$	single crystal	irradiation, illumination	V_O	[50]
$g_{\parallel}=1.9948$, $g_{\perp}=1.9963$	single crystal	irradiation, illumination	V_O	[51]
$g_{\parallel}=1.9951$, $g_{\perp}=1.9956$	single crystal	irradiation, illumination	V_O-Li_{Zn}	[51]
$g_{\parallel}=1.9945$, $g_{\perp}=1.9960$	single crystal	irradiation, illumination	V_O	[22]

Table 5.1 also contains many references reporting a $g \approx 1.96$ EPR line, also assigned to oxygen vacancies in ZnO. However, scant evidence for this was offered. For instance, no hyperfine interactions were observed for the $g \approx 1.96$ line. It is much more likely that the $g \approx 1.96$ signal is associated with electrons in the conduction band or in a donor band, as originally proposed by Müller and Schneider[35] and most recently confirmed by Garces *et al.*[47] and Evans *et al.*[23] The tendency for authors to assign the $g \approx 1.96$ line to V_O was probably largely based on the prevailing hypothesis that oxygen vacancies were the donors

responsible for the unintentional n-type conductivity – a hypothesis that we now know to be incorrect.

An overview of experimental observations of the $g \approx 1.96$ line up to 1970 was given by Sancier,[38] who also favored assigning the $g \approx 1.96$ line to electrons in the conduction band. A critical review of results up to 1981 was given by Neumann,[52] who observed that doping with Al, Ga, or In increases the intensity of the $g \approx 1.96$ signal. This behavior is consistent with the $g \approx 1.96$ signal being due to delocalized electrons, but would be hard to reconcile with oxygen vacancies as the source for the $g \approx 1.96$ line.

It should be emphasized that lines around $g \approx 1.96$ may actually be due to different types of centers, as pointed out by Geisler and Simmons.[36] We also note that the $g \approx 1.96$ line has also been reported to be photosensitive; in particular, the signal is enhanced after UV illumination.[33,35–38,40,41,44] This observation is of course consistent with the $g \sim 1.96$ line corresponding to electrons in conduction-band states, since UV light can promote electrons into these states.

5.3.5.2 Green Luminescence

ZnO often exhibits a weak and broad luminescence band in the green, centered between 2.4 eV and 2.5 eV.[53–55] This green luminescence has been observed in samples prepared with a variety of growth techniques, and it is important to point out that there may not be a single source for this luminescence. For instance, Cu has been suggested as a potential cause.[56,57] Still, not all ZnO samples contain copper. Native defects have also been suggested as a potential source of the green luminescence. Experimental and theoretical results have suggested that the Zn vacancy can give rise to green luminescence.[6,7,53,54] Indeed, the calculated transition level between the –1 and –2 charge states of V_{Zn} shown in Figure 5.1 occurs around 0.9 eV above the VBM,[6] and therefore a transition between the conduction band (or a shallow donor) and the V_{Zn} acceptor level would give rise to luminescence around 2.5 eV, in reasonable agreement with the observed transition energy. In addition to the agreement with the observed emission energy, the Zn vacancy is also a likely candidate because it is an acceptor-type defect: acceptor defects are more likely to occur in n-type material, and most ZnO material exhibits unintentional n-type conductivity. This proposed explanation for the green luminescence is similar to the proposal that gallium vacancies are source of the yellow luminescence in GaN.[58]

Oxygen vacancy has also been suggested as source of green luminescence in ZnO.[8,59,60] Vanheusden *et al.* have even reported a correlation between the intensity of the green luminescence and the concentration of oxygen vacancies.[45,61] However, their assessment of the presence of oxygen vacancies was based on the observation of a line with $g \approx 1.96$ in EPR, and, as explained in Section 5.3.5.1, this line has been erroneously associated with oxygen vacancies, undermining their arguments.

The green luminescence has also been assigned to oxygen vacancies based on ODEPR signals of a S=1 center.[62,63] However, it has been argued that oxygen vacancy is unlikely to be related to a S=1 signal.[6] The calculated configuration coordinate diagrams for V_O in Janotti and Van de Walle[14] also do not support the existence of any transition that is consistent with green luminescence.

Finally, Sekiguchi *et al.* have reported strong passivation of the green luminescence by hydrogen plasma treatment.[64] This observation is consistent with the green luminescence being caused by zinc vacancies, which act as acceptors and are likely to be passivated by hydrogen, as discussed in Section 5.4.3.

Table 5.2 *Properties of dopant impurities in ZnO. n_{max} or p_{max} indicates the highest carrier concentration experimentally observed to date*

Impurity	Character	Ionization energy (meV)	n_{max} or p_{max} (cm^{-3})
Al	donor	120 [8]	8×10^{20}[65]
Ga	donor	–	1.1×10^{20}[66]
		–	3.7×10^{20}[67]
In	donor	–	[68,69]
F	donor	80 [8]	5×10^{20}[70]
H	donor	35 [88]	–
Li	acceptor	–	
Cu	acceptor	–	
N	acceptor	100 [109]	9×10^{16}[109]

5.4 Donor Impurities

Properties of dopant impurities in ZnO are summarized in Table 5.2.

5.4.1 Aluminum, Gallium and Indium

The group-III impurities, when substituted on the Zn site, act as shallow donors. For Al, Zhang *et al.*[8] calculated an ionization energy of 120 meV and a low formation energy, referenced to the elemental Al phase. This formation energy would be raised if equilibration with Al_2O_3 were taken into account. Hu and Gordon[65] obtained carrier concentrations up to $8 \times 10^{20}\,cm^{-3}$ with Al doping in chemical vapor deposition of ZnO. For Ga, Ko *et al.*[66] found that high Ga doping (up to $10^{20}\,cm^{-3}$) did not degrade the structural quality of the film, but a degradation in the photoluminescence intensity was observed for Ga concentrations exceeding $2.6 \times 10^{19}\,cm^{-3}$. Indium, finally, has also been used as a donor in ZnO.[68] An EPR study for In in ZnO was carried out by Block *et al.*[69]

5.4.2 Fluorine

Fluorine in ZnO is a shallow donor. Zhang *et al.* calculated an ionization energy of 80 meV and a low formation energy, indicative of easy incorporation into the lattice.[8] Intentional fluorine doping can produce electron concentrations up to $5 \times 10^{20}\,cm^{-3}$.[70] The atomic structure of the fluorine donor is unexpected: for F^+, a large displacement of one of the neighboring Zn atoms is found. This Zn atom moves away from the F atom by 25% of the bond length; the F atom itself moves off-site by 12% of the bond length.[10] The bond between the F atom and one of its Zn neighbors is thus effectively broken. Such large relaxations are usually thought of as giving rise to deep (localized) states, but F in ZnO still behaves as a shallow donor.

5.4.3 Hydrogen

Recent first-principles calculations have drawn attention to the role of hydrogen as a shallow donor in ZnO.[71,72] It was found that interstitial hydrogen behaves as a shallow donor: only the positive charge state (H_i^+) is thermodynamically stable. An electron can be bound to the H_i^+ center in an extended state, of course, characteristic of a shallow donor. The behavior of hydrogen in ZnO is highly unusual. In all other semiconductors studied to date, interstitial hydrogen has been found (theoretically as well as experimentally) to act as an *amphoteric* impurity:[73–75] in p-type material, hydrogen incorporates as H_i^+, and in n-type material as H_i^-, always counteracting the prevailing conductivity of the material. This amphoteric behavior precludes hydrogen from acting as a dopant, i.e., from being a *source* of conductivity of the material. In ZnO, however, interstitial hydrogen occurs exclusively in the positive charge state, i.e., it always acts as a donor.

Interstitial hydrogen in ZnO can be located at the bond-center (BC) site, or the antibonding (AB_O) site, with comparable energies. Hydrogen in ZnO thus prefers sites where it can strongly bind to an oxygen atom, forming an O-H bond with a length of 0.99–1.01 Å. Large lattice relaxations occur around the hydrogen interstitial; in particular, for the BC configuration the Zn atom moves outward over a distance equal to 40% of the bond length (0.8 Å), to a position slightly beyond the plane of its nearest neighbors as shown in Figure 5.5(a). Simultaneously, the O atom moves outwards by 11% of the bond length. For the ABO configuration, the relaxation of both Zn and O amounts to about 20% of the bond length as shown in Figure 5.5(b). As we noted in Section 5.4.2, such large relaxations are not unique to hydrogen; they also occur for fluorine in ZnO.[10]

In addition to the interstitial positions, it was recently found that hydrogen can also replace oxygen in ZnO (H_O), forming a multicenter bond in which H is equally bonded to the four Zn nearest neighbors,[72] as shown in Figure 5.5(c). Substitutional hydrogen H_O is also a shallow donor in ZnO, occurring exclusively in the positive charge state H_O^+.[72] The hydrogen multicenter bond can be understood as a coupling between the fully symmetric state of V_O and the H 1s state, and is located at ~7 eV below the VBM as shown in Figure 5.6(a). The electronic charge density of the hydrogen multicenter bond in ZnO is shown in Figure 5.6(b).

The substitutional and interstitial forms of hydrogen have low formation energies in ZnO as shown in Figure 5.7, indicating that they can occur in significant concentrations. Hydrogen is obviously not the only possible donor in ZnO, but it is a very attractive candidate for an impurity that can be unintentionally incorporated and can give rise to background n-type conductivity. Hydrogen is present in many of the growth techniques used to produce ZnO, either intentionally or unintentionally. These techniques include vapor-phase transport,[76] hydrothermal growth,[77]metal oxide chemical vapor deposition

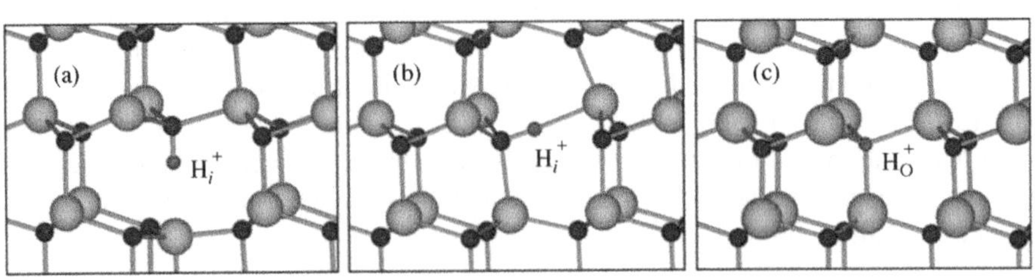

Figure 5.5 *Ball and stick model of the relaxed atomic positions of intersititial hydrogen at the bond-center site parallel to the c-axis in (a), antibonding site perpendicular to the c-axis in (b), and substitutional hydrogen in (c)*

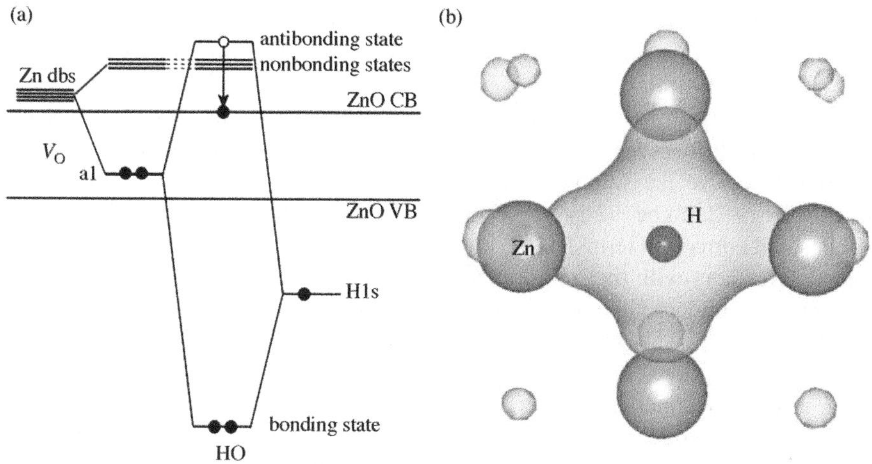

Figure 5.6 *Coupling between the H 1s orbital and the Zn 4s dangling bonds (Zn dbs) to form the hydrogen multicenter bond in ZnO in (a). The H 1s orbital combines with the a1 state and results in a fully symmetric bonding state in the valence band, and an antibonding state in the conduction band. The electron that would occupy this antibonding state is then transferred to the conduction-band minimum, making the substitutional hydrogen H_O a shallow donor. Electronic charge density distributions of the lowest-energy fully symmetric bonding state of the hydrogen multicenter bond in ZnO in (b). The isosurfaces are at 0.05 electrons Å^{-3}. Reprinted from A. Janotti and C. G. Van de Walle, Hydrogen multicentre bonds, Nat. Mat. 6, 44–47. Copyright (2007) Macmillan Publishers Ltd*

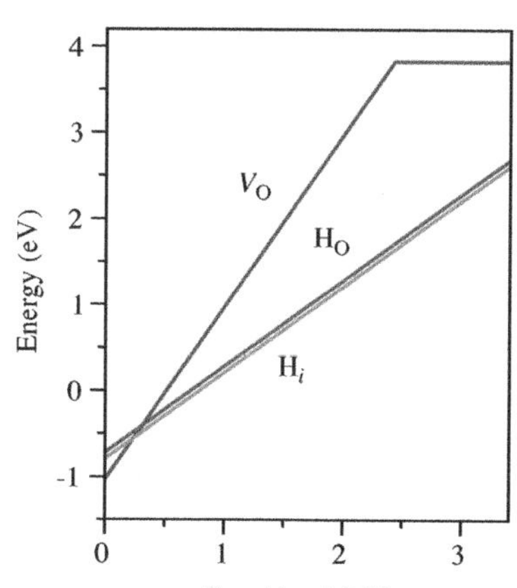

Figure 5.7 *Formation energies as a function of the Fermi level position for substitutional hydrogen H_O, interstitial hydrogen H_i, and the oxygen vacancy V_O in ZnO, under Zn-rich conditions. The slopes correspond to the charge of the stable states. H_O and H_i are both stable in the +1 charge state. The oxygen vacancy is stable only in the +2 and 0 charge states. The Fermi level is referenced to the valence-band maximum. Reprinted from A. Janotti and C. G. Van de Walle, Hydrogen multicentre bonds, Nat. Mat. 6, 44–47. Copyright (2007) Macmillan Publishers Ltd*

(MOCVD),[78] laser ablation[79] and sputtering.[80] In addition, processing steps such as wet etching or annealing in forming gas can easily introduce hydrogen into the material.

Experimental indications for hydrogen's behavior as a donor in ZnO were already reported in the 1950s.[81–83] Thomas and Lander[82] observed an increase in n-type conductivity when H diffuses into ZnO. They used the measured conductivity as a function of temperature to derive the solubility of H in ZnO, and found the heat of the reaction $H_2(gas) \rightarrow 2H^+ + 2e$ to be 3.2 eV, or 1.6 eV per hydrogen. This value should correspond (to within small correction terms) to the formation energy of interstitial H^+ in ZnO, and it is in good agreement with the first-principles calculations.[71,72]

An increase in conductivity upon exposure to H_2 was also observed by Baik *et al.*[84] and by Kohiki *et al.*[85] who introduced hydrogen by proton implantation followed by annealing at 200 °C. Low-energy H implantation of ZnO surfaces also results in formation of strong electron accumulation layers, consistent with H acting as a donor.[86]

The nature of hydrogen as a donor impurity has been microscopically established in recent experiments. Muon spin rotation is a technique similar to EPR, based on muonium, which is a pseudo-isotope of hydrogen. Muonium in ZnO was observed to exhibit all characteristics of a shallow donor (including ionization behavior consistent with a level close to the conduction band, and a delocalized wavefunction), confirming that H acts as a shallow donor.[87] EPR has also resulted in a direct observation of hydrogen in ZnO, with behavior consistent with a shallow donor.[88] Hydrogen was identified as one of two residual donors in commercial ZnO samples; the presence of two donors in this material is consistent with Hall measurements. Hydrogen was found to correspond to the donor labeled *D*1, which has an ionization energy of 35 meV. The involvement of hydrogen in the structure of the *D*1 donor was confirmed by electron-nuclear double resonance (ENDOR). The *D*1 shallow donor produces a line with *g* values $g_{\parallel} = 1.9569$ and $g_{\perp} = 1.9552$, consistent with the shallow donor signals discussed in Section 5.3.5.1. It is an interesting question whether these experiments observe the interstitial and/or the substitutional species. Because of the way the muons are introduced in the ZnO material, muon spin rotation probably observes the equivalent of isolated interstitial hydrogen.

Infrared (IR) spectroscopy has identified two lines corresponding to interstitial hydrogen in hydrogenated ZnO single crystals: one at 3326 cm^{-1} associated with the stretching H-O mode with H at the antibonding site,[89,90] shown in Figure 5.5(b); and another at 3611 cm^{-1} associated with H at the bond center site.[91] Shi *et al.* recently showed that samples that were grown by a pressurized melt growth technique showed a strong 3326 cm^{-1} line and a weak 3611 cm^{-1} line, and vice versa for samples grown by seeded vapor transport technique.[92] The origin of these differences has been discussed by McCluskey and Jokela.[93] On the other hand, the identification of substitutional hydrogen by IR spectroscopy is much more difficult. The H_O vibration modes occur in the range of 750–1000 cm^{-1};[72] in this spectral region the current ZnO samples are opaque due to strong IR absorption by free carriers.

The experiments of Shi *et al.* indicate that hydrogen is stable in ZnO up to ~500 °C,[92] a result that is not consistent with the presence of the highly mobile interstitial hydrogen.[94,95] However, the temperature dependence is consistent with the calculated migration barrier of 2.5 eV for substitutional hydrogen.[72] Moreover, H_O can also explain the observed dependence of the electrical conductivity on oxygen partial pressure, a dependence that, as we now know, has been erroneously attributed to the presence of oxygen vacancies.[96]

5.5 Acceptor Impurities

5.5.1 Lithium

Lithium may behave both as a donor and as an acceptor in ZnO.[97] The donor behavior arises when lithium occurs as an interstitial impurity; the acceptor behavior is exhibited when lithium substitutes on a Zn site. Kolb and Laudise[98] reported that incorporation of Li during growth can compensate the background n-type conductivity of hydrothermally grown ZnO, provided a post-growth heat treatment is applied. The heat treatment was considered necessary to achieve outdiffusion of zinc interstitials, but may also result in outdiffusion of Li interstitials as well as hydrogen donors. EPR studies of Li in ZnO were carried out by Kasai[34] and by Schirmer.[99] For the axial configuration, the measured g values were $g_{\parallel} = 2.0028$ and $g_{\perp} = 2.0253$.[99] Lithium has been used to compensate n-type doping; no p-type doping using Li has been reported. Computational investigation of substitutional and interstitial Li have been reported in the literature.[100–103]

5.5.2 Copper

Copper acts as a deep acceptor in ZnO. It can be used to reduce the carrier concentration of n-type ZnO.[104] The electronic structure of the Cu impurity is complicated by the fact that a hole can be formed in the Cu 3d shell.[105] Copper has been mentioned as a source of green luminescence.[56,57]

5.5.3 Nitrogen

Nitrogen is a shallow acceptor in other II–VI semiconductors[1] and has been considered as a suitable p-type dopant for ZnO for some time.[106] Minegishi *et al.*[107] have reported p-type doping of ZnO by chemical vapor doposition, using NH_3 as the nitrogen source, resulting in a carrier concentration of $1.5 \times 10^{16}\,cm^{-3}$. They estimated an ionization energy of 100 meV. The authors pointed out that hydrogen may play a role in the nitrogen incorporation. Thonke *et al.*[108] have reported an acceptor binding energy of 195 meV, based on an analysis of donor–acceptor pair transitions observed in photoluminescence on ZnO crystals that were not intentionally doped (but are likely to contain N). Carlos *et al.*[46] have reported EPR signals for N in ZnO, with g values $g_{\parallel} = 1.9953$ and $g_{\perp} = 1.9633$. These seem to be in reasonable agreement with the observations of Garces *et al.*[47] who reported $g_{\parallel} = 1.9948$ and $g_{\perp} = 1.9632$.

Look *et al.*[109] have investigated the electrical and optical properties of nitrogen-doped ZnO grown by molecular beam epitaxy on a Li-diffused semi-insulating ZnO substrate. Their results indicated a hole density of $9 \times 10^{16}\,cm^{-3}$ and mobility of $2\,cm^2\,V^{-1}\,s^{-1}$, and suggested an acceptor activation energy of 170–200 meV. This relatively high activation energy is expected from the fact that N is less electronegative than O. More recently, Tsukazaki *et al.*[110] reported blue light emission from a p-n ZnO homojunction using nitrogen as the acceptor in the p-type layer. However, it is important to note that the band gap of ZnO is 3.4 eV and, as such, one would expect UV light emission from a ZnO p-n junction. The reliability and reproducibility of p-type doping in ZnO are still major issues. Compensation of nitrogen substitutional acceptors by intrinsic defects as well as by nitrogen incorporated in different configurations has been addressed in Lee *et al.*[11]

5.5.4 Phosphorous, Arsenic and Antimony

Phosphorous and arsenic have been found to introduce *deep* acceptor states in ZnSe,[111] so the prospects for these impurities to act as shallow acceptors in ZnO are not good. Antimony is expected to result in even deeper states. Aoki *et al.*[112] have reported ZnO diodes in which a p-type layer was created by excimer laser doping from a Zn_3P_2 compound. No details about the acceptors were reported. Ryu *et al.* have investigated arsenic as an acceptor in ZnO grown on GaAs(001).[113] Hall measurements indicated p-type conductivity, but it was pointed out that large uncertainties resulted due to contributions from interference layers between the ZnO film and the GaAs substrate. From optical measurements an ionization energy of 100 meV was deduced. Xiu *et al.* reported results of Sb-doped ZnO grown on Si, with a relatively high hole concentration of $2 \times 10^{18}\,cm^{-3}$.[114] The same group recently reported results for a p-n homojunction using Sb as p-type dopant, with a turn-on voltage of 6 eV.[115] Based on first-principles calculations, Limpijumnong *et al.* proposed that these large anions (As, Sb) occupy Zn lattice sites and form complexes with two nearby Zn vacancies.[116] A similar model has also been proposed for P in ZnO by Lee *et al.*[117] These calculations indicate that these complexes are shallow donors with ionization energies less than 100 meV. However, the formation of these complexes would not be favorable by entropic considerations, as discussed in Janotti and Van de Walle.[6]

5.5.5 Co-Doping

Co-doping has been suggested to be an effective method of achieving p-type conductivity in ZnO. The term co-doping means that, along with the acceptors that are incorporated to produce holes, donors are also incorporated during the growth. At first sight, this would lead merely to compensation. In fact, compensation during the growth is actually quite desirable, since it shifts the Fermi level away from the VBM towards the middle of the gap. This results in a lowering of the formation energy of acceptors (and hence in an increase of the acceptor solubility), as well as an increase in the formation energy of compensating donor-type native defects (such as V_O). However, the compensation by the intentionally introduced donor will persist after growth, and the material will not exhibit p-type conductivity.

One potential strategy for overcoming this limitation is to remove the donor after growth. This is only possible with donor impurities that are not strongly bound and that exhibit a sufficiently high diffusivity, so that they can be removed from the vicinity of the acceptors during an anneal at modest temperatures (to avoid formation of other compensating defects). It has been proposed that hydrogen may be a candidate for such a donor.[8,10] Indeed, hydrogen plays such a role in p-type doping of GaN, where it enhances the solubility of Mg acceptors and suppresses compensation by nitrogen vacancies.[118]

Whether this type of dopant engineering also works in the case of ZnO will depend on the binding and dissociation energies of N-H complexes, and on the barriers that need to be overcome to remove H from the vicinity of the acceptors during a post-growth anneal. The potential beneficial aspects of simultaneous incorporation of hydrogen along with nitrogen acceptors were noted by Minegishi *et al.*[107]

Another type of co-doping has been proposed by Yamamoto and Katayama-Yoshida.[119] This proposal is based on the incorporation of complexes consisting of two acceptors and one donor. The authors claimed that such complexes would result in higher hole concentrations due to an enhancement of the solubility and a lowering of the ionization energy. These conclusions were qualitatively based on first-principles calculations but formation energies and ionization energies have not been calculated. Explicit calculations for similar situations, i.e., for acceptor doping of GaN[120] and of CdTe[121] have in fact indicated that co-doping with such complexes is unlikely to produce favorable results.

Experimentally, Joseph *et al.* have reported co-doping of ZnO using N (in the form of N_2O) with Ga as a co-dopant.[122] These results have not been reproduced. Yan *et al.* have proposed an alternative interpretation of the results, in terms of the chemical potentials of the gases used during growth.[123]

5.6 Isoelectronic Impurities

Addition of isoelectronic impurities (such as Mg or Cd, which substitute on the Zn site) is mainly pursued in the context of alloy formation for band-gap engineering, a topic that is beyond the scope of the currrent chapter. An interesting case that has received a lot of attention is that of Mn. Indeed, Dietl *et al.* have predicted ferromagnetism with a very high Curie temperature in p-type Mn-doped ZnO.[124] Many experimental and theoretical investigations of Mn-doped ZnO have been reported, however, the conclusions about room temperature ferromagnetism are still highly controversial.[125–129]

Acknowledgements

Thanks are due to C. Litton, L. Halliburton, J. McCaldin, M. D.McCluskey, J. Neugebauer, T. Ive, J. Speck and G. D. Watkins for valuable discussions. This work was supported by the NSF MRSEC Program under award No. DMR05-20415 and by the UCSB Solid State Lighting and Energy Center.

References

[1] C. G. Van de Walle, D. B. Laks, G. F. Neumark and S. T. Pantelides, *Phys. Rev. B* **47**, 9425 (1993).
[2] C. G. Van de Walle and J. Neugebauer, *J. Appl. Phys.* **95**, 3851 (2004).
[3] J. A. Dean, *Lange's Handbook of Chemistry*, 14th Edn, McGraw–Hill, Inc., New York, 1992.

[4] P. Hohenberg and W. Kohn, *Phys. Rev.* **136**, B864 (1964); W. Kohn and L. J. Sham, *Phys. Rev.* **140**, A1133 (1965).
[5] M. Bockstedte, A. Kley, J. Neugebauer and M. Scheffler, *Comp. Phys. Commun.* **107**, 187 (1997).
[6] A. Janotti and C. G. Van de Walle, *Phys. Rev. B* **76**, 165202 (2007).
[7] A. F. Kohan, G. Ceder, D. Morgan and C. G. Van deWalle, *Phys. Rev. B* **61**, 15019 (2000).
[8] S. B. Zhang, S.-H. Wei and A. Zunger, *Phys. Rev. B* **63**, 075205 (2001).
[9] F. Oba, S. R. Nishitani, S. Isotani, H. Adachi and I. Tanaka, *J. Appl. Phys.* **90**, 824 (2001).
[10] C. G. Van de Walle, *Physica B* **308–310** 899 (2001).
[11] E.-C. Lee, Y.-S. Kim, Y.-G. Jin and K. J. Chang, *Phys. Rev. B* **64**, 085120 (2001).
[12] P. Erhart, A. Klein and K. Albe, *Phys. Rev. B* **72**, 085213 (2005).
[13] P. Erhart and K. Albe, *Phys. Rev. B* **70**, 115207 (2006).
[14] A. Janotti and C. G. Van de Walle, *Appl. Phys. Lett.* **87**, 122102 (2005).
[15] A. Janotti and C. G. Van de Walle, *J. Cryst. Growth* **287**, 58 (2006).
[16] C. H. Patterson, *Phys. Rev. B* **74**, 144432 (2006).
[17] S. Lany and A. Zunger, *Phys. Rev. B* **72**, 035215 (2005).
[18] S. Lany and A. Zunger, *Phys. Rev. Lett.* **98**, 045501 (2007).
[19] T. R. Paudel and W. R. L. Lambrecht, *Phys. Rev. B* **77**, 205202 (2008).
[20] F. Oba, A. Togo, I. Tanaka, J. Paier and G. Kresse, *Phys. Rev. B* **77**, 245202 (2008).
[21] A. Janotti, D. Segev and C. G. Van deWalle, *Phys. Rev. B* **74**, 045202 (2006).
[22] L. S. Vlasenko and G. D. Watkins, *Phys. Rev. B* **71**, 125210 (2005).
[23] S. M. Evans, N. C. Giles, L. E. Halliburton and L. A. Kappers, *J. Appl. Phys.* **103**, 043710 (2008).
[24] D. G. Thomas, *J. Phys. Chem. Solids* **3**, 229 (1957).
[25] A. R. Hutson, *Phys. Rev.* **108**, 222 (1957).
[26] D. C. Look, J. W. Hemsky and J. R. Sizelove, *Phys. Rev. Lett.* **82**, 2552 (1999).
[27] D. C. Look, G. C. Farlow, P. Reunchan, S. Limpijumnong, S. B. Zhang and K. Nordlund, *Phys. Rev. Lett.* **95**, 225502 (2005).
[28] A. L. Taylor, G. Filipovich and G. K. Lindeberg, *Solid State Commun.* **8**, 1359 (1970).
[29] D. Galland and A. Herve, *Phys. Lett.* **33A**, 1 (1970).
[30] G. Neumann,in Current Topics in Materials Science, edited by E. Kaldis, North Holland, Amsterdam, 1981. Vol. 7, p. 279.
[31] G.W. Tomlins, J. L. Routbort and T. O. Mason, *J. Appl. Phys.* **87**, 117 (2000).
[32] G. W. Tomlins and J. L. Routbort, *J. Am. Ceram. Soc.* **81**, 869 (1998).
[33] J. Schneider and A. Räuber, *Z. Naturforsch.* **16a**, 712 (1961).
[34] P. H. Kasai, *Phys. Rev.* **130**, 989 (1963).
[35] K. A. Müller and J. Schneider, *Phys. Lett.* **4**, 288 (1963).
[36] C. H. Geisler and G. L Simmons, *Phys. Lett.* **11**, 111 (1964).
[37] R. B. Lal and G. M. Arnett, *J. Phys. Soc. Jpn.* **21**, 2743 (1966).
[38] K. M. Sancier, *Surf. Sci.* **21**, 1 (1970).
[39] A. Hausmann, *Z. Phys.* **237**, 86 (1970).
[40] G. K. Born, A. B. Hofstaetter, A. O. Scharmann, G. M. Arnett, R. L. Kroes and U.E. Wegner, *Phys. Status Solidi A* **4**, 675 (1971).
[41] T. Mookherji, Phys. *Status Solidi A* **13**, 293 (1972).
[42] K. Hoffmann and D. Hahn, *Phys. Status Solidi A* **24**, 637 (1974).
[43] A. Pöppl and G. Völkel, *Phys. Status Solidi* **115**, 247 (1989).
[44] A. Pöppl and G. Völkel, *Phys. Status Solidi* **121**, 195 (1990).
[45] K. Vanheusden, C. H. Seager, W. L. Warren, D. R. Trallant, J. Caruso, M. J. Hampden-Smith and T. T. Kodas, *J. Lumin.* **75**, 11 (1997).
[46] W. E. Carlos, E. R. Glaser and D. C. Look, *Physica B* **308–310** 976 (2001).
[47] N. Y. Garces, N. C. Giles, L. E. Halliburton, G. Cantwell, D. B. Eason, D. C. Reynolds and D. C. Look, *Appl. Phys. Lett.* **80**, 1334 (2002).
[48] J. M. Smith and W. E. Vehse, *Phys. Lett. A* **31**, 147 (1970).
[49] C. Gonzalez, D. Galland, and A. Herve, *Phys. Status Solidi B* **72**, 309 (1975).
[50] V. Soriano and D. Galland, *Phys. Status Solidi B* **77**, 739 (1976).
[51] K. Tarkpea, A. Ots and V. Nikitenko, *J. Phys. Chem. Solids* **55**, 1353 (1994).

[52] G. Neumann, in *Current Topics in Materials Science*, edited by E. Kaldis, North Holland, Amsterdam, 1981, Vol. 7, p. 269.
[53] D. C. Reynolds, D. C. Look, B. Jogai and H. Morkoç, *Solid State Commun.* **101**, 643 (1997).
[54] D. C. Reynolds, D. C. Look, B. Jogai, J. E. Van Nostrand, R. Jones and J. Jenny, *Solid State Commun.* **106**, 701 (1998).
[55] R. B. Lauer, *J. Phys. Chem. Solids* **34**, 249 (1973).
[56] R. Dingle, *Phys. Rev. Lett.* **23**, 579 (1969).
[57] K. C. Mishra, P. C. Schmidt, K. H. Johnson, B. G. DeBoer, J. K. Berkowitz and E.A. Dale, *Phys. Rev. B* **42**, 1423 (1990).
[58] J. Neugebauer and C. G. Van de Walle, *Appl. Phys. Lett.* **69**, 503 (1996).
[59] F. A. Kröger and H. J. Vink, *J. Chem. Phys.* **22**, 250 (1954).
[60] S. A. Studenikin, N. Golego and M. Cocivera, *J. Appl. Phys.* **84**, 2287 (1998).
[61] K. Vanheusden, W. L. Warren, C. H. Seager, D. R. Trallant and J. A. Voigt, *J. Appl. Phys.* **79**, 7983 (1996).
[62] F. H. Leiter, H. R. Alves, A. Hofstaetter, D. M. Hofmann and B. K. Meyer, *Phys. Status Solidi B* **226**, R4 (2001).
[63] F. Leiter, H. Alves, D. Pfisterer, N. G. Romanov, D. M. Hofmann and B. K. Meyer, *Physica B* **340–342** 201 (2003).
[64] T. Sekiguchi, N. Ohashi and Y. Terada, *Jpn. J. Appl. Phys.* **36**, **L289** (1997).
[65] J. Hu and R. G. Gordon, *J. Appl. Phys.* **71**, 880 (1992).
[66] H. J. Ko, Y. F. Chen, S. K. Hong, H. Wenisch, T. Yao and D. C. Look, *Appl. Phys. Lett.* **77**, 3761 (2000).
[67] J. Hu and R. G. Gordon, *J. Appl. Phys.* **72**, 5381 (1992).
[68] J. Hu and R. G. Gordon, *Mater. Res. Soc. Symp. Proc.* **283**, 891 (1993).
[69] D. Block, A. Hervé and R. T. Cox, *Phys. Rev. B* **25**, 6049 (1982).
[70] J. Hu and R. G. Gordon, *Solar Cells* **30**, 437 (1991).
[71] C. G. Van de Walle, *Phys. Rev. Lett.* **85**, 1012 (2000).
[72] A. Janotti and C. G. Van de Walle, *Nature Mater.* **6**, 44 (2007).
[73] J. I. Pankove and N. M. Johnson(Eds), *Hydrogen in Semiconductors, Semiconductors and Semimetals*, Academic Press, Boston, 1991, Vol. **34**.
[74] C. G. Van de Walle and N. M. Johnson, in *Gallium Nitride (GaN) II,* edited by J. I. Pankove and T. D. Moustakas, *Semiconductors and Semimetals*, Academic Press, Boston, 1998, Vol. **57**, p. 157.
[75] C. G. Van de Walle and J. Neugebauer, *Annu. Rev. Mater. Res.* **36**, 179 (2006).
[76] D. C. Look, D. C. Reynolds, J. R. Sizelove, R. L. Jones, C. W. Litton, G. Cantwell and W. C. Harsch, *Solid State Commun.* **105**, 399 (1998).
[77] M. Suscavage, M. Harris, D. Bliss, P. Yip, S.-Q. Wang and D. Schwall, *MRS Internet J. Nitride Semicond. Res.* **4S1**, G3.40 (1999).
[78] S. Y. Myong, S. J. Baik, C. H. Lee, W. Y. Cho and K. S. Lim, *Jpn. J. Appl. Phys.* **36**, L1078 (1997).
[79] H. Kordi Ardakani, *Thin Solid Films* **287**, 280 (1996).
[80] A. Valentini, F. Quaranta, M. Penza and F. R. Rizzi, *J. Appl. Phys.* **73**, 1143 (1993).
[81] E. Mollwo, *Z. Physik* **138**, 478 (1954).
[82] D. G. Thomas and J. J. Lander, *J. Chem. Phys.* **25**, 1136 (1956).
[83] J. J. Lander, *J. Phys. Chem. Solids* **3**, 87 (1957).
[84] S. J. Baik, J. H. Jang, C. H. Lee, W. Y. Cho and K.S. Lim, *Appl. Phys. Lett.* **70**, 3516 (1997).
[85] S. Kohiki, M. Nishitani, T. Wada and T. Hirao, *Appl. Phys. Lett.* **64**, 2876 (1994).
[86] G. Yaron, J. Levy, A. Many, Y. Goldstein, S. Z. Weisz and O. Resto, *J. Phys. Chem. Solids* **47**, 401 (1986).
[87] S. F. J. Cox, E. A. Davis, S. P. Cottrell, P. J. C. King, J. S. Lord, J. M. Gil, H. V. Alberto, R. C. Vilo, J. P. Duarte, N. A. de Campos, A. Weidinger, R. L. Lichti and S. J. C. Irvine *Phys. Rev. Lett.* **86**, 2601 (2001).
[88] D. M. Hofmann, A. Hofstaetter, F. Leiter, H. Zhou, F. Henecker, B. K. Meyer, S. B. Orlinskii, J. Schmidt and P. G. Baranov, *Phys. Rev. Lett.* **88**, 045504 (2002).
[89] M.D. McCluskey, S.J. Jokela, K.K. Zhuravlev, P.J. Simpson and K.G. Lynn, *Appl. Phys. Lett.* **81**, 3807 (2002).

[90] S.J. Jokela and M.D. McCluskey, *Phys. Rev. B* **72**, 113201 (2005).
[91] E.V. Lavrov, J. Weber, F. Boörrnert, C.G. Van de Walle and R. Helbig, *Phys. Rev. B* **66**, *165205* (2002).
[92] G. A. Shi, M. Stavola, S.J. Pearton, M. Thieme, E.V. Lavrov and J. Weber, *Phys.Rev. B* **72**, 195211 (2005).
[93] M. D. McCluskey and S. J. Jokela, *Physica B* **401402**, 355 (2007).
[94] M. G. Wardle, J. P. Goss and P. R. Briddon, *Phys. Rev. Lett.* **96**, 205504 (2006).
[95] D. G. Thomas and J. J. Lander, *J. Chem. Phys.* **25**, 1136 (1956).
[96] F. A. Kröger, *The Chemistry of Imperfect Crystals*, North-Holland, Amsterdam, 1974.
[97] J. J. Lander, *J. Phys. Chem. Solids* **15**, 324 (1960).
[98] E. D. Kolb and R. A. Laudise, *J. Elctrochem. Soc.* **48**, 344 (1965).
[99] O. F. Schirmer, *J. Phys. Chem. Solids* **29**, 1407 (1968).
[100] C. H. Park, S. B. Zhang and S. H. Wei, *Phys. Rev. B* **66**, 073202 (2002).
[101] M. G. Wardle, J. P. Goss and P. R. Briddon, *Phys. Rev. B* **71**, 155205 (2005).
[102] E.-C. Lee and K. J. Chang, *Physica B* **376**, 707 (2006).
[103] Y. J. Zeng, Z. Z. Ye, J. G. Lu, W. Z. Xu, L. P. Zhu, B. H. Zhao and S. Limpijumnong, *Appl. Phys. Lett.* **89**, 042106 (2006).
[104] G. Müller, *Phys. Status Solidi B* **76**, 525 (1976).
[105] R. E. Dietz, H. Kamimura, M. D. Sturge and A. Yariv, *Phys. Rev.* **132**, 1559 (1963).
[106] A. Kobayashi, O. F. Sankey and J. D. Dow, *Phys. Rev. B* **28**, 946 (1983).
[107] K. Minegishi, Y. Koiwai, Y. Kikucji, K. Yano, M. Kasuga and A. Shimizu, *Jpn. J. Appl. Phys.* **36**, **L1453** (1997).
[108] K. Thonke, Th.Gruber, N. Teofilov, R. Schöfelder, N. Kerwien, A. Waag and R. Sauer, *Physica B* **308**, 945 (2001).
[109] D. C. Look, D. C. Reynolds, C. W. Litton, R. L. Jones, D. B. Eason and G. Cantwell, *Appl. Phys. Lett.* **81**, 1830 (2002).
[110] A. Tsukazaki, A. Ohtomo, T. Onuma, M. Ohtani, T. Makino, M. Sumiya, K. Ohtani, S. F. Chichibu, S. Fuke, Y. Segawa, H. Ohno, H. Koinuma and M. Kawasaki, *Nat. Mater.* **4**, 42 (2005).
[111] D. J. Chadi, *Appl. Phys. Lett.* **59**, 3589 (1991).
[112] T. Aoki, Y. Hatanaka and D. C. Look, *Appl. Phys. Lett.* **76**, 3257 (2000).
[113] Y. R. Ryu, S. Zhu, D. C. Look, J. M. Wrobel, H. M. Yeong and H. W. White, *J. Cryst. Growth* **216**, 330 (2000).
[114] F. X. Xiu, Z. Yang, L. J. Mandalapu, D. T. Zhao and J. L. Liu, *Appl. Phys. Lett.* **87**, 152101 (2005).
[115] S. Chu, J. H. Lim, L. J. Mandalapu, Z. Yang, L. Li and J. L. Liu, *Appl. Phys. Lett.* **92**, 152103 (2005).
[116] S. Limpijumnong, S. B. Zhang, S.-H. Wei and C. H. Park, *Phys. Rev. Lett.* **92**, 155504 (2004).
[117] W. J. Lee, J. Kang and K. J. Chang, *Phys. Rev. B* **73**, 024117 (2006).
[118] J. Neugebauer and C. G. Van de Walle, *Appl. Phys. Lett.* **68**, 1829 (1996).
[119] T. Yamamoto and H. Katayama-Yoshida, *Jpn. J. Appl. Phys.* **38**, **L166** (1999).
[120] C. G. Van de Walle, S. Limpijumnong and J. Neugebauer, *Phys. Rev. B* **63** 245205 (2001).
[121] S. B. Zhang, S.-H. Wei and Y. Yan, *Physica B* **302–303** 135 (2001).
[122] M. Joseph, H. Tabata and T. Kawai, *Jpn. J. Appl. Phys.* **38**, L1205 (1999).
[123] Y. Yan, S. B. Zhang and S. T. Pantelides, *Phys. Rev. Lett.* **86**, 5723 (2001).
[124] T. Dietl, H. Ohno, F. Matsakura, J. Clibert and D. Ferrand, *Science* **287**, 1019 (2000).
[125] K. Sato and H. Katayama-Yoshida, *Jpn. J. Appl. Phys.* **39**, L555 (2000).
[126] T. Fukumura, Z. Jin, M. Kawasaki, T. Shono, T. Hasegawa, S. Koshihara and H. Koinuma, *Appl. Phys. Lett.* **78**, 958 (2001).
[127] P. Sharma, A. Gupta, K. V. Rao, F. J. Owens, R. Sharma, R. Ahuja, J. M. O. Guillen, B. Johansson and G. A. Gehring, *Nat. Mater.* **2**, 673 (2003).
[128] G. Lawes, A. S. Risbud, A. P. Ramirez and R. Seshadri, *Phys. Rev. B* **71**, 045201 (2005).
[129] L. Petit, T. C. Schulthess, A. Svane, Z. Szotek, W. M. Temmerman and A. Janotti, *Phys. Rev. B* **73**, 045107 (2006).

6

Spectral Identification of Impurities and Native Defects in ZnO

B.K. Meyer[1], *D.M. Hofmann*[1], *J. Stehr*[1] *and A. Hoffmann*[2]
[1]*Physics Institute, Justus Liebig University Giessen, Giessen, Germany*
[2]*Institute of Solid State Physics, Technical University Berlin, Berlin, Germany*

6.1 Introduction

This chapter deals with the spectral identification of impurities and defects in ZnO. High resolution luminescence and magnetic resonance spectroscopy are employed to study the role of extrinsic and intrinsic defects in ZnO. The neutral donor bound exciton transitions are studied and identified. The most common donors stem from the group-III elements Al, Ga and In. Hydrogen as a shallow donor – quite uncommon in compound semiconductors – has the lowest binding energy. The doping with nitrogen and arsenic is discussed and the different roles of the acceptors in the optical recombinations are presented. Magnetic resonance adds valuable information to the role of the shallow donors in ZnO and often allows for an unambiguous identification. The situation is more controversial for the deep, intrinsic centers such as the oxygen and zinc vacancy, their contribution to the deep luminescence bands in the green yellow spectral range is discussed. For the acceptors N, Li and Na magnetic resonance gives clear evidence that they act as deep centers, and there is up to now no evidence for shallow acceptors, at least from magnetic resonance.

Zinc Oxide Materials for Electronic and Optoelectronic Device Applications, First Edition.
Edited by Cole W. Litton, Donald C. Reynolds and Thomas C. Collins.

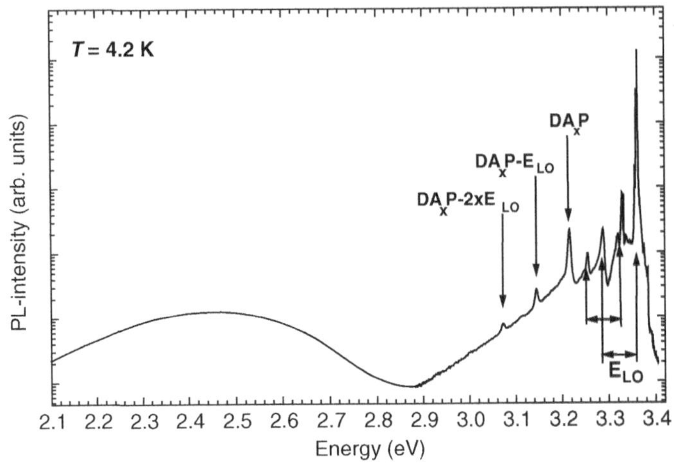

Figure 6.1 *Photoluminescence spectrum of bulk ZnO showing excitonic, donor acceptor pair (DAP) and deep level emission. The corresponding phonon replica with longitudinal optical phonons (LO) are indicated. Reprinted from B.K. Meyer, et al., Bound exciton and donor–acceptor pair recombinations in ZnO, phys. stat. sol. (b) 241, 231–260. Copyright (2004) with permission from WILEY-VCH Verlag GmbH & Co. KGaA, Weinheim*

6.2 Optical Spectroscopy

6.2.1 Excitons Bound to Shallow Donors

ZnO is a direct band gap semiconductor which crystallizes in the wurtzite symmetry. The valence band is split by crystal field and spin orbit interaction into three states named A, B and C. The symmetry of the upper valence subband (A-subband) in ZnO has been the subject of controversy (Γ_9 or Γ_7 character) for more than 40 years.[1–11] Based on the polarization properties of the free exciton transitions, most of the authors assume that the symmetry of the A-valence subband is Γ_7.[1,2,4–8,12–14] At low cryogenic temperatures bound exciton emission is the dominant radiative channel, whereas at higher temperatures free exciton emission usually takes over. Up to eleven excitonic recombinations where excitons bind to ionized, neutral donors and/or acceptors have been observed.[1,2,15–17] As there was no definite assignment of the bound exciton recombinations to the chemical nature of the donors or acceptors, they are numbered I_0 to I_{11} in the early work of Reynolds *et al.*[15] There was considerable dispute on these assignments. Based on magneto-photoluminescence experiments, Gutowski *et al.*[12] attributed all recombinations from I_5 to I_{10} to acceptor bound excitons in conflict with the early magneto-absorption data of Loose *et al.*[14] and with recently published experiments[17] which showed that I_4 to I_8 are neutral donor bound excitons.

The luminescence from high quality bulk ZnO crystals extends from the band edge to the green/orange spectral range (an overview spectrum is shown in Figure 6.1). Very common is a broad band centered around 2.45 eV extending from the blue into the green range, where different impurities and defects contribute (see below). The lines dominating originate from bound exciton (BE) recombinations [excitons bound to ionized (D^+X),

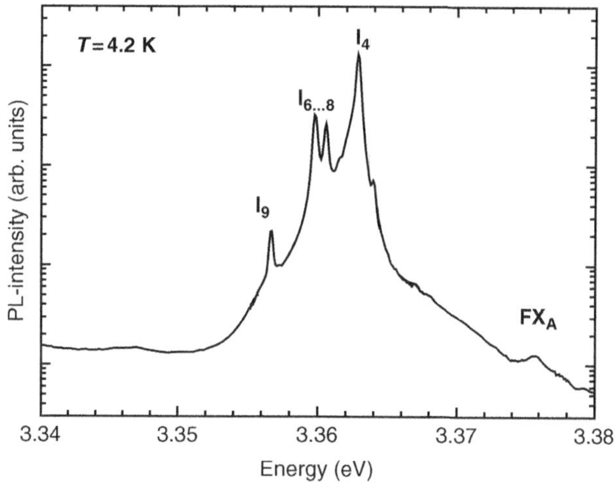

Figure 6.2 *Photoluminescence spectrum in the excitonic range.* FX_A *is the transversal free exciton,* I_4*–*I_9 *are bound exciton lines*

neutral donors (D^oX) and/or acceptors (A^oX)] followed by longitudinal optical (LO) phonon replicas with an energy separation of 72 meV. In some samples a donor-acceptor-pair (DAP) transition is found, the chemical identity of the acceptor A_x is unknown. The transition energy is around 3.22 eV again followed by phonon replicas (see Figure 6.1). The prominent lines in the bulk ZnO are the bound excitons positioned at 3.3628, 3.3608 and 3.3598 eV (labeled in Figure 6.2 as I_4, $I_{6..8}$). At 3.357 eV another bound exciton (I_9) can be observed. At lower energies from 3.34 to 3.31 eV further recombination lines appear. It is the region where the two-electron satellite (TES) recombination lines of the neutral donor bound excitons show up.[18,19] During the recombination of an exciton bound to a neutral donor the donor final state can be the 1s state (normal D^oX line) or the 2s, 2p state (TES line). The energetic distance between the D^oX and its TES is consequently the difference between the donor energies in the 1s and 2p states, which is three-quarters of the donor binding energy (E_D) in the hydrogenic effective mass approach (EMA). Therefore, by determining the position of the TES the related donor binding energy is obtained with high precision.

From luminescence experiments on ZnO the localization energies E_L are determined as the difference between the recombination energy of the respective bound exciton lines and the free transversal exciton line (see Figure 6.1). In many semiconductors the localization energies E_L of excitons bound to ionized (D^+) and neutral shallow donors (D^0) and acceptors (A^0) follow a certain ordering e.g. $E_L(D^+X) < E_L(D^0X) < E_L(A^0X)$ as found in CdS[20] and is the case for ZnO[21] or $E_L(D^0X) < E_L(D^+X) < E_L(A^0X)$ as found in ZnSe[22] and proposed for GaN.[23] This sequence can be understood as a result of the electron to hole mass ratio $\sigma = m^*_e/m^*_h$ which has a characteristic value for each semiconductor. From this mass ratio one can theoretically predict the localization energies of the bound excitons as a function of the binding energies of the shallow donors E_D and acceptors E_A. In addition it also answers the question on the stability of the ionized donor bound exciton complexes.

Magnetic resonance showed that one of the shallow donors (I_4) is hydrogen. Based on electron nuclear double resonance (ENDOR) experiments the hyperfine interaction with a single H-nucleus was resolved.[24] Experiments showed that I_4 annealed out completely at temperatures between 650 °C and 800 °C depending on the annealing time. Electron paramagnetic resonance (EPR) showed a distinct reduction in the signal amplitude of the neutral donor resonance together with a disappearance of the hydrogen-related ENDOR signal. I_4 and the hydrogen donor are typical for ZnO bulk crystals grown by the hydrothermal and seeded vapor transport methods. It is, however, absent in crystals grown from the vapor phase.

As to $I_{6/6a}$ we can refer to the implantation studies of Schilling *et al.*[25] They showed (see Figure 6.3) that upon implantation of Al and successive annealing the $I_{6/6a}$ line gained in intensity and was the strongest for Al concentrations above $9 \times 10^{16}\,cm^{-3}$. Aluminum seems to be an omnipresent impurity in vapor grown ZnO as already outlined by Gonzales *et al.*[26]

In Ga-doped epitaxial films, the prominent excitonic recombinantion is I_8. There has been a report published on Ga-doped ZnO by the Sendai group.[27] In agreement with the findings in Meyer *et al.*[17] and Reuss *et al.*[28] they showed that the exciton bound to neutral Ga-donor recombination occurs at 3.359 eV. I_8 always dominates in epitaxial films on GaN templates. Secondary ion mass spectrometry (SIMS) experiments showed a severe interdiffusion of Ga from the GaN template into the ZnO epitaxial film.

I_9 could be identified with the donor In based on diffusion[17] and implantation experiments.[29] Müller *et al.* implanted radioactive ^{111}In into Zn bulk single crystals, and In occupies substitutional Zn lattice sites after annealing at 700 °C. They monitored the photoluminescence while the donor In decayed into the stable and isoelectronic Cd;[29] the decay of the I_9 line is shown in Figure 6.4.

For the donors which are responsible for the I_4 to I_{10} recombinations we can deduce the 1s to 2p transition energy from the location of the two-electron satellites. Transitions to higher excited states e.g. $n=3$, $n=4$, ... were not observed (see Figure 6.5).

In a simple hydrogen-like EMA the energy separation between $n=1$ and $n=2$ states would be equal to three-quarters of the donor binding energy R^*. A short range chemical potential of the impurity would affect only the states of the S symmetry thus leading to the chemical shift of the real donor binding energy E_D from the effective mass value R^* and to the chemical shift of the 2S state. In this case, the donor energy E_D can be estimated as $(E_{2p}-E_{1s}) + {}^1\!/_4 R^*$. However, the 2S and 2P states in polar hexagonal semiconductors are additionally split due to the effects of anisotropy (into 2S, $2P_z$ and $2P_{x,y}$ states, where the hexagonal axis c is directed along z) and the polar interaction with optical phonons. The latter is different for the states of S and P symmetry and may also modify the chemical shift corrections on all states. The results of numerical calculations on the binding energies of the different donors are given in Meyer *et al.*[17]

It is interesting to test whether the donor binding energies E_D show a linear relationship to the bound exciton localization energies $E_{loc} = (E_{FX} - E^0_{DX})$ known as Haynes rule, $E_{loc} = \alpha E_D$.[30] More generally, E_{loc} is given by:

$$E_{loc} = A' + B' E_D$$

and A' and B' must be determined by experiment.

The localization energies can be measured with rather high precision, whereas the binding energies could be in error of ±5% depending on the choice of the band parameters (electron effective masses, dielectric constants including the anisotropy of both). The

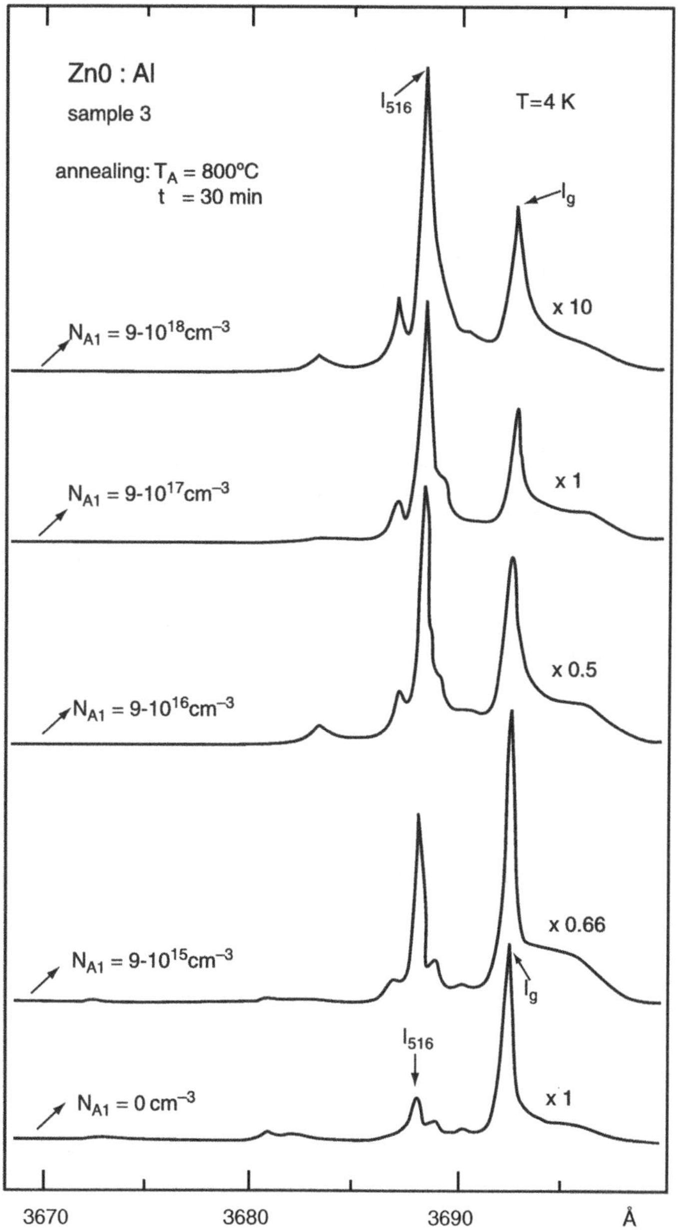

Figure 6.3 *Luminescence spectra as a function of wavelength of Al-implanted ZnO taken after implantation and annealing at 800 °C for 30 min. Reprinted from M. Schilling, R. Helbig, and G. Pensl, Bound exciton luminescence of Ar- and Al-implanted ZnO, Journal of Luminescence 33, 201–212. Copyright (1985) with permission from Elsevier*

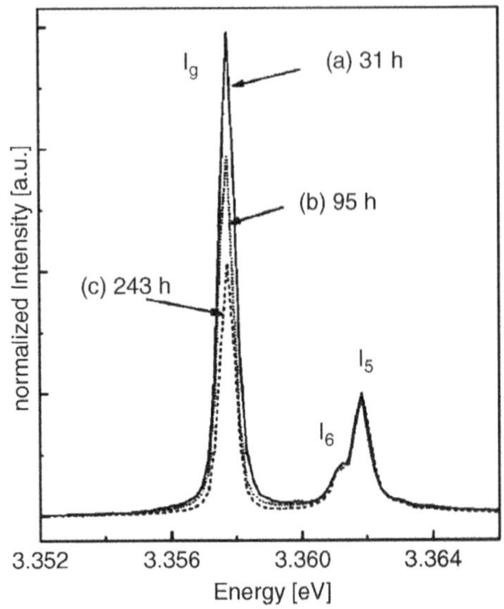

Figure 6.4 Luminescence spectra of the bound exciton range of radioactive ^{111}In-implanted ZnO (a) 31 h, (b) 95 h and (c) 243 h after the implantation and annealing process. Reprinted from S. Müller, D. Stichtenoth, M. Uhrmacher, H. Hofsäss, C. Ronning, and J. Röder, Unambiguous identification of the PL-I9 line in zinc oxide, App. Phys. Lett. 90, 012107. Copyright (2007) American Institute of Physics

fitting of the data in Figure 6.6 gives $A' = -3.8$ meV, $B' = 0.365$ (note that the Haynes rule actually does not demand for $A' = 0$). The good linear relationship gives further support that I_{10} is also donor related.

We demonstrated that the localization energies of the neutral and ionized donor bound excitons scale with the respective donor binding energies.[21] Figure 6.7 shows the

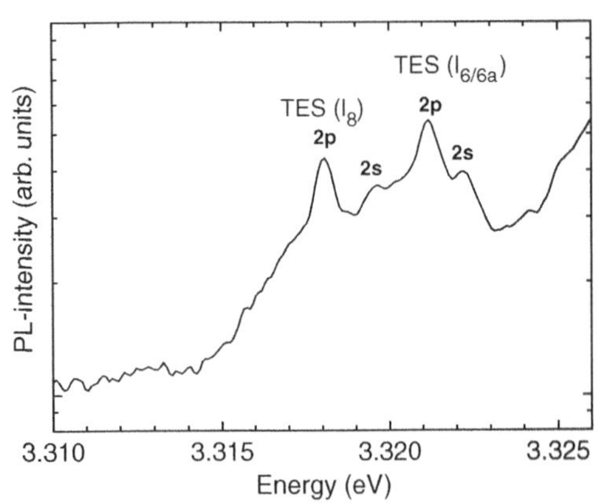

Figure 6.5 Two-electron-satellite transitions of different donor bound excitons. The splitting of the excited states into the 2s and 2p states is indicated. Reprinted from B.K. Meyer, et al., Bound exciton and donor–acceptor pair recombinations in ZnO, phys. stat. sol. (b) 241, 231–260. Copyright (2004) with permission from WILEY-VCH Verlag GmbH & Co. KGaA, Weinheim

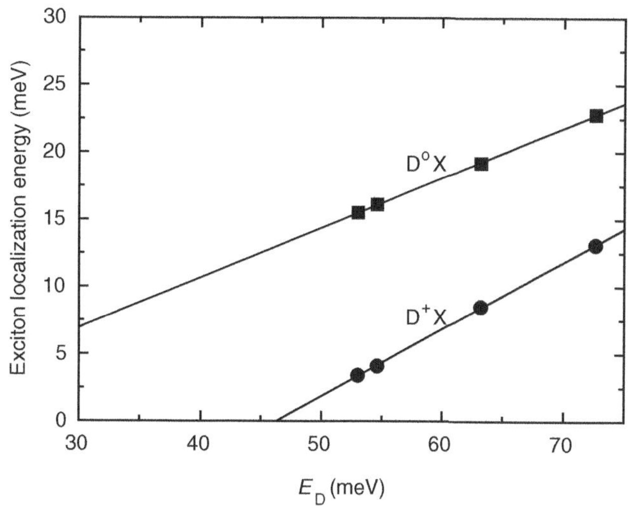

Figure 6.6 *Localization energies of neutral D^0 and ionized D^+ bound excitons in ZnO as a function of the donor binding energies E_D. Reprinted from B.K. Meyer, J. Sann, S. Lautenschlaeger, M.R. Wagner, and A. Hoffmann, Ionized and neutral donor-bound excitons in ZnO, Phys. Rev.B 76, 184120. Copyright (2007) with permission from American Physical Society*

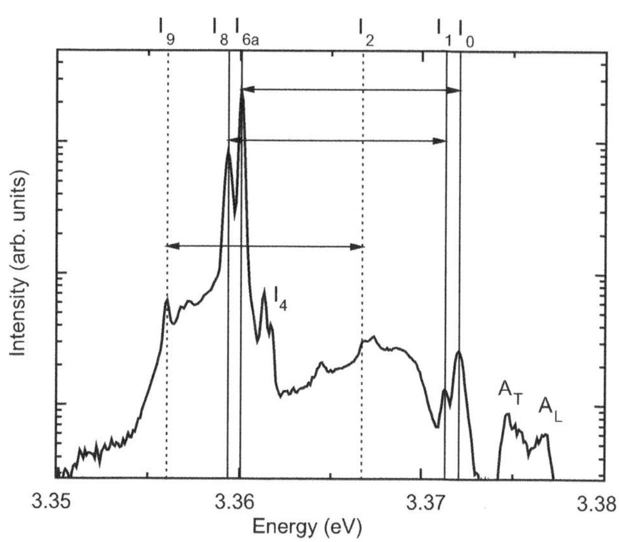

Figure 6.7 *Photoluminescence spectrum of a homoepitaxial ZnO thin film taken at $T = 4.2$ K showing neutral donor bound exciton recombination I_{6a}, I_8 and I_9 and the corresponding ionized donor bound exciton lines I_0, I_1 and I_2. A_T and A_L are the transversal and longitudinal free A-exciton recombinations. Reprinted from B.K. Meyer, J. Sann, S. Lautenschlaeger, M.R. Wagner, and A. Hoffmann, Ionized and neutral donor-bound excitons in ZnO, Phys. Rev.B 76, 184120. Copyright (2007) with permission from American Physical Society*

photoluminescence spectra of a homoepitaxial sample with prominent I_{6a} (Al-donor) and I_8 (Ga-donor) neutral donor bound exciton recombinations. The I_9 (In-donor) recombination is rather weak. They are accompanied by high energy transitions at 3.3726 and 3.3718 eV labeled I_0 and I_1, respectively. These two lines have the same intensity ratio as I_8 and I_{6a}. Moreover, in samples with prominent I_8 recombination as found e.g. for the samples grown on GaN templates (interdiffusion of Ga into the ZnO layer) the I_8 and I_1 lines show up almost exclusively. For samples grown on sapphire substrates (Al_2O_3), Al is the dominant impurity in the ZnO films, and I_{6a} and I_0 are observed as correlated pairs. Another correlated pair of lines exists for I_2 and I_9. In Na-doped ZnO crystals due to compensation the neutral donor bound exciton recombination I_9 related to In had the highest luminescence intensity (see Figure 2 in Meyer *et al.*)[21] followed by I_2. The I_{6a} and I_8 recombinations were much lower in intensity as was the I_{10} line. A similar intensity ratio of I_9 to I_2 was found in a Li-doped sample, here the I_{10} recombination was more pronounced.[21]

As to the nature of the I_0 to I_2 recombinations magneto-optical experiments provided a conclusive picture. In Figure 6.8 the magnetic field behavior of I_0 and I_1 (for I_2/I_3 see the published data in the literature)[15,31] is presented. Both lines show the appearance of an

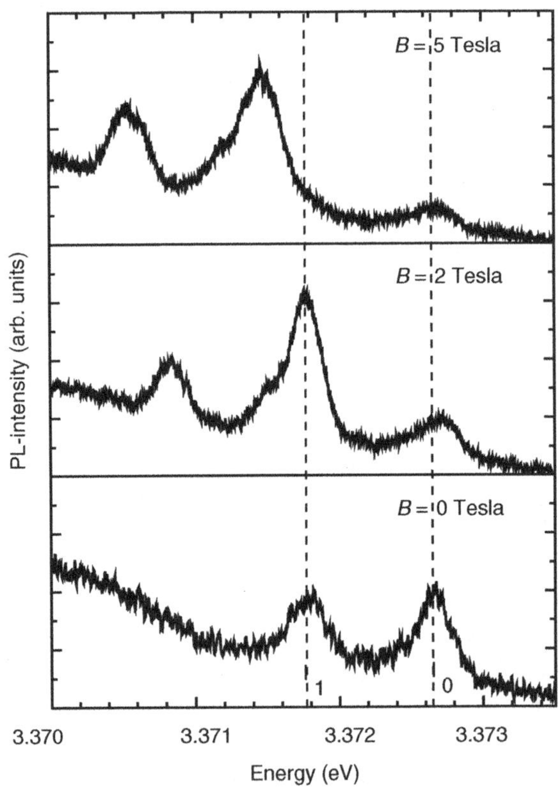

Figure 6.8 *Photoluminescence measurements at T = 1.6 K on a homoepitaxial ZnO thin film showing the behavior of the I_0 and I_1 recombinations in an applied magnetic field (Voigt configuration). Reprinted from B.K. Meyer, J. Sann, S. Lautenschlaeger, M.R. Wagner, and A. Hoffmann, Ionized and neutral donor-bound excitons in ZnO, Phys. Rev.B 76, 184120. Copyright (2007) with permission from American Physical Society*

additional line at the low energy side of I_0 and I_1 in an applied magnetic field. In the case of I_0 this component is superimposed by the I_1 bound exciton transition. Without an applied magnetic the lines are not visible (see Figure 6.8). The extrapolation of the position of the low energy Zeeman component to $B = 0$ Tesla reveals the presence of a zero field splitting. This is strong evidence for ionized donor bound excitons.[31] Additionally, one does not observe a linear splitting in Voigt configuration as would be expected for a neutral bound exciton. Therefore, there is convincing evidence that I_0 to I_2 are caused by ionized donor bound exciton recombinations.

In Figure 6.6 the localization energies of the ionized and neutral donor bound excitons as a function of the donor binding energy are presented. Two aspects are remarkable: The localization energy E_L has a linear dependence on the donor binding energy E_D for both neutral and ionized bound excitons, although with a different slope, and for donor binding energies of $E_D < 47$ meV, excitons will not be bound to the ionized donors.

The shallow donor impurities in ZnO seem to be of extrinsic origin, hydrogen, aluminum, gallium and indium in order of increasing binding energy. However, for many years it was common sense that intrinsic defects dominate the n-type conductivity of ZnO.[32,33] Interstitial zinc and oxygen vacancies were the natural choice. Interstitial zinc as well as oxygen vacancies should be double donors, and in order to contribute to the n-type conduction they should have shallow levels, and low formation energies to be abundant. Theoretical calculations predict indeed that the oxygen vacancies are the main intrinsic defects in zinc-rich ZnO whereas Zn_i have higher formation energies, i.e. are less abundant.[34] However, the oxygen vacancy theory predicts a negative-U behavior with a transition from the neutral to the twofold positive charge state around $E_{VB} + 0.5$ eV, i.e. the vacancy does not induce a shallow level.[34]

From electrical measurements (Hall, DLTS, admittance spectroscopy) on various ZnO single crystals from different sources as well as on epitaxial films there is general consensus that apart from the shallow donor a deep donor state exists which is located around 130 ± 20 meV below the conduction band. Its concentration falls into the $10^{16}\,cm^{-3}$ range thus being relevant for the conductivity as well as for the compensation of acceptors.[35–40]

The Zn interstitial is a double donor in ZnO. It should exist in three charge states, 2+, +, and 0, and thus have two energy levels in the gap. If we assume the level at $E_C \sim 130$ meV is the 2+/+ level of the zinc interstitial the transition +/0 would fall in the range where the binding energies of the extrinsic shallow donors are i.e. from 46 to 53 meV. Look *et al.*[41] reported on Hall effect measurements of electron irradiated ZnO. They concluded with the presence of a native donor in ZnO whose binding energy would be around 35 meV.

In Figure 6.9 we compare three films grown under different Zn/O ratios.[42] In the film grown (# 1) under oxygen-deficient conditions the prominent recombination occurs at 3.366 eV (in Reynolds *et al.*[15] and Meyer *et al.*[17] named I_{3a}). It has comparable intensity as the neutral donor bound exciton recombination with Ga as shallow donor (I_8 at 3.359 eV). For sample # 1 the additional NO_2 flux was 100 sccm, for # 2 it was 200 sccm and for # 3 it was 300 sccm. One notes that with increasing NO_2 flux the I_{3a} recombination decreases in intensity more and more. Finally, for the sample # 3 the neutral donor bound exciton recombination I_8 is the main radiative excitonic recombination. In sample # 3 the I_8 recombination is shifted to lower energy by 0.7 meV, it is connected to a change in the c-lattice constant as revealed by X-ray diffraction measurements. Also the line width increased which is consistent with the findings in Ko *et al.*[43] When a film grown under Zn-rich conditions was annealed in an O_2 atmosphere at high temperatures e.g. $T = 900\,°C$

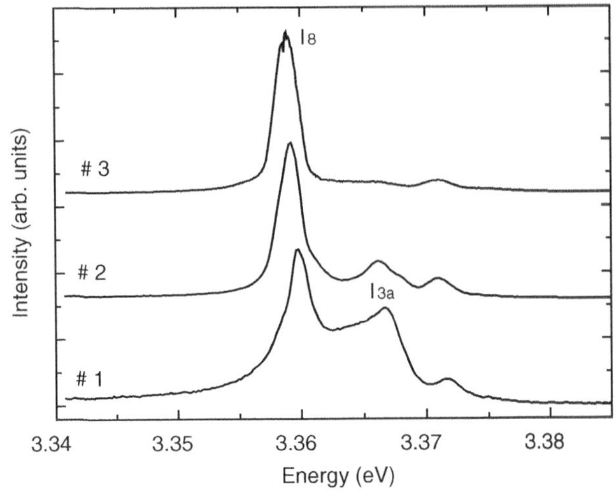

Figure 6.9 *Photoluminescence spectra of ZnO epitaxial films grown under different Zn/O ratios, the additional NO_2 flux was 100 (# 1), 200 (# 2) and 300 (# 3) sccm, respectively ($T = 4.2$ K, HeCd excitation). Reprinted from B.K. Meyer, S. Lautenschlaeger, S. Graubner, C. Neumann, and J. Sann, Mater. Res. Soc. Symp. Vol. 891 , 0891-EE08-02.1. Copyright (2006) Materials Research Society*

with O_2 at atmospheric pressures, and for 30 min the I_{3a} recombination disappears (see Figure 6.10). This can be explained by the fact that excess Zn is oxidized to form ZnO. Both experiments provide evidence that under Zn-rich conditions a shallow center is introduced that binds excitons and gives rise to the I_{3a} recombination. The localization energy of the I_{3a} recombination is 9.9 meV and is hence different from the recombinations I_2 and I_3, which are caused by ionized donor bound excitons (see above). If one assumes a neutral donor bound exciton transition as the origin with $E_{loc} = 9.9$ meV a donor binding energy of 37.5 meV will result (using the data shown in Figure 6.6), which is in close agreement with the data from electrical measurements of 35 meV by Look *et al.*[41] Such a

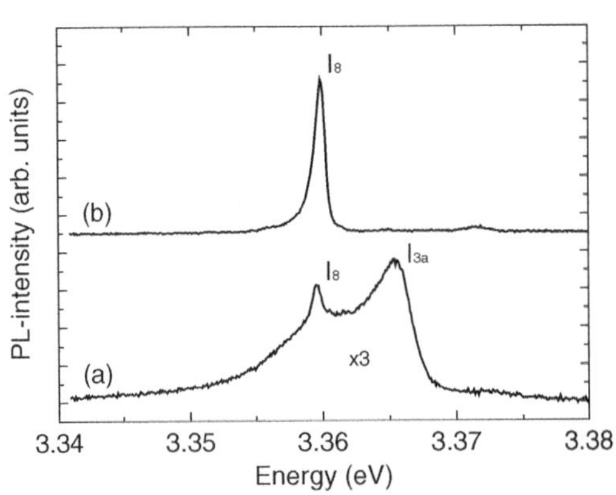

Figure 6.10 *Photoluminescence spectra of an as-grown Zn-rich ZnO film (a) and after subsequent annealing in oxygen atmosphere (b) (for details see text; $T = 4.2$ K, HeCd excitation). Reprinted from B.K. Meyer, S. Lautenschlaeger, S. Graubner, C. Neumann, and J. Sann, Mater. Res. Soc. Symp. Vol. 891 , 0891-EE08-02.1. Copyright (2006) Materials Research Society*

shallow donor level will contribute to the electrical transport properties significantly. The electrical properties of the samples grown under Zn- and O-rich conditions were analyzed by Hall effect measurements in the Van der Pauw configuration at room temperature. The Zn-rich film had almost a factor of ten higher carrier concentration than the O-rich film (n-type carrier density of $6.2 \times 10^{17}\,cm^{-3}$ and a mobility of $110\,cm^2\,V^{-1}\,s^{-1}$ vs $6.7 \times 10^{16}\,cm^{-3}$ and $194\,cm^2\,V^{-1}\,s^{-1}$).

In a recent work on MBE grown ZnO[35] it was demonstrated that the level ET2 ($E_C-0.12\,eV$) is present in Zn-rich and stoichiometric ZnO but appears to be less abundant in O-rich ZnO (concentration $2 \times 10^{16}\,cm^{-3}$ in Zn-rich and $2 \times 10^{15}\,cm^{-3}$ in O-rich). Oh *et al.* further concluded that the trap L1 found in bulk ZnO corresponds to ET2 and may be assigned to the Zn interstitial.[35] Its second ionization energy (2+/+) was estimated to be 0.2 eV below the conduction band. Oh *et al.*[35] observed another trap ET1 in ZnO with thermal activation energies of 33 meV under Zn-rich flux conditions and stoichiometric flux conditions and with 46 meV under O-rich flux conditions. It was a factor 10 lower in concentration than ET2 but showed a similar trend i.e. being less abundant in O-rich ZnO. This level is commonly observed[44–48] and Hagemark and Chacka[48] suggested that the defect causing this shallow donor level is due to the singly ionized Zn interstitial.

The evidence for an intrinsic defect is supported by SIMS experiments. In the films grown either Zn-rich or O-rich the extrinsic impurity content did not change. Since in the growth experiments beside the Zn/O ratio all other parameters remained unchanged (growth temperature, film thickness, etc.) extrinsic donors cannot be made responsible for the change in carrier density upon change in stoichiometry. Whether the defect giving rise to the I_{3a} recombination is an isolated point defect or a more complex species remains to be established.

6.2.2 Recombinations Caused by Nitrogen and Arsenic Doping

In wide-band-gap semiconductors bipolar doping has been a challenge of research for many years. ZnO obviously has a doping asymmetry, since it can easily be doped n-type into concentrations where it shows metallic conduction, but reports on successful, reproducible, stable and homogeneous p-type doping are very rare.[49–56] In the late 1950s lithium and sodium doping into ZnO was tested but the group-I elements act as self-compensating centers – Li (Na) at interstitial sites is a donor and incorporated substitutionally on Zn sites it behaves as an acceptor.[57–59] Interstitial Li (Na) donors in ZnO have so far escaped detection. However, on the basis of magnetic resonance experiments it could be demonstrated convincingly that Na and Li are on Zn sites. In contrast to other II–VI semiconductors they are deep (600–800 meV) acceptors[57–59] caused by a substantial lattice distortion.

Nitrogen on an oxygen site can be considered to be the best candidate as demonstrated by its success in p-type doping ZnSe[60] and ZnS.[61] There are many theoretical reports on the behavior of N in ZnO revealing a complex behavior by the interaction with intrinsic defects, passivation by hydrogen and formation of nitrogen molecules.[62–64] Experiments used ion implantation,[17,28,65] doping during growth[49,51,55,66] and diffusion.[67] There are only a few reports as to the optical properties of the materials in terms of DAP recombination involving a shallow nitrogen acceptor level.[51,65–69] The optical activity of nitrogen was investigated in samples prepared by three different techniques: Nitrogen doping of ZnO was carried out by ion implantation (10^{14} ions cm^{-2}) at room temperature,

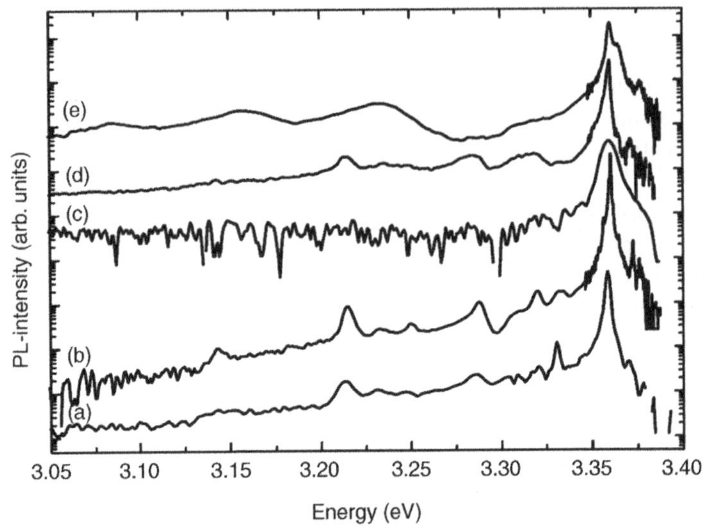

Figure 6.11 *Photoluminescence spectra of bulk ZnO as-implanted (a) and after annealing of the as-implanted sample at 900 °C (b), dose: 10^{13} cm^{-2} N; (c) as-implanted, annealing of the as-implanted sample at 800 °C (d) and at 900 °C (e), dose: 10^{14} cm^{-2} N, Zn co-implant. Reprinted from B.K. Meyer, et al., Shallow donors and acceptors in ZnO, Semicond. Sci. Technol. 20, S62. Copyright (2005) with permission from IOP Publishing*

by diffusion from a thermal decomposition of ammonium or lithium nitrate, and during growth of epitaxial films using ammonia as nitrogen source.[70]

In Figure 6.11 we compare two samples implanted at a total dose of 10^{13} cm^{-2} in the as-received state [Figure 6.11(a)] and after rapid thermal annealing at 900 °C [Figure 6.11(b)]. The spectra differ only in small details and are representative of the recombinations seen in the as-grown ZnO substrate. Due to the small mass of nitrogen the implantation damage results in a reduction of the overall luminescence intensity by a factor 5–10. Similar results were obtained when the dose was changed to 10^{14} cm^{-2}. Therefore, Zn and N were subsequently implanted into ZnO within the same box profile. As seen in Figure 6.11(c) the implantion damage is now substantial, the band edge emission due to bound excitons is hardly observable. At temperatures above 800 °C [see Figure 6.11(d)] a recovery of the luminescence occurs, and for a temperature of 900 °C nitrogen was activated as seen in the appearance of the DAP recombination located at 3.23 eV followed by phonon replica [see Figure 6.11(e)]. The findings about the recovery of the luminescence are very similar to the results of Schilling *et al.*[25] in Ar- or Al-implanted ZnO crystals. The luminescence [see Figure 6.11(e)] is still dominated by the neutral donor bound exciton recombinations (Al is the major trace impurity in the bulk substrates), and there are no indications as to the appearance of a neutral acceptor bound exciton line connected with the incorporation of N into ZnO.

Following the approach of Rommeluère *et al.*[67] the incorporation of N into ZnO from a thermal decomposition of ammonium and lithium nitrate salts was studied. The diffusion of N from the gas phase into ZnO results in the very same DAP recombination as found for the implanted samples [see Figure 6.12(b); Figure 6.12(a) shows the luminescence of the

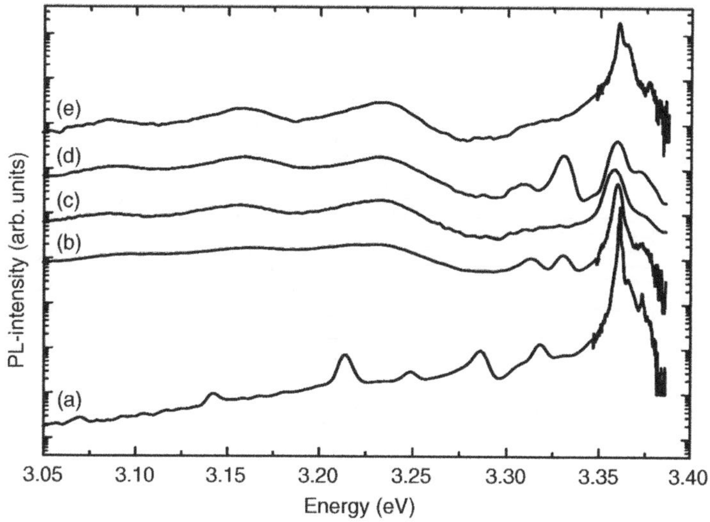

Figure 6.12 *Photoluminescence spectra of ZnO samples: (a) bulk crystal as received, and after annealing in the presence of ammonium nitrate (b) and lithium nitrate (c). Spectrum (d) shows a ZnO epitaxial film with in situ doping by ammonia and (e) is for a comparison the implanted and annealed sample shown in Figure 6.1. Reprinted from B.K. Meyer, et al., Shallow donors and acceptors in ZnO, Semicond. Sci. Technol. 20, S62. Copyright (2005) with permission from IOP Publishing*

substrate before treatment]. Similar results were obtained for $LiNO_3$ [see Figure 6.12(c)}. Figure 6.12(d) shows the luminescence spectrum of an epitaxial ZnO film doped with N during growth using NH_3 as a nitrogen source. The observation of the same DAP band in samples where nitrogen was introduced by three different techniques, implantation, *in situ* doping and diffusion from the gas phase, provides strong evidence that nitrogen on oxygen site is the acceptor.[70]

Temperature-dependent measurements could be performed up to 150 K before the signal disappears in the noise (see Figure 6.13). The transition energy is not very temperature dependent. The band gap energy decreases as a function of temperature $[E(T) = E(T=0) - (5.05 \times 10^{-4} T^2)/(900 - T)]$ which amounts to ~16 meV for 150 K. With increasing temperature a line emerges at higher energies and takes over in intensity which is typical for a change to a conduction band to acceptor transition when the donors start to become ionized. Correcting for the shift of band gap on temperature the band-acceptor transition shifts to higher energies with approximately ½ $k_B T$.

The acceptor binding energy can be estimated from the peak position of the zero-phonon line (ZPL) of the DAP transition at 3.23 eV:

$$E_A = (E_g - E_D) - [E(D^o, A^o) - \alpha N_D^{1/3}]$$

where E_D is the shallow donor binding energy, N_D is its concentration and $\alpha = 3 \times 10^{-5}$ meV cm. In Zeuner *et al.*[66] time resolved luminescence was used to determine N_D and more importantly the acceptor binding energy E_A of 165 ± 40 meV. Similar conclusions on E_A have been derived by Reuss *et al.*[28] on N-implanted ZnO.

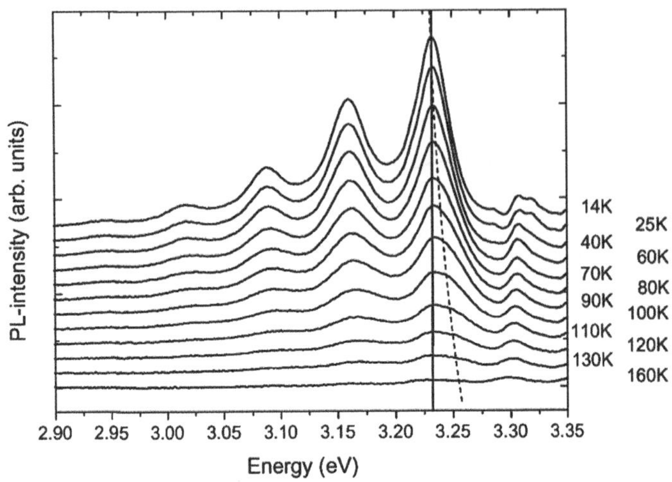

Figure 6.13 *Temperature-dependent photoluminescence measurements on the DAP band in a ZnO epitaxial film doped with nitrogen. The straight line gives the line position at low temperatures; the dashed line indicates the shift to higher energies at high temperatures*

There are a few reports on the optical activity of N in ZnO.[51,65–69] Reuss *et al.*[28] studied N-implanted ZnO and agreed in all major points with the findings presented elsewhere.[17,65,66,71] Yamauchi *et al.*[68] showed in ZnO layers grown by plasma-assisted epitaxy in mixed oxygen-nitrogen gas plasma a DAP emission at 3.27 eV with practically identical phonon coupling. They derived an acceptor binding energy of 135 meV assuming a shallow donor binding energy of 40 meV. Tamura *et al.*[69] reported on a DAP band in ZnO films, the films were doped *in situ* by a radio frequency plasma activated N species. They observed a more or less structureless band which for nitrogen concentrations of $2 \times 10^{18}\,cm^{-3}$ was centered around 3.16 eV, and determined a binding energy of 266 meV for the acceptor. Wang and Giles[72] determined the ionization energy of nitrogen acceptors in bulk ZnO in as-grown, thermally annealed and N_2 added samples. They assigned a DAP line at 3.216 eV to the nitrogen acceptor, and estimated a binding energy of 209 meV. It appears that there is a discrepancy in the data of the acceptor binding energy of N in ZnO which should be resolved in the future.

SIMS was applied to determine the concentration of nitrogen and unintentional dopants such as hydrogen in samples containing different nitrogen concentrations.[71] Nitrogen was detected as $^{14}N^{16}O^-$ and hydrogen as $^{64}Zn^1H^-$ clusters. The given absolute concentrations are accurate to within half an order of magnitude. Nitrogen related vibrations[70] were studied by Raman-scattering experiments in backscattering geometry. Wurtzite ZnO belongs to the C_{6v} symmetry group and, therefore, there are the Raman-active phonon modes E_2(low), E_2(high), A_1(TO), A_1(LO), E_1(TO) and E_1(LO). The B_1 modes are silent. According to the well-known selection rules we expect to observe the E_2 modes and the A_1(LO) mode in unpolarized Raman spectra taken in backscattering geometry. Their respective frequencies are 101, 437 and 574 cm^{-1}.[73]

Figure 6.14 shows Raman spectra of nitrogen-doped ZnO. Beside the expected E_2(low) and E_2(high) mode of ZnO we find a variety of other modes, which need to be explained.

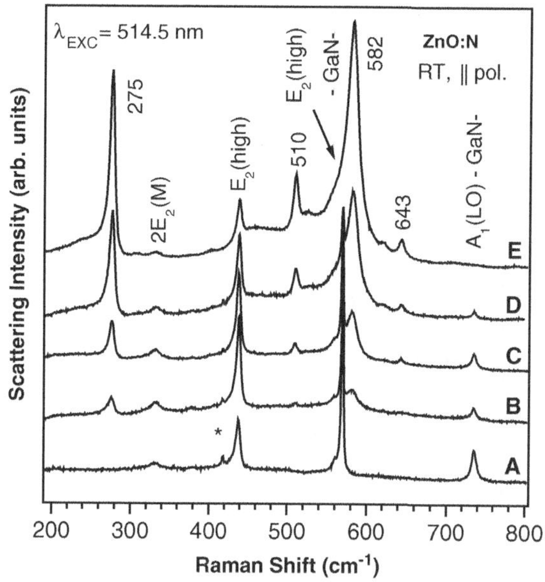

Figure 6.14 Room temperature Raman spectra of five ZnO samples. Sample A is undoped, whereas the nitrogen concentration increases from sample B to E. The peak marked by an asterisk originates from the sapphire substrate. Reprinted from A. Kaschner, et al., Nitrogen-related local vibrational modes in ZnO:N, Appl. Phys. Lett. 80, 1909. Copyright (1969) with permission from American Institute of Physics

At 569 cm^{-1} one finds the E_2(high) mode of the GaN template. The feature at 332 cm^{-1} is a second-order structure of ZnO, which was interpreted as $2E_2$(M) by Calleja and Cardona.[74] There are five more modes with frequencies of 275, 510, 582, 643 and 856 cm^{-1} which do not belong to first- or second-order structures of ZnO or the GaN template material and are nitrogen-related.

In total five samples with different nitrogen concentration as estimated from the growth parameters were investigated (see Figure 6.14). Sample A is an undoped ZnO reference, and the estimated nitrogen concentration increases from sample B to sample E. Neither of the additional modes is found in undoped material. Also in ZnO:Ga (not shown) none of these Raman modes was observed. The intensity of the five modes increases with nitrogen concentration. Furthermore, the intensity of the A_1(LO) mode from the GaN template subsequently decreases from A to E. This results from the higher absorption in the visible spectral range with increasing nitrogen concentration.[75] The correlation of the mode intensities and the nitrogen concentration leads to the conclusion that all the additional modes are local vibrational modes (LVMs) related to nitrogen in ZnO. For a quantitative correlation of the nitrogen concentration and the LVMs SIMS measurements (Figure 6.15) were performed. The nitrogen concentration was deduced from the signal intensity of the NO^- cluster consisting of ^{14}N and ^{16}O isotopes. The detection limit of this method is around 10^{17} cm^{-3}. As expected from the growth parameters the nitrogen concentration increases from sample A to E with a maximum concentration of about 10^{19} cm^{-3}. Simultaneously, the hydrogen concentration as detected by the $^{64}Zn^1H^-$ cluster linearly increases with the nitrogen content. Since a calibration standard for hydrogen was not

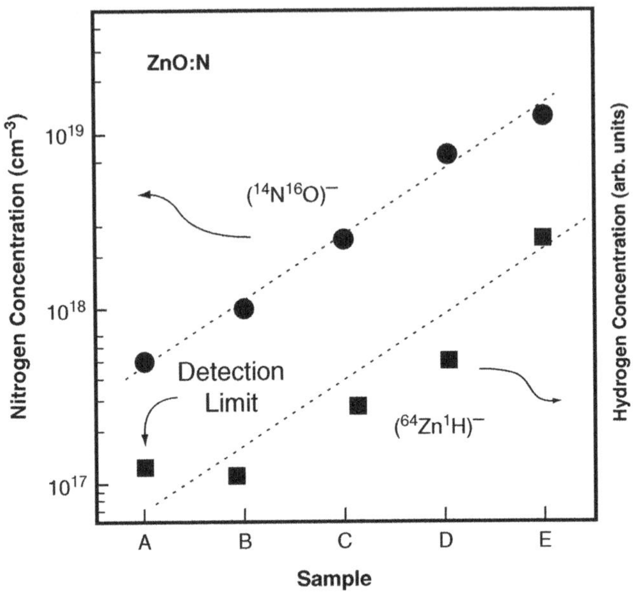

Figure 6.15 *Nitrogen and hydrogen concentration in samples A–E as determined from SIMS. The nitrogen concentration increases from sample A to E. The hydrogen concentration linearly correlates with the nitrogen concentration. Dashed lines are a guide to the eye only. Reprinted from A. Kaschner, et al., Nitrogen-related local vibrational modes in ZnO:N, Appl. Phys. Lett. 80, 1909. Copyright (1969) with permission from American Institute of Physics*

available, a quantitative scaling of the hydrogen concentration is unfortunately not possible. Nevertheless, a qualitative comparison of the samples can be given. A higher nitrogen concentration seems to drive the material to build in a higher concentration of compensating hydrogen (see Figure 6.15). A similar mechanism as for GaN, where the concentration of magnesium acting as an acceptor and compensating hydrogen are linearly correlated,[76] may be the reason for this behavior. This fact may have a strong impact on the compensation mechanism in ZnO.

The LVMs with frequencies of 275, 510, 582 and 643 cm^{-1} have a linear dependence on the nitrogen concentration in the sample, which is the final evidence for their origin being related to vibrations of nitrogen in ZnO.[71] The different slopes may indicate different formation probabilities of the nitrogen-containing complexes.[71] The exact configuration of the vibrating nitrogen-related complex cannot be deduced from these experiments. First-principles calculations show that the formation of N_O and related complexes is very likely in ZnO.[77] The vibrational properties and the correlation between the acceptor and hydrogen seems very similar in ZnO:N and GaN:Mg.[78]

The research on p-type doping was stimulated when in 1997 Minegishi *et al.*[49] reported on p-type behavior of chemical vapor deposition (CVD) grown ZnO films doped with nitrogen. Indeed the first light-emitting device based on a ZnO p-n homojunction used nitrogen as acceptor dopant.[55] The authors demonstrated that the solubility of nitrogen is high only at low growth temperatures ($<400\,°C$). Substrate temperatures in CVD and metal organic chemical vapor decposition are typically 200 and 400 °C higher limiting the incorporation of nitrogen. Alternatives have been searched and found by using P and As

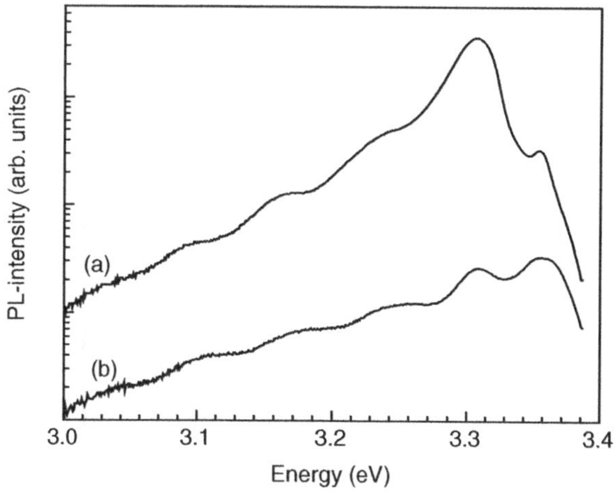

Figure 6.16 *Temperature-dependent photoluminescence measurements in a ZnO epitaxial film doped with arsenic at 4 K (a) and 110 K (b)*

despite the fact that there is a considerable size mismatch (in terms of ionic radii) of P and As compared with O [assuming P (As) replaces O].

In comparison with ZnO:N As doping during growth by CVD[79] leads to a fundamentally different PL spectrum (see Figure 6.16). D^oX recombinations are completely absent. The first transition is located at 3.354 eV followed by the most intense line at 3.307 eV. A sequence of transitions separated by 72 meV (LO phonon frequency) follows which can be seen more clearly at $T = 110$ K (see Figure 6.16). The line at 3.307 eV is assigned to a free to bound transition, i.e. from the conduction band to an acceptor level (eA^o). This assignment is supported by the temperature-dependent studies, which allowed the acceptor binding energy of 135 ± 5 meV to be determined. Meanwhile this characteristic transition centered around 3.31 eV has been observed in a variety of ZnO materials (bulk crystals, thin films and nanostructures) often most prominent in ZnO:P, ZnO:As, ZnO:Sb and ZnO:N (for a compilation see Schirra *et al.*).[80] It is also present in undoped material and Schirra *et al.*[80] have made an extensive study on the behavior of this line by temperature-dependent cathodoluminescence with high spectral resolution, by scanning electron microscopy and by transmission electron microscopy. A detailed temperature dependence of the band-acceptor transition and replicas is shown in Figure 6.17. From a line shape analysis at different temperatures the acceptor binding energy of 130 ± 3 meV was determined in very close agreement with the value obtained in ZnO:As mentioned above.

The first attempt of doping ZnO with As goes back to Ryu *et al.*[53] growing ZnO on GaAs substrates. They showed by SIMS the diffusion from As into ZnO which was high close to the interface GaAs/ZnO but maintained a level of 10^{18} cm^{-3} throughout the rest of the film. They claimed the recombination at 3.32 eV seen in their films is caused by an acceptor bound exciton. In 2003 Ryu *et al.*[54] reported on As-doped p-type ZnO grown by hybrid beam deposition. The lightly doped films with N_{As} in the low 10^{18} cm^{-3} range [see Figure 2(b) in Ryu *et al.*)[54] showed recombinations at 3.359 eV attributed to neutral acceptor bound excitons (A^oX), at 3.32 and 3.273 eV attributed to free to bound transitions (eA^o), and at 3.204 eV attributed to DAP recombination. Thus two acceptors with binding

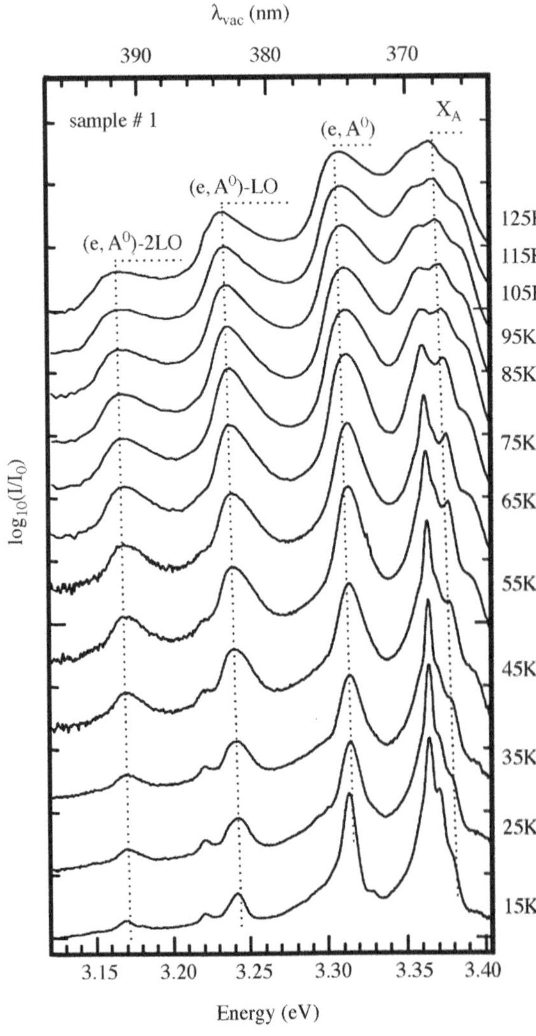

Figure 6.17 *Temperature-dependent photoluminescence measurements in an undoped ZnO epitaxial film grown by vapor phase epitaxy. Reprinted from M. Schirra, et al., Stacking fault related 3.31-eV luminescence at 130-meV acceptors in zinc oxide, Phys. Rev.B 77, 125215. Copyright (2008) with permission from American Physical Society*

energies of 115 and 164 meV would contribute to the measured p-type conduction. In 2004 Jeong *et al.*[81] reported on As-implanted bulk crystals and assigned a peak at 3.3589 eV to an acceptor bound exciton, and transitions at 3.3159 and 3.31859 eV to eA^o and DAP, respectively.

Results on phosphorous doping in ZnO came from Hwang *et al.*[82] They obtained a bound exciton recombination at 3.355 eV, a free to bound transition at 3.31 eV, and a DAP recombination at 3.241 eV, and deduced a binding energy of the P-acceptor in ZnO of 127 meV in agreement with the results from Ye *et al.*[83] who obtained a value of 123 meV.

Antimony doping was presented in the work of Xiu *et al.*[84] The luminescence line at 3.353 eV was assigned to neutral donor bound exciton recombination. Two emissions at 3.296 and 3.222 eV were attributed to eA^o and DAP transitions, respectively. The acceptor binding energy was calculated to be 140 meV.

In the work of Schirra *et al.*[80] the eA^o transition is correlated with the structural data of the film as provided by electron microscopy. They concluded that the acceptors are located in the basal plane stacking faults. Doping with group-V elements could lower the threshold for stacking fault formation, where the stacking faults act as acceptors. The observation and identification of the eA^o transition at 3.31 eV is not necessarily an indication for successful p-type doping of ZnO.

The high p-type carrier densities for ZnO:As, ZnO:Sb and ZnO:P require doping concentrations of the order of 10^{20}–10^{21} cm^{-3}. The solubility limits for dilute doping with the group-V elements are not known. There is experimental evidence[85] that secondary phases within the systems Zn-P-O and Zn-As-O form easily and limit the doping efficiency.

6.3 Magnetic Resonance Investigations

EPR spectroscopy has the ability to resolve hyperfine interactions of electrons or holes related to doped atoms, impurities, or defects with their own nuclear core and/or the nuclei of the constituting lattice atoms, e.g. Zn and O. This connection between the nuclear and the electronic system allows on one hand an unambiguous identification of species via the isotopes and their abundances, and on the other hand statements on their electrical activity and wave function distribution. This makes magnetic resonance spectroscopy a valuable tool for the characterization of materials. The spectra are typically analyzed in the framework of a Spin-Hamiltonian formalism which we briefly introduce in the following:[86]

$$\mathcal{H} = \mu_B \cdot \vec{S} \cdot \tilde{\mathbf{g}} \cdot \vec{H} + \vec{S} \cdot \tilde{\mathbf{A}} \cdot \vec{I} \tag{6.1}$$

The first term on the right-hand side of the equation describes the Zeeman interaction of an electron characterized by its electron spin S with an external magnetic field H via the **g**-tensor, and the second term gives the hyperfine interaction (**A**-tensor) with the spin I of the nuclei. For electronic centers with a single unpaired spin ($S = ½$) in axial symmetry the Hamiltonian is given by:

$$\mathcal{H} = \mu_B[g_\parallel S_z H_z + g_\perp(S_x H_x + S_y H_y) + A_\parallel S_z I_z + A_\perp(S_x I_x + S_y I_y)] \tag{6.2}$$

For $S > ½$ an additional fine structure interaction has to be taken into consideration:

$$\mathcal{H} = \vec{S} \cdot \tilde{\mathbf{D}} \cdot \vec{S} \tag{6.3}$$

which in the principal axis system leads to

$$\mathcal{H} = D\left[S_z^2 - \frac{1}{3}S(S+1)\right] + E\left(S_x^2 - S_y^2\right) \tag{6.4}$$

where $\vec{S}, \vec{I}$ are the electron spin nuclear spin, respectively and their components S_x, S_y, S_z, I_x, I_y, I_z, $\tilde{\mathbf{A}}$ is the hyperfine interaction tensor and its components $A_\parallel$, $A_\perp$, A_{xx}, A_{yy}, A_{zz}, $\tilde{\mathbf{g}}$ is the **g**-tensor and its components $g_{\parallel}$, $g_{\perp}$, g_{xx}, g_{yy}, g_{zz}, $\tilde{\mathbf{D}}$ is the fine structure tensor and its components, D is the axially symmetric part and E is the asymmetry parameter.

It is helpful to recall that the hyperfine interaction **A** can be split into an isotropic part (a) and an anisotropic fraction (b):

$$a = 1/3(A_{xx} + A_{yy} + A_{zz}), b = 1/2(A_{zz} - a) \tag{6.5}$$

The isotropic part which is a measure of the spin density on the position of the nucleus (Fermi contact term)is given by:

$$a = 16\pi/3\mu_e\mu_N g_e g_N \eta^2 \alpha^2 \Psi^2 \tag{6.6}$$

and the anisotropic part which is due to dipole–dipole interactions is given by:

$$b = 4/5\mu_e\mu_N g_e g_N \eta^2 \beta^2 \langle r^{-3} \rangle \tag{6.7}$$

We will use these equations to show that Na_{Zn} is a deep acceptor rather than a shallow one (see below).

To structure the magnetic resonance data on ZnO we have organized the rest of this section as follows: we start with the results on shallow donors, than describe deep center properties, especially the oxygen vacancy, and finally present the results on acceptors. Table 6.1 summarizes the data on the individual centers.

6.3.1 Shallow Donors

In ZnO shallow donors are characterized by a single line spectrum at a field position corresponding to a *g*-value of $g \sim 1.96$. In ZnO single crystals the resonance is slightly anisotropic due to the wurzite symmetry. The deviation of the shallow donor *g*-value from the free electron value of $g = 2.0023$ is in first order a result of the admixture of valence band states (p-type) to the s-type conduction band.[87–89]

In most cases the EPR linewidth of shallow donors varies between 1 G and 10 G, which depends on the concentration of the shallow donors and the details of the measurement conditions. A hyperfine interaction is rarely resolved, thus direct information on the chemical nature of the donors cannot be given. The reason is that the donor electron is in an effective mass-like state with a Bohr radius of about 1.5 nm, thus it is distributed over many lattice sites. Consequently the electron density on the nucleus of the donor atom is low, typically less than a percent compared with the atomic values, thus the hyperfine interaction is often smaller than the instrumental resolution in the EPR experiment.

Two fortunate cases have been found where the hyperfine interaction was large enough to be resolved. In the case of the In-donors a 10 line spectrum was observed reflecting the $I = 9/2$ nuclear spin of the ^{115}In isotope (95% abundance)[26,90] and Ga-donors gave a four line spectrum due to the nuclear spin $I = 3/2$.[26,91] For the third common group-III donor in ZnO, namely Al, the enhanced resolution of the ENDOR had to be used to resolve the interaction.[92] Experiments of the same kind were successful to identify H as a shallow donor.[24] To act as a shallow donor in semiconductors is quite unusual behavior for H. We just recall here the passivating properties of H on dangling bonds in amorphous Si.[93,94] However, meanwhile also complex centres were observed in which H plays a role as a passivating

Table 6.1 *Spin Hamilton parameters (spin state, g-values, hyperfine interactions) and defect assignments*

Spin *g*-values	Hyperfine interaction	Fine structure	Defect assignment	Reference
$S=\frac{1}{2}$ $g_{\parallel}=1.955$	Not observed		Shallow donor	[87]
$g_{\perp}=1.953$				
$S=\frac{1}{2}$ $g_{\parallel}=1.957$	Not observed		Shallow donor	[88]
$g_{\perp}=1.956$				
$S=\frac{1}{2}$ $g=1.956$	Not observed		Shallow halogen donors	[88]
$S=\frac{1}{2}$ $g_{\parallel}=1.957$	$A(^{115}In)=36.6\ G$		Shallow In donor	[26,90]
$g_{\perp}=1.956$	$A(^{69}Ga)=4.2\ G$		Shallow Ga donor	[26]
	$A(^{69}Ga,\ ^{72}Ga)=6.7\ G$		Shallow Ga donor	[91]
			Shallow Ga donor	[89]
$S=\frac{1}{2}$ $g=1.9595$	$A(^{27}Al)=1.45\ MHz$		Shallow Al donor	[92]
$S=\frac{1}{2}$ $g_{\parallel}=1.9569$	$A(^{1}H)=1.4\ MHz$		Shallow H donor	[24]
$g_{\perp}=1.9552$				
$S=1/2$ $g_{\parallel}=1.9605$			Shallow Zn interstitial	[97]
$g_{\perp}=1.9595$				
$S=\frac{1}{2}$ $g_{\parallel}=1.9945$	Axial center:		Oxygen vacancy, F^{+a}	[101]
$g_{\perp}=1.9960$	$A_{\parallel}=57.34\ MHz$			[99,100]
	$A_{\perp}=42.3\ MHz$			[102,103]
	Nonaxial center:			
	$A_{xx}=76.6\ MHz$			
	$A_{yy}=75.9\ MHz$			
	$A_{zz}=94.8\ MHz$			

(continued)

Table 6.1 *(Continued)*

Spin *g*-values	Hyperfine interaction	Fine structure	Defect assignment	Reference
Axial at room temperature:			P-center in ref. [131]	[132]
$S = 1/2$ $g_{zz} = 2.0038$			Zinc-vacancy related	[131]
$g_{yy} = 2.018$			Di- or trivacancy centers[a]	[133]
$g_{xx} = 2.0191$				[102,103]
$g_{\parallel} = 2.0149$				
$g_{\perp} = 2.0134$				
$S = 1/2$ $g_{zz} = 2.0038$			Zinc-vacancy related[a]	[131]
$g_{yy} = 2.0018$				
$g_{xx} = 2.0217$				
$S = 1$ $g_{zz} = 2.0095$		$\|D\| = 1465\,MHz$	Zinc-vacancy related[a]	[131]
$g_{yy} = 2.0141$		$\|E\| = 58\,MHz$		[133]
$g_{zz} = 2.019$				[102,103]
$S = 1/2$ Axial center:	Axial center:		Hole on an oxygen atom adjacent to a Li atom; deep Li acceptor	[88]
$g_{\parallel} = 2.0028$	$A_{\parallel} = 0.22\,G$			[114]
$g_{\perp} = 2.0253$	$A_{\perp} = 1.81\,G$			[115]
Nonaxial center:	Nonaxial center:			
$g_{zz} = 2.004$	$A_{zz} = 0.29\,G$			
$g_{yy} = 2.0254$	$A_{yy} = 1.8\,G$			
$g_{xx} = 2.0223$	$A_{xx} = 1.8\,G$			

$S = 1/2$ Axial center:	Axial center:		Hole on an oxygen atom adjacent to a Na atom; deep Na acceptor	[116]
$g_{\parallel} = 2.0024$	$A_{\parallel} = 2.89$ G			
$g_{\perp} = 2.0298$	$A_{\perp} = 1.58$ G			
Nonaxial center:	Nonaxial center:			
$g_{zz} = 2.0032$	$A_{zz} = 4.15$ G			
$g_{yy} = 2.0302$	$A_{yy} = 1.42$ G			
$g_{xx} = 2.0241$	$A_{xx} = 1.61$ G			
$S = 1/2$ $g_{\parallel} = 1.9953$	$A_{\parallel} = 81.3$ MHz		Deep nitrogen acceptor	[108]
$g_{\perp} = 1.9633$	$A_{\perp} = 9.5$ MHz			[91,127]
$S = 1/2$ $g_{\parallel} = 2.0036$	$A_{\parallel} = 9.8$ MHz		N_2^- acceptor	[127,128]
$g_{\perp} = 1.9935$	$A_{\perp} = 20.1$ MHz			
$S = 1$ $g_{\parallel} = 1.971$		$D = 0.763$ GHz	None	[108]
$g_{\perp} = 2.0224$				[106]
$g_{\parallel} = 1.984$				
$g_{\perp} = 2.025$		$D = 260 \times 10^{-4}$ cm^{-1}	Oxygen vacancy; F-center	[106]
$S = 1/2$ $g_{\parallel} = 2.020$			$(Zn_i^+ - N_0^-)$	[129]
$g_{\perp} = 2.006$				

[a]After e^- or n^o irradiation.

defect (see below). While the understanding of hydrogen and the group-III donors in ZnO has greatly developed over recent years much less is known about the halogen donors.

Not only the extrinsic donors may contribute to the n-type conductivity of nominally undoped ZnO, but also intrinsic defects are often of great importance and relevance. Very often the exact stoichiometry of the ZnO samples is not known, and defects accommodating nonstoichiometry such as the donor Zn interstitials and oxygen vacancies have to be considered with equal importance. In order to study the role of intrinsic defects Gorelkinskii and Watkins[95] and Vlasenko and Watkins[96,97] used optically detected magnetic resonance (ODMR) in combination with low-temperature, high-energy electron irradiation to create and possibly identify the relevant defects. By choosing the appropriate electron energy they created and characterized the primary radiation defects and were able to show convincingly that Zn interstitial atoms are shallow donors. Due to the low abundance of Zn with nuclear spin (~5%) hyperfine interactions were not observed, but the *g*-value for the Zn interstitials is clearly that of a shallow donor. The oxygen vacancy turned out to be a deep level defect. Son *et al.* also performed ODMR experiments on high energy electron irradiated samples – this time at room temperature – and discovered similar defects but with different thermal stabilities.[98]

6.3.2 Deep Level Defects

By the 1970s the paramagnetic charge state of the oxygen vacancy was observed by EPR.[99–103] Very recently the deep level behavior was confirmed, and the V_O^+ level position was determined to be about 0.9 eV above the valence band.[104] This diverges somewhat from theoretical calculations which locate the level closer to the conduction band.[105] Also the negative correlation energy predicted for the oxygen vacancy demands an experimental confirmation. Another open question is the relationship of the oxygen vacancy to the "green" luminescence band in ZnO. At least three models are under discussion: (i) it is related to the presence of Cu impurities[91] – a less likely model – the pronounced phonon structured luminescence band can easily be distinguished; (ii) a shallow donor (D^0) to oxygen vacancy (V_O^0) recombination;[96,104] and (iii) it originates from an internal triplet recombination of the neutral oxygen vacancy ($S = 1$), similar to color center emissions in ionic crystals.[106] New aspects in the discussion of the properties of the oxygen vacancies came recently from the theoretical calculations.[107] Their calculations predict that oxygen vacancies have high formation energies in n-type material, hence have negligible abundance, and play no role in the optical recombinations in ZnO. They further predict that Zn vacancies are the most abundant intrinsic defects and recombinations in the green spectral range are caused by Zn vacancies and associates. They assigned the spin triplet ($S = 1$) recombination investigated in the work of Leiter *et al.*,[106] Carlos *et al.*[108] and Vlasenko and Watkins[109] to a pair defect consisting of a singly negatively charged Zn vacancy ($S = 1/2$) and a neutral shallow donor ($S = 1/2$) which by exchange interaction couple to a $S = 1$ state ($EM^0 + V_{Zn}^- \rightarrow EM^+ + V_{Zn}^{2-}$). While such a possibility cannot be ruled out, it is in conflict with the available experimental data and arguments are given in the following.

In the model of Leiter *et al.*[106] the $S = 1$ signal originates from the two-electron state of the oxygen vacancy (V_O). In its neutral charge state two electrons are paired with antiparallel spins, a singlet ground state. Photo-excitation does not involve band states but occurs as an internal atomic excitation into a singlet excited state. From there inter-system crossing occurs to the emitting states, a spin- and orbital-triplet state. Here the radiative recombination into the singlet ground state takes place ("green emission"). In an atomic model this is a ^{3}P to ^{1}S type recombination, the band states of ZnO are not involved. For a comparison see the results on the two-electron centres in CaO (color centers).[110–112]

The parameters of the S = 1 center are $g_{\parallel} = 1.984$, $g_{\perp} = 2.025$, and the fine structure splitting constant is D = +0.78 GHz [see Table 6.1 and Equation (6.3)].

Janotti and Van de Walle[107] suggest as a model an exchange coupled pair consisting of a singly negatively charged Zn vacancy (S = 1/2) and an effective mass type donor (S = 1/2) for the recombination. Coupling the two spins to a total S = 1 system and assuming the fine structure splitting D to be zero, we expect the *g*-value of this center to be:[113]

$$g_{\text{pair}} = (g_{V_{Zn}} + g_{\text{donor}})/2 \tag{6.8}$$

Shallow donors have *g*-values of $g_{\parallel} = 1.957$ and $g_{\perp} = 1.955$ and the *g*-values for V_{Zn}^{-} are $g_{\parallel} = 2.0024$ and $g_{\perp} = 2.0193$, thus the *g*-values should be $g_{\parallel} = 1.979$ and $g_{\perp} = 1.987$. Such *g*-values are typical for the centers observed by Vlasenko and Watkins[109] but they are considerably different from the experimental values of $g_{\parallel} = 1.971$ and $g_{\perp} = 2.022$ (see Table 6.1). Vlasenko and Watkins[109] indeed observed several Zn vacancy related centers which show an effective mass donor to Zn vacancy acceptor type of recombination ($EM^0 + V_{Zn}^{-} \rightarrow EM^{+} + V_{Zn}^{2-}$), however, none of these had this particular *g*- and D-value. Furthermore, the trigonal symmetry of V_{Zn}^{-} should show up in the angular dependence of the magnetic resonance signals.

The S = 1 center detected in the green band has a *positive* fine structure or zero field splitting D. The D of spin-exchange coupled pairs is mediated by the dipole–dipole interaction D_{ss} and is *negative*.[112]

$$D_{ss} = -(\mu_o/4\pi)(3g_1g_2\mu_B/2r^3) \tag{6.9}$$

where g_1 and g_2 are the *g*-values of the two spins and *r* is the distance between the spins. For anion vacancies the zero field splitting became positive because there are two contributions to $D = D_{ss} + D_{so}$, the dipole–dipole interaction as given in Equation (6.3) and D_{so} is the spin-orbit interaction in the triplet states which are not spin-only states but have angular momentum.[112]

In the ODMR experiments Leiter *et al.*[106] were in the fortunate situation to observe the level anticrossing (LAC) in the circular polarization of the emission (Figure 6.18). The LAC allows an unambiguous determination of the sign of D. The intensity of the circular polarization of the emission is measured as the difference of the left- and right-circular polarized emission components:

$$\Delta I(B,T) = I_{\sigma-} - I_{\sigma+} \tag{6.10}$$

In case of a positive zero field splitting D the $|0\rangle$ sub-state is located below the $|\pm1\rangle$ sub-states of the S = 1 triplet. With increasing magnetic field *B* the MCPE intensity increases due to partial thermalization in the sub-states. At the field position of LAC the level mixing of the $|-1\rangle$ radiative state and the more populated nonradiative $|0\rangle$ results in an increase of the σ^{-} emission, i.e. a positive signal ΔI. For a negative D the opposite holds, the intensity of the MCPE would decrease.

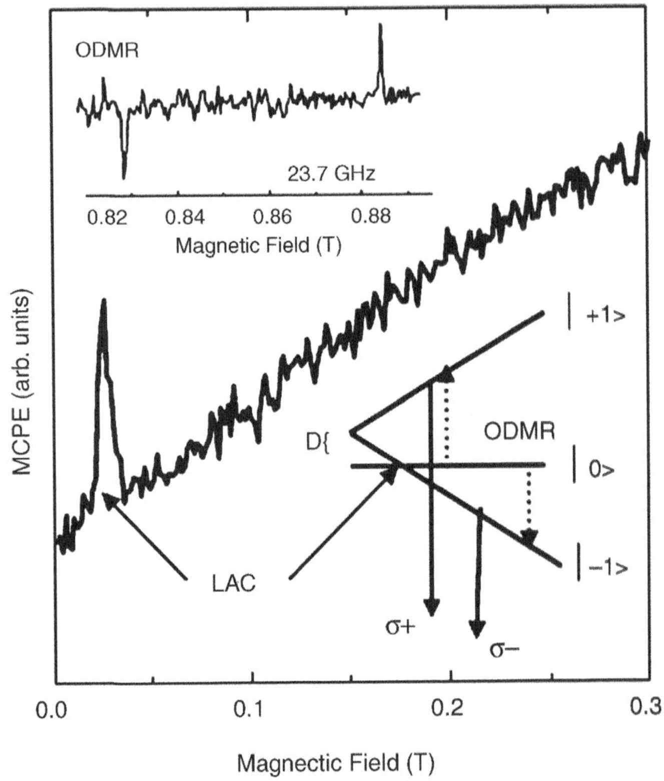

Figure 6.18 *Observation of the level-anti-crossing in the magnetic circular polarized emission (MCPE) of the S = 1 center in ZnO. The insets show the magnetic resonance as signal decrease and increase, and a schematic model of the zero-field splitting of the S = 1 states. Reprinted with permission from F.H. Leiter, Thesis, Gieβen, 2002*

A typical situation in many ZnO samples is that the emission in the visible spectral range is a superposition of various recombinations. A broad vibronic coupled band with almost no structure is the consequence. Therefore it was of importance to investigate the spectral dependence of the LAC signal and the ODMR resonances responsible for our S = 1 center (see Figure 6.19). The dependence of the ODMR (LAC) signals on the emission wavelength reproduces the photoluminescence spectrum (see Figure 6.19), thus the recombination band centered at 2.45 eV is caused by this particular spin triplet recombination.

This does not mean that other recombinations and hence other defects cannot participate in the emission in the visible spectral range but for the samples used by Carlos *et al.*,[108] Leiter *et al.*[106] and Vlasenko and Watkins[109] the broad unstructured green luminescence band was caused by a defect with a S = 1 configuration in the excited state. However, the relationship between the "green" emission and the oxygen vacancies is still under discussion. Rather more settled is the identification of the oxygen vacancy in its paramagnetic single ionized charge state.[99–103] Further experiments are needed to get solid information on the neutral charge state and if possible also on the 2+ charge state. Furthermore, the predicted negative correlation energy for the oxygen vacancy needs confirmation.[105,107]

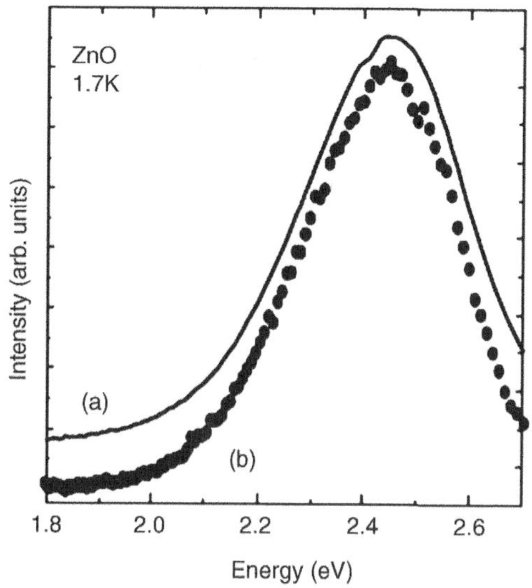

Figure 6.19 (a) Photoluminescence spectrum and (b) ODMR excitation spectrum of the $S=1$ defect in ZnO. Reprinted with permission from F.H. Leiter, Thesis, Gießen, 2002

6.3.3 Extrinsic Acceptors: Li, Na and N

For the group-I elements it is well established that they introduce deep acceptor states. With binding energies of 800 meV (Li) and 600 meV (Na) they compensate the background shallow donors and produce high resistive material.[114–116] For "shallow" effective mass like acceptors in ZnO one would expect in an atomic limit g-values around $g \sim 1$ (a hole localized in a p-orbital) which is clearly not the case for the deep Li and Na centers (see Table 6.1).

The hyperfine constants of the centers can be interpreted in a linear combination of atomic orbitals (LCAO) approach where one assumes that the electron wavefunction is made up from Na 3s and 3p orbitals:

$$\Psi = \alpha\Psi(3s) + \beta\Psi(3p) \qquad (6.11)$$

$\alpha^2 + \beta^2 = \eta^2 < 1$ allows for admixtures of non-sodium orbitals for example from surrounding oxygen atoms. The term β^2/α^2 reveals the hybridization of the investigated atom which is also reflected in the bond angle φ. Using the relationship:[117]

$$\tan(\varphi) = -[2(1 + \beta^2/\alpha^2)]^{1/2} \qquad (6.12)$$

and the measured angle φ we can estimate the term to be 4.7 times more p-like than in an ideal sp^3 hybrid.

Inserting the measured values of a and b and the wavefunction values of the free atoms given by Morton and Preston[118] one finds $\alpha^2 = 0.12$, $\beta^2 = 0.59$, and $\eta^2 = 0.71$ for the nonaxial Na centers. For the axial Na centers the values become: $\alpha^2 = 0.11$, $\beta^2 = 0.29$, and $\eta^2 = 0.4$. The η^2 value shows different admixtures of oxygen states for the axial type and nonaxial type of centers. The hybridization also varies, being closer to sp^2 in the case of the nonaxial centers while the orbital is much more p-like for the axial center.

This leads to a distorted center in which the sodium atom is shifted from the regular Zn position. A geometrical interpretation of the bond angle gives a value of $r_{nax} = 2.05$ Å for the nonaxial case if the shift is along the normal bond direction. The normal length of this bond is 1.97 Å. One can also find an estimate for the bond length r_{ax} in the axial case by considering that b reflects a dipole-type of state. The lengths in both cases should relate to each other like the sixth root of the relationship of the corresponding anisotropic hyperfine interaction parameters:

$$r_{ax}/r_{nax} = (b_{ax}/b_{nax})^{1/6} \tag{6.13}$$

Using this equation one finds a bond length $r_{ax} = 2.2$ Å, that means the bond is longer in the axial case (which is parallel c).

The g-tensor allows an estimate of the energy separation between the acceptor state and the valence band. The g-shift is related to this energy separation and the spin-orbit coupling of the atom that contributes to the state.[113] The spin-orbit coupling of sodium is close to zero, therefore the contribution comes from the oxygen ligand. Using the admixture factor η^2 one obtains for the energy separation:

$$\Delta E = -2\lambda(1-\eta^2)/(g_{\|}-g_{\perp}) \tag{6.14}$$

The values of 730 meV (axial centers) and 640 meV (nonaxial centers) are in agreement with the results of the photoluminescence measurements (see below). A detailed study of the EPR temperature dependence confirms these results on the energy barrier between both types of centers.[119]

The sodium-doped samples showed a strong luminescence line in the excitonic range followed by a broad, unstructured luminescence band in the visible spectral range (see Figure 6.20). This band can be decomposed into two Gaussian bands with peak energies of 2.48 eV (marked as II) and 2.15 eV (marked as I). Temperature-dependent measurements[120] of the half-width of band I revealed a Huang–Rhys factor S of about 16 ± 1.5 and participation of local phonons of $\hbar\omega = (45 \pm 5)$ meV. This leads to a (hypothetical) zero phonon transition at 2.87 eV (peak position at 2.15 eV plus $S\hbar\omega$), that is, 580 meV below the band gap energy in fair agreement with the data analysis of the g-anisoptropy. The corresponding values for the Li acceptor are: peak energy at 2.1 eV, a Huang–Rhys factor S of about 13 ± 1 and participation of local phonons of $\hbar\omega = (50 \pm 5)$ meV, zero phonon transition at 2.75 eV (peak position plus $S\hbar\omega$), that is, 700 meV below the band gap energy. The values of 580 and 700 meV for the Na and Li acceptors are equivalent to the acceptor binding energies since the recombinations investigated in ODMR were of the shallow donor to deep acceptor type.

A direct correlation to the Na related photoluminescence band was obtained by ODMR excitation spectroscopy (Figure 6.20).[120] The sample shows overlapping broad emission bands in the visible; the Na related magnetic resonance signals were detected only on the deeper lying emission I, while emission II showed no magnetic resonance response.

The intense research to find suitable acceptors for the p-type doping has led to a reinvestigation of the behavior of the group-I elements over recent years. While in the old

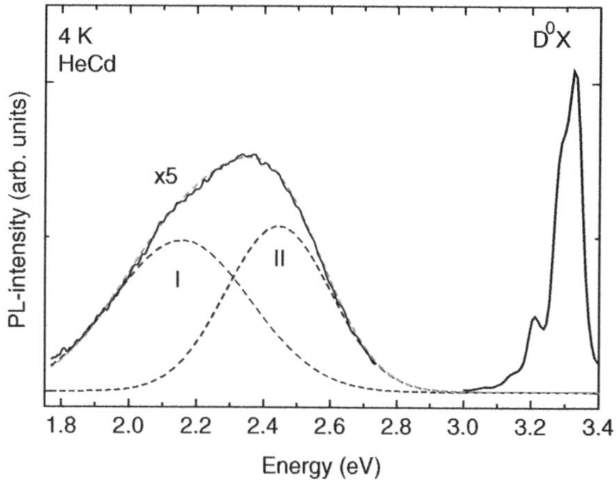

Figure 6.20 *Luminescence spectrum of a Na-doped ZnO single crystal showing the band edge excitonic recombinations and the deep centre emission decomposed into two Gaussian contributions. The ODMR spectrum of the Na acceptor is only detected on band I. Reprinted with permission from F.H. Leiter, Thesis, Gießen, 2002*

work bulk crystals and high diffusion temperatures were used, now moderate temperatures (<800 °C) and epitaxial films came into play. Evidence for a Li related center causing a DAP recombination with a ZPL at 3.05 eV was found by Sann *et al.*[121]

Figure 6.21 shows a low-temperature photoluminescence measurement of a sample where Li was introduced during growth. A newly observed luminescence starts at around 3.1 eV, and shows pronounced coupling to longitudinal optical (LO) phonons separated by

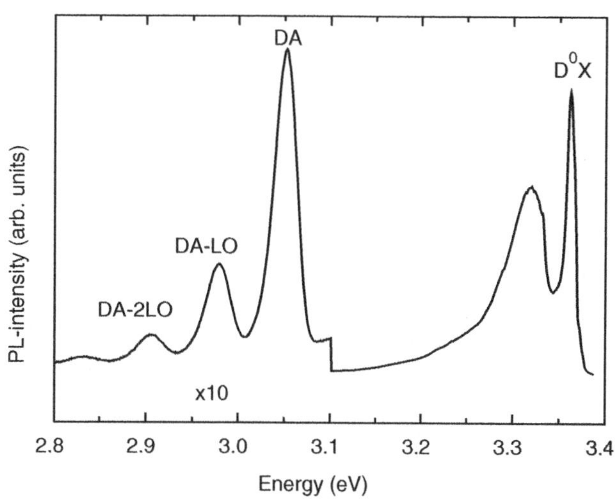

Figure 6.21 *Luminescence spectrum at $T = 4.2$ K of a Li-doped ZnO epitaxial film grown by CVD. The position of the DAP band at 3.05 eV and its LO replica are indicated. Reprinted from J. Sann, A. Hofstaetter, D. Pfisterer, J. Stehr, and B.K. Meyer, phys. stat. sol.(c) 3, 952. Copyright (2006) with permission from WILEY-VCH Verlag GmbH & Co. KGaA, Weinheim*

72 meV. The shape is typical for DAP recombination. The coupling strength given by the Huang–Rhys factor S is estimated from the Poisson distribution $I = I_0 S^n/n!$ where $n = 0, 1, 2, \ldots$ and I_0 is the intensity of the first, the ZPL. For S one obtains a value of 0.35 ± 0.05 which can be compared with the value of 0.55 ± 0.05 obtained for the nitrogen acceptor with a binding energy of 165 meV.[66] If we assume a value of 50 meV for the donor binding energy, and neglect the coulombic interaction one obtains an acceptor binding energy around 300 meV, and one may wonder why the phonon coupling strength is so small.

In 1960 Lander[122] reported on the donor and acceptor properties of Li and gave first indications about the possible defect structure. He proposed the interplay between interstitial Li donors and substitutional Li_{Zn} acceptors. Lander[122] observed a decrease in conductivity which was dependent on the oxygen pressure during doping and concluded that for the realization of p-type ZnO:Li very high (and not achievable) oxygen pressures were needed. The main reason for the conversion from n-type to high resistive ZnO:Li was presented in the work of Zwingel.[123] He showed that the conductivity decreased in inverse proportion to the square of the Li concentration. Only 1% of the total Li concentration was effective in the compensation of the donors already present in the crystal. He assumed equal concentrations for the interstitial Li donors and Li_{Zn} acceptors and postulated the existence of another Li acceptor, needed to model the reaction kinetics. It should arise from the pairing of two Li centers. The luminescence/EPR investigations[121] do not allow to conclude whether one or two Li atoms are part of the 300 meV acceptor. For the epitaxial films Li and Na doping in all cases enhanced significantly the luminescence in the yellow spectral range (compared with undoped samples where the Cu emission with well resolved phonon structure is observed). This can be used as an indication that also the deep Li (Na) acceptor was formed, and that there is a branching ratio between the shallow and deep Li centers. A defect model of shallow Li with nearby H donors was proposed.[121] A similar model was proposed by Halliburton *et al.*[124] based on a combined investigation of the infrared absorption and magnetic resonance.

Novel aspects on the group-I elements come from the synthesis of ZnO nanocrystals. Here LiOH or NaOH are used as oxygen precursors and the materials are prepared at room temperature. ENDOR investigations show that under such conditions Na and Li can exist in interstitial positions thus being shallow donors. Together with the Li_{Zn} acceptors charge neutrality is obtained in the nanocrystals by this type of extrinsic-self-compensation process.[125]

The most crucial point for the future of ZnO as material for electronic and optoelectronic devices is the ability to dope it reliable p-type and thus to find an efficient shallow acceptor which is able to release holes in sufficient concentration to the valence band. Meanwhile, many doping techniques and various growth processes have been applied to achieve this goal, however, with limited success. From the point of characterization by means of magnetic resonance methods to our knowledge only information on two types of species was obtained. One is the Zn vacancy and related defects, and the other one is nitrogen. Nitrogen can be expected to occupy an oxygen site considering the small size mismatch, and thus has the potential to be an acceptor. Indeed a nitrogen acceptor center compatible was detected in EPR, originally in single crystals, later also in nitrogen-treated ZnO powders.[108,126] Figure 6.22 shows the EPR signal of the nitrogen acceptor for different orientations of the magnetic field with respect to the *c*-axis of the ZnO crystal.

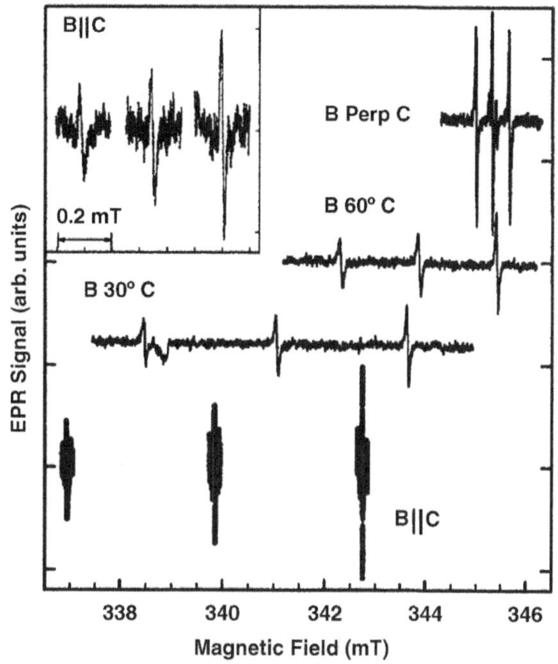

Figure 6.22 The EPR signal of the N acceptor at selected orientations. Reprinted from W. E. Carlos, E. R. Glaser, and D. C. Look, Magnetic resonance studies of ZnO, Physica B 308–310, 976. Copyright (2001) with permission from Elsevier

The g-values are not what one expects for shallow effective-mass-type acceptors (see Table 6.1 and Figure 6.23), also the observation of hyperfine interaction with the nitrogen N = 1 nuclear spin rules against a shallow center. Carlos *et al.*[108] remarked that the large anisotropy in the hyperfine interaction suggests that a p-orbital (p-hole) is involved. From

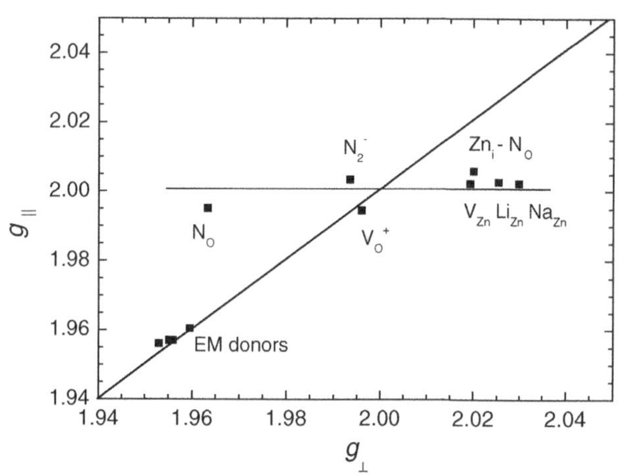

Figure 6.23 Compilation of the g-values of shallow and deep defects in ZnO. The diagonal line refers to isotropic g-values; the horizontal line is a guide for the eye

an atomic orbital analysis follows that the center has 96% p and 4% s character, and has a 4% localization of the nitrogen atom. It is obvious that from the spin Hamilton parameters and their analysis one deals with a deep and not a shallow acceptor. Carlos *et al.*[108] pointed out the similarity to the deep states of As and P in ZnSe – a p-orbital on the impurity atom which is Jahn–Teller distorted. While this is certainly correct one may notice: As and P in ZnSe also introduce shallow effective-mass-type acceptor states, and depending on the growth method and growth parameters (melt grown or low temperature epitaxial growth, II/VI ratio, etc.) there can be preference which acceptor state is favored. The shallow acceptor state introduced by nitrogen doping, by diffusion and implantation (see above) obviously has so far escaped detection by magnetic resonance techniques.

Apart from this deep N^- acceptor also molecular nitrogen (N_2^-) acceptors were found.[127,128] Studying the annealing behavior of these two N centers,[127] compensation (probably by hydrogen) was noticed as well as the deactivation of the centers at temperatures above 800 °C. Here we also would like to draw attention to the work of Aliev *et al.*[129] who obtained evidence that compensation of the nitrogen acceptors may also happen via the pairing with zinc interstitials. Such defects models were also found in ammonia-treated ZnO powder samples by Sann *et al.* [130]

6.3.4 Intrinsic Acceptors

The first information on Zn vacancies was obtained in the 1970s. Room temperature electron- and neutron-irradiated samples were studied.[131–133] More recently the work of Vlasenko and Watkins[109] revealed the creation and annealing behavior of V_{Zn} by low temperature electron irradiation. Frenkel pairs of the Zn vacancy with the Zn interstitial were observed and the annealing stages below 170 K were related to the migration of the Zn interstitials. In correlating these ODMR results to the photoluminescence bands present in the material and comparing the results with the better studied case of ZnSe, the authors obtained evidence that the Zn vacancy levels are at least 350 meV above the valence band or even higher.

In summary, while the identification of the impurities, dopants and intrinsic defects in ZnO by magnetic resonance methods is in a good state there is a lack of data which relates this information to electrical or optical properties of the material.

Note, the magnetic resonance data of the transition metal elements Sc.....Cu were not considered in this chapter. They are defects in their own right and would demand an independent compilation.[134–139] While they are common defects in ZnO bulk crystals they play a minor role in thin films relevant for optoelectronic applications. Using the same argument, defects such as Mo[140] and Pb[141] in ZnO have not been reviewed.

References

[1] D.G. Thomas, *J. Phys. Chem. Solids* **15**, 86 (1960).
[2] J.J. Hopfield, *J. Phys. Chem. Solids* **15**, 97 (1960).
[3] Y.S. Park. C.W. Litton, T.C. Collins and D.C. Reynolds, *Phys. Rev.* **143**, 512 (1966).
[4] B. Segall, *Phys. Rev.* **163**, 769 (1967).
[5] W.Y. Liang and A.D. Yoffe, *Phys. Rev. Lett.* **20**, 59 (1968).
[6] K. Hümmer, *Phys. Status Solidi.* **56**, 249 (1973).

[7] M. Rosenzweig, Diploma thesis, TU Berlin, 1975.
[8] G. Blattner, G. Kurtze, G. Schmieder and C. Klingshirn, *Phys. Rev. B* **25**, 7413 (1982).
[9] B.C. Reynolds, D.C. Look, B. Jogai, C.W. Litton, G. Cantwell and W.C. Harsch, *Phys. Rev. B* **60**, 2340 (1999).
[10] B. Gil, *Phys. Rev. B* **64**, *201310 (R)* (2001); see also B. Gil, A. Lusson, V. Sallet, R. Triboulet and P. Bigenwald, *Jpn J. Appl. Phys. Lett.* (to be published).
[11] W.L.R. Lambrecht, A.V. Rodina, S. Limpijumnong, B. Segall and B.K. Meyer, *Phys. Rev. B* **65**, 075207 (2002).
[12] J. Gutowski, N. Presser and I. Broser, *Phys. Rev. B* **38**, 9746 (1988).
[13] G. Blattner, C. Klingshirn, R. Helbig and R. Meinl, *Phys. Status Solidi B* **107**, 105 (1981).
[14] P. Loose, M. Rosenzweig and M. Wöhlecke, *Phys. Status Solidi B* **75**, 137 (1976).
[15] D.C. Reynolds, C.W. Litton and T.C. Collins, *Phys. Rev.* **140**, A1726 (1965).
[16] D.C. Reynolds and T.C. Collins, *Phys. Rev.* **185**, 1099 (1969).
[17] B.K. Meyer, H. Alves, D.M. Hofmann, W. Kriegseis, D. Forster, F. Bertram, J. Christen, A. Hoffmann, M. Straßburg, M. Dworzak, U. Haboeck and A.V. Rodina, *Phys. Status Solidi B* **241**, 231 (2004).
[18] K. Thonke, Th. Gruber, N. Teofilov, R. Schönfelder, A. Waag and R. Sauer, *Physica B* **308–310** 945 (2001).
[19] D.C. Reynolds, D.C. Look, B. Jogai, C.W. Litton, T.C. Colins, W. Harsch and G. Cantwell, *Phys. Rev. B* **57**, 19 (1998).
[20] K. Nassau, C.H. Henryand J.W. Shiever,in *Proceedings of the Tenth International Conference of the Physics of Semiconductors*, NBS, Springfield, 1970, p. 629.
[21] B.K. Meyer, J. Sann, S. Lautenschlaeger, M.R. Wagner and A. Hoffmann, *Phys. Rev. B* **76**, 184120 (2007).
[22] J.L. Merz, H. Kukimoto, K. Nassau and J.W. Shiever, *Phys. Rev. B* **6**, 545 (1972).
[23] B. Šantic, C. Merz; U. Kaufmann, R. Niebuhr, H. Obloh and K. Bachem, *Appl. Phys. Lett.* **71**, 1837 (1997).
[24] D.M. Hofmann, A. Hofstaetter, F. Leiter, H. Zhou, F. Henecker and B.K. Meyer, *Phys. Rev. Lett.* **88**, 045504 (2002).
[25] M. Schilling, R. Helbig and G. Pensl, *J. Lumin.* **33**, 201 (1985).
[26] C. Gonzales, D. Block, R.T. Cox and A. Hervé, *J. Crystal Growth* **59**, 357 (1982).
[27] H.J. Ko, Y.F. Chen. S.K. Hong, H. Wenisch and T. Yao, *Appl. Phys. Lett.* **77**, 3761 (2000).
[28] F. Reuss, C. Kirchner, Th. Gruber, R. Kling, S. Maschek, W. Limmer, A. Waag and P. Ziemann, *J. Appl. Phys.* **95**, 3385 (2004).
[29] S. Müller, D. Stichtenoth, M. Uhrmacher, H. Hofsäss, C. Ronning and J. Röder, *Appl. Phys. Lett.* **90**, 012107 (2007).
[30] J.R. Haynes, *Phys. Rev. Lett.* **4**, 361 (1960).
[31] A.V. Rodina, M. Strassburg, M. Dworzak, U. Haboeck, A. Hoffmann, A. Zeuner, H.R. Alves, D.M. Hofmann and B.K. Meyer, *Phys. Rev. B* **69**, 125206 (2004).
[32] D.G. Thomas, *J. Phys. Chem. Solids* **3**, 229 (1957).
[33] K.I. Hagemark, *J. Solid State Chem.* **16**, 293 (1976).
[34] A.F. Kohan, G. Ceder, D. Morgan and C.G. Van de Walle, *Phys. Rev. B* **61**, 15019 (2000).
[35] D.C. Oh, T. Suzuki, J.J. Kim, H. Makino, T. Hanada, M.W. Ch,and T. Yao, *Appl. Phys. Lett.* **86**, 032909 (2005).
[36] N. Shohata, T. Matsumura and T. Ohno, *Jpn. J. Appl. Phys.* **19**, 1793 (1980).
[37] A. Rohatgi, S.K. Pang, T.K. Gupta and W.D. Straub, *J. Appl. Phys.* **63**, 5375 (1988).
[38] W.I. Lee, R.L. Young and W.K. Chen, *J. Appl. Phys., Part 2* **35**, L1158 (1996).
[39] N. Ohashi, J. Tanaka, T. Ohgaki, H. Haneda, N. Ozawa and T. Tsurumi, *J. Mater. Res.* **17**, 1529 (2002).
[40] F.D. Auret, S.A. Goodman, M. Legodi, W.E. Meyer and D.C. Look, *Appl. Phys. Lett.* **80**, 1340 (2002).
[41] D.C. Look, J. W. Hemsky and J. R. Sizelove, *Phys. Rev. Lett.* **82**, 2552 (1999).
[42] B.K. Meyer, S. Lautenschlaeger, S. Graubner, C. Neumann and J. Sann, *Mater. Res. Soc. Symp.* **891**, 0891-EE08-02.1 (2006).
[43] Hang-Ju Ko, Takafumi Yao, Yefan Chen and Soon-Ku Hong, *J. Appl. Phys.* **92**, 8 (2002).

[44] A.P. Roth and D.F. Williams, *J. Appl. Phys.* **52**, 6685 (1982).
[45] E. Ziegler, A. Heinrich, H. Oppermann and G. Stoever, *Phys. Status Solidi A* **66**, 635 (1981).
[46] Y. Natsume, H. Sakata and T. Hirayama, *Phys. Status Solidi A* **148**, 484 (1995).
[47] Y. Natsume, H. Sakata, T. Hirayama and H. Yanagida, *J. Appl. Phys.* **72**, 4203 (1992).
[48] K.I. Hagemark and L.C. Chacka, *J. Solid State Chem.* **15**, 261 (1975).
[49] K. Minegishi, Y. Koiwai, Y. Kikuchi, K. Yano, M. Kasuga and A. Shimizu, *Jpn. J. Appl. Phys.* **36**, L1453 (1997).
[50] M. Joseph, H. Tabata and T. Kawai, *Jpn. J. Appl. Phys.* **38**, L1205 (1999).
[51] D.C. Look, D.C. Reynolds, C.W. Litton, R.L. Jones, D.B. Eason and G. Gantwell, *Appl. Phys. Lett.* **81**, 1830 (2002).
[52] Y.R. Ryu, S. Zhu, D.C. Look, J.M. Wrobel, H.M. Jeong and H.W. White, *J. Crystal Growth* **216**, 330 (2000).
[53] Y.R. Ryu, W.J. Kim and H.W. White, *J. Crystal Growth* **219**, 419 (2000).
[54] Y.R. Ryu, T.S. Lee and H.W. White, *Appl. Phys. Lett.* **83**, 87 (2003).
[55] A. Tsukazaki, A. Ohtomo, T. Onuma, M. Oktami, T. Makino, M. Sumiya, K. Oktami, S.F. Chichibu, S. Fuke, Y. Segawa, H. Ohno, H. Koinuma and M. Kawasaki, *Nat. Mater.* **4**, 42 (2005).
[56] K.-K. Kim, H.-S. Kim, D.-K. Hwang, J.-H. Lim and S.-J. Park, *Appl. Phys. Lett.* **83**, 63 (2003).
[57] O.F. Schirmer, *J. Phys. Chem. Solids* **29**, 1407 (1968).
[58] D. Zwingel and F. Gärtner, *Solid State Commun.* **14**, 45 (1974).
[59] R.T. Cox, D. Block, A. Hervé, R. Picard and C. Santier, *Solid State Commun.* **25**, 77 (1978).
[60] J. Qiu, J.M. De Puydt, H. Cheng and M.A. Haase, *Appl. Phys. Lett.* **59**, 2992 (1991).
[61] L. Svob, C. Thiandoume, A. Lusson, M. Bouanani, Y. Marfaing and O. Gorochov, *Appl. Phys. Lett.* **76**, 1696 (2000).
[62] C.H. Park, S.B. Zhang,and S.-H. Wei, *Phys. Rev. B* **66**, 073202 (2002).
[63] S.B. Zhang, S.-H. Wie and A. Zunger, *Phys. Rev. B* **63**, 075205 (2001).
[64] E.-C. Lee, K. Y.-G. Jin and K.J. Chang, *Phys. Rev. B* **64**, 085120 (2001).
[65] G. Xiong, K.B. Ucer, R.T. Williams, J. Lee, D. Bhattacharyya, J. Metson and P. Evans, *J. Appl. Phys.* **97**, 043528 (2005).
[66] A. Zeuner, H. Alves, D.M. Hofmann, B.K. Meyer, A. Hoffmann, U. Haboeck, M. Strassburg and M. Dworzak, *Phys. Status Solidi B* **234**, R7 (2002).
[67] J.F. Rommeluère, L. Svob, F. Jomard, J. Mimila-Arroyo, A. Lusson. V. Sallet and Y. Marfaing, *Appl. Phys. Lett.* **83**, 287 (2003).
[68] S. Yamauchi, Y. Goto and T. Hariu, *J. Cryst. Growth* **260**, 1 (2004).
[69] T. Tamura, T. Makino, A. Tsukazaki, M. Sumiya, S. Fuke, T. Furumochi, M. Lippmaa, C.H. Chia, Y. Segawa, H. Koinuma and M. Kawasaki, *Solid State Commun.* **127**, 265 (2003).
[70] B.K. Meyer, J. Sann, D.M. Hofmann, C. Neumann and A. Zeuner, *Semicond. Sci. Technol.* **20**, S62 (2005).
[71] A. Kaschner, U. Haboeck, Martin Strassburg, Matthias Strassburg, G. Kaczmarczyk, A. Hoffmann, C. Thomsen, A. Zeuner, H.R. Alves, D.M. Hofmann and B.K. Meyer, *Appl. Phys. Lett.* **80**, 1909 (2002).
[72] Lijun Wang and N.C. Giles, *Appl. Phys. Lett.* **84**, 16 (2004).
[73] T. C. Damen, S. P. S. Porto and B. Tell, *Phys. Rev.* **142**, 570 (1966).
[74] J. M. Calleja and M. Cardona, *Phys. Rev. B* **16**, 3753 (1977).
[75] X. Wang, S. Yang, J. Wang, M. Li, X. Jiang, G. Du, X. Liu and R. P. H. Chang, *J. Cryst. Growth* **226**, 123 (2001).
[76] L. Sugiura, M. Suzuki and J. Nishio, *Appl. Phys. Lett.* **72**, 1748 (1998).
[77] E.-C. Lee, Y.-S. Kim, Y.-G. Jin and K. J. Chang, *Phys. Rev. B* **64**, 085120 (2001).
[78] A. Kaschner, H. Siegle, G. Kaczmarczyk, M. Straßburg, A. Hoffmann, C. Thomsen, U. Birkle, S. Einfeldt and D. Hommel, *Appl. Phys. Lett.* **74**, 3281 (1999).
[79] S. Lautenschläger,unpublished, Giessen (2007).
[80] M. Schirra, R. Schneider, A. Reiser, G.M. Prinz, M. Feneberg, J. Biskupek, U. Kaiser, C.E. Krill, K. Thonke and R. Sauer, *Phys. Rev. B* **77**, 125215 (2008).
[81] T.S. Jeong, M.S. Han, C.J. Youn and Y.S. Park, *J. Appl. Phys.* **96**, 175 (2004).

[82] D.-K. Hwang, H.-S. Kim, J.-H. Lim, J.-Y. Oh, J.-H. Yang, S.-J. Park, K.-K. Kim, D.C. Look and Y.S. Park, *Appl. Phys. Lett.* **86**, 151917 (2005).
[83] J.D. Ye, S.L. Gu, F. Li, M. Zhu, R. Zhang, Y. Shi, Y.D. Zheng, X.W. Sun, G.Q. Lo and D.L. Kwang, *Appl. Phys. Lett.* **90**, 152108 (2007).
[84] F.X. Xiu, Z. Yang, L.J. Mandalapu, D.T. Zhao and J.L. Liu, *Appl. Phys. Lett.* **87**, 252102 (2005).
[85] N. Volbers, S. Lautenschläger, T. Leichtweiss, A. Laufer, S. Graubner, B.K. Meyer, K. Potzger and S. Zhou, *J. Appl. Phys.* **103**, 123106 (2008).
[86] A. Abragam and B. Bleany, *Electron Paramagnetic Resonance of Transition Ions,* Oxford University Press, Oxford, 1970.
[87] J. Schneider and A. Räuber, *Z. Naturforsch.* **16a**, 712 (1961).
[88] P.H. Kasai, *Phys. Rev.* **130**, 989 (1963).
[89] M. Schulz, *Phys. Status Solidi A* **27**, K5 (1975).
[90] D. Block, A. Hervé and R.T. Cox, *Phys. Rev. B* **25**, 6049 (1982).
[91] N.Y. Garces, N.C. Giles, L.E. Halliburton, G. Cantwell, D.B. Eason, D.C. Reynolds and D.C. Look, *Appl. Phys. Lett.* **80**, 1334 (2002).
[92] S. Orlinskii, J. Schmidt, P. G. Baranov, V. Lormann, I. Riedel, D. Rauh and V. Dyakonov, *Phys. Rev. B* **77**, 115334 (2008).
[93] C. G. Van de Walle and J. Neugebauer, *Nature* **423**, 626 (2003).
[94] S. F. Cox, E. A. Davis, S. P. Cottrell, P. J. King, J. S. Lord, J. M. Gil, H. V. Alberto, R. C. Vilão, J. Piroto Duarte, N. Ayres de Campos, A. Weidinger, R. L. Lichti,and S. J. Irvine, *Phys. Rev. Lett.* **86**, 2601 (2001).
[95] Yu. V.Gorelkinskii and G.D. Watkins, *Phys. Rev. B* **69**, 115212 (2004).
[96] L.S. Vlasenko and G.D. Watkins, *Phys. Rev B* **71**, 125210 (2005).
[97] L.S. Vlasenko and G.D. Watkins, *Phys. Rev B* **72**, 035203 (2005).
[98] N.T. Son, I.G. Ivanov, A. Kuznetsov, B.G. Svensson, Q.X. Zhao, M. Willander, N. Morishita, T. Oshima, H. Itoh, J. Isoya, E. Janzén and R. Yakimova, *J. Appl. Phys.* **102**, 093504 (2007).
[99] J.M. Smith and W.E. Vehse, *Phys. Lett.A* **31**, 147 (1970).
[100] V. Soriano and D. Galland, *Phys. Status Solidi B* **77**, 739 (1976).
[101] C. Gonzales, D. Galland and A. Herve, *Phys. Status Solidi B* **72**, 309 (1975).
[102] B. Schallenberger and A. Hausmann, *Z. Phys. B* **23**, 177 (1976).
[103] B. Schallenberger and A. Hausmann, *Z. Phys. B: Condens. Matter* **44**, 143 (1981).
[104] R. Laio, L.S. Vlasenko and P.M. Vlasenko, *J. Appl. Phys.* **103**, 12379 (2008).
[105] S.B. Zhang, S.-H. Wei and A. Zunger, *Phys. Rev. B* **63**, 075205 (2001).
[106] F.H. Leiter, H.R. Alves, A. Hofstaetter, D.M. Hofmann and B.K. Meyer, *Phys. Status Solidi B* **226**, R4 (2001).
[107] A. Janotti and C.G. Van de Walle, *Phys. Rev. B* **76**, 165202 (2007).
[108] W.E. Carlos, E.R. Glaser and D.C. Look, *Physica B* **308–310** 976 (2001).
[109] L.S. Vlasenko and G.D. Watkins, *Phys. Rev. B* **71**, 125210 (2005).
[110] P. Edel, F.J. Ahlers, C.M. Mc Donagh, B. Henderson and J.M. Spaeth, *J. Phys. C* **15**, 4913 (1982).
[111] P. Edel, C. Hennies, Y. Merle d'Aubigné, R. Romestain and Y. Twarowski, *Phys. Rev. Lett.* **28**, 19 (1972).
[112] B. Henderson and G.F. Imbusch,in *Optical Spectroscopy of Inorganic Solid,* Clarendon Press, Oxford, 1989, p. 563.
[113] J.J. Davies, G.N. Aliev, S.J. Bingham, D. Wolverson, S. Stepanov, B. Yavich and W.N. Wang, *Phys. Rev. B* **67**, 035203 (2003).
[114] J. Schneider and O. Schirmer, *Z. Naturforsch.* **18a**, 20 (1963).
[115] O.F. Schirmer, *J. Phys. Chem. Solids* **29**, 1407 (1968).
[116] D. Zwingel and F. Gärtner, *Solid State Commun.* **14**, 45 (1974).
[117] O.F. Schirmer, H.A. Müller and J. Schneider, *Phys. Kondens. Mater.* **3**, 323 (1965).
[118] J.R. Morton and K.F. Preston, *J. Magn. Reson.* **30**, 577 (1978).
[119] B.K. Meyer, A. Hofstaetter and V.V. Laguta, *Physica B* **376**, 682 (2006).
[120] F.H. Leiter,PhD thesis, Gießen, 2002.

[121] J. Sann, A. Hofstaetter, D. Pfisterer, J. Stehr and B.K. Meyer, *Phys. Status Solidi C* **3**, 952 (2006).
[122] J.J. Lander, *J. Phys. Chem. Solids* **15**, 324 (1960).
[123] D. Zwingel, *J. Lumin.* **5**, 385 (1972).
[124] L.E. Halliburton, Lijun Wang, Lihua Bai, N.Y. Garces, N.C. Giles, M.J. Callahan and Buguo Wang, *J. Appl. Phys.* **96**, 7168 (2004).
[125] S.B. Orlinskii, J. Schmidt, P.G. Baranov, D.M. Hofmann, C.M. Donega and A. Meijerink, *Phys. Rev. Lett.* **92**, 047603–1 (2004).
[126] D. Pfisterer, J. Sann, D.M. Hofmann, M. Plana, A. Neumann, M. Lerch and B.K. Meyer, *Phys. Status Solidi B* **243**, R1 (2006).
[127] N.Y. Garces, Lijun Wang, N.C. Giles, L.E. Halliburton, G. Cantwell and D.B. Eason, *J. Appl. Phys.* **94**, 519 (2003).
[128] S. Larach and J. Turkevich, *J. Phys. Chem. Solids* **29**, 1519.
[129] G.N. Aliev, S.J. Bingham, D. Wolverson, J.J. Davis, H. Makino, H.J. Ko and T. Yao, *Phys. Rev. B* **70**, 115206 (2004).
[130] J. Sann, J. Stehr, A. Hofstaetter, D.M. Hofmann, A. Neumann, M. Lerch, U. Haboeck, A. Hoffmann and C. Thomsen, *Phys. Rev. B* **76**, 195203 (2007).
[131] K. Leutwein and J. Schneider, *Z. Naturforsch.* **26a**, 1236 (1971).
[132] A.L. Taylor, G. Filipovich and G.K. Lindeberg, *Solid State Commun.* **8**, 1359 (1970).
[133] D. Galland and A. Herve, *Solid State Commun.* **14**, 953 (1974).
[134] P.B. Dorain, *Phys. Rev.* **112**, 1058 (1958).
[135] W.W. Walsh and L.W. Rupp, *Phys. Rev.* **126**, 952 (1962).
[136] A. Hausmann, *Phys. Status Solidi* **31**, K131 (1969).
[137] W.C. Holton, J. Schneider and T.L. Estle, *Phys. Rev.* **133**, A1638 (1964).
[138] R.E. Dietz, H. Kamimura, M.D. Sturge and A. Yariv, *Phys. Rev.* **132**, 1559 (1963).
[139] A. Hausmann and E. Blaschke, *Z. Phys.* **230**, 255 (1970).
[140] J. Stehr, C. Knies, A. Hofstaetter, D.M. Hofmann, W. Xu, Y. Zhou and X. Zhang, *Solid State Commun.* **145**, 95 (2008).
[141] G. Born, A. Hofstaetter and A. Scharmann, *Z. Phys.* **240**, 163 (1970).

7

Vapor Transport Growth of ZnO Substrates and Homoepitaxy of ZnO Device Layers

Gene Cantwell[1], *Jizhi Zhang*[1] *and J.J. Song*[1,2]
[1]*ZN Technology, Inc., Brea, CA, USA*
[2]*Department of Electrical and Computer Engineering, University of California at San Diego, La Jolla, CA, USA*

7.1 Introduction

Vapor growth methods have been used to produce high quality single crystals of compound semiconductors in the II–VI family for many years.[1–4] This is primarily due to the high vapor pressures of the components of these compounds well below their melting points along with their typically high melting points. The vapor growth methods are usually either physical vapor transport (PVT) or chemical vapor transport (CVT). PVT, in this case, refers to the sublimation of the compound into components volatile at the growth temperatures and has often been favored for the more volatile II–VI compounds such as CdS, CdTe, ZnTe, CdSe, and ZnSe. The less volatile II–VI compounds, primarily ZnS and ZnO, often require a transporting agent to react with the compound and accomplish the vaporization of the compound's components at practical growth temperatures; thus chemical vapor transport or CVT. The potential transporting agents that have been used with ZnO form an extensive list including hydrogen, carbon, and group VII compounds.[5]

Zinc Oxide Materials for Electronic and Optoelectronic Device Applications, First Edition.
Edited by Cole W. Litton, Donald C. Reynolds and Thomas C. Collins.

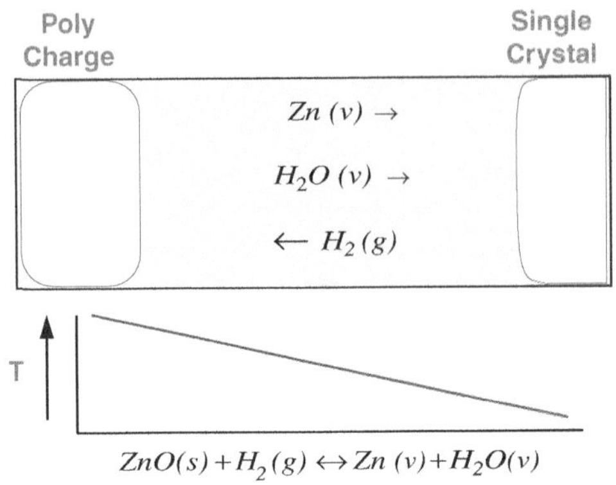

Figure 7.1 *Schematic illustration of the chemical vapor transport of ZnO in an enclosed ampoule*

CVT growth of ZnO is illustrated in Figure 7.1. In this configuration, a quartz ampoule is used to hold a polycrystalline charge of ZnO at one end with a seed crystal at the other end. By placing the ampoule in a temperature gradient such that the polycrystalline charge is at a higher temperature than the seed, the charge will be vaporized and transported to the cooler end where it recrystallizes on the seed crystal. To form a single crystal, the seed would ideally be of the same material as the growing crystal. However, in some cases, single crystal seeds composed of other materials have been used.[6] "Self-seeding" (no seed is used; the ampoule is usually configured to localize the initial growth to a few small single crystals) is also possible but usually requires a tapered ampoule such that the first deposited material will form at the tip in such a small size that it will naturally form a very small single crystal. The rest of the deposit then grows epitaxially on this seed and, as it gets thicker, expands in diameter as the ampoule tapers out. This can require an extended growth time for the growing crystal to expand to a usable diameter. A more economical method has been developed by ZN Technology in which a full diameter seed is used with a straight wall ampoule. This method has been shown to produce high quality, large area, single crystals of CdS, CdTe,[7] ZnTe, ZnSe,[8,9] and many ternaries. These crystals are typically twin free without low angle grain boundaries and have been grown at 2- and 3-in. diameters by the PVT method. The same methodology was adapted for use in the CVT of ZnO and 2-in. diameter single crystals have been grown. The addition of a full diameter seed to the PVT method results in the process being named the Seeded Physical Vapor Transport (SPVT) method and for CVT, the Seeded Chemical Vapor Transport (SCVT) method.

7.2 Transport Theory and Comparison with Growth Data

While PVT of ZnO is possible, the sublimation vapor pressures of zinc and oxygen over ZnO are very low at practical growth temperatures: $\sim 3 \times 10^{-6}$ atm at 1000 °C and $\sim 1 \times 10^{-4}$ atm at 1200 °C for the Zn partial pressure.[10] Typical Zn vapor partial pressures in the SCVT system are calculated[10] to be $\sim 1.5 \times 10^{-2}$ atm at 1000 °C and $\sim 1.9 \times 10^{-2}$ atm at 1200 °C. Thus the use of the chemical transporting agent should increase the transport and growth rates by a factor of approximately 10^4 at 1000 °C and 10^2 at 1200 °C. Most vapor growth is done in quartz ampoules and the typical upper limit for use of quartz is in the 1200–1300 °C range. Experimental measurements by ZN Technology of the transport rate of ZnO using PVT have typically resulted in rates lower than 1×10^{-6} mol $cm^{-2}\,h^{-1}$ which is too low to be practical for crystal growth. In these experiments, a buffer gas of helium at a pressure of 370 Torr was used with a temperature differential between the charge and the seed of ~50 °C. Even this low rate may have been enhanced by residual moisture or other impurities in the growth charge and system that may have acted as transporting agents and increased the transport rate beyond that of sublimation only.

The SCVT method provides the potential for much higher transport rates of ZnO at significantly lower temperatures. In this method, the transporting agent reacts with the source material causing a decrease in the partial pressure of the transporting agent and production of high vapor pressure products. The vapor products formed by the reaction then flow towards the growing crystal where the reverse reaction takes place. At the growth interface, the reverse reaction releases the transporting agent causing an increase in the transporting agent's partial pressure at that point. Therefore, the pressure differentials of the transporting agent and the crystal forming products force the net fluxes of the zinc and oxygen containing vapors to be from source to crystal while that of the transporting agent is in the opposite direction. This counter flow of gases does not normally exist in PVT and results in a slightly more complex flow system inside the ampoule.

Although not the only possibility, the transporting agent chosen for ZnO is hydrogen. The reversible chemical reaction of hydrogen with ZnO proceeds as:

$$\mathrm{ZnO(s)} + \mathrm{H_2(g)} \leftrightarrow \mathrm{Zn(v)} + \mathrm{H_2O(v)} \tag{7.1}$$

At the source (polycrystalline charge) end, the reaction proceeds from left to right (see Figure 7.1) resulting in vaporization of the solid ZnO through usage of the hydrogen. At the growth end, the reaction proceeds in the opposite direction with solid ZnO formed and molecular hydrogen released. Transport rates of $\sim 5 \times 10^{-4}$ mol $cm^{-2}\,h^{-1}$ were achieved by ZN Technology using a 20% mixture of hydrogen in helium.

While it would be possible to use pure hydrogen to increase the transport rate, growth of high quality crystals often requires reduced growth rates. To control the growth rate, the SCVT growth of ZnO has been conducted with a mixture of inert gas and hydrogen. This not only reduces the hydrogen partial pressure, a controlling factor in the transport rate, but also provides a medium through which the vapors must diffuse to reach the growing crystal. If conditions are set such that the transport rate is limited by the diffusion of the vapors from one end of the ampoule to the other, independent control of the growth rate and growth temperature can be achieved resulting in improved reproducibility and long term stability of the system.

Faktor *et al.* developed a one-dimensional model for vapor transport and applied it to crystal growth of CdS by PVT.[11] In their analysis, they used a combination of drift and diffusion to describe the flux of a vapor per unit cross-sectional area given by:

$$J_{\mathrm{vap}} = \frac{U}{RT} p_{\mathrm{vap}} - \frac{D}{RT} \frac{\mathrm{d}}{\mathrm{d}x} p_{\mathrm{vap}} \tag{7.2}$$

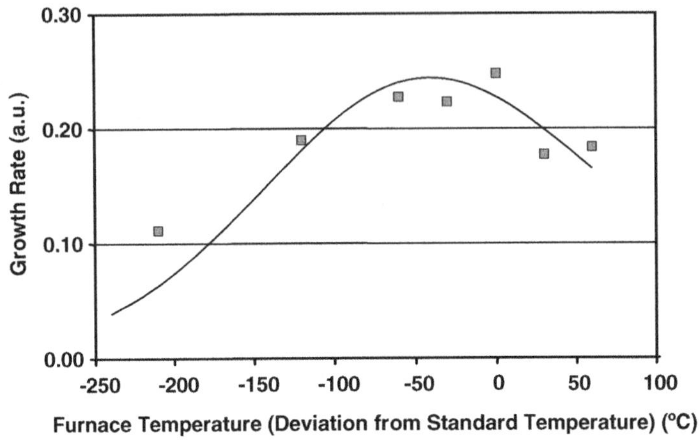

Figure 7.2 *Comparison of experimental growth rates (points) with the growth rate (solid line) as calculated by the thermodynamic model described in the text. Each point is the average growth rate for that growth temperature*

where the term $\frac{U}{RT}p_{\mathrm{vap}}$ represents the drift of the vapor and $\frac{D}{RT}\frac{\mathrm{d}}{\mathrm{d}x}p_{\mathrm{vap}}$ represents the diffusion term. Boone[12] originally applied these equations to the SCVT system to determine the transport and growth rates as functions of the growth parameters. This work was extended by ZN Technology to more closely model the current growth system. The results of these calculations along with experimental data are shown in Figure 7.2 where the growth rate of ZnO is plotted vs furnace temperature. The experimental values are averages over all growth runs conducted at each temperature. Though the fit of the calculated curve is not exact, it does exhibit the maximum in growth rate seen in the data.

In addition to growth rate, it is important to maintain an optimum stoichiometry in the vapor at the crystal growth interface for vapor growth of high quality II–VI compound crystals. In the case of the SCVT ZnO growth process, this means controlling the ratio of water, the oxygen bearing compound, to zinc vapor. A set of experiments by Reisman *et al.*[13] involving the gas phase reaction of Zn and H_2O vapors in an helium carrier gas indicated the conditions for producing smooth and transparent layers involved a $\frac{P_{H_2O}}{P_{Zn}}$ ratio of 10–50. Similar observations have been made for metal organic chemical vapor deposition (MOCVD) growth processes[14,15] except the ratios were in the 4–8 range. To accomplish vapor stoichiometry control, an external water vapor source was added to the SCVT ZnO growth system. This source permits control of the water vapor pressure in the growth ampoule. The one-dimensional growth model utilized for this growth system includes the external water vapor source. In conjunction with the crystal growth temperature the influence of the external water vapor source can be used to change the ratio of water vapor to zinc vapor partial pressures at the growing crystal interface from ~1.5 at high growth temperatures and low external water vapor pressure to ~80 at low growth temperatures and high external water vapor pressure. This is illustrated in Figure 7.3 by the two curves representing the lowest and highest temperatures used for ZnO crystal growth.

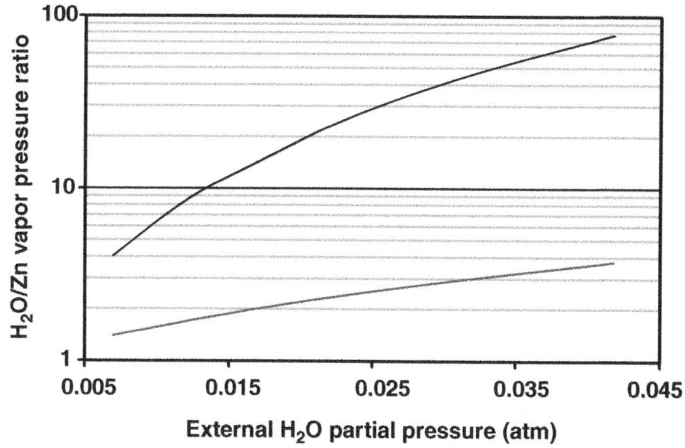

Figure 7.3 The ratio of water vapor to zinc vapor partial pressures at the growth interface vs the applied external water vapor pressure for two different growth temperatures. Upper curve: –240 °C; lower curve: +60 °C

7.3 Characterization

7.3.1 Crystallinity

High resolution X-ray omega rocking curves were used to check the crystallinity of the SCVT ZnO. In this procedure, the crystal is rocked along the omega axis through the Bragg peak and the detector is unmasked such that all of the diffracted radiation is collected. The values measured are the full width at half-maximum (FWHM), the peak height relative to the incident beam intensity, and the omega position of the Bragg peak. The rocking curves resulting from this procedure are similar to those in Figure 7.4. The

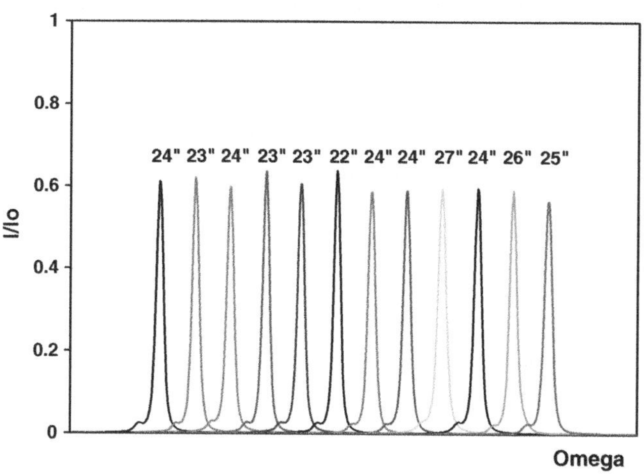

Figure 7.4 X-ray rocking curves (omega) for one of the best SCVT ZnO crystals. The measurements shown are for different spots along the diameter of the wafer and cover a length of approximately 40 mm

FWHM (shown in Figure 7.4 as the number above each peak in arc-sec) is measured as the width of the peak at a vertical position half the total height of the peak. The rocking curves shown in Figure 7.4 are from several positions across a single 40 mm diameter SCVT ZnO wafer. The uniformity of the FWHM values and the peak heights are indicative of both the high crystalline quality and excellent uniformity of the material. Although lower (better) FWHM values have been measured in individual areas on a wafer, the crystalline quality indicated by the FWHM values in Figure 7.4 should be sufficient for excellent homo-epitaxial results and subsequent device fabrication.

7.3.2 Purity

One characteristic of the SCVT ZnO growth process is the extremely high purity of the resulting ZnO crystals. This can be very important for homoepitaxial growth of ZnO since any impurities in the substrate may diffuse into the homoepitaxial film and alter its electrical characteristics. Glow discharge mass spectrometry (GDMS) was used to characterize the impurities in the SCVT ZnO crystals. Table 7.1 shows the results for the impurities detected in three samples from different growth runs. All other elements were either not detectable down to a few parts per billion atomic (ppba) or were a part of the sample matrix or sample mount such as zinc, oxygen, and tantalum. The majority of the impurity content is silicon and nitrogen. Silicon is expected since the crystals are grown in quartz ampoules. There has been no indication this level of silicon doping changes the electrical properties of SCVT ZnO. The amount of nitrogen present in the three samples shows significant variation, but is consistently present. This nitrogen presence may be related to imperfect sealing of the growth apparatus resulting in a minor component of nitrogen in the growth atmosphere. Of the remaining detected impurities, only aluminum and boron are known to be n-type dopants and their total concentration of ~20–50 ppba is not likely to impact the doping of any epitaxial films grown on these substrates by outdiffusion from the substrate to the film. This high purity is also evidenced in the optical data discussed in a later section.

Table 7.1 *Results of GDMS of one of the SCVT ZnO crystals. Only detected elements are shown. Note that the data are shown in parts per million atomic (ppma)*

	Sample A (ppma)	Sample B (ppma)	Sample C (ppma)
B	0.012	0.012	0.028
C	0.040	0.004	0.091
N	0.028	0.180	1.200
Na	0.015	ND	ND
Al	0.009	0.007	0.020
Si	0.330	0.280	0.660
Ti	0.001	ND	ND
Sn	ND	0.077	ND
Pb	0.002	ND	ND

ND, not detected.

7.3.3 Electrical

All undoped ZnO grown by the SCVT method has been n-type conductive. Typical room temperature electrical characteristics are ~0.5 ohm cm with a mobility of ~200 $cm^2 V^{-1} s^{-1}$ and a net donor concentration of ~$6 \times 10^{16} cm^{-3}$. Growth parameters have been found to cause the net donor concentration to change, though not drastically. Figure 7.5 illustrates the decrease observed in the net donor concentration as the growth temperature decreases. Although the direct cause of this has not been investigated, as was shown in an earlier section, the thermodynamic model describing the transport process predicts that the H_2O/Zn ratio at the growth interface will increase with decreasing growth temperature. Thus, changes in the stoichiometry at the growth interface for different growth temperatures may be the cause of this trend. Changes in mobility are also observed, but they appear to be associated with the decrease in the net donor concentration. Figure 7.6 plots the mobilities and net donor concentrations for all samples measured showing the apparent correlation.

The conductivity of the bulk SCVT ZnO has been attributed to three donors by Look *et al.*:[16] one with a ~30 meV activation energy proposed to be due to a native defect complex donor, the second at ~44 meV activation energy associated with hydrogen interstitials, and the third at ~75 meV that is probably associated with a group III impurity (most likely Ga or Al) on a Zn site. From Hall vs temperature measurements, the density of each of these was estimated to be $4.5 \times 10^{15} cm^{-3}$ for the 30 meV donor level, $3.0 \times 10^{16} cm^{-3}$ for the 44 meV donor level, and $2.0 \times 10^{16} cm^{-3}$ for the 75 meV level. The acceptor density was calculated to be $1.3 \times 10^{15} cm^{-3}$. Therefore, the room temperature n-type conductivity of the SCVT ZnO is likely due to a combination of the three sources and the compensation level is low with acceptor densities more than a factor of 20 lower than the donor densities. In addition, the peak electron mobility has been measured at $> 2000 cm^2 V^{-1} s^{-1}$ at low temperatures after a thermal anneal that is thought to remove hydrogen.[17]

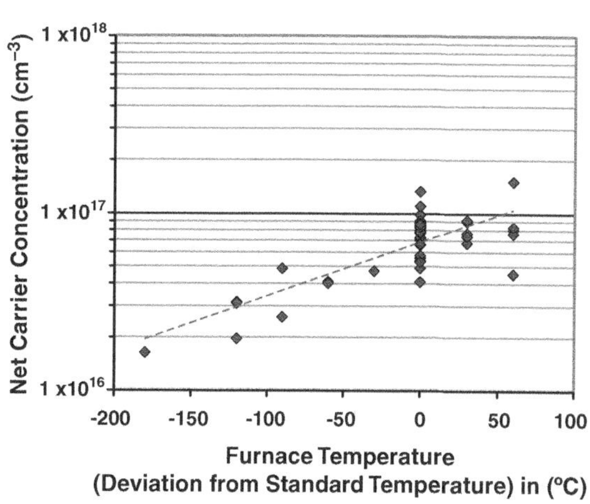

Figure 7.5 Plot of the net carrier concentrations for SCVT ZnO samples grown over a range of furnace set point temperatures

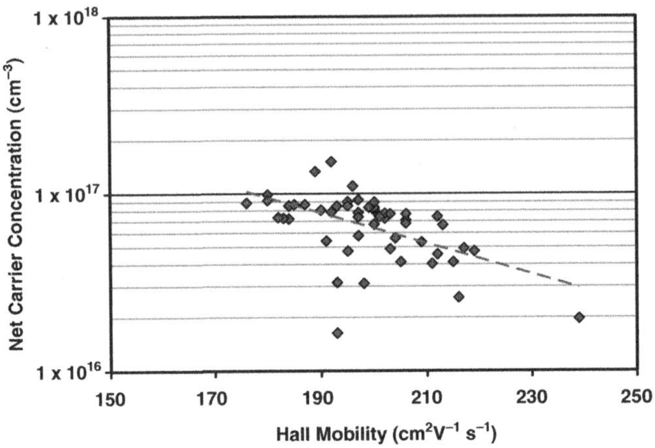

Figure 7.6 Plot of average net carrier concentrations of SCVT ZnO grown over a range of furnace set point temperatures

The donor level associated with hydrogen has been eliminated by annealing the SCVT ZnO in air at temperatures around 800 °C. Look *et al.*[16] report elimination of this donor level using a 715 °C anneal. This resulted in a reduction of the room temperature net donor concentration by a factor of ~1.7. The results of annealing in air at 1000 °C have been found to reduce the room temperature net donor concentration by a factor of two to three.[10]

7.3.4 Optical

SCVT grown ZnO has very good optical properties as measured by photoluminescence and cathodoluminescence. At room temperature, the typical photoluminescence response shows a strong UV peak at 378.6 nm along with a very low intensity visible (green) luminescence.[10] Cathodoluminescence also shows the extremely low value of the visible as compared with the UV luminescence. This is illustrated in Figure 7.7[18] where, in this experiment, the cathodoluminescence visible peak height is less than 1% of the UV peak height. Figure 7.7 also includes a comparison of the cathodoluminescence of SCVT ZnO, hydrothermal ZnO, and melt grown ZnO showing that the SCVT ZnO ratio of visible/UV luminescence is much better than the samples grown by the other two techniques.

Using low temperature cathodoluminescence, the large number of observable phonon replicas in Figure 7.8[18] also indicates the extremely high optical quality of the SCVT ZnO sample. The uniformity of the optical properties was also tested on this sample, showing less than 0.1% variation in the cathodoluminescence when mapped across the face of the wafer.[18] Similar luminescence spectra exhibiting phonon replicas are observed using low temperature photoluminescence.[19] Room temperature, optically pumped stimulated emission has also been observed in SCVT ZnO samples.[10] Figure 7.9 shows the spectra of the stimulated emission. The luminescence spectra are shown for various pump densities from below to above the stimulated emission threshold. This observation of room temperature stimulated emission in an SCVT ZnO substrate with no special optical cavity preparation demonstrates the exceptional optical quality of the SCVT ZnO and the efficiency of ZnO's radiative processes.

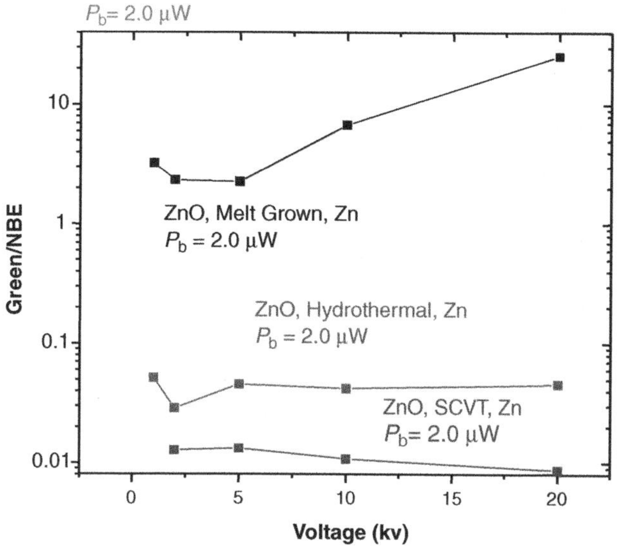

Figure 7.7 Ratio of green luminescence to near band edge (NBE) luminescence as a function of electron beam energy for SCVT grown bulk ZnO (lower plot) and those of two other methods showing that the SCVT ZnO has very low green luminescence. Reprinted from L. J. Brillson et al., Nanoscale depth-resolved cathodoluminescence spectroscopy of ZnO surfaces and metal interfaces, Superlattices and Microstructures 45, 206–213. Copyright (2009) with permission from Elsevier

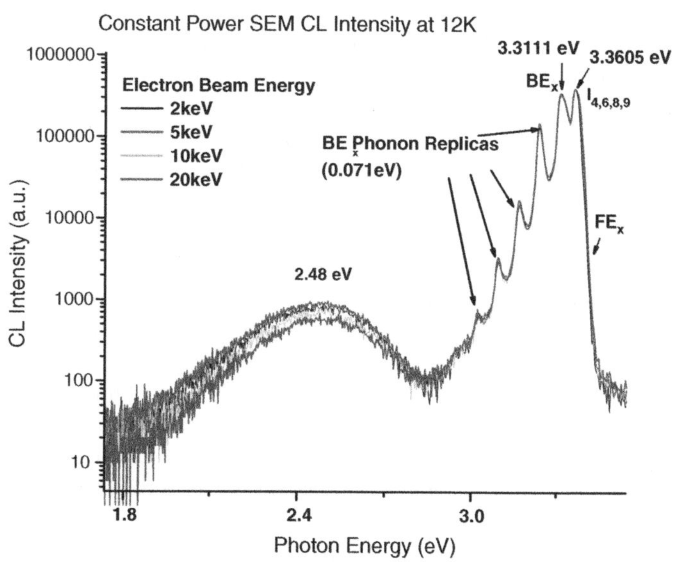

Figure 7.8 Cathodoluminescence (conducted at Ohio State University) of an (0001) SCVT ZnO substrate. The sample shows no discernible dependence on electron beam energy indicating that surface damage was not detectable. Reprinted from H. L. Mosbacker, et al., Role of subsurface defects in metal-ZnO(000) Schottky barrier formation, J. Vac. Sci. Technol. 25, 1405. Copyright (2007) with permission from American Vacuum Society

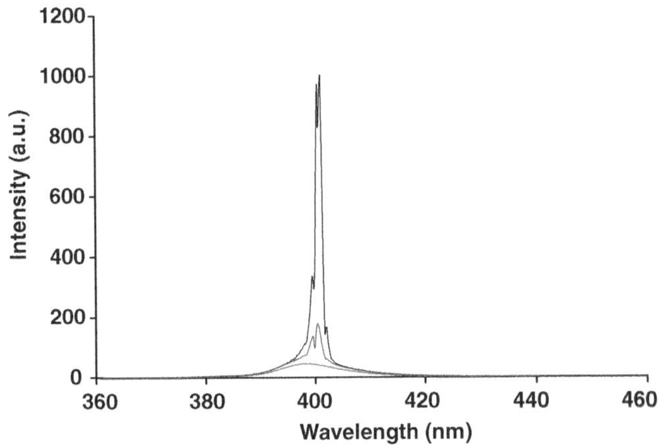

Figure 7.9 *Room temperature, optically pumped, stimulated emission from a centimeter square, standard chemo-mechanically polished SCVT ZnO wafer. At low pumping power, the emission peaks are broad*

7.4 In-situ Doping

Even though the SCVT ZnO is electrically conductive as grown, a higher conductivity would be beneficial for its use as a substrate for light-emitting diodes (LEDs) and lasers. By adding n-type dopants to the growth system, it is possible to incorporate them during growth. The key is to find dopants or dopant compounds that have appropriate volatility in the growth system with normal growth parameters. In addition, a method of control of the doping rate must be devised to keep its vapor concentration constant for the entire growth run to assure uniform doping. Group VII dopants such as fluorine, chlorine and bromine are potential candidates for use. However, they may also act as transporting agents, complicating the growth system; not to mention that they are hazardous and can severely deteriorate the growth equipment during an accidental leak. Possible group III dopants are boron, aluminum, gallium and indium. These are more benign but elemental forms are not volatile at the growth temperatures typically employed in the SCVT system. Most reports of highly doped, high conductivity ZnO utilize gallium as the dopant.[20–22] However, calculation of the probable dopant compound vapor pressures in the SCVT growth system showed that those most suited to maintaining a desirable and controllable vapor pressure were compounds of boron and indium. Both have previously been reported for use in thin film deposition methods to increase the conductivity of ZnO films.[23–25] Figure 7.10 shows plots of the calculated ratios of In/Zn and B/Zn in the vapor for a range of growth temperatures. As the ratio affects the degree of doping, there will be an optimum dopant to Zn vapor pressure ratio. This is usually determined experimentally but one could estimate that the optimum ratio should be in the 10^{-5}–10^{-3} range as a starting point. The optimum value will also depend upon how effectively the dopant is incorporated into the growing interface. The In/Zn ratio is higher than what is considered optimum but can be lowered somewhat by maintaining a high growth temperature. The B/Zn vapor ratio is much lower than the In/Zn ratio for the same temperatures. This ratio is also temperature dependent with the lowest ratio being at the highest growth temperatures.

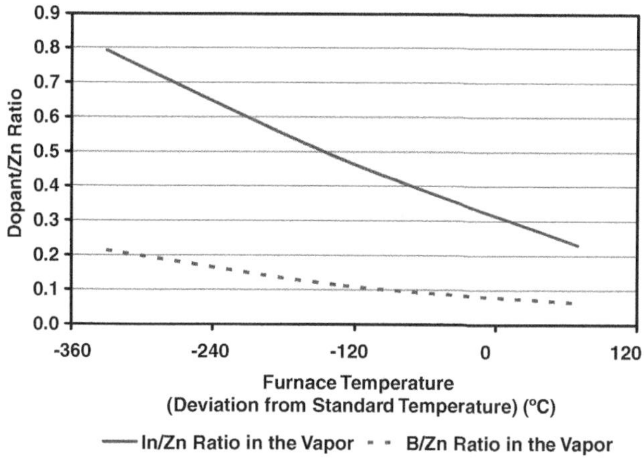

Figure 7.10 *Calculated ratios of dopant atoms to Zn atoms in the vapor of the SCVT growth system for boron and indium*

ZnO crystals were doped *in situ* by this method with indium and boron. Both dopants were successfully incorporated into ZnO ingots and the room temperature Hall measurement data for the resulting substrates are shown in Table 7.2. While the indium seems to be the more effective dopant of the two, giving a resistivity of the order of 10^{-3} ohm cm with a mobility of the order of $40\,cm^2\,V^{-1}\,s^{-1}$, the indium-doped ZnO was completely opaque for this doping level. The boron-doped ZnO crystals showed a much larger decrease in the room temperature Hall mobility ($6–10\,cm^2\,V^{-1}\,s^{-1}$) than did the indium-doped crystals but the transparency of the ZnO was much better. Further work is needed to find the correct dopant and concentration to give the increased conductivity without having a dramatic impact on the transparency of the ZnO.

7.5 ZnO Homoepitaxy

The primary use of SCVT ZnO is projected to be substrates for homoepitaxy of ZnO-based optoelectronic devices, particularly LEDs and laser diodes (LDs). Therefore, the ultimate test of the quality of the substrates is their ability to be used to produce high quality epitaxial layers. The results of the homoepitaxy (defined here as growth of ZnO-based

Table 7.2 *Room temperature Hall measurement results for boron- and indium-doped SCVT ZnO. A typical undoped result is included for comparison*

	Resistivity (ohm-cm)	Mobility ($cm^2/V\cdot s$)	Net carrier concentration (cm^{-3})
In Doped	7.9×10^{-3}	41.5	1.9×10^{19}
B Doped	1.2×10^{-1}	9.5	5.4×10^{18}
	2.6×10^{-1}	5.9	4.1×10^{18}
Typical Undoped	5.0×10^{-1}	200	6.0×10^{16}

compounds on ZnO substrates) work at ZN Technology are evidence of the quality of the ZnO substrates produced by the SCVT process.

To achieve high quality ZnO films, not only does the substrate and its surface need to be high quality and effectively prepared, the growth parameters of the epitaxial film must be highly optimized to produce device structures that will lead to efficient, durable ZnO-based devices. It is commonly known that ZnO grows fastest along the <0001> direction in contrast to other orientations, resulting in typical growth being mostly in a three-dimensional (3D) or Volmer–Weber mode. This 3D growth mode has been observed for heteroepitaxy at ZN Technology based on both molecular beam epitaxy (MBE) and MOCVD results. Similar 3D nucleation has also been reported for ZnO heteroepitaxy on GaN, sapphire, and Si.[26–28] ZnO epitaxial films grown in the 3D mode have a high density of defects with overall surface area significantly larger than that of a film grown in layer-by-layer mode. This leads to a very high density of surface states that significantly affect the film properties. For most ZnO-based device applications, such low quality films usually degrade device performance and thus high quality ZnO films grown in a layer-by-layer or two-dimensional (2D) mode are highly preferred.

Homoepitaxy is the most logical way to achieve the high quality ZnO films. Unlike heteroepitaxy, ZnO homoepitaxy is ideally performed with no difference in lattice constants and thermal properties between the film and the substrate. Zn or O adatoms at the growth interface can be incorporated in the same lattice as the substrate with minimal defects. In spite of that, achieving 2D homoepitaxial growth of ZnO is still difficult. At ZN Technology, it has been found that control of growth parameters and proper preparation of the ZnO substrates are crucial to obtain 2D growth. As a result, the ZnO homoepitaxial films have been grown in the 2D mode on in-house produced SCVT ZnO substrates by both MOCVD and MBE.[29] This not only shows that the two widely used epitaxial tools have high potential in the growth of high performance device structures, but also that the SCVT ZnO substrates have sufficiently high quality for device applications.

Most of the films have been grown on <0001> oriented SCVT ZnO substrates. However, film growth has also been conducted on nonpolar, a-plane ZnO substrates. This orientation is important because it may be used in the near future to help achieve LEDs with high internal quantum efficiency. Some significant features of these homoepitaxial films will be described in the following sections.

7.5.1 Substrate Preparation

A high quality, well prepared substrate is necessary to succeed in producing homoepitaxial films with low defect densities, good morphology, and the desired electronic properties. The ZnO substrates are prepared using a long developed wafering, lapping, and chemo-mechanical polishing procedure that has been optimized to remove damage and leave a pristine, near stoichiometric surface. In this manner, an ideal surface used for epitaxy should be free of defects with the near surface region having the same crystalline quality as the bulk ZnO. Figures 7.7 and 7.8 illustrate that properly prepared SCVT ZnO surfaces have detected defect densities similar to the bulk material. In both of these figures, the data represent cathodoluminescence results with variable incident beam energies, the lowest of which samples within nanometers of the surface. In both figures, the cathodoluminescence spectra do not change appreciably with electron beam energy (or sampling depth). In both of the other samples of different growth methods, as shown in Figure 7.7, depth variation of the green or visible luminescence/near band edge emission ratio is present. Although testing of the surfaces of the SCVT ZnO substrates by this technique helps to assure that the surfaces are ready for epitaxy, the positive results of homoepitaxial growth discussed in the next section are evidence of good bulk crystal and substrate surface properties.

7.5.2 Homoepitaxial Films on c-plane SCVT ZnO Substrates

The ZnO homoepitaxial films grown on (0001) SCVT ZnO substrates by MOCVD appear smooth and clear to the eye. Observation using a Nomarski microscope shows that they have a crack-free surface without visible orange-peel. Atomic force microscopy (AFM) was used to further distinguish atomic scale morphology of the films. Figure 7.11 is a 1 μm × 1 μm 3D AFM picture taken from a 0.5 μm thick ZnO film, showing the high quality of the film. One significant feature is the "flow pattern" composed of well defined atomic steps on the surface. This indicates that the epitaxial growth progressed with a 2D growth mode, rather than columnar, 3D growth. The rms roughness of the film is only ~0.23 nm. X-ray rocking curve measurements indicated that the 2D-grown homoepitaxial films are of high crystallinity. An omega-scan X-ray rocking curve is illustrated in Figure 7.12 using a logarithmic vertical scale. This rocking curve is a symmetric, single peak without any appreciable side lobes. The FWHM of these rocking curves have been measured as narrow as 82 arc-sec, equal to the value measured from the substrate. It is possible that most of the diffraction signal is from the ZnO substrate (the CuK_α X-rays can penetrate through the ZnO epitaxial film). However, no significant deterioration or broadening of the curve shape can be observed, suggesting the epitaxial film is of high crystallinity.

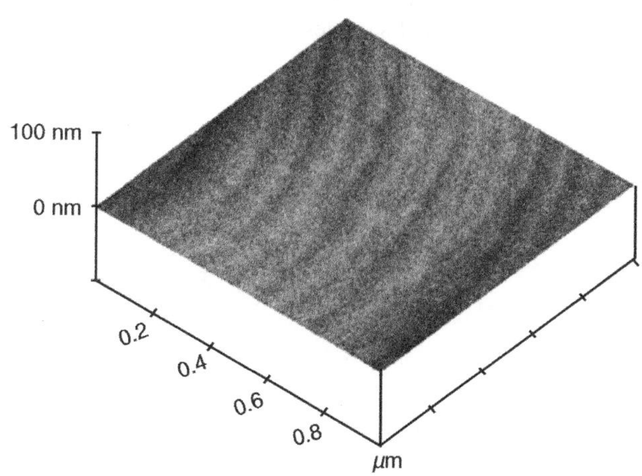

Figure 7.11 *AFM picture taken from a ZnO film that is homoepitaxially grown in 2D mode by MOCVD on an SCVT ZnO substrate*

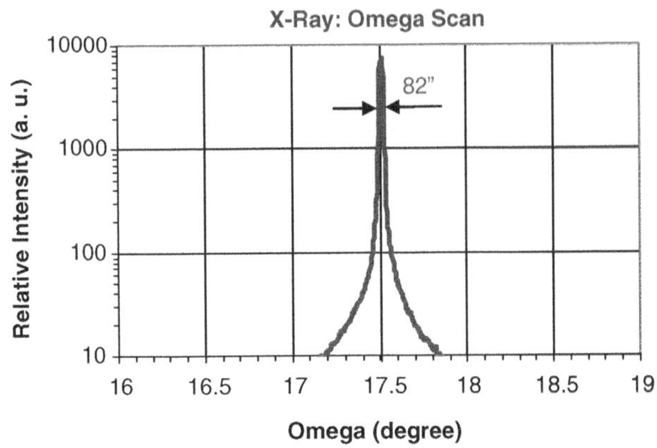

Figure 7.12 *Omega-scan X-ray rocking curve taken from a homoepitaxial film by MOCVD*

ZnO homoepitaxial films grown in 2D mode by MBE are featureless under the Nomarski microscope at 1000x magnification. A typical morphology of these films is shown in the AFM image in Figure 7.13, generated for a 1 μm × 1 μm area of a film grown on an <0001> oriented SCVT ZnO substrate. The image illustrates the full coverage of the surface by triangular facets. These are different from the "flow pattern" morphology of MOCVD-grown homoepitaxial films but still represent a 2D growth mode. The step heights at the edges of the triangular facets are either one or two monolayers. According to these characteristics, the 3D growth mode often observed in ZnO epitaxial growth has been completely suppressed. The rms roughness across the 1 μm × 1 μm area is ~0.15 nm. The homoepitaxial films have also been evaluated by X-ray diffraction rocking curve measurements. The (0002) omega-scan X-ray rocking curves (using CuK_{α} radiation) show a single peak at the Bragg diffraction position. Comparing the rocking curves of the ZnO substrates prior to growth and the ZnO samples after growth (ZnO substrate with a ZnO

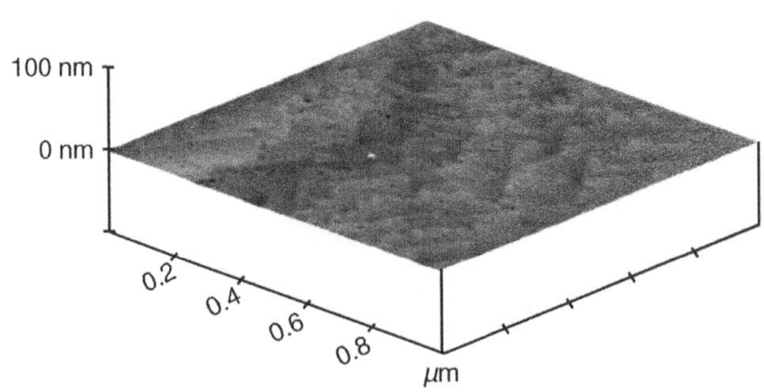

Figure 7.13 *AFM image taken from a homoepitaxial film grown by MBE showing the growth was in a 2D mode*

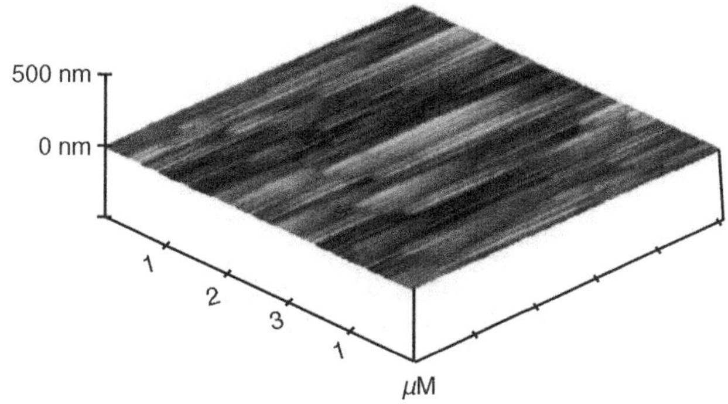

Figure 7.14 AFM image of a homoepitaxial film grown on an a-plane ZnO substrate by MOCVD

homoepitaxial film), no significant deterioration/broadening of the curve shapes were observed, suggesting these epitaxial films are also of high crystallinity.

7.5.3 ZnO Homoepitaxial Films on a-plane SCVT ZnO Substrates

Nonpolar growth of ZnO homoepitaxial films was performed on a-plane SCVT ZnO substrates using MOCVD. Figure 7.14 is an AFM image taken from one of the nonpolar films showing a typical morphology for films of this orientation. It features an in-plane line structure along the $<0001>$ direction, indicating that ZnO has a selective growth characteristic along the *c*-axis.

7.6 Summary

CVT and the SCVT process in particular are capable of producing high crystalline quality, extremely high purity, conductive, n-type ZnO substrates. In addition, the diameter of the SCVT growth process is easily scalable for higher volume applications. The surface preparation techniques and proper homoepitaxial growth parameters have enabled high quality ZnO epilayers to be grown on the SCVT ZnO substrates. These epilayers are the precursors to highly efficient optoelectronic devices.

The low cost of starting materials, the ability to grow high quality substrate material at low cost by SCVT and the benign nature of ZnO all contribute to making ZnO a prime candidate for high volume production of environmentally friendly optoelectronic devices. In addition, ZnO has long been touted as a potentially excellent bulk substrate for GaN epitaxy, especially for GaN-based LEDs and LDs and the techniques to produce high quality GaN epitaxial films on ZnO substrates have continued to improve.[30–33] With the interest in semi-polar and nonpolar orientations of GaN for increased emission efficiency, the need for bulk substrates for GaN devices has intensified.[34–37] While GaN bulk substrates have recently become available, their potential production cost is very high compared with that of bulk ZnO substrates. This makes nonpolar ZnO substrates for nonpolar GaN epitaxy very attractive.

Acknowledgement

This work was supported in part by DOE, AFOSR, and ARO.

References

[1] W.W. Piper and S.J. Polich, *J. Appl. Phys.* **32**, 1278 (1961).
[2] V.A. Akimov, M.P. Frolov, Yu.V. Korostein, V.K. Kozlovsky, A.K. Landman, Yu.P. Podmarkov and A.A. Voronov, *Phys. Status Solidi C* **3**, 1213 (2006).
[3] C.-H. Su, *Pro. SPIE* **3123**, 7 (1997).
[4] A. Micielski and A.J. Szadkowski, *Jpn. J. Cryst. Growth* **27**, 63 (2000).
[5] R. Tena-Zaera, C. Martinex-Tomas, C.J. Gomez-Garcia and V. Munoz-Sanjose, *Cryst. Res. Technol.* **41**, 742 (2006).
[6] J.T. Mullins, B.J. Cantwell, A. Basu1 Q. Jiang, A. Choubey, A.W. Brinkman and B.K. Tanner, *J. Electron. Mater.* **37**, 1460 (2008).
[7] J.L. Boone, G. Cantwell, W.C. Harsch, J.E. Thomas and B.A. Foreman, *J. Cryst. Growth* **139**, 27 (1994).
[8] X.H. Yang, J. Hays, W. Shan, J.J. Song, E. Cantwell and J. Aldridge, *Appl. Phys. Lett.* **59**, 1681 (1991).
[9] H.L. Cotal, B.G. Markey, S.W.S. McKeever, G. Cantwell and W.C. Harsch, *Physica B* **185**, 103 (1993).
[10] ZN Technology, unpublished internal data.
[11] M.M. Faktor, I. Garrett and R. Heckingbottom, *J. Cryst. Growth* **9**, 3 (1971).
[12] J. Boone, ZN Technology, unpublished internal report.
[13] A. Reisman, M. Berkenblib, S. A. Chan and J. Angilello, *J. Electron. Mater.* **2**, 177 (1973).
[14] F. T. J. Smith, *Appl. Phys. Lett.* **43**, 1109 (1983).
[15] P. Souletie, S. Bethke, B. W. Wessels and H. Pan, *J. Cryst. Growth* **86**, 248 (1988).
[16] D.C. Look, G.C. Farlow, PakPoom Reunchan, Sukit Limpijumnong, S.B. Zhang and K. Nordlund, *Phys. Rev. Lett.* **95**, 225502 (2005).
[17] D.C. Look, B.B. Claflin, G. Cantwell, S.-J. Park and G.M. Renlund, in *Zinc Oxide – A Material for Micro- and Optoelectronic Applications*, edited by N.H. Nickel and E. Terukov, Springer, Dordrecht, 2005, Ch. 4.
[18] L. Brillson,Ohio State University, unpublished data.
[19] W. Shan, W. Walukiewicz, J.W. Ager III, K.M. Yu, H.B. Yuan, H.P. Xin, G. Cantwell and J.J. Song, *Appl. Phys. Lett.* **86**, 191911 (2005).
[20] V. Bhosle, J. T. Prater, F. Yang and D. Burk, *J. Appl. Phys.* **102**, 023501 (2007).
[21] V. Assuncao, E. Fortunato, A. Marques, A. Goncalves, I. Ferreira, H. Aguas and R. Martins, *Thin Solid Films* **442**, 102 (2003).
[22] H. Kim, Y. Lee, Y. Roh, J. Jung, M. Lee and H. Kwon, *1998 IEEE Ultrasonics Symp. Prc.* **1**, 323 (1998).
[23] B. Kotlyarchuk, V. Savchuk and M. Oszwaldowski, *Cryst. Res. Technol.* **40**, 1118 (2005).
[24] C.-H. Lu, Yu-H. Huang and A. Dhobale, *Jpn. J. Appl. Phys.* **45**, 228 (2006).
[25] Y. Cao, L. Miao, S. Tanemura, M. Tanemura, Y. Kuno, Y. Hayashi and Y. Mori, *Jpn. J. Appl. Phys.* **45**, 1623 (2006).
[26] J. Zhong, H. Chen, G. Saraf, Y. Lu, C. K. Choi, J. J. Song, D. M. Mackie and H. Shen, *Appl. Phys. Lett.* **90**, 203515 (2007).
[27] B. P. Zhang, N. T. Binh, K. Wakatsuki, Y. Segawa, Y. Yamada, N. Usami, M. Kawasaki and H. Koinuma, *Appl. Phys. Lett.* **90**, 4098 (2004).
[28] Weizhong Xu, Zhizhen Ye, Liping Zhu, Yujia Zeng, Liu Jiang and Binghui Zhao, *J. Cryst. Growth* **277**, 490 (2005).
[29] J. Zhang, G. Cantwell, C. K. Choi, J. J. Song, J. Zhong, H. Cheng and Y. Lu, in *Procedings of GOMACTECH-07*, Lake Buena Vista, FL, 2007, paper #9-3.

[30] T. Ohgaki, N. Ohashi, T. Ohnishi, H. Haneda, M. Lippmaa and A. Yasumori, *J. Ceram. Soc. Jpn.* **115**, 414 (2007).
[31] A. Kobayashi, S. Kawano, K. Ueno, J. Ohta, H. Fujika, H. Amanai, S. Nagao and H. Horie, *Appl. Phys. Lett.* **91**, 191905 (2007).
[32] A. Kobayashi, S. Kawano, Y. Kawaguchi, J. Ohta and H. Fujioka, *Appl. Phys. Lett.* **90**, 041908 (2007).
[33] K Ueno, A. Kobayashi, J. Ohta and H. Fujioka, *Appl. Phys. Lett.* **90**, 141908 (2007).
[34] H. Masui, H. Yamada, K. Iso, J.S. Speck, S. Nakamura and S.P. DenBaars, *J. Soc. Inf. Display* **16**, 571 (2008).
[35] Y. Tsuda, M. Ohta, P.O. Vaccaro, S. Ito, S. Hirukawa, Y. Kawaguchi, Y. Fujishiro, Y. Takahira, Y. Ueta, T. Takakura and T. Yuasa, *Appl. Phys. Express* **1**, 011104 (2008).
[36] A. Tyagi, H Zhong, N.N. Fellows, M. Iza, J.S. Speck, S.P. DenBaars and S. Nakamura, *Jpn. J. Appl. Phys.* **46**, L129 (2007).
[37] B.A. Haskell, S. Nakamura, S.P. DenBaars and J.S. Speck, *Phys. Status Solidi B* **244**, 2847 (2007).

8

Growth Mechanisms and Properties of Hydrothermal ZnO

M. J. Callahan[1,3], *Dirk Ehrentraut*[2], *M. N. Alexander*[3] *and Buguo Wang*[4]
[1]*Teleos Solar, Hanson, MA, USA*
[2]*Institute of Materials Research (IMR), Tohoku University, Aoba-ku, Sendai, Japan*
[3]*Air Force Research Laboratory, Sensors Directorate, Hanscom AFB, MA, USA*
[4]*Solid State Scientific Corp., Nashua, NH, USA*

8.1 Introduction

Because of its direct band gap at 3.4 eV and large excitonic binding energy (60 meV), zinc oxide (ZnO) has attracted considerable interest for possible applications in emitter and detector applications in the blue to ultraviolet (UV) portion of the spectrum.[1] ZnO also has many other unique properties such as the ability to alloy with Mg or Cd in fabrication of quantum wells, resistance to radiation damage, high voltage breakdown strength, and high electron saturation velocity.[2] In addition, ZnO also has large piezoelectric and acousto-optic coefficients,[3] and with appropriate doping it can have ferroelectric, magnetic, and nonlinear electrical properties. It exhibits strong two-photon absorption and a very high optical damage threshold, rendering it attractive for saturable absorber applications.[4]

ZnO powder, polycrystalline, and single crystal material have been and will be integral elements in such applications as low voltage varistors, surge protectors, photocatalysts, chemical detectors, biological and medical processing, phosphors, visible-UV detectors and emitters, and high power electronics.

A large, low cost ZnO substrate would speed development of ZnO-based optoelectronics. ZnO boules up to 50 mm in diameter and 10 mm thick have been grown by vapor phase

Zinc Oxide Materials for Electronic and Optoelectronic Device Applications, First Edition.
Edited by Cole W. Litton, Donald C. Reynolds and Thomas C. Collins.

transport,[5] and melt-grown wafers 50 mm diameter are now also available.[6] Both processes typically yield wafers with dislocation densities of 10^4–10^5 cm^{-2}.

Hydrothermal growth has certain advantages over other methods in producing high purity, low defect-density, and small uniform grain size crystallites (for powder) because of lower growth temperature and slower growth rates.[7–11] Hydrothermal ZnO wafers have been produced with dislocation densities as low as 100 cm^{-2} and carrier concentrations as low as 2×10^{12}.[12] Early interest in single crystal ZnO focused on piezoelectric transducers; for this purpose, large hydrothermal crystals were grown in the 1960s.[13–15] Renewed interest arose because of ZnO's potential as an isostructural, nearly lattice-matched substrate for group III nitride semiconductor device structures.[16] Advances in fabricating ZnO and ZnMgO quantum wells on sapphire that exhibit strong optically stimulated UV emission have further driven demand for development of ZnO substrates.[17]

This review focuses on the hydrothermal growth of single crystals of ZnO, including pertinent information on microcrystalline growth when it provides insight on the growth kinetics of larger crystals. Phase stability, solubility, growth kinetics, and impurity incorporation are discussed. Electrical and optical properties are reviewed with special interest in the effects of crystalline anisotropy. Phenomena related to etching and polishing are also discussed.

8.2 Overview of Hydrothermal Solution Growth

Byrappa and Yoshimura define hydrothermal growth as "any heterogeneous chemical reaction in the presence of a solvent (whether aqueous or non-aqueous) above room temperature at an pressure greater than 1 atm in a closed system".[8] Hydrothermal crystal growth offers several advantages over better-known methods such as melt growth. Hydrothermal growth is a low temperature process, which often makes possible the growth of materials that are difficult or impossible to melt, or materials which, in solidifying from a melt and cooling down, would undergo phase changes (because of such changes, α-quartz cannot be grown from the melt). Low temperature hydrothermal growth can also minimize or eliminate the incidence of temperature-induced point defects, as illustrated by growth of $Bi_{12}SiO_{20}$,[18] and it can produce large amounts of material (over 4000 kg of single crystal quartz has been produced in a single run).[8] Some disadvantages of hydrothermal growth are the low growth rate and initial capital equipment costs, but these are offset by the ability to grow multiple crystals in a single run and the long lifetimes of the autoclaves. Additionally, the labor requirements are minimal due to the simplicity of the autoclaves where closed system crystal growth occurs for long periods of time at steady-state temperatures and pressures. An in depth review of hydrothermal technology is covered in several references.[7,8]

8.3 Thermodynamics of Hydrothermal Growth of ZnO

8.3.1 Solubility of ZnO in Various Aqueous Media

In hydrothermal growth a "nutrient", often the compound one intends to grow, is dissolved in an aqueous medium. The dissolved compound may form intermediate complexes or species such as hydrides, hydrates, hydroxides, or oxides in solution. In hydrothermal growth, convective circulation and/or diffusion of these species throughout the aqueous medium provides the primary mechanism for synthesizing crystalline compounds. A hydrothermal growth system is designed to bring the soluble species into a region of the aqueous medium where a change in conditions – e.g. in temperature, solvent composition, pH, or pressure – promotes crystal growth by bringing the soluble species into supersaturation. Supersaturation is defined as the amount of material in solution above the equilibrium concentration for a substance under a specific set of conditions. A supersaturated species must eventually come out of solution – hopefully as the desired compound (e.g. ZnO) – until equilibrium state (saturation) is achieved.

The first requirement for hydrothermal growth follows immediately from the foregoing discussion: the solubility of nutrient in the solvent medium must be adequate for growth. Low solubility will result in low growth rates; excessively high solubility will result in polycrystalline growth or spontaneous nucleation, which prohibits growth of low defect-density single crystals. Typically, the nutrient should be 1–10 wt% soluble in the solvent. Solubility can be increased by adding a proper complexing agent (mineralizer). Most mineralizers used in hydrothermal growth change the pH of the solvent, making the solvent either acidic or alkaline, i.e. increasing the number of H^+ or OH^- ions that attack the nutrient material in aqueous solutions. For ZnO, a solubility of approximately 5 wt% in alkaline solution yields high quality single crystals at reasonable growth rates.

The majority of compounds that can be dissolved in solutions will have a change in solubility when the temperature of the solution is raised or lowered. Therefore, most solution growth methods, including the hydrothermal method, use a change in temperature at the seed interface to create supersaturation. Thus, the second requirement of hydrothermal growth is that the solubility of the compound has an adequate range of temperature dependence for high growth rates. The solubility of ZnO in an OH^- alkaline medium is shown in Figure 8.1.[19] Note that solubility increases with temperature. This is called *normal solubility or forward solubility.*

ZnO is an amphoteric oxide, meaning it acts as an acid in alkaline solutions and as a base in acidic solutions. It is possible to grow hydrothermal ZnO in an acidic solution as well as the alkaline solutions shown in Figure 8.1. McCandlish and Uhrin recently studied the solubility of ZnO in an acidic medium and grew ZnO at 100–250 °C, with growth rates up to 0.25 mm day^{-1}.[20] Figure 8.2 illustrates the solubility in acidic regimes. The squares signify 2 molal aqueous nitric acid and the circles signify a proprietary acidic solution. Note that the nitric acid solution exhibits normal solubility, whereas the proprietary solution exhibits decreasing solubility with increasing temperature (*retrograde solubility*). To grow ZnO crystals under conditions of normal solubility the seed is placed in a colder region than the source material (see Section 8.4.2); to grow under conditions of retrograde solubility, the seed is placed in the hotter region.

8.3.2 ZnO Phase Stability in H_2O System

A third requirement for hydrothermal growth is that the desired compound is thermodynamically favored at the growth interface. If an oxide like ZnO is desired, the hydride, hydroxide or hydrate cannot be thermodynamically favored for a given mineralizer,

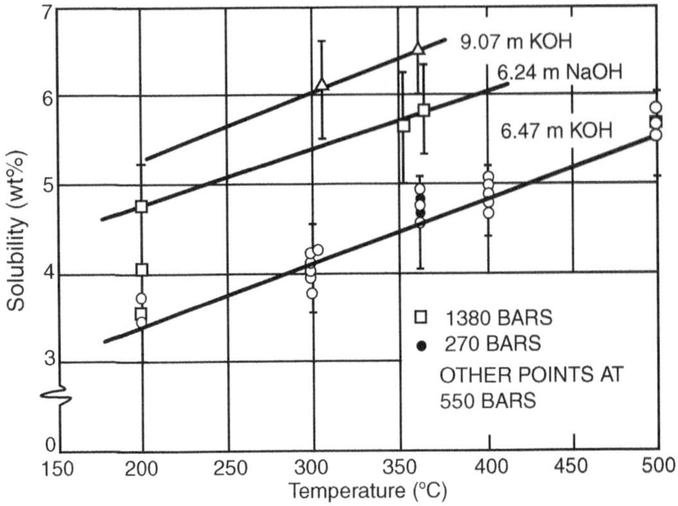

Figure 8.1 Solubility of ZnO vs temperature in aqueous NaOH and KOH solutions[19]. Reprinted from R.A. Laudise, E.D. Kolb, American Mineralogist (USA) 48 [3] 642. Copyright (1963) with permission from the Mineralogical Society of America

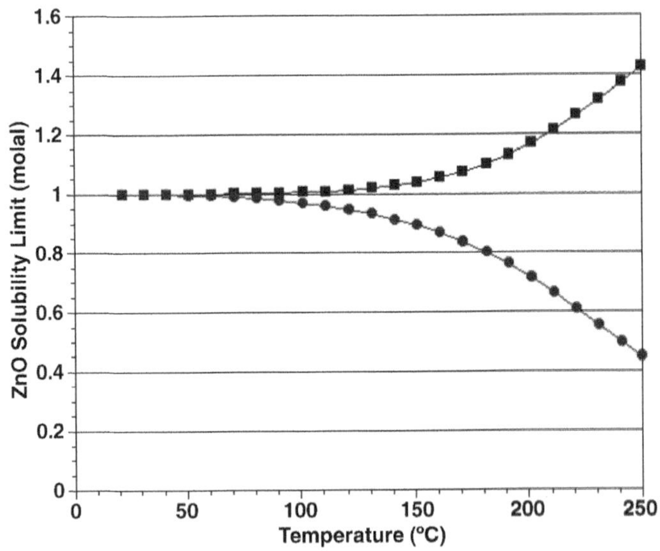

Figure 8.2 Normal solubility of ZnO in 2 molal nitric acid (squares) and retrograde solubility of ZnO in a proprietary acid solution (circles). Reprinted from Handbook of Crystal Growth, Editors Govindhan Dhanaraj, Kullaiah Brrappa, Vishwanath Prasad, and Michael Dudley, Publisher Springer 2010, ISBN: 978-3-540-74182-4, Chapter 19 Hydrothermal and Ammonothermal Growth of ZnO and GaN pp 655–689, Figure 19.6

temperature and pressure. Also, the mineralizer must not form stable solid compounds at the growth interface.

Hüttig and Mőldner studied the phase equilibrium of the ZnO–H_2O system to 40 °C and found ZnO to be the stable solid phase at pressures above 50 Torr and temperatures above 35 °C.[21] Lu and Yeh experimentally showed that ZnO is the stable product up to pH = 12.5 in an aqueous ammonia solution at 100 °C.[10] Laudise and Ballman grew large ZnO crystals in alkaline media and found ZnO to be the stable product from 200 to 400 °C in 1.0 M NaOH.[22] ZnO has been grown in hydroxide solutions up to 10 M. Recent advances in thermodynamic modeling of aqueous solution chemistry can aid in choosing conditions that achieve crystal growth of ZnO and other materials. A thermodynamic model of aqueous-based chemistry has been developed that computes the stability of ZnO in different aqueous regimes. The model uses commercial software (OLI systems Inc., Morris Plains, NJ, USA) and is detailed in several publications.[23–25] McCandlish and Uhrin initially modeled ZnO in the hydroxide system to validate the model against experimental data. Subsequently, a thermodynamic model was created for the growth of ZnO in acidic environments. Figure 8.3 exhibits the computed stability of ZnO at 150 °C as a function of pH with HNO_3 as the mineralizer.[20]

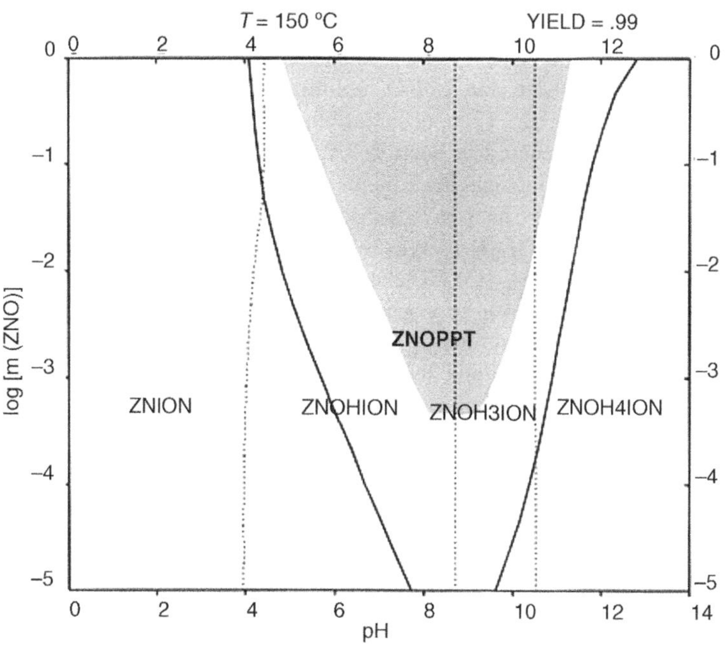

Figure 8.3 *Yield diagram for the precipitation of ZnO in 2 molal nitric acid at 150 °C as a function of pH. Reprinted from Handbook of Crystal Growth, Editors Govindhan Dhanaraj, Kullaiah Brrappa, Vishwanath Prasad, and Michael Dudley, Publisher Springer 2010, ISBN: 978-3-540-74182-4, Chapter 19 Hydrothermal and Ammonothermal Growth of ZnO and GaN pp 655–689, Figure 19.6*

8.4 Hydrothermal Growth Techniques

8.4.1 Hydrothermal Growth of ZnO Powder

Hydrothermal growth of microcrystalline ZnO can improve particle size uniformity and suppress agglomeration of crystallites.[10] A desire for higher purity, low cost, microcrystalline material for varistors,[26] spurred research in microcrystalline ZnO synthesis in various aqueous media by several groups.[9,10,27–29] Recently, an intense interest in nanocrystalline ZnO for a variety of new applications has resulted in an increase in the use of the hydrothermal technique for the synthesis of microcrystalline and nanocrystalline ZnO. Since many experiments can be performed in a short period of time, growth of microcrystalline ZnO can provide insights that may be helpful in optimizing future production methods for bulk ZnO crystals. Although an in-depth review of hydrothermal synthesis of micro- and nano-ZnO structures is not presented, much of the initial work on ZnO powder synthesis will be reviewed as it provides insight into the growth kinetics of large ZnO crystals grown by the hydrothermal method.

Various hydrothermal media have been used to prepare ZnO powder. Chen *et al.*[9] used $Zn(OH)_2$ colloids in aqueous HCl at pH = 5–8 and growth temperatures of 100–220 °C to produce ZnO powder; they also investigated several organic and inorganic solvents. Lu and Yeh[10] used zinc nitrate as a source in aqueous ammonia to investigate product yield of ZnO powder. Komareni *et al.*[30] formed ZnO by using microwave irradiation to decompose zinc nitrate in aqueous sodium hydroxide. Wang *et al.*[27,28] and Li *et al.*[29] used $Zn(OH)_2$ colloids and $Zn(CH_3COO)_2$ as source materials, dissolving them in pure H_2O and aqueous KOH, KBr, and $NaNO_2$ solutions at temperatures up to 350 °C and pressures approaching 400 bar. Their observations of changes in ZnO morphology as a function of pH will be discussed in Section 8.5.

Recently considerable effort has been made by dozens of research groups on synthesizing nanocrystalline ZnO structures. Zinc acetate, zinc nitrate and zinc chloride have been used as zinc sources. CTAB (cetyltrimethylammonium), PPA (poly acrylic acid), SDS (sodium dodecyl sulfate), HMT (hexamethylenetramine), ethanol and PVP (polyvinylpyrrolidone) have all been used as chemical additives, surfactants, and modifiers to alter the morphology of the ZnO nanostructures. The intense effort on controlling ZnO nanostructures can provide additional insight on altering the kinetics of large ZnO crystals.

8.4.2 Hydrothermal Crystal Growth of ZnO Single Crystals

Hydrothermal growth of single crystals is performed in steel vessels called autoclaves. The autoclaves are usually made out of some form of stainless steel, or a high strength nickel-based alloy if higher growth temperatures or pressures are required. The vessel must be corrosion-resistant and able to withstand the temperature and pressure requirements for long periods of time. Large ZnO crystals are typically grown in 2–10 molal alkali solutions at 200–500 °C and 500–2000 bar.[12,13,19,31–34]

Corrosive solutions employing concentrated KOH mineralizer are required to obtain acceptable growth rates for ZnO crystals. Therefore, to protect the autoclave, a noble metal liner (e.g. silver, gold or platinum) must be used. Two techniques can be employed. The first is to insert a lipped noble metal liner against the wall of a Bridgman autoclave and then put a

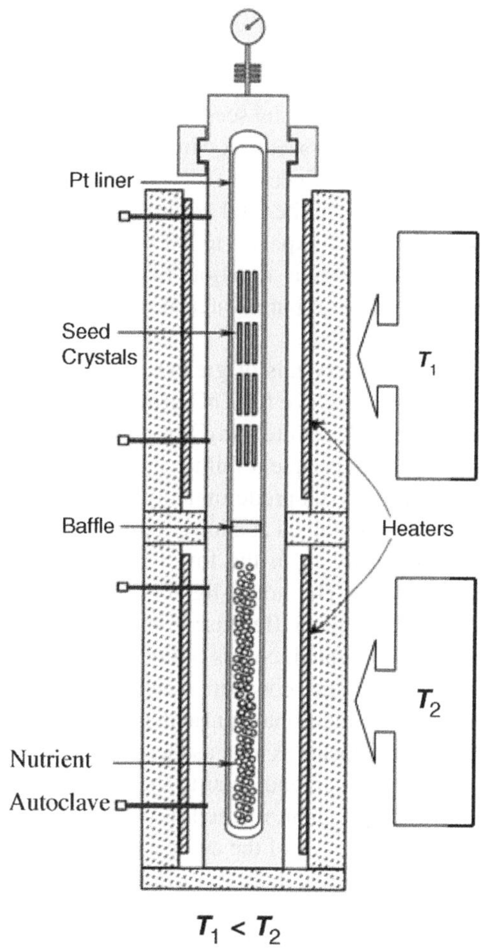

Figure 8.4 *Hydrothermal autoclave used for growth of ZnO. Adapted from Zinc Oxide Materials and Devices II, edited by Ferechtec Hosseini Teherani, Cole W. Litton, Proc. of SPIE Vol. 6474, 647412, Fig. 1 on page 647412-2. Copyright (2007) with permission from SPIE*

noble metal disk between the lip and the plunger. The second method is to contain the experiment in a complete welded noble metal "can" inside the autoclave. In this case a specific amount of mineralizer solution is put inside the container ("can") and a less corrosive solution outside the container so that, upon heating to the growth temperature, the pressure caused by the solution inside the container is approximately balanced by the pressure generated by the solution outside the container. Such pressure balancing prevents the noble metal container from rupturing. See the literature[7,8] for details on autoclave designs and apparatus. Figure 8.4 schematically shows the cross-section of a large hydrothermal autoclave used by Tokyo Denpa LLC for the production of ZnO. The autoclave has an inner diameter (ID) of 200 mm with an inner length (IL) of 3–4 m.

The growth of ZnO crystals is typical of the hydrothermal growth of many other crystals, e.g., α-ZnO,[35,36] $Bi_{12}SiO_{20}$[37] and $LiGaO_2$.[38] ZnO nutrient, which is either crystalline material (free nucleated material from previous hydrothermal runs) or high purity ZnO powder that has been sintered or hot pressed at high temperature (usually 1000–1300 °C) to obtain dense material[39,40] is processed into 1 mm^3–1 cm^3 pieces and put into the bottom of the liner. Single crystal seeds are cut in the desired orientation, and a small hole is drilled near the top of the seeds. A wire is inserted through the hole so seeds can be hung in the upper half of the liner.

The aqueous solvents fill 60–90% of the total volume (the liner, and also the rest of the autoclave). When heated, the liquid expands to 100% of the total volume; continued heating generates pressure, which is a function of temperature, degree of fill, and composition of the liquid. For data on pure water at different percent fills, see the *PVT* curves generated by Kennedy.[41]

Lithium in the solvent produces crystals of greater perfection, as will be discussed in Section 8.6. Before using a new autoclave for ZnO growth, quartz growth is performed to coat the walls of the autoclave with acmite, $Na_2O{\cdot}Fe_2O_3{\cdot}SiO_2$. The acmite helps protect the autoclave from corrosion and subsequent failure in case the platinum container leaks.

Hydrothermal growth of large crystal predominately employs the temperature gradient method. The upper region T_1 is maintained at a lower temperature than the lower region T_2 as depicted in Figure 8.4. The solvent in the lower region of the vessel dissolves the nutrient until it reaches saturation. The hotter-lighter-solvated species is transported by fluid convection to the colder seed region. Because of the lower temperature at the seed the solvated species becomes supersaturated, comes out of solution, and deposits on the seed (normal saturation conditions). Fluid convection returns the cooler-heavier-depleted solution to the hot zone, where additional nutrient is dissolved to regain equilibrium solubility (saturation). The cycle repeats as long as there is nutrient in the lower zone. ZnO growth rates of up to 0.3 mm day^{-1} perpendicular to the basal plane can be maintained for months by using the temperature gradient method. For large crystals of ZnO, dissolution temperatures in the lower region or zone of the autoclave are typically above 350 °C; with a temperature gradient of at least 10 °C between the lower region and the upper region where crystallization takes place. Various conditions for hydrothermal growth of ZnO single crystals over the past 40 years have been tabulated elsewhere.[42]

Detailed computational fluid dynamics simulations have been performed in various hydrothermal systems to analyze the flow and temperature gradients in large autoclaves. Uniform flow and temperature profiles are critical for the growth of large hydrothermally grown crystals of high perfection. Figure 8.5 shows a computer simulation of the fluid flow and thermal profile of an autoclave used to grow ZnO crystals.[43] The simulations were only performed on one half of the autoclave because the system is assumed to be axisymmetric. The simulation shows that the solution in the upper region primarily flows up through the center and down in close proximity to the autoclave walls in a clockwise motion. The fluid passes through the baffle primarily through diffusion and an opposite counter-clockwise motion is seen in the lower region. Several simulations were done to see the effect a change in viscosity had on the flow in the system. The authors found that changes in the thermal conductivity of the fluid had only a small effect on fluid flow, whereas changes in the heat capacity and viscosity of the fluid had much greater effects. Because of the difficulty in getting accurate fluid properties such as density, viscosity, specific heat, thermal conductivity, and thermal expansion coefficient in high pressure closed systems, the models may not accurately

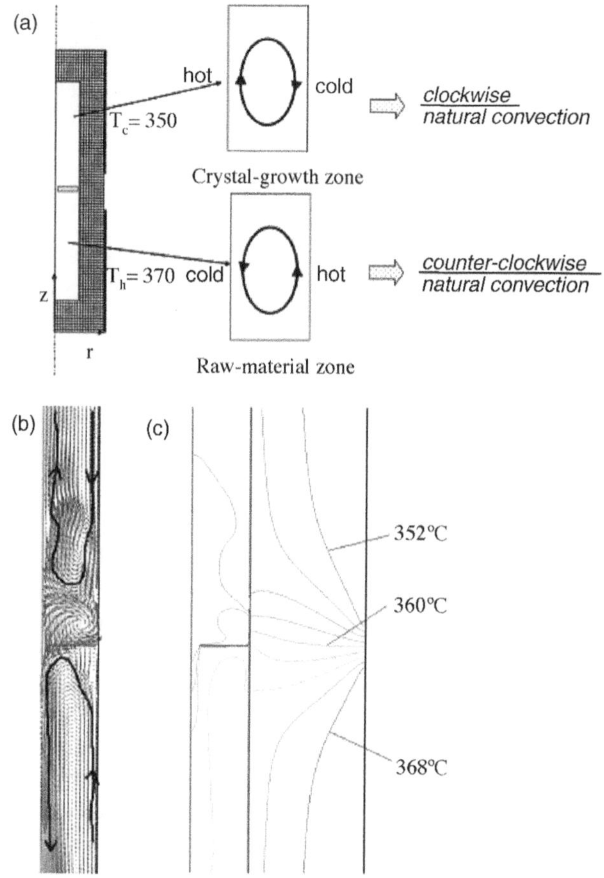

Figure 8.5 *(a) Main flow direction of natural convection in a raw-material zone and a crystal-growth zone. Flow (b) and temperature (c) fields using reference values as follows: thermal conductivity = 0.6143 $W m^{-1} K^{-1}$; heat capacity = 4605 $J kg^{-1} K^{-1}$; and reference viscosity = 9.574×10^{-5} Pa s. Reprinted from Yoshio Masuda et al., Numerical simulation of natural convection heat transfer in a ZnO single-crystal growth hydrothermal autoclave—Effects of fluid properties, Journal of Crystal Growth, 311. Copyright (2009) with permission from Elsevier*

represent real systems. Nonetheless, fluid flow and thermal profile simulation can still yield viable insights in optimizing autoclave design to increase the yield and uniformity of hydrothermal ZnO crystals.

8.4.3 Industrial Growth of Large ZnO Crystals

The minimum wafer diameter size typically required by semiconductor device manufacturers is 50 mm, therefore ZnO crystals with diameters in excess of 50 mm must be grown and afterwards processed to ship as epi-ready wafers. The first ZnO crystals weighing upwards of 200 g were reported in 1997 and 2001.[44,45] Fifty millimeter diameter (0001) ZnO bulk crystals were first reported in 2004 using the hydrothermal

Figure 8.6 Almost 100 ZnO crystals of 50 mm diameter in size displayed after growth at Tokyo Denpa. Reprinted from D. Ehrentraut et al., Solvothermal growth of ZnO, Prog. Cryst. Growth Charact. Mater. vol. 52. Copyright (2006) with permission from Elsevier

growth technique.[46] Furthermore, to yield reasonable prices for 50 mm wafers, large quantities of ZnO crystals are required, which therefore stipulates growth processes with high throughput. Despite the slow growth rate of ZnO (about one-fifth the growth rate of quartz) and the need to use noble metal liners, hydrothermal technology is predestined to yield large quantities of high-quality ZnO crystals in a single growth run: currently around 100 crystals of ZnO[47] or even >1000 crystals as in the case of quartz (SiO_2).[48]

Figure 8.6 illustrates the result from one growth run at Tokyo Denpa.[47,49] Almost 100 crystals approximately 50 mm in diameter have been grown on (0001) and $(10\bar{1}0)$ oriented seeds. Smooth facets were formed and the average crystal thickness was approximately 1 cm for those crystals grown on (0001) ZnO seeds. The aqueous growth solution was composed of 3 M KOH + 1 M LiOH at pressures of 80–100 MPa with temperatures approaching 400 °C.

A 75 mm diameter (0001) ZnO crystal and a (0001) wafer processed from it are shown in Figure 8.7(a) and (b), respectively. The crystal is almost 1 cm in thickness. Note the homogeneously pale-yellow coloration which points to lower impurity levels than ZnO crystals grown previously. Figure 8.7(c) shows a typical crystal. Clearly visible are the newly formed, almost colorless crystal fractions which are surrounding the seed crystal.

Contrary to the crystal growth from the melt with its higher growth rates of the order of mm–cm h^{-1} as in the case of silicon, the slow growth rates of 200–300 μm day^{-1} for the $\langle 0001 \rangle$ direction of ZnO requires time-consuming scale up of crystal size in two steps. After the size has been increased, the subsequent step is to greatly improve the structural quality. Next, new seeds are fabricated from this high-quality crystal and the cycle starts again until the desired size has been obtained.

Recently, several more reports on 75 mm size ZnO single crystals grown by the hydrothermal method have been published which employed autoclaves with volumes of up to 500 liters to grow 100–200 crystals per growth cycle.[50,51] These developments are encouraging and enhance the likelihood that large diameter ZnO wafers, i.e. $\geq$50 mm will soon enter the market in high qualities if commercial homoepitaxial ZnO light-emitting diodes (LEDs) on ZnO substrates can be developed in the near future.

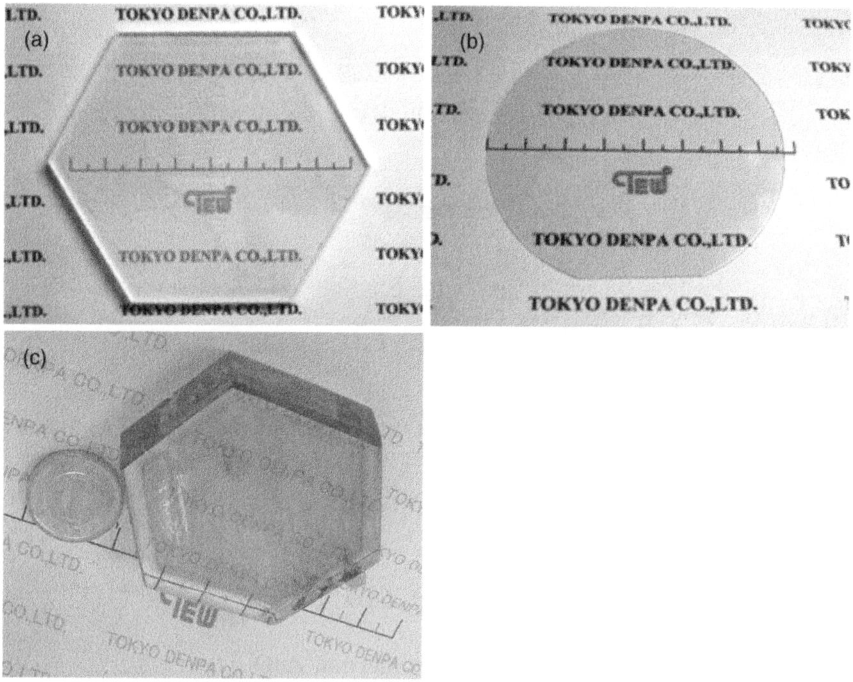

Figure 8.7 A 75 mm diameter (0001) ZnO crystal as grown recently (a), a 75 mm diameter epi-ready (0001) ZnO wafer prepared from such a crystal (b) and a 75 mm diameter (0001) ZnO crystal obtained from an earlier growth cycle (c). Reprinted from E. Ohshima, et al., Growth of the 2-in-size bulk ZnO single crystals by the hydrothermal method, Journal of Crystal Growth 260, 166–170, Issue 1–2. Copyright (2004) with permission from Elsevier

The major challenge to be solved for large industrial ZnO crystals is the lowering of impurity concentrations of the alkali metals such as Li, Na and K, which gives rise to acceptor compensation, and Al and Fe, both of which are electronic donors. Non-radiative recombination centers are observed. Growth sectors are typically formed due to differences in the growth rate and surface chemistries for faces with different crystallographic orientations. The thermodynamically stable hexagonal phase of ZnO incorporates impurities such as Li, Al and Fe with quite different concentrations in the basal, prismatic and pyramidal growth sectors which will be described in more detail later in this chapter. Growth in the $\langle 0001 \rangle$ direction is preferred since the impurity levels are strikingly lower and growth rates are higher than the $\langle 000\bar{1} \rangle$ direction. The formation of $\{10\bar{1}1\}$ facets should be suppressed for economic reasons Figure 8.7(a) and (c) display the proper morphology to ensure that all wafers cut from one crystal are of the same size and have uniform impurity concentrations across the wafer. The preceding discussions show that understanding and controlling the kinetics of large hydrothermal ZnO crystals is critical for the production of low cost ZnO wafers with uniform properties. Therefore, growth kinetics will be discussed in the next section and impurities and their influence on kinetics will be discussed in detail in Section 8.6.

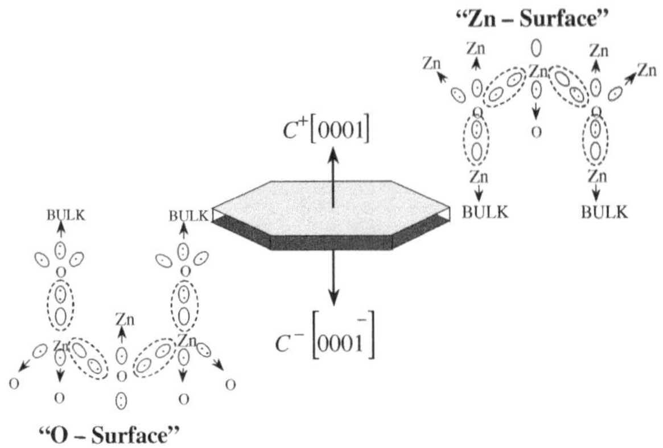

Figure 8.8 Electronic charge distribution of ZnO basal faces. Reprinted from M. Suscavage, et al., MRS Internet J. Nitride Semiconductor Res. (USA) vol. 4, p. 294., Figure 1 Page 295. Copyright (1999) with permission from MRS

8.5 Growth Kinetics of Hydrothermal ZnO

8.5.1 Crystallographic Structure of Hydrothermal ZnO

The stable phase of ZnO has the wurtzite crystal structure, which is hexagonal with a space group of $P6_3mc$. The noncentrosymmetric structure of wurtzite ZnO produces an anisotropy in which the opposite sides of a basal plane wafer have different atomic arrangements at their surfaces. The C^+ side of the basal plane is comprised of a Zn-rich layer, and the C^- is comprised of an O-rich layer, as illustrated in Figure 8.8.[12,52]

The resulting distribution of electric charge causes disparities, among the various growth planes, in hydrothermal growth rates, as well as in impurity incorporation, chemical etching, optical, and electrical properties.

8.5.2 Growth Rates of the Crystallographic Facets of Hydrothermal ZnO

Figure 8.9 shows the growth planes for hydrothermal ZnO crystals. Hydrothermal crystals are highly faceted due to the slow growth rates and lack of confinement during growth. Hydrothermal ZnO in an alkaline medium grows with the following facets: (0001) and $(000\bar{1})$ monohedra (C^+ and C^- planes, respectively), the six $(10\bar{1}0)$ prismatic faces (M planes), and the six $(10\bar{1}1)$ pyramid faces (P planes). Laudise and Ballman first observed the anisotropic growth rate on both spontaneous crystallites and crystals grown on seeds in 1 M NaOH.[22]

Growth on the C^+ face is always faster than on the C^- face in hydroxide solutions above 1 M. Typical growth rates for 6 M KOH and 1 M LiOH are approximately 0.45 mm day^{-1} in the C^+ direction, and 0.20 mm day^{-1} in the C^- direction for growth on C-plane seeds.[31] Growth rates on M-plane seeds average 0.20 mm day^{-1} in the direction normal to the M-plane.[32] ZnO crystals grown in KOH solutions are of higher structural quality but more highly faceted than those grown in NaOH solutions.[34] At AFRL-Hanscom AFB, a

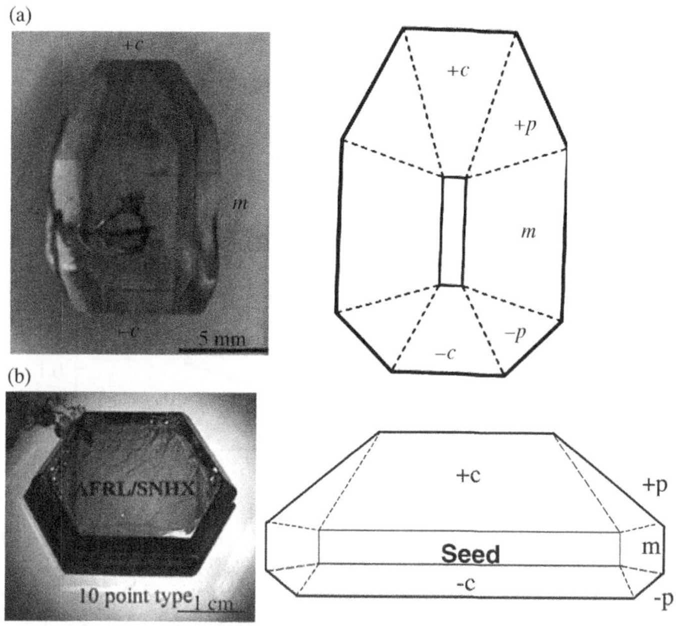

Figure 8.9 ZnO crystal grown on M-plane seed (a) and ZnO crystal grown on C-plane seed at AFRL-Hanscom (b). Image in (a) reproduced from Sekiguchi et al.[52]. Image in (b) provided by M. J. Calahan while at AFRL-Hanscom AFB. Reprinted from T. Sekiguchi, et al., Hydrothermal growth of ZnO single crystals and their optical characterization, J. Cryst. Growth, vol. 214/215, Copyright (2000) with permission from Elsevier

3:1 NaOH: KOH solution produced crystals with low defect densities and less faceting than KOH-grown crystals. The mixed NaOH–KOH solvent had the added benefit of being less corrosive than KOH solutions. ZnO crystals grown on a C-plane seed and a M-plane seed are shown in Figure 8.9.[53]

Demianets *et al.* measured the growth kinetics on the different crystallographic faces by varying the type and concentration of mineralizer, growth temperature, and temperature difference between the dissolution and crystallization zones.[54] Figure 8.10 shows more detailed kinetics of ZnO growth as a function of temperature. Note the effect the addition of lithium, which improves the perfection of the ZnO crystal, has in decreasing the growth rate.

The authors went on to determine the elementary surface layers for the possible growth facets of ZnO. The elementary surface layers were then used to determine the relative theoretical growth velocities under ideal conditions (*the absence of any additional components in the crystallization medium*) for the different crystallographic faces of ZnO. The relationships of the velocities are:

$$v(10\bar{1}0) < v(000\bar{1}) \sim v(0001) < v(10\bar{1}1) < v(10\bar{1}2) < v(1\bar{1}20)$$

The sequence would be reversed to characterize the prevalence of the faces in the formed crystal. The ideal velocities above would form simple shapes such as monohedra

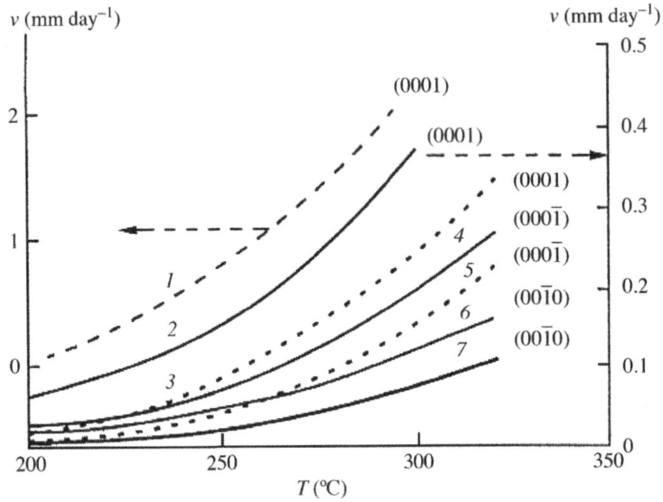

Figure 8.10 *Growth rates of the faces of the monohedra and the {10$\bar{1}$0} prism of ZnO single crystals in alkaline solutions as a function of temperature: (1) 5 M KOH; and (2–7) 5.15 M KOH + 1.2 M LiOH. Solid lines correspond to $\Delta T = 75\,°C$; dashed lines correspond to $\Delta T = 50\,°C$. Reprinted from L.N. Demianets, et al., Crystallography Reports vol. 47, Supp 1, S86, Fig. 4, pg S94. Copyright (2002) with permission from Springer*

and prisms but water, a polar solvent, adds a great deal of complexity and anisotropy to the kinetics of crystal growth.

Growth kinetics in solutions with forced convection can be simplified into two basic components: supersaturation and surface kinetics.[55] Growth rates in solutions are linearly dependent on supersaturation on all faces for compounds whose supersaturations are large, e.g. in the percent range.[56] Surface kinetics has a much more complicated effect on growth rates and can be broken down into several molecular processes: surface diffusion, adsorption, desorption, desolvation, and finally molecular exchanges between various surface atoms or molecules with each other and in and out of solution.[56]

The polar nature of water, ZnO, and the intermediate complexes in solution further complicates growth rates for the various crystal facets. Each family of a crystallographic plane has a unique set of activation energies in the molecular processes discussed above. A charge distribution is also associated with each plane which will interact with the charge distribution and surface composition of the ions in solution. Thus to gain insight into the growth kinetics of hydrothermal ZnO crystals, one needs not only to understand the atomic composition and charge of each crystallographic surface, but also the atomic composition and charge of intermediate species at or near the growth interface.

Several researchers have studied the solubility and thermodynamic parameters of aqueous Zn species for hydrothermal systems. Wesolowski *et al.*[57] and Bénézeth *et al.*[58] investigated the solubility of ZnO in 0.03–1.0 M sodium trifluoromethanesulfonate solutions to determine thermodynamic properties of the transport species in dilute acidic and alkaline solutions. The Gibbs free energy of formation, entropy, and enthalpy at

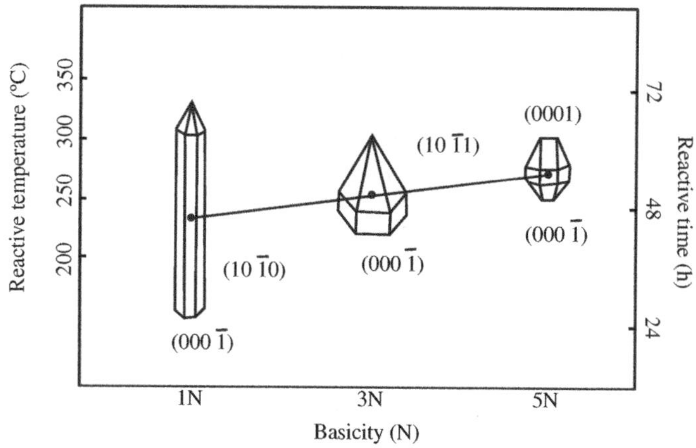

Figure 8.11 *Morphological changes of ZnO crystallites with increasing pH of the growth solution. Reprinted from B.G. Wang, E. W. Shi and A. Z. Zhong, Understanding and Controlling the Morphology of ZnO Crystallites under Hydrothermal Conditions, Cryst. Res. Technol. 32,5 659–667. Copyright (1997) WILEY-VCH Verlag GmbH & Co. KGaA*

25 °C and 1 atm and were determined for Zn^{2+}, $Zn(OH^+)$, $Zn(OH)_2^0$, and $Zn(OH)_3^-$ by employing a hydrogen electrode concentration cell and periodic sampling of cell potentials. Solubility data at temperatures up to 200 °C for $Zn(OH^+)$, $Zn(OH)_2^0$, and $Zn(OH)_3^-$, and at temperatures up to 290 °C for Zn^{2+}, were also obtained. The authors concluded that $Zn(OH)_4^{2-}$ was the predominant species in OH^- solutions above 0.1 M NaOH. Khodakovsky and Yelkin also concluded that $Zn(OH)_4^{2-}$ was the dominant species in alkaline solution at the high temperatures and high pH values at which bulk crystals are grown.[59]

Wang[27,28] and Li *et al.*[29] performed systematic studies of the morphology and growth rates of ZnO powder in alkaline solutions. Starting with $Zn(OH)_2$ colloids as nutrient, they investigated morphological changes as a function of pH (see Figure 8.11).[27]

The study assumed that $Zn(OH)_4^{2-}$ growth units were the predominant species in solution when hydroxide colloids were dissolved. These growth units have a tetrahedral form and charge distribution similar to those of ZnO, which itself is a series of tetrahedra (Figure 8.12).[29] The corner (point) of a ZnO tetrahedron can bind with three hydroxide growth units, the edge with two growth units, and the face of the tetrahedron with only one growth unit. The viscosity of the hydrothermal growth solution is low, so the crystal interface structure plays a large role in kinetics. Thus analysis of Figure 8.12 would predict the following relative growth rates under ideal conditions:

$$v\langle 0001\rangle > v\langle 01\bar{1}\bar{1}\rangle > v\langle 0\bar{1}10\rangle > v\langle 01\bar{1}1\rangle > v\langle 000\bar{1}\rangle$$

The hydroxide growth units cluster together by dehydration. In strong alkali solution these clusters are shielded by ions such as $Na{\cdot}O^-$, shielding the growth units and slowing down growth. Wang contends that these mechanisms account for the growth rates and shape of bulk ZnO crystals in strong alkali solutions.

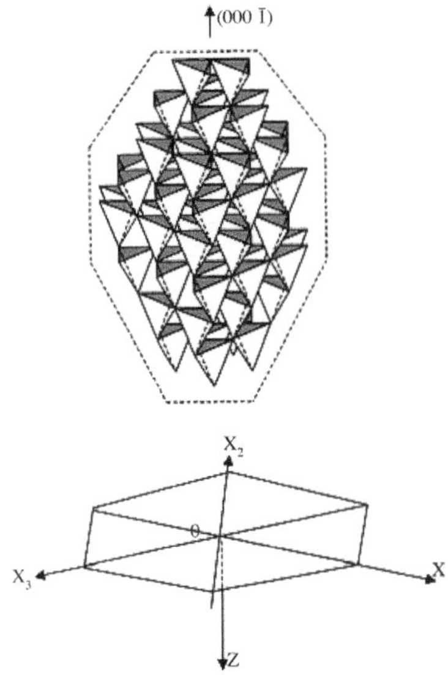

Figure 8.12 ZnO crystal structure image represented in the form of the coordination tetrahedron along the x-direction (C^- surface made up of flat faces at the top). Reprinted from W. J. Li, et al., Growth mechanism and growth habit of oxide crystals, J. Crystal Growth 203, 186. Copyright (1999) with permission from Elsevier

Demianets and Kostomarov proposed the same mechanism, but argued that $Zn(OH)_4^{2-}$ dissociates into $ZnO_2^{2-} + 2H^+$, and that the ZnO_2^{2-} concentration increases with increasing pH.[33] Reaction of one ZnO_2^{2-} with the zinc surface of the crystal allows two ZnO units to form, whereas reaction on the oxygen surface allows only one ZnO unit because of charge compensation.

Impurities and dopants further complicate growth kinetics. Addition of lithium to the solution, as hydroxide or carbonate, improves the quality of the bulk ZnO crystals but also reduces the growth rate in the (0001) facet while increasing the rate on the $(10\bar{1}0)$ facets.[60,61] This may be due to the shielding mechanism mentioned above.

Sakagami found that hydrothermal ZnO crystals have tens of ppm excess zinc.[32] He therefore added H_2O_2 as an oxidizer; excess zinc was reduced to 1–2 ppm. The addition of an oxidizer like H_2O_2 slows the growth rate on all faces, especially the C^- facet.[34] Manganese and nickel had no effect on the kinetics but did color the crystals red and green, respectively. No effect of these dopants on the electrical resistance could be discerned.[34] Addition of NH_4^+ increased the growth rate on the $(10\bar{1}0)$ facets, but crystal quality was degraded.[34] Demianets *et al.* published a more detailed paper on the effect of Li^+ and several of the divalent and trivalent metals (Co^{2+}, Fe^{2+}, Mn^{2+}, Fe^{3+}, Mn^{3+}, Sc^{3+}, In^{3+}) on the growth kinetics and morphology of hydrothermal ZnO.[54] Figure 8.13 shows as metallic impurities are increased in the ZnO growth medium, P-plane facets are formed and C-axis growth rates decrease. The decrease in growth rates can be explained by the shielding effects impurities can have on the matrix compound (i.e. ZnO) as discussed above by Wang.

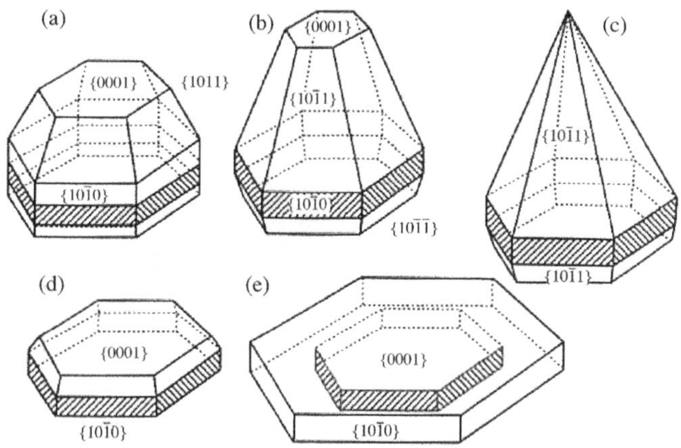

Figure 8.13 *Impurity effects on the morphology of hydrothermal ZnO crystals: (a) Li^+; (b, c) Fe^+; (d) Mn^{2+}; and (e) In^{3+}*

8.6 Properties of Bulk Hydrothermal ZnO

8.6.1 Extended Imperfections (Dislocations, Voids, etc.) and Surface Studies

Because hydrothermal ZnO and quartz are both amphoteric single-component oxides, many insights into hydrothermal ZnO can be obtained from studies of hydrothermal quartz growth, which has been intensely investigated during the last 50 years. Laudise and Barnes[35] and Armington[36] have published excellent reviews on the growth of high perfection quartz and on dislocation mechanisms. Extended imperfections that can be formed in both ZnO and quartz include:

- *Seed veils and etch channels.* Small holes or channels filled with voids, water vapor, or liquid, caused by etch tracks that form on seeds during initial growth.
- *Voids.* Small holes filled with air, water vapor, or liquid; can occur whenever growth conditions change abruptly at the growth surface (impurity clusters, cracks, or crystalline particles from nutrient brought to growth interface by fluid flow).
- *Crevice flaws.* Equivalent to dendritic growth in metals. Uneven or rough growth caused by surface kinetics of growth faces. In extreme cases can cause gaps, cracks, and large numbers of dislocations.
- *Dislocations.* Equivalent to those in melt-grown bulk crystals. Strain-induced, because of impurity incorporation or intersection of growth planes, dislocations often propagate from seed into crystal.
- *Vertical etch channels.* Equivalent to micropipes that form in vapor-grown crystals. Dislocations decorated with impurities causing cylindrical voids that can reach from the seed to the surface of the crystal.

All these imperfections have been observed in hydrothermal ZnO bulk crystals. Addition of lithium, use of low dislocation-density seeds, and use of high purity nutrient reduce the

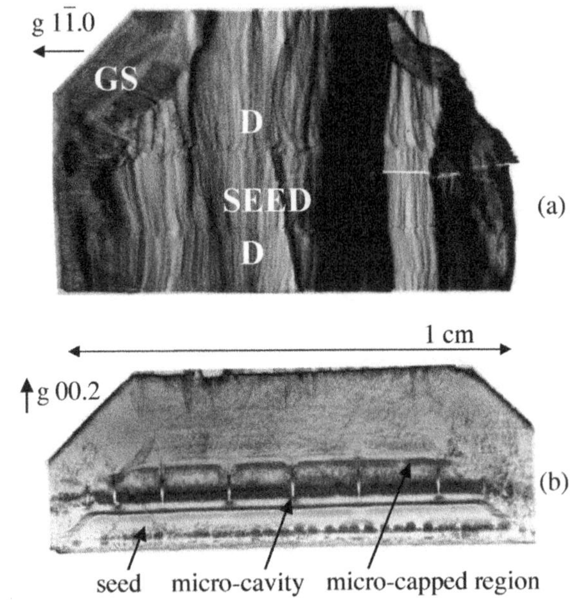

Figure 8.14 *Synchrotron white-beam X-ray topographs of ZnO seed-crystal interfaces: (a) normal seed interface (D, dislocations; GS, growth sector boundary); (b) ZnO micro-capped region, exhibiting micro-cavities and low dislocation generation above micro-capped region in the C^+-direction. Adapted from G. Dhanaraj, et al., Growth and process induced dislocations in Zinc Oxide crystals. Journal of Crystal Growth vol. 297, 74. Copyright (2006) with permission from Elsevier*

concentrations of most imperfections. Lithium may reduce imperfections by decreasing the surface free energy when lithium ions incorporate at the growth interface. Lithium may also limit the incorporation of H_2O and OH^- into the crystal lattice at the growth interface.[14,15,62]

Synchrotron white-beam X-ray topography (SWBXT) in Laue configuration was performed on a series of crystals grown at AFRL-Hanscom AFB.[63] To trace the growth history of the crystal, (10$\bar{1}$0) crystal plates containing both the seed crystal and the bulk region were imaged. One set of topographs showed the usual propagation of edge dislocations from the seeds [Figure 8.14(a)]. The growth sector boundary can also clearly be seen, marking a change of growth morphology.

On several other crystals the topographs revealed a capping phenomenon similar to that observed in KH_2PO_4.[64] Figure 8.14(b) shows that the dislocation density is very high near the seed/crystal interface, revealing strain associated with growth initiation rather than dislocations propagating from the seed into the bulk. A growth band, possibly because of a fluctuation in the growth conditions, stops many of these dislocations. In some cases micro-cavities originating at the seed/crystal interface were observed; the cavities heal during subsequent growth, nucleating dislocations.

The cause of micro-capping has not been established. One possible mechanism is seed etch-back during initial growth, due to fluctuations in the temperature gradient coupled with seed misalignment from C-plane orientation.

Surface studies on hydrothermal crystals showed the C^+ surface to be smooth and specular, with spiral hexagonal growth pyramids.[34] The C^- surface is more three-dimensional, with layer-like growth. The M-planes also have layered growth; the P-planes have a series of terraces. No in-depth study has addressed the mechanisms responsible for these various morphologies on the growth surfaces.

X-ray data from polished ZnO wafers cut from industrial hydrothermal ZnO have recently been published. The quality of (0001) substrates was investigated by X-ray rocking curve (XRC) measurements (Rigaku RINT-2000 diffractometer, CuK_α, four crystal Ge (220) channel monochromator, beam divergence 12 arc-sec, scan speed $0.01\,^\circ\,min^{-1}$, step width $10^{-4}\,^\circ$) using the (0002) reflection. The full width at half-maximum (FWHM) ranges between 19 arc-sec and 30 arc-sec after chemical–mechanical polishing (CMP) compared with values reported for ZnO wafers prepared from pressure-melt grown ZnO, 49 arc-sec,[65] and rocking curves measured on ZnO fabricated by vapor growth with FWHM (~30 arc-sec).[66] X-ray reciprocal space mapping using the (0002) symmetric reflection reveled a highly symmetric single peak with the FWHM from the ω scan of 15 arc-sec measured on hydrothermal ZnO substrates.[67]

X-ray topography (CuK_α, 40 kV, 10 mA, detected by a film IX80; Berg–Barrett geometry) was measured on an epi-ready (0001) ZnO wafer 50 mm in diameter and 500 µm in thickness, supplied by Tokyo Denpa. The XRC FWHM from the (0002) reflection of the sample was 19 arc-sec. Approximately 2500 scans have been assembled to yield the contrast-enhanced image of Figure 8.15(a). The (114) reflection at $2\theta = 98.6^\circ$ and $\omega = 49.3^\circ$ was used to obtain the image. The wafer appears very homogeneous over the entire area. Slight contrast effects are seen, presumably due to slightly different lattice parameters caused by fluctuations in the impurity concentration or stoichiometry.

The observation of an epi-ready (0001) 50 mm ZnO wafer under crossed polarizer is shown in Figure 8.15(b). Only a small fraction of random strain was observed, which

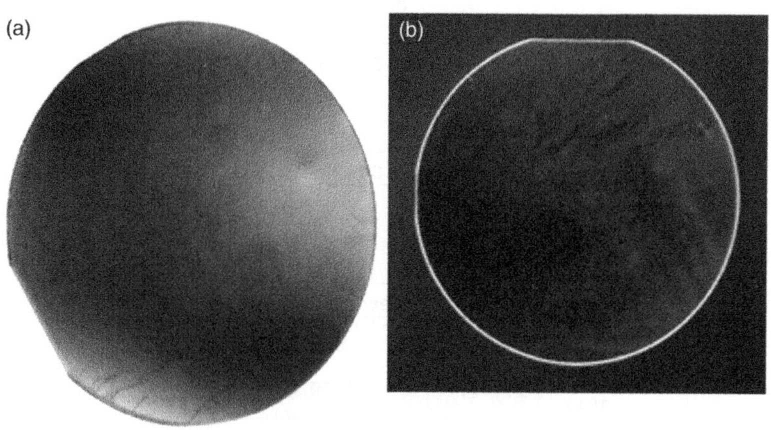

Figure 8.15 *(a) Transmission X-ray topograph of a 50 mm epi-ready (0001) ZnO wafer produced by Tokyo Denpa. (b) Observation under crossed polarizer reveals little random strain in a 50 mm (0001) ZnO wafer. Reprinted from D. Ehrentraut et al., Solvothermal growth of ZnO, Prog. Cryst. Growth Charact. Mater. vol. 52. Copyright (2006) with permission from Elsevier*

supports the result from the X-ray topography that the homogeneity of the wafer is very high.

Measurements employing contact-mode atomic force microscopy (AFM) under ambient air conditions have revealed a root mean square (rms) roughness of 0.285 nm and 0.155 nm for the (0001) and (000$\bar{1}$) face, respectively, for ZnO wafers which are treated by CMP. Thermal annealing further reduces surface roughness to about 0.12 nm if proper conditions are applied.[67]

The etch pit density (EPD) was determined for the (0001) and (000$\bar{1}$) polar face of several wafers. An aqueous solution of concentrated H_3PO_4 was applied for 5 min at 25 °C. The EPD was about 300 cm^{-2} after CMP and is further lowered to less than 80 cm^{-2} by annealing.[68] To our knowledge these wafers are the best hydrothermal ZnO wafers in terms of structural perfection that are currently commercially available.

8.6.2 Impurities

We introduce the occurrence and distribution of impurities in hydrothermal ZnO by referring to crystal ZnO34A, grown at AFRL-Hanscom AFB, and samples cut from the crystal.[12] Figure 8.16 illustrates the faceted crystal morphology of ZnO34A and labels the three slices that were analyzed. Slice ZnO34A3, from the C^- sector, was a darker green than slices ZnO34A1 and ZnO34A2, which were taken from the C^+ sector. In the C^+ sector, coloration became less intense as distance from the seed increased.

The differences in coloration, together with the discussion of Section 8.5 and the fact that the crystal growth rate in the C^+ direction was ~3 times the growth rate in the C^- direction, may mean that impurity incorporation differs between the C^+ and C^- sectors.

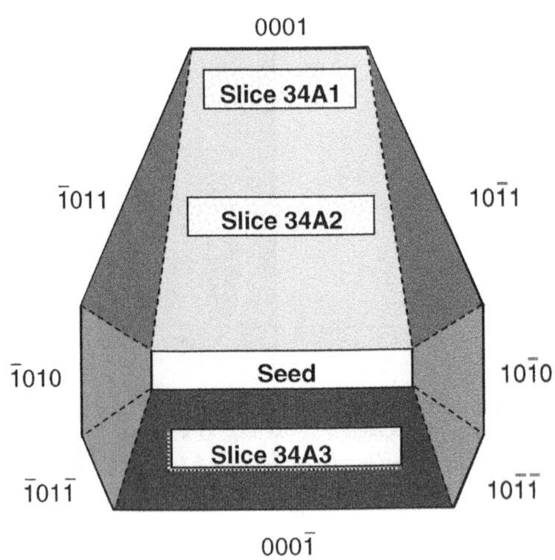

Figure 8.16 Schematic of AFRL-Hanscom AFB crystal ZnO34A. Reprinted from M. Suscavage, et al., MRS Internet J. Nitride Semiconductor Res. (USA) vol. 4, p.294. Copyright (1999) with permission from MRS

Chemical and electrical analyses support such an inference, but also point to a more complex story. Glow discharge mass spectrometry (GDMS) showed the dark sector contains 3–5 ppm of iron, about twice as much as in the lightest colored sector, in general agreement with Croxall *et al.*[31] This is qualitatively consistent with electrical measurements, which showed that ZnO34A3 had a free electron concentration of 3×10^{18} cm^{-3} at room temperature, whereas ZnO34A1 had donor and acceptor concentrations of only $N_D = 7 \times 10^{15}$ and $N_A = 1 \times 10^{15}\,cm^{-3}$. However, slice ZnO34A1 contained about twice as much Al (known to be a shallow donor)[69] and Si (5–6 ppm) as ZnO34A3. (All other impurities occurred at <1 ppm in this crystal.) The anisotropic incorporation of impurities was also conferred by Sekiguchi *et al.*, who reported impurity concentration variations among crystals that were grown under various conditions of temperature and pressure but that otherwise were nominally identical.[53]

ZnO seems to have a high affinity for iron in hydrothermal growth. The ZnO nutrient Croxall *et al.* employed contained <1ppm Fe, but green sections of their crystal contained ~50 ppm. In our laboratory, ZnO crystals contained parts per *thousand* of iron if platinum liners leaked, thereby allowing the diluted growth solution to leach iron from the autoclave wall.

Secondary ion mass spectrometry (SIMS) was employed to study the impurity distribution in the depth of a commercial 50 mm diameter hydrothermally grown ZnO wafer from Tokyo Denpa Corporation which had been CMP treated and afterwards annealed. The primary beam species was Cs^+ (5 kV, 350 nA) and the sputter speed was around 120–150 nm min^{-1}. The impurity levels remain constant with increasing scan depth and concentrations have been revealed for Li ($2 \times 10^{16}\,cm^{-3}$), Na ($8 \times 10^{15}\,cm^{-3}$), K ($3 \times 10^{15}\,cm^{-3}$), Mg ($10^{16}\,cm^{-3}$), Al ($4 \times 10^{15}\,cm^{-3}$), Si ($7 \times 10^{17}\,cm^{-3}$), Fe ($8 \times 10^{15}\,cm^{-3}$) and Cd ($10^{17}\,cm^{-3}$).

Impurity concentrations of Fe, Al, Li and K using inductively coupled plasma mass spectrometry (ICP-MS) from 50 mm diameter ZnO wafers cut from the same hydrothermal crystalline boule have also been investigated.[48,68] Specimens grown and cut from the $(000\bar{1})$ face and from the (0001) face of the seed crystal show different impurity levels. Lower impurity levels are generated in the grown crystal at increasing distance from the $(000\bar{1})$ face of the seed. This was particularly obvious for the case of Li whereas the concentration of K remains nearly unchanged for both faces. Both Fe and Al show higher concentrations in wafers grown on the $(000\bar{1})$ face of the seed crystal: 11 and 1 ppm for Fe and 8 and 0.5 ppm for Al for the $(000\bar{1})$ and (0001) face, respectively. Figure 8.17 shows that the concentration of Al can be reduced by a factor of up to three upon proper annealing.[47] There is a similar result for Li, which can be reduced by about one order of magnitude to currently $10^{16}\,cm^{-3}$.

To summarize, impurity concentrations in hydrothermal ZnO depend not only upon the purity of the starting materials but also upon growth conditions (solution chemistries, growth temperature, etc.) in ways that have yet to be fully characterized. Fe, Ag, Si, Na, Li, K and Al have been found in hydrothermal ZnO crystals; the concentrations vary from run to run. Nonetheless, sub-ppm impurity and sub-$10^{16}\,cm^{-3}$ donor/acceptor concentrations have been achieved, demonstrating that hydrothermal ZnO crystals can have purities that rival or exceed the purities of bulk ZnO grown by other methods – purities, in fact, that rival or exceed those of III–V semiconductors such as InP and GaN.

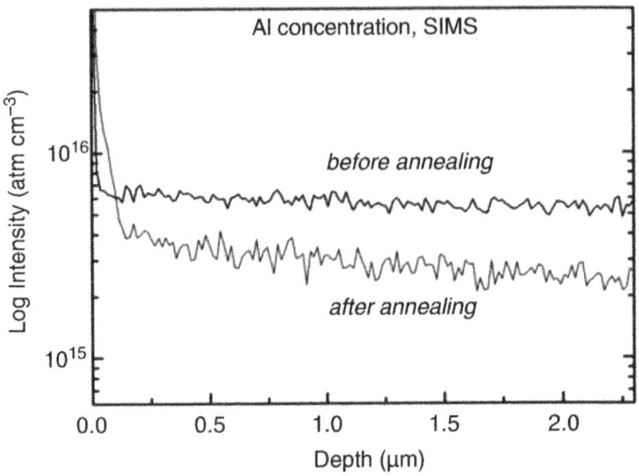

Figure 8.17 *SIMS depth scans in the surface-near region demonstrating the effect of annealing on the Al concentration*

8.6.3 Electrical Properties

Studies of the electronic and optical properties of hydrothermal ZnO are similar to studies of ZnO grown by other techniques. Differences in emphasis relate primarily to issues associated with solution growth, with incorporation of Li, and with opportunities created by the production of highly faceted crystals.

Li is often added to hydrothermal solutions as hydroxide or carbonate because it improves crystallinity and morphology.[70] Li can therefore occur in concentrations of >10 ppm in hydrothermal ZnO;[32,53,71] it has been employed to achieve resistivities as high as 10^{10} ohm cm by compensating native donors.[71] Li is anathema, however, to most electronic and optical device fabricators, who fear that Li – typically a fast diffuser – will incorporate into devices and thereby "poison" them. This may not be an insurmountable obstacle to device applications, for Li apparently can be removed from ZnO by annealing in a zinc atmosphere.[15] Also, using appropriate mineralizer solutions, it is possible to obtain high quality as-grown hydrothermal ZnO crystals that have sub-ppm Li concentrations.[12]

In, Ga and Al are shallow donors in ZnO.[69,72] As mentioned above, Li occupying the Zn site is believed to be an acceptor (interstitial Li is believed to be a donor);[15] addition of Li or Cu increases the resistivity of ZnO after annealing in air or Zn,[13–15,31] probably by compensating donors. The donor/acceptor properties of Fe^{2+} and Fe^{3+} in ZnO are not known. The role of hydrogen in ZnO is controversial: recent theoretical calculations predict it should be a shallow donor,[73] in overall agreement with experimental measurements performed in the 1950s that associated increases in electrical conductivity with hydrogen incorporation (reviewed in Van de Walle);[73] however, in recent work on ZnO films grown by metal organic chemical vapor deposition, the conductivity increase was attributed to passivation of acceptors.[74] Hydrothermal growth of ZnO in an effective overpressure of H_2 was achieved by adding Zn powder to the growth solution; unfortunately only the carrier concentrations after annealing in vacuum or air ($2–5 \times 10^{15}\,cm^{-3}$), not carrier concentrations in as-grown ZnO, were reported.[15]

Native defects such as oxygen vacancies or zinc interstitials have long been regarded as donor centers in ZnO (see, e. g., Littbarski[70] and Look *et al.*[75] and references therein). In many cases, they may constitute the most numerous donor sites. In sample ZnO34A3 (discussed in Section. 8.6.2), for example, impurity concentrations are in the ppm range, but the free electron concentration at room temperature is about two orders of magnitude higher, $3 \times 10^{18}\,cm^{-3}$. This does not by itself exclude hydrogen as a possible donor, but Sakagami's observation that the ZnO electrical resistance increased when an oxidizing agent was added to the hydrothermal growth solution (equivalent to growth in an oxygen overpressure)[32] is indirect evidence that many donor defects result from imperfect zinc–oxygen stoichiometry. Positron annihilation spectroscopy on hydrothermal ZnO from Tokyo Denpa has revealed about $10^{16}\,cm^{-3}$ zinc vacancies (V_{Zn}) as neutral defect complexes and about $10^{17}\,cm^{-3}$ oxygen vacancies (V_O) as neutral oxygen.[76] Clear effects of V_{Zn} and V_O and Li-related defects can be seen in the low temperature photoluminescence (PL) spectrum obtained from excitation with a 325 nm laser ($P_{out} = 1.6\,mW$).[77] Clearly the role of native defects and their relationship to extrinsic point defects is a complicated issue that needs to be elucidated further.

Semi-insulating behavior in hydrothermal ZnO has, as already noted, been achieved through Li doping[72] and growth in an effective oxygen overpressure.[32] Semi-insulating behavior, with a net room temperature free electron concentration of $\sim 2 \times 10^{12}\,cm^{-3}$, and electron mobility of $\sim 175\,cm^2\,V^{-1}\,s^{-1}$, was observed. The cause of this behavior was found to be a donor center located 340 meV below the conduction band;[12] the microscopic nature of this donor is not understood.

The temperature-dependent Hall-effect technique with Van der Pauw geometry was used to examine the carrier concentration N, carrier mobility μ_H, and electrical resistivity R from a $10 \times 10\,mm^2$ specimen cut from a high-quality Tokyo Denpa ZnO crystal. The surface was polished to an rms roughness of about 0.2 nm[47] and Ti/Au contacts were produced by thermal evaporation. A melt-grown sample was measured for comparison.

The carrier concentration is substantially lower in the hydrothermal sample than in the melt-grown sample and decreased with increasing $1/T$ from $N = 2 \times 10^{16}\,cm^{-3}$ at 500 K ($10^3/T = 20$) to $N = 4 \times 10^{13}$ at 100 K. Polyakov *et al.*[78] have obtained similar results of $N = 1.3–4.6 \times 10^{13}$ and $N = 6.4 \times 10^{11}$ at 300 K and 77 K, respectively, for measurements on four samples of ZnO as purchased from Tokyo Denpa. The value of N is clearly the effect of low impurity, point defect, and dislocation concentrations, which substantiates the results previously discussed from X-ray diffraction and impurity analysis. The slight hysteresis slope at $10^3/T < 4$ was obtained from the measurement during heating up and cooling down and is likely the effect related to surface conductivity.[79,80] It has been noted in Graubner *et al.*[81] that annealing at 1150 °C would convert a ZnO substrate from highly resistive to n-type conductive. The higher numbers of shallow donors reduce the concentration of acceptors by compensation and are clearly the dominate cause of the high conductivity.

The Hall mobility peaks at 100 K, $\mu_H = 530\,cm^2\,V^{-1}\,s^{-1}$ and drops down to about $40\,cm^2\,V^{-1}\,s^{-1}$ at 580 K. The higher mobility in comparison with the melt-grown sample, that peaks at $480\,cm^2\,V^{-1}\,s^{-1}$ at 80 K and $430\,cm^2\,V^{-1}\,s^{-1}$ at 100 K, is due to the lower impurity concentration as indicated by the lower carrier concentration. An

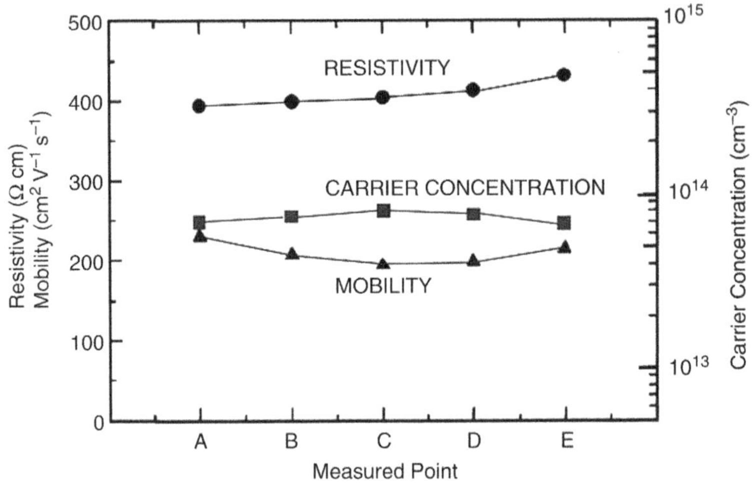

Figure 8.18 *Resistivity, carrier concentration, and mobility across a hydrothermal 50 mm epi-ready ZnO wafer*

earlier hydrothermal grown sample[2] had already showed very similar results for μ_H for the measured temperature range of 200–400 K, where the μ_H decreased from about 300 to $100\,cm^2\,V^{-1}\,s^{-1}$. The vapor grown ZnO shows a higher μ_H up to almost $2000\,cm^2\,V^{-1}\,s^{-1}$ at 40 K. These results are expected, for ZnO compared with GaN has a lower mobility because of a higher effective mass and larger optical phonon scattering parameter.[2]

The electrical resistivity of the hydrothermal ZnO sample is about two orders of magnitude higher than the sample grown from the melt, with a minimum of 20 Ω cm at 60 K and 0.1 Ω cm at 200 K, respectively. Hydrothermal ZnO from the Russian producer SPC GoodWill[82] shows a higher electrical resistivity of 500–1000 Ω cm than the TEW material. Other results on TEW ZnO[78] reported a large variation of R between 96 Ω cm and $5 \times 10^5\,\Omega\,cm$, which was speculated to come from the Li content in the samples investigated. It is possible that different growth sectors were present in the specimens and therefore Li was incorporated in quite different concentrations there.

The uniformity of resistivity R, carrier concentration N and mobility μ_H over a 50 mm wafer from Tokyo Denpa was measured (Figure 8.18) and the following values obtained: $R = 380\,\Omega\,cm \pm 15\%$, $N = 8 \times 10^{13}\,cm^{-3} \pm 20\%$ and $\mu_H = 200\,cm^2\,V^{-1}\,s^{-1} \pm 10\%$.[68]

Hydrothermal conducting indium-doped ZnO was grown using sintered ZnO powder that was mixed with a small percentage of indium oxide.[83] As previously stated, growth rates were dramatically reduced in the C-axis (see Figure 8.13). The resistivity measurement for a $5 \times 5\,mm^2$ In:ZnO sample was $\sim 2.1 \times 10^{-2}$ ohm cm. The conductivity achieved is adequate for most semiconductor device applications that would benefit from conducting substrates.

The presence of large growth facets on hydrothermal ZnO crystals (Figures 8.9, 8.10 and 8.15) facilitates the study of electronic properties as a function of crystallographic orientation and surface polarity. Urbieta *et al.* employed scanning tunneling microscopy and found clear differences in surface electronic structure that distinguished the C^+

(zinc-rich), C^- (oxygen-rich), and nonpolar ZnO faces.[84] Sakagami *et al.* measured current–voltage and capacitance–voltage characteristics on the crystallographic C^+, C^-, and m faces.[85] Along all crystallographic axes studied, they observed nearly ohmic behavior when surfaces were zinc-rich and rectifying behavior when surfaces were oxygen-rich. The authors' judgement that m-sectors are more suitable than C^+ and C^- sectors for making electrical contacts, if confirmed, has potential significance for device fabrication that could stimulate interest in ZnO crystal growth on nonbasal-plane seeds.

8.6.4 Optical Properties

To date, the overwhelming majority of optical measurements of hydrothermal ZnO have been at room temperature or liquid nitrogen temperature. Such measurements may usefully compare the luminescence efficiencies of samples prepared under conditions in which a researcher varies a growth parameter, and they provide qualitative indications of crystal quality or the presence of impurities. However, the most sensitive indications of crystal quality often come from optical measurements at near-liquid-helium temperatures, where excitonic and other bands indicate the underlying quality (or lack thereof) of the material. Broadband spectra at both high and low temperatures can provide useful information about impurities and native defects.

Near-band-edge PL spectra, obtained at 4.2 K, are shown in Figure 8.19, which compares hydrothermally grown and vapor-phase-grown ZnO (from AFRL-Hanscom AFB and Eagle Picher Technologies, respectively).[12] The narrow-line spectra – emissions from decays of excitons bound to a neutral donor and to some unidentified impurity or defect – demonstrate that carriers in both samples have the long lifetimes characteristic

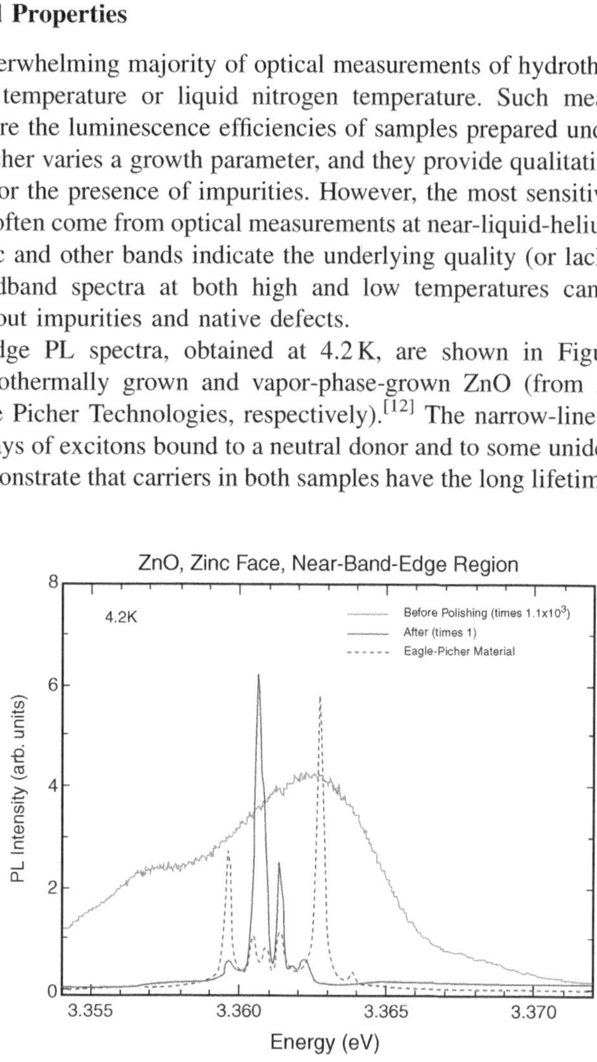

Figure 8.19 *Zinc-face PL spectra of AFRL-Hanscom AFB hydrothermal ZnO and Eagle Picher Technologies vapor-grown ZnO. The two narrow-line spectra are for samples polished by Eagle PicherTechnologies; the broad weak band is for the AFRL-Hanscom AFB sample after conventional polishing. With kind permission of Dr. David Look*

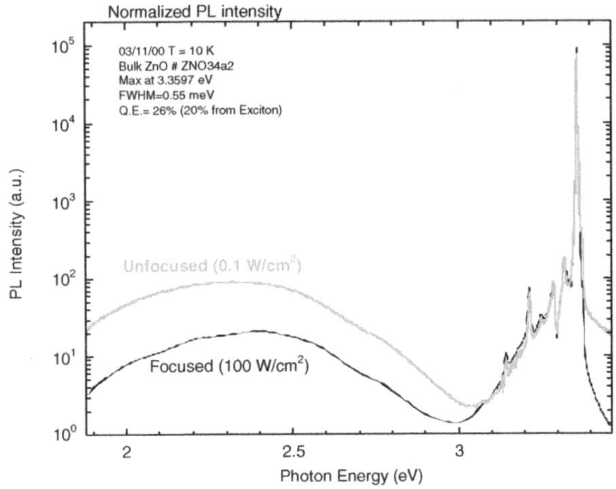

Figure 8.20 *Low temperature broadband PL spectrum of AFRL-Hanscom AFB sample ZnO34A2. Courtesy of M. Reshchikov. With Kind Permission of M. Reshchikov*

of excellent crystallinity. The weak, broad band exhibited in Figure 8.19, caused by damage introduced by a mechanical polishing, contrasts starkly with the intense narrow bands that characterize undamaged surfaces (which were prepared by a proprietary Eagle Picher Technologies chemical–mechanical polish). Near-helium-temperature PL is thus seen to be a sensitive indicator of both crystal and surface quality for ZnO.

Higher temperature PL spectra tend to be less sensitive to surface quality. Room temperature PL from the polish-damaged C^- face was approximately one-third as intense as PL from the undamaged surface.[12] Thus an investigator who does not know a priori what PL intensity should be expected from an undamaged surface might conclude, erroneously, that ZnO sample surfaces were well-prepared. (PL on the C^+ surface, on the other hand, might have raised a red flag.)[12]

The broadband low temperature PL spectrum of AFRL-Hanscom AFB sample ZnO34A2 is shown in Figure 8.20. In addition to the excitonic and donor–acceptor bands at energies above 3 eV, there is a broad band, centered at ~2.3 eV (it is probably not a composite of sub-bands, since its shape remains essentially unchanged over three orders of magnitude of excitation intensity). The intensity of the broad band (measured at both 10 K and 300 K) increases from sample ZnO34A1 to ZnO34A3, i.e., as the material becomes more heavily colored. Similar broad extrinsic "orange" and "green" bands have been reported in several PL and cathodoluminescence (CL) measurements of hydrothermal and vapor-grown ZnO.

Conflicting mechanisms have been advanced for the origin of the green band. Reynolds *et al.* compared the ZnO green band to the "yellow band" in GaN (a wurtzitic crystal having a band gap similar to that of ZnO), whose origin remains a matter of debate, and concluded that both bands arise from a transition between a shallow donor and a deep level.[86] However, Garces *et al.* concluded, from electron paramagnetic resonance (EPR) and PL spectra of unannealed and annealed vapor-phase-grown ZnO, that ZnO green bands are caused by emission from Cu^+ and Cu^{2+} ions, respectively.[69]

Orange and green CL have been observed in the m, C, and p sectors of hydrothermal ZnO crystals. Strong green emission (in addition to near-band-edge emission, which we ignore for the purposes of this discussion) was observed from C^+ (Zn-terminated) surfaces, and weak orange emission from C^- (oxygen-terminated) surfaces; green emission was observed from P^- surfaces (Zn-terminated) and orange emission from P^+ (oxygen-terminated) surfaces; and there was weak orange emission from the (nonpolar) M face.[53] Combining these and related observations with the growth kinetic models presented in Section 8.5.2, Sekiguchi *et al.*[53] and Urbieta *et al.*[87,88] associated the occurrence of ZnO green and orange CL bands with impurity incorporation efficiencies (during crystal growth) that the growth model attributes to the polarization state of ZnO sectors (C^+, C^-, M, etc.); similarly, UV and broadband luminescence efficiencies were associated with presumed incorporation rates of nonradiative recombination centers. In this vein, Sekiguchi *et al.* noted that orange emission was strongest in their sample that had the highest Li concentration and lower in a flux-grown crystal that contained virtually no Li; they also noted that use of H_2O_2 in the hydrothermal growth solution, which presumably lowered the concentration of oxygen vacancies, significantly reduced the orange emission.[53]

The broad emission band from an industrial ZnO wafer was measured. The band went from 1.7 to 2.8 eV with a maxium intensity at approximately 2.3 eV. The nature of this broad emission involves donor–acceptor pair recombination due to Li.[89,90] As already above reported, V_{Zn} as neutral defect complexes and V_O as neutral oxygen are typically observed in hydrothermal ZnO.[76] A recent overview of visible luminescence in ZnO has more comprehensive information on the possible mechanisms of broadband emission in ZnO bulk crystals and thin films.[91]

We complete our summary of the optical properties of hydrothermal ZnO with transmission spectra (Figure 8.21).[92] Insulating ZnO is transparent from the near-ultraviolet almost to 10 μm (spectrum in Figure 8.21 labeled "Clear C+ Sector"). Electrically conducting samples (e.g., spectrum labeled "Dark C- sector") exhibit a long-wavelength free-carrier absorption tail. The optical transparency shown in Figure 8.21, together with the high laser breakdown strength of ZnO,[44] have made ZnO a leading candidate for transparent conducting electrodes for high-power near-infrared laser beam-steering devices.[93]

8.6.5 Etching and Polishing

Because the literature[52,70,94–96] provides in-depth discussions of the etching behavior of ZnO, we offer a very brief summary. Oxygen-rich surfaces etch faster (e.g., in phosphoric acid) than zinc-rich surfaces; etch pyramids and hillocks develop. The Zn surface becomes smooth, but small pits develop where dislocations intersect the surface. The prism face exhibits triangular etch pits that look like arrowheads pointing in the $\langle 0001 \rangle$ direction. Thus, etching can be a quick and simple method for obtaining information about crystal orientation and the polarity of ZnO facets.

Upon annealing in air, the C^+ surface becomes smoother, and dimples gradually disappear. On the other hand, the C^- surface roughens, but the size of bumps decreases.

ZnO is a very soft material whose surface is easily damaged by polishing. In Section 8.6.4 we showed that near-band-gap PL spectra measured at temperatures $\leq$15 K

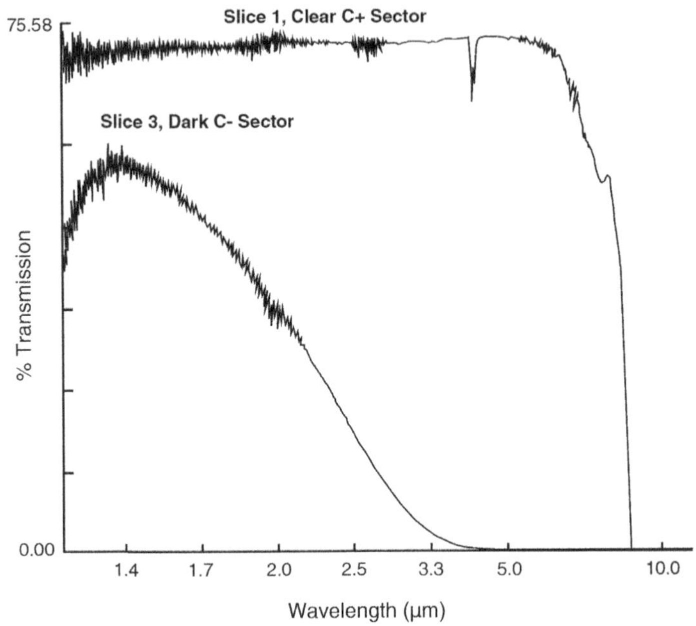

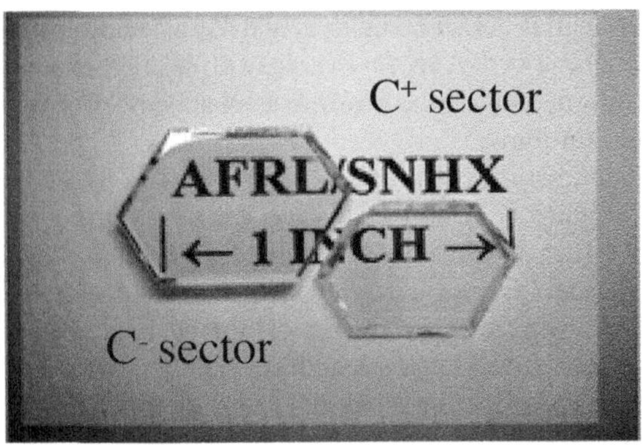

Figure 8.21 *Infrared transmission spectra of ZnO slices cut from C^+ and C^- growth sectors of crystal ZnO37A. Reprinted from D. F. Bliss, Encyclopedia of Materials: Science and Technology. Copyright (2008) with permission from Elsevier*

are sensitive to surface damage. This sensitivity arises from the fact that the above-band-gap excitation employed in such measurements penetrates only a few tens of nanometers into the material. Near-band-gap PL at higher temperatures tends to be less sensitive to surface damage because the emission is due to transitions of thermally dissociated excitons and of carriers that have been thermally excited into the conduction band. In the absence of information about emission from undamaged surfaces, higher temperature PL from damaged surfaces might erroneously be interpreted as "good enough". The same holds true for measurements of the widths of X-ray rocking curves.

Table 8.1 Comparison of X-ray rocking curves of AFRL-Hanscom AFB ZnO after application of different polishing techniques[12]

Sample number and surface termination	ω Scan after abrasive polish, in arc-sec (002)	ω Scan after chemical–mechanical polish, in arc-sec (002)
34A1 C–	52	17.5
34A1 C+	78	23.2
34A2 C–	143	18.6
34A2 C+	214	20.3
34A3 C–	62	13.2
34A3 C+	92	18.0

Most X-ray diffraction measurements sample to depths of 1–2 μm, which may include comparatively undamaged material. In the data of Table 8.1, for example, rocking curve widths of 50–100 arc-sec might be mistakenly attributed to mediocre crystal quality rather than to damaged surfaces in an otherwise excellent crystal.

8.7 Conclusion

Hydrothermal growth can produce very high quality ZnO crystals. Nearly dislocation-free growth can be produced in zinc-terminated growth sectors and on nonpolar prism faces. Semi-insulating crystals have been grown, as well as crystals exhibiting superior low temperature PL characteristics.

The various crystallographic faces of hydrothermal ZnO exhibit pronounced differences not only in growth rates, but also in structural, electrical, and optical properties. The growth mechanisms, the incorporation of impurities, and the generation of native defects are (at least in part) connected to the polarity of the growth surfaces; however, the precise mechanisms – and especially the properties of impurities and native defects – have not been fully elucidated. Exploiting faceted morphology of hydrothermal crystals may facilitate studies of these phenomena, many of which surely have counterparts in ZnO thin film growth and ZnO bulk growth by other methods. The hydrothermal technique has been used for the economical growth of large, low-defect-density ZnO crystals. Advances in homoepitaxy P-doping, and device design and fabrication will drive demand for hydrothermal ZnO substrates in the future.

Acknowledgements

We give special thanks to collaborators whose work either influenced or was explicitly incorporated into this article. Finally, we thank the Air Force Office of Scientific Research for its support of our research on ZnO and other wide-band-gap semiconductors.

References

[1] D. Bagnall, Y. Chen, Z. Zhu, T. Yao, M. Shen and T. Goto, *Appl. Phys Lett.* **73**, 1038 (1998).
[2] D. C. Look, *Mater. Sci. Eng., B* **80**, 383 (2001).
[3] T. Sciosaki, N. Kitamura and A. Kawabata, *Proc. IEEE Ultrasonics Symp.* 296 (1991).
[4] F. J. Aranda, M. T. Harris, M. J. Callahan, S. S. Bailey, M. J. Suscavage, D. F. Bliss, B. R. Campbell, M. Nakashima, B. DeCristofans and D.V.L.N. Rao, *SPIE Proc.* **3798**, 22 (1999).
[5] D.C. Look, D.C. Reynolds, J.R. Sizelove, R.L. Jones, C.W. Litton, G. Cantwell and W.C. Harsch, *Solid State Commun.* **105**, 399 (1998).
[6] J. Nause, *III-Vs Rev.* **12**, 28 (1999).
[7] R.A. Laudise, *The Growth of Single Crystals*, Prentice-Hall, Inc., Upper Saddle River, NJ, 1970, p. 257.
[8] K. Byrappa and M. Yoshimura, *Handbook of Hydrothermal Technology*, William Andrew Publishing, New York, 2001, Ch. 1.
[9] D. Chen, X. Jiao and G. Cheng, *Solid State Commun.* **113**, 363 (2000).
[10] C.H. Lu and C.H. Yeh, *Ceram. Int.* **26**, 351 (2000).
[11] A.S. Khishinuma, *Bull. Mater. Sci.* **18**, 811 (1995).
[12] M. Suscavage, M. Harris, D. Bliss, P. Yip, S.Q. Wang, D Schwall, L. Bouthillette, J. Bailey, M. Callahan, D.C. Look, D.C. Reynolds, R.L. Jones and C.W. Litton, *MRS Internet J. Nitride Semicond. Res.* **4**, 294 (1999).
[13] R. R. Monchamp, R. C. Puttbach and J.W. Nielson, AFML Technical Report AFML-TR-67-144 (USA), 1967.
[14] E. D. Kolb and R.A. Laudise, *J. Am. Ceram. Soc.* **49**, 302 (1966).
[15] E.D. Kolb, S. Coriell, R.A. Laudise and A.R. Hutson, *Mater. Res. Bull.* **2**, 1099 (1967).
[16] F. Hamdani, A. Botchkarev, W. Kim, H. Morkoç, M. Yeadon, J.M. Gibson, S.-C. Y. Ten, D. J. Smith, D.C. Reynolds, D.C. Look, K. Evans, C.W. Litton, W.C. Mitchel and P. Hemenger, *Appl. Phys. Lett.* **70**, 467 (1997).
[17] H.D. Sun, Y. Segawa, M. Kawasaki, A. Ohtomo, K. Tamura and H. Koinuma, *Appl. Phys. Lett.* **91**, 6457 (2002).
[18] M.T. Harris, J.J. Larkin and J.J. Martin, *Appl. Phys. Lett.* **60**, 2162 (1992).
[19] R.A. Laudise and E.D. Kolb, *Am. Mineral.* **48**, 642 (1963).
[20] L.E. McCandlish and R. Urhin, Presentation at 5th International Conference on Solvothermal Reactions, East Brunswick, NJ, July 2002; and personal communications with authors.
[21] G.F. Hüttig and H. Möldner, *Z. Anorg. Chem.* **211**, 368 (1933).
[22] R.A. Laudise and A.A. Ballman, *J. Phys. Chem.* **64**, 688 (1960).
[23] M. M. Lencka and R. E. Riman, *J. Am. Cer. Soc.* **76**, 2649 (1993).
[24] M. M. Lencka, A. Anderko and R. E. Riman, *J. Am. Cer. Soc.* **78**, 2609 (1995).
[25] M. M. Lencka and R.E. Riman, *Chem. Mater.* **5**, 61 (1993).
[26] M. Takehana, T. Nishino, K. Sugawara and T. Sugawara, *Mater. Sci. Eng., B* **41**, 186 (1996).
[27] B.G. Wang, E. W. Shi and A. Z. Zhong, *Cryst. Res. Technol.* **32**, 659 (1997).
[28] B.G. Wang, *Cryst. Res. Technol.* **33**, 937 (1998).
[29] W. J. Li, E.W. Shi, W.Z. Zhong and Z.W. Yin, *J. Cryst. Growth* **203**, 186 (1999).
[30] S. Komareni, M. Bruno and E. Mariani, *Mater. Res. Bull.* **35**, 1843 (2000).
[31] D. F. Croxall, R.C.C. Ward, C. A Wallace and R.C. Kell, *J. Cryst. Growth* **22**, 117 (1974).
[32] N. Sakagami, *J. Cryst. Growth* **99**, 905 (1990).
[33] L. Demianets and D. Kostomaro, *Ann. Chim. Sci. Mater.* **26**, 193 (2001).
[34] I.P. Kuz'mina, A.N. Lobachev and N.S. Triodina, *Crystallization Process Under Hydrothermal Conditions*, Nauka Press, Moscow, 1973, p. 27.
[35] R.A. Laudise and R.L. Barnes, *IEE T. Ultrason. Ferr.* **35**, 277 (1998).
[36] A. F. Armington, *Prog. Cryst. Growth Charact.* **21**, 97 (1990).
[37] J. Larkin, M. Harris, E. Cormier and A. Armington, *J. Cryst. Growth* **128**, 871 (1993).
[38] E.D. Kolb, A.J. Caporaso and R.A. Laudise, *J. Cryst. Growth* **8**, 354 (1971).
[39] S. K. Dutta and R. M. Spriggs, *Mater. Res. Bull.* **4**, 797 (1969).
[40] T. K. Gupta and R. L. Coble, *J. Am. Ceram. Soc.* **51**, 521 (1968).

[41] G. C. Kennedy, *Am J. Sci.* **248**, 540 (1950).
[42] D. Ehrentraut, H. Sato, Y. Kagamitani, H. Sato, A. Yoshikawa and T. Fukuda, *Prog. Cryst. Growth Charact. Mater.* **52**, 287 (2006).
[43] Y. Masuda, A. Suzuki, Y. Mikawa, C. Yokoyama and T. Tsukada, *J. Cryst. Growth* **311**, 675 (2009).
[44] E.V. Kortounova and V.I. Lyutin, *Ann. Chim. Sci. Mater.* **22**, 647 (1997).
[45] E.V. Kortounova, V.I. Lyutin, V.D. Dubovskaya and P.P. Chvanski, *High Press. Res.* **20**, 175 (2001).
[46] E. Ohshima, H. Ogino, I. Niikura, K. Maeda, M. Sato, M. Ito and T. Fukuda, *J. Cryst. Growth* **v260**, 166 (2004).
[47] D. Ehrentraut, H. Sato, Y. Kagamitani, H. Sato, A. Yoshikawa and T. Fukuda, *Prog. Cryst. Growth Charact. Mater.* **52**, 280 (2006).
[48] T. Fukuda and D. Ehrentraut, *J. Cryst. Growth* **305**, 304 (2007).
[49] Public data and personal discussions with Tokyo Denpa Co. Ltd, Tokyo (http://www.tew.co.jp).
[50] L.N. Dem'yanets and V.I. Lyutin, *J. Cryst. Growth* **310**, 993 (2008).
[51] E.V. Kortunova, P.P. Chvanski and N.G. Nikolaeva, *J. Phys. IV* **126**, 39 (2005).
[52] A.N. Mariano and R. E. Hanneman, *J. Appl. Phys.* **34**, 384 (1963).
[53] T. Sekiguchi, S. Miyashita, K. Obara, T. Shishido and N. Sakagami, *J. Cryst. Growth* **214–215**, 72 (2000).
[54] L.N. Demianets, D.V. Kostomarov, I.P. Kuz'mina and S.V. Pushko, *Crystallogr. Rep.* **47** (Suppl. 1), S86 (2002).
[55] F. Abbona, in *Crystal Growth of Technologically Important Electronic Materials*, edited by K. Byrappa, H. Klapper, T. Ohachi and R. Fornari, Allied Publishers PVT Limited, New Delhi, p. 89.
[56] G.H. Gilmer, R. Ghez and N. Cabrera, *J. Cryst. Growth* **8**, 79 (1971).
[57] D. Wesolowski P. Bénézeth and D. Palmer, *Geochim. Cosmochim. Acta* **62**, 971 (1998).
[58] P. Bénézeth, D. Palmer and D. Wesolowski, *Geochim. Cosmochim. Acta* **63**, 1571 (1999).
[59] I.L. Khodakovsky and A. Y. Yelkin, *Geokhimiya* **10**, 1490 (1975).
[60] M. M. Lukina, M. V. Lelekova and V. E. Khadzhi, *Sov. Phys. Cystallogr.* **15**, 530 (1970).
[61] R.A. Laudise, E.D. Kolb and A.I. Caporaso, *J. Am. Ceram. Soc.* **47**, 9 (1964).
[62] I. P. Kuz'mina, *Sov. Phys. Cystallogr.* **13**, 803 (1969).
[63] G. Dhanaraj, M. Dudley, D. Bliss, M. Callahan and M. Harris, *J. Cryst. Growth* **297**, 74 (2006).
[64] H. Youping, Z. Jinbo, W. Dexang, S. Genbo and Y. Mingshan, *J. Cryst. Growth* **169**, 196 (1996).
[65] J. Nause and B. Nemeth, *Semicond. Sci. Technol.* **20**, S45 (2005).
[66] J.-M. Ntep, S. Said Hassani, A. Lusson, A. Tromson-Carli, D. Ballutaud, G. Didier and R. Triboulet, *J. Cryst. Growth* **207**, 30 (1999).
[67] D. Ehrentraut, H. Sato, M. Miyamoto, T. Fukuda, M. Nikl, K. Maeda and I. Niikura, *J. Cryst. Growth* **287**, 367 (2006).
[68] K. Maeda, M. Sato, I. Niikura and T. Fukuda, *Semicond. Sci. Technol.* **20**, S49 (2005).
[69] N. Y. Garces, L. Wang, L. Bai, N. C. Giles, L. E. Halliburton and G. Cantwell, *Appl. Phys. Lett.* **81**, 622 (2002).
[70] R. Littbarski, in *Current Topics in Materials Science*, edited by E. Kaldis, North-Holland, Amsterdam, 1981, Vol. 7, p. 212.
[71] R. Laudise, in *Crystal Growth: an Introduction*, edited by P. Hartman, North-Holland, Amsterdam, 1973, p. 190.
[72] E. D. Kolb and R. A. Laudise, *J. Am. Ceram. Soc.* **49**, 302 (1966).
[73] C. G. Van de Walle, *Phys. Rev. Lett.* **85**, 1012 (2000).
[74] B. Theys, V. Sallet, F. Jomard, A. Lusson, J. Rommeluère and Z. Teukam, *J. Appl. Phys.* **91**, 3922 (2002).
[75] D. C. Look, J. W. Hemsky and J. R. Sizelove, *Phys. Rev. Lett.* **82**, 2552 (1999).
[76] F. Tuomisto, *Proc. SPIE* **6474**, 647413 (2007).
[77] T. Moe Børseth, B.G. Svensson, A. Yu. Kuznetzov, P. Klason, Q. X. Zhao and M. Willander, *Appl. Phys. Lett.* **89**, 262112 (2006).

[78] A.Y. Polyakov, N.B. Smirnov, A.V. Govorkov, E.A. Kozhukhova, S.J. Pearton, D.P. Norton, A. Osinsky and A. Dabiran, *J. Electron. Mater.* **35**, 663 (2006).

[79] I.V. Markevich, V.I. Kushnirenko, L.V. Borkovska and B.M. Bulakh, *Solid State Commun.* **135**, 475 (2005) and references therein.

[80] O. Schmidt, P. Kiesel, C.G. Van de Walle, N.M. Johnson, J. Nause and G.H. Döhler, *Jpn. J. Appl. Phys.* **44**, 7271 (2005); O. Schmidt, P. Kiesel, D. Ehrentraut, T. Fukuda and N. M. Johnson, *Appl. Phys. A* **88**, 71 (2007).

[81] S. Graubner, C. Neumann, N. Volbers, B.K. Meyer, J. Bläsing and A. Krost, *Appl. Phys. Lett.* **90**, 042103 (2007).

[82] D. Look, Personal communication; M. Callahan and B. Wang, unpublished data.

[83] B. Wang, M. J. Callahan, Chunchuan Xu, L.O. Bouthillette, N. C. Giles and D. F. Bliss, *J. Cryst. Growth* **304**, 73 (2007).

[84] A. Urbieta, P. Fernández, J. Piqueras and T Sekiguchi, *Semicond. Sci. Technol.* **16**, 589 (2001).

[85] N. Sakagami, M. Yamashita, T. Sekiguchi, S. Miyashita, K. Obara and T. Shishido, *J. Crys. Growth* **229**, 98 (2001).

[86] D. C. Reynolds, D. C. Look, B. Jogai and H. Morkoç, *Solid State Commun.* **101**, 643 (1997).

[87] A. Urbieta, P. Fernández, J. Piqueras, Ch. Hardalov and T. Sekiguchi, *J. Phys. D: Appl. Phys.* **34**, 2945 (2001).

[88] A. Urbieta, P. Fernández, Ch. Hardalov, J. Piqueras and T. Sekiguchi, *Mater. Sci. Eng., B* **91–92**, 345 (2002).

[89] O.F. Schirmer and D. Zwingel, *Solid State Commun.* **8**, 1559 (1970).

[90] D. Zwingel, *J. Lumin.* **5**, 385 (1972).

[91] J. Mass, M. Avella, J. Jiménez, M. Callahan, E. Grant, K. Rakes, D. Bliss and B. Wang, in *New Materials and Processes for Incoming Semiconductor Technologies*, edited by S. Dueñas and H. Castán. Research Signpost, Kerala, 2006, p. 113.

[92] D. Bliss, in *Encyclopedia of Materials: Science and Technology*, edited by K.H.J. Buschow, R. Cahn, M. Flemings, B. Ilschner, E. Kramer, S. Mahajan and P. Veyssiere, Elsevier Science, New York, 2001, p. 9888.

[93] C. Woods and A. J. Drehman, Presentation at the National Space and Missile Materials Symposium, Monterey, CA, June 2001.

[94] G.A. Wolff, B.N. Das and F.H. Cocks, *J. Appl. Cryst.* **4**, 379 (1971).

[95] G. Heland, P. Kunstmann and H. Pfister, *Z. Phys.* **176**, 485 (1963).

[96] S. Krebs and R Littbarski, in *Current Topics in Materials Science*, edited by E. Kaldis, North-Holland, Amsterdam, 1981, Vol. 7, p. 181.

9

Growth and Characterization of GaN/ ZnO Heteroepitaxy and ZnO-Based Hybrid Devices

Ryoko Shimada and Hadis Morkoç
Department of Electrical and Computer Engineering, Virginia Commonwealth University, Richmond, VA, USA

9.1 Introduction

It is a well known fact that GaN-based heterostructures have been playing a major role in optoelectronic devices such as blue-green light-emitting diodes (LEDs), laser diodes (LDs), color displays, moving signals, low level lighting, traffic signals, data storage, UV and solar blind detectors, and high temperature high-power electronics such as field effect transistors and Schottky devices.[1] The semiconductors, which show a higher electron drift velocity, high mobility and large dielectric breakdown field, are essential for high frequency power devices in communication and other systems. The nitride semiconductor family satisfies many of these requirements. A mobility of 160 000 $cm^2 V^{-1} s^{-1}$ and sheet density of $1.5 \times 10^{12} cm^{-2}$ at 4.2 K for AlGaN/GaN have been recently reported.[2] Almost on a daily basis new milestones are attained. Sapphire has been used as a substrate for GaN growth, owing to demonstration of reasonable quantum efficiency from the LEDs and pulsed laser devices grown on sapphire. For example, Nichia Company, Japan and others have produced high efficiency blue LEDs with lifetimes in excess of 60 000 h, output powers near 300 mW with efficiencies near 200 lumens per watt ($lm W^{-1}$),[3] despite of high density of defects occurring in the layers. The origin of this high density of defects is

Zinc Oxide Materials for Electronic and Optoelectronic Device Applications, First Edition.
Edited by Cole W. Litton, Donald C. Reynolds and Thomas C. Collins.

due to poor lattice match and thermal match of substrates, typically sapphire, to GaN. They affect both the electrical and optical properties of the devices. The origin of the defects has been the topic of many investigations, as detailed in reference [4]. There is still a need to achieve more efficient violet lasers, green lasers and LEDs, which could be operated continuously at room temperature for long lifetimes.

The role of the buffer layer is predominant for better wetting of the growing layer with the substrate, i.e. from three-dimensional growth to layered growth. A suitable buffer layer, either homo or hetero, is essential to grow nitride compounds to promote nucleation centers and to reduce defect level concentration, hence, to increase the quality of the growth. Various buffer layers have been explored to grow high quality GaN such as AlN, GaN and ZnO. Early on AlN was identified as a good buffer layer for GaN and AlGaN on sapphire substrates, which led to improved optical and electrical properties. As further development took place GaN buffer layers also began to be used successfully.[5]

It is preferable to employ a nearly lattice-matched and also atomic stacking match buffer layers for follow up epitaxial layers, in order to avoid defect formation and to yield quality epitaxial growth. Among these types of strategies, ZnO has been reported to be a suitable buffer layer for GaN epitaxial layers and templates, owing to low lattice mismatch, about less than 2% between ZnO and GaN. An additional advantage of ZnO is that it has a common stacking order with GaN.

Pulsed laser vapor deposition (PLD), molecular beam epitaxy (MBE), hydride vapor phase epitaxy (HVPE), and metal organic vapor phase epitaxy (MOVPE) techniques have been employed to grow GaN using ZnO buffer on *c*-plane (0001) sapphire. It was found that the growth of ZnO is in the *c*-direction, which is perpendicular to the sapphire substrate. ZnO perfectly matches with $In_{0.22}Ga_{0.78}N$. In many respects, ZnO may also be considered as a suitable buffer for GaN because of the similarity in its physical properties to those of GaN, such as its wide band gap leading to transparency in the visible region. Moreover, ZnO can be grown on *c*-, *a*-, and *r*-plane sapphire.

9.2 Growth of GaN/ZnO

Several techniques have been employed and are being developed to grow nitride-based compounds on different substrates. However, we focus our attention on reviewing growth of GaN on ZnO by these techniques and the role of ZnO as either a buffer layer or a substrate. The PLD technique has been extensively used for deposition of oxide compounds such as superconducting, ferrite, and magnetic materials and has been successfully extended to grow nitride compounds. Vispute *et al.*[6,7] prepared GaN layers using ZnO as a buffer layer by the PLD technique. A KrF excimer laser with a wavelength of 248 nm, and pulse duration of 25 ns was used for ablation of a ZnO (99.99 %) target. The focused beam energy and pulse repetition rate were 2–3 J cm^{-2} and 10–15 Hz, respectively (Figure 9.1). The substrate temperature for the deposition of ZnO was in the range of 300–800 °C. The background oxygen pressure was in the range of 10^{-5}–10^{-2} Torr. In the same fashion, GaN films were deposited on ZnO/sapphire templates at 850 °C under a background NH_3 pressure of 10^{-6}–10^{-5} Torr.

Note that the growth temperature is very important. If it is too low, the obtained smaller grains or number of grain boundaries and defect levels lead to poor quality of the growth.

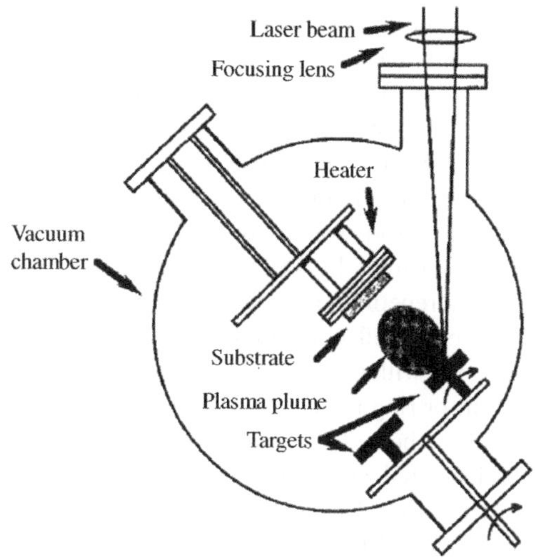

Figure 9.1 *A schematic diagram of a pulsed laser deposition system. Reprinted from R. D. Vispute, V. Talyansky, R. P. Sharma, S. Choopun, M. Downes, T. Venkatesan, K. A. Jones, A. A. Iliadis, M. Asif Khan, and J. W. Yang, Growth of epitaxial GaN films by pulsed laser deposition, Applied Physics Letters, 71, 102. Copyright (1997), American Institute of Physics*

At high temperatures, the ZnO growth is preferable. In fact, it has been concluded that the growth temperature for both ZnO buffer and GaN should be the same.[8] In this case, the ZnO films have been grown by pulsed laser ablation of a solid ZnO target in oxygen ambient on fused silica substrates. An ArF excimer laser with a wavelength of 193 nm, a 250 mJ pulse energy and a 10 Hz repetition was used to deposit films. The substrate temperature was kept at 600 °C.[9] On this ZnO buffer layer, the GaN has been deposited by a liquid-target pulsed laser technique. The flow rate of NH_3 and pressure in the chamber were 20 sccm and 1 mbar, respectively. The same substrate temperature for the deposition of GaN was maintained. If the substrate temperature is less than 550 °C, the grown GaN films are either amorphous or polycrystalline. Columnar growth has proved to be highly important for device applications.[3] An attempt to deposit GaN in N_2 atmosphere instead of NH_3 was unsuccessful.[9] The deposition temperature of the PLD technique is much lower than that of MOVPE, hence, it can be speculated that the optical grade GaN film can be grown by this technique for device applications.

Wang *et al.*[10] prepared GaN layers on sapphire using ZnO buffer layers by the PLD technique. The ZnO and GaN targets were prepared using commercially available powders. The ZnO powder was pre-sintered at 900 °C in O_2 ambient for 2 h, from which a 2 cm diameter pallet was made and the GaN powder was also sintered at 800 °C in the presence of N_2 flow in a similar manner. A Nd:YAG pulsed laser, with a wavelength of 266 nm, a fluence of 0.8 J cm^{-2} per pulse and a repetition rate of 5 Hz, was employed for deposition of ZnO and GaN layers at 800 °C on sapphire substrates.

Room temperature layer by layer GaN growth by PLD on atomically flat ZnO $(000\bar{1})$ substrate was demonstrated by Kobayashi *et al.*[11] To obtain an atomically flat ZnO surface, ZnO $(000\bar{1})$ was annealed at 1150 °C in air. During annealing, the ZnO substrates were kept in a ceramic ZnO box in order to suppress desorption of Zn atoms from the ZnO surface and maintain the surface stoichiometry. GaN growth by PLD at room temperature was conducted with the background pressure 5×10^{-10} Torr, and Ga metal target (99.9999% purity) was irradiated with a KrF excimer laser ($\lambda = 248$ nm, $t = 20$ ns), having an energy density of 3.0 J cm^{-2} and pulse repetition rate of 10 Hz. As a nitrogen source, a radio frequency (RF) plasma radical generator, which was operated at 360 W with a N_2 pressure of 2.0×10^{-5} Torr, was used.

Ueda *et al.*[12] prepared ZnO films by PLD using a 248 nm KrF excimer laser with a pulse duration of 20 ns and a repetition rate of 5 Hz. The partial pressure of oxygen was 0.01 Torr at 550 °C. The streaky reflection high energy electron diffraction (RHEED) pattern indicated a smooth layer with good atomic ordering rather than a spotty (indicative of three-dimensional growth) or ring pattern (indicative of polycrystalline growth) of RF sputtered ZnO, as reported by others. Thick GaN layers were grown by using the HVPE technique on pulsed laser deposited ZnO buffer layers on sapphire. $GaCl_3$ and NH_3 were the source materials. In order to form $GaCl_3$, HCl was passed over liquid Ga held at a high temperature. The production rate of $GaCl_3$ was controlled by the temperature of the Ga boat and HCl flow rate. The growth was performed at 1000 °C at a V/III ratio of 500 and the growth rate was 10 μm h^{-1}.[12] As soon as the required substrate temperature was reached, GaN growth was started immediately to prevent thermal desorption of the ZnO buffer layer from the sapphire substrate. The large difference in thermal expansion coefficient between these compounds (ZnO, 2.9×10^{-6} K^{-1}; GaN, 5.6×10^{-6} K^{-1}; and Al_2O_3, 7.5×10^{-6} K^{-1}) must be kept in mind.

Detchprohm *et al.*[13] used sputtered ZnO on sapphire as buffer layers for GaN epitaxy. A 10 cm diameter ZnO ceramic disk was used as target and Ar or O_2 was employed as sputtering gas. Prior to deposition of ZnO, the chamber was evacuated to 10^{-5} Torr and then the chamber pressure was maintained at 0.13 and 0.4 Torr for Ar and O_2, respectively. The discharging input and reflecting powers were 200 and 20 W, respectively, for both gases. The ZnO growth rates were about 150 and 5 nm h^{-1} for Ar and O_2, respectively. The ZnO growth was performed at room temperature. The sputtered ZnO was not annealed prior to deposition of GaN otherwise it would have decomposed according to the authors. The GaN films were deposited by the HVPE technique. The Ga boat temperature was kept at 850 °C for the formation of $GaCl_3$.

Molnar *et al.*[14] also used the HVPE technique to grow GaN thick templates by employing ZnO as a buffer layer (Figure 9.2). The HVPE is useful to grow thick but not necessarily large area GaN samples. GaN samples grown by this technique are of considerable interest for follow up epitaxial growth because of reducing defect concentration, as the film thickness increases. HCl used in the preparation process of GaN is corrosive and affects the deposition system adversely. For this growth, GaCl, AlN and ZnO were used to treat the substrates prior to growth of GaN. GaCl was synthesized by mixing HCl and liquid Ga metal at 800–900 °C and then it was passed down the reactor tube to the substrate where it was allowed to react with NH_3. Either ZnO or GaCl treated GaN on sapphire samples showed highly transparent films, fewer hexagonal islands, and did not show any polycrystalline nature.

Gu *et al.*[15] extensively studied the role of ZnO while depositing GaN by the HVPE method. They observed that the thin ZnO prelayer disappeared between the sapphire and GaN. The reason is that ZnO is thermodynamically unstable in the presence of reducing

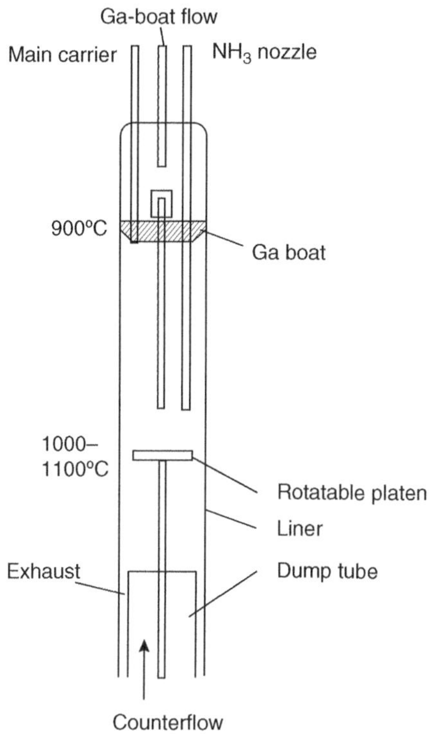

Figure 9.2 *HVPE reactor for GaN growth. Reprinted from R. J. Molnar, et al., Growth of gallium nitride by hydride vapor-phase epitaxy. J. Cryst. Growth, 178, 147. Copyright (1997) with permission from Elsevier*

agents coming from the source materials in the HVPE system. Gu *et al.* have used two 20 nm-thick ZnO samples deposited on sapphire, one annealed at 1050 °C for 10 min in N_2 ambient and the other unannealed. The unannealed sample was dipped into HCl to dissolve any ZnO film. The X-ray photoelectron spectroscopy (XPS) analysis showed no residual ZnO or any other compound formed with sapphire. However, the annealed one, after dipping in HCl, showed traces of $ZnAl_2O_4$ by XPS analysis, which is the same as standard compound $ZnAl_2O_4$. One can conclude that $ZnAl_2O_4$ might form and may act as nucleation centers for GaN growth. The authors have attempted to confirm the aforementioned results by using a supportive technique such as X-ray diffraction (XRD); however, it is very difficult to trace this compound in thin layers so they did the analysis on a 200 nm-thick sample annealed in air for 30 h that avoided any ambiguity. The rocking curve line width of (0002) of GaN grown on sapphire was 700 arcsec, compared with 400–500 arcsec for samples grown with ZnO buffer layer. The carrier concentration was $1 \times 10^{16}\,cm^{-3}$ for 20 nm-thick ZnO buffered GaN samples. The $ZnAl_2O_4$ may promote nucleation centers, in order to get growth and reduce lattice mismatch to GaN.

Building on the earlier investigations of Hamdani *et al.* on as-received ZnO substrates,[16] Gu *et al.*[17] thoroughly studied ZnO substrates with O and Zn polarity prior to GaN thin film deposition and developed a method to render the surfaces atomically flat and damage free. ZnO substrates were annealed at 1050 °C in air for 3 h in order to improve the quality of the surface, which showed full width at half-maximum (FWHM) of 0.576 arcmin for the (0002) X-ray rocking curve, FWHM of 0.7 meV for the photoluminescence exciton peak, and surface roughness of 0.6 nm using atomic force microscopy (AFM). The unannealed and annealed ZnO substrates at temperature less than 1050 °C showed rough surfaces, terraces, and scratches on the surface. GaN thin films were prepared on both O and Zn face ZnO substrates using the RF plasma-enhanced nitrogen MBE technique.[18] GaN films were also grown on ZnO substrates by reaction of Ga and NH_3 at 760 °C by employing either GaN or AlN buffer layer grown at lower temperature by use of RF nitrogen source. For the growth of GaN films on ZnO substrates with the MBE technique, Ga was supplied from a Knudsen cell, and filtered and purified NH_3 was fed through the chamber by an injector retained at 300 °C to prevent liquid NH_3 formation. During the growth, the chamber pressure was maintained at $2–5 \times 10^{-5}$ Torr.[18]

Similar approaches for ZnO substrate treatment[19,20] were reported using higher annealing temperature in air. Ohgaki *et al.* annealed four different orientations of ZnO substrates covered with single crystalline Y-stabilized ZrO_2 (YSZ) plate to prevent evaporation of Zn.[19] After annealing at 1200 °C for 2 h with the YSZ plate, all the scratches were removed. Particularly step-and-terrace in the *c*(+) and (–) - face were observed, having surface roughness of 0.5 nm using AFM. On the other hand, *a*- and *m*-ZnO showed a stripe pattern having surface roughness of 6~7 nm using AFM. GaN films were grown on ZnO and low-temperature nearly lattice matched 30 nm-thick InGaN buffer layer by MBE. The growth was initiated by simultaneously exposing the ZnO surface to incident Ga (Ga and In) and N fluxes, which prevented nitridation and metallization of the ZnO surface. Most of the growth experiments with GaN/ZnO structures were conducted at low temperatures since the grown GaN films at higher temperatures peel off from the ZnO substrate.[21] GaN on low-temperature InGaN buffer layer was grown at 800 °C. To prevent phase separation of the InGaN layer, the Ga and N fluxes were supplied to the film surface while increasing the temperature from 550 to 800 °C. In the case of epi-ZnO films, Heinze *et al.* performed *in-situ* annealing following a low-temperature ZnO buffer layer growth at 900 °C under *t*-BuOH and N_2O atmosphere with N_2 carrier gas to achieve smooth and clean surfaces and then high temperature ZnO growth was performed increasing the growth temperature to 800~900 °C.[20] Oxygen-face ZnO substrates grown by the liquid solution method (CrysTec GmbH, Berlin, Germany) were annealed at 1100 °C for 15 min in the ZnO box with a high amount of ZnO powder. After the annealing process, surface roughness using AFM was 0.5 nm, which is the height of a bilayer along the *c*-axis length. FWHM of the XRD rocking curve at the (0002) reflection of the untreated samples was 220 arcsec, while the annealed one was ~35 arcsec. In terms of electrical characterization, the charge carrier density increased from 3.1×10^{13} cm^{-2} in the as-grown sample to $3.8 \times 10^{14}\,cm^{-2}$ in the annealed sample. The carrier mobility slightly decreased from 96.5 to 77.8 $cm^2\,V^{-1}\,s^{-1}$, respectively.[20]

Matsuoka *et al.*[22] used degreased and etched ZnO substrates to grow thin GaN films by MOVPE. Until attaining the required growth temperature, a nitrogen flow was maintained in the chamber, then ammonia and group III sources were allowed into the MOVPE chamber. A vertical cold walled MOVPE reactor was used for the growth of InGaN and InGaAlN epitaxial layers. The sources were trimethylindium (TMI), triethylgallium (TEG), trimethylaluminum (TMA), and purified ammonia. The incorporation of In was increased by reducing the substrate temperature from 800 to 500 °C. This is due to the much higher vapor pressure of In compared with that of Ga. The typical ammonia flow rate was 20 l min^{-1}. The growth temperature was 500–700 °C and the V/III flow rate ratio was maintained at 20 000. At 800 °C, the flow rate of TEG was kept at 17 µmol min^{-1}, which resulted in a V/III ratio of 4000. The InGaN layers were grown on ZnO single crystal substrates, which were grown by the flux method using lead fluoride as a flux. A typical size of the substrates was several millimeters thick and 10 mm long on each side.

There is a challenging issue related to the MOVPE growth process, which is back-etching of ZnO under the NH_3 and/or H_2 environments, that either damages the ZnO surface or removes ZnO buffer layers during the growth temperature ramp up. To overcome these problems, the MOVPE growth was performed at relatively low substrate temperature using N_2 gas as carrier and dimethylhydrazine (DMHy) as a precursor of atomic nitrogen.[23] Using this MOVPE system, high-quality GaN by MOVPE after only 140 nm thickness using ZnO buffer layer deposited by PLD technique was demonstrated.[24,25] The growth temperature ranged from 550 to 690 °C and the growth pressure was 450 Torr. The carrier gas was N_2, and TMG and DMHy were used as sources of gallium and nitrogen, respectively. The V/III ratio was equal to 11. After the GaN growth, the root mean square (rms) surface roughness was about 2 nm over an area $5 \times 5\,\mu m^2$.

In order to promote future integration of optical function possible with GaN to Si devices and make use of low cost, large area Si substrates, GaN growth on Si is an important technology. Since, the properties of GaN are expected to improve by using buffer layer grown on Si substrates, ZnO buffer layers on Si(001) have been studied by various techniques such as electron beam evaporation, PLD and RF sputtering.[26–28] Polycrystalline GaN on Si with ZnO buffer layer was deposited by radio frequency plasma-enhanced chemical vapor deposition (RF-PECVD).[26] Some 100 nm-thick ZnO buffer layers were grown on Si (001) substrates by the electron beam evaporation technique at 250 °C and annealed at 400 °C for 1 h. A GaN layer was grown on ZnO/Si by RF-PECVD at 650 °C. Triethylgallium (TEGa) with N_2 carrier gas as Ga source and inductively coupled plasma of N_2 generated at 13.56 MHz as nitrogen source were used for the growth. Kim *et al.* deposited ZnO on Si (001) at 600 °C by the PLD technique using a KrF excimer laser ($\lambda = 248$ nm, $t = 25$ ns).[27] After post annealing at 300 °C under an O_2 ambient pressure of 300 Torr, GaN layers were grown on ZnO/Si by metal organic chemical vapor deposition (MOCVD) using TEGa and NH_3 as precursors and N_2 as a carrier gas. The growth temperature was varied from 600 to 850 °C to prevent the thermal decomposition of ZnO buffer layers and the growth pressure was fixed at 100 Torr. The flow rates of TEGa and NH_3 were 50 and 1600 sccm, respectively, to give a V/III ratio of 3500. The thickness of the ZnO buffer layers was an important parameter to obtain high quality GaN layers. GaN on Si (001) with ZnO buffer layers by RF magnetron sputtering was carried out at room temperature in a 30 sccm Ar gas flow by supplying a RF power of 13.56 MHz.[28] The ZnO buffer layer was grown at a RF power of 100 W and a pressure of 5.0×10^{-2} Torr and GaN growth was at a RF power of 50 W under the same pressure. GaN on a 300 nm-thick ZnO buffer layer was obtained with XRD FWHM of 0.22 ° and rms surface roughness of 2.2 nm as determined by AFM.

In GaN epitaxial growth, the polarity control of GaN is very important, since Ga-polar GaN could have a better quality. Suzuki *et al.* demonstrated Ga-polar GaN layers grown on epi-ZnO templates with NH_3 pre-exposure.[29] The GaN growth by ammonia-assisted MBE was conducted on ZnO/MgO/*c*-sapphire template grown by oxygen plasma-assisted

MBE, where solid Ga and NH_3 gas were used as Ga and nitrogen sources. The substrate temperature for GaN growth was 800 °C and the Ga beam pressure was 2.8×10^{-7} Torr and the NH_3 flow rate was 10 sccm. Before GaN growth, NH_3 was provided onto the O-polar ZnO template at 800 °C to grow a Zn_3N_2 layer, which could allow the Ga-polar GaN on O-polar ZnO. By surface etching in 0.5 M KOH solution, the surface polarity was confirmed. Another way to control the GaN polarity is by controlling the growth temperature in the PLD technique.[30] Ga-polar GaN layers were grown at room temperature, while those grown at 700 °C had N-polarity. However, Ga-polar GaN could be grown at 700 °C by introducing room-temperature GaN buffer layers.

To achieve high performance and reliable GaN-based devices, freestanding GaN substrates are desirable. Butcher *et al.* demonstrated a GaN freestanding layer by chemical wet etching using ZnO buffer/soda lime glass as a sacrificial layer.[31] The 50 nm-thick ZnO buffer layers were grown on soda lime glass substrates by atomic layer epitaxy. Metallic Zn and water vapor from liquid source was used as precursors and the optimized substrate temperature was 430 °C. From the XRD spectrum, the ZnO film was not found to be oriented along the *c*-axis. After a ZnO buffer layer growth, GaN was grown by remote plasma-enhanced laser-induced chemical vapor deposition (RPE-LICVD) at a low temperature (570 °C). Following the growth of ZnO, GaN on soda lime glass with ZnO buffer was intended to eliminate the glass substrate and ZnO buffer layer by chemical wet etching without any damage on GaN layers. Although the GaN layer was polycrystalline, 6 μm-thick and 5 mm × 10 mm area free-standing GaN films were obtained. Lee *et al.* demonstrated strain-free thick GaN films grown on ZnO buffer layers with two-step HVPE growth and *in situ* lift-off.[32] Figure 9.3 shows a schematic diagram of the preparation sequence of free-standing GaN substrate preparation by the *in situ* lift-off process. A 200 nm-thick ZnO layer with Zn polarity was grown on *c*-sapphire using 8 nm-thick MgO buffer layer by plasma-assisted MBE. The polarity of ZnO can be controlled by the thickness of the MgO buffer layer through modifying the crystal structure of the MgO buffer layer.[33] A 1 μm-thick GaN layer with Ga polarity was grown on ZnO/MgO/*c*-sapphire by plasma-assisted MBE. Following GaN/ZnO growth, thick GaN layers were grown by HVPE. During HVPE growth at 900 °C in order to prevent decomposition of GaN at high temperatures, the ZnO layer was gradually etched from the side walls by NH_3 and HCl gases, which were also used for the growth of GaN layers. The *in situ* etching rate of ZnO was controlled by the growth temperature and flow rates of NH_3 and HCl gases. During the growth of the over 100 μm-thick GaN layer by HVPE at 900 °C, the ZnO layer between sapphire and GaN layers was completely etched away. Note that the GaN buffer should be thick enough to prevent cracking of the GaN buffer before ZnO removal. Then, thick GaN layers were grown at 1040 °C. The free-standing GaN surface roughness was 1.44 nm for a $10 \times 10\,\mu m^2$ area with no microcracks (Figure 9.4). The calculated lattice parameters from XRD data were 5.185 Å for the *c*-axis and 3.189 Å for the *a*-axis, which indicated strain-free GaN layers. The FWHM of the (0002) ω-rocking curve by high resolution XRD analysis ranges from 300 to 450 arcsec, which is comparable with the typical value of HVPE-grown GaN on ZnO/*c*-sapphire. The FWHM of the (0002) ω-2θ scan lies at about 50 arcsec, indicating uniform strain in the free-standing GaN layer.

The growth and fabrication of GaN-based LEDs using nonpolar (*m*-plane or *a*-plane) semiconductors has attracted much attention due to the potential for avoiding undesirable built-in field effects and achieving higher efficiencies than conventional devices using *c*-plane

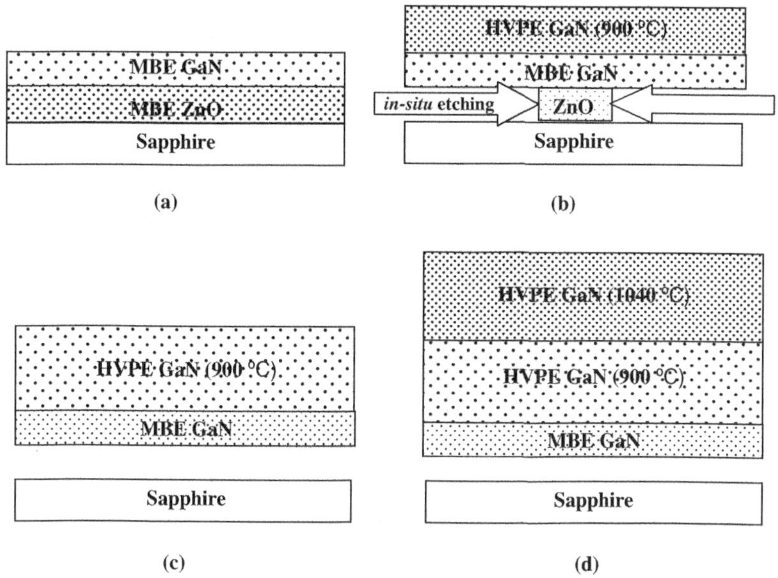

Figure 9.3 *Schematic diagram of the growth sequence for free-standing GaN substrates. A 200 nm-thick single crystal ZnO layer with Zn polarity was first grown on c-plane sapphire using an 8 nm thick MgO buffer layer deposited by plasma assisted molecular beam epitaxy followed by the growth of a 1 μm-thick GaN layer (a). During the low temperature growth of a relatively thick (100 μm) GaN by HVPE, commencing at 700 °C and ending at 900 °C, the ZnO buffer was gradually etched from the sides by NH_3 and HCl gases (b). The ZnO layer between sapphire and GaN layers was completely etched off during growth of low-temperature GaN buffer (c). This was followed by growth of the final high temperature GaN layer (d). Reprinted from S. W. Lee, et al., Strain-free GaN thick films grown on single crystalline ZnO buffer layer with in situ lift-off technique, Appl. Phys. Lett., 90, 061907. Copyright (2007) American Institute of Physics*

nitrides. Kobayashi *et al.* demonstrated nonpolar (*m*- and *a*-plane) GaN growth on nonpolar (*m*- and *a*-plane) ZnO substrates by PLD.[34,35] Atomically flat *m*-plane ZnO substrates were obtained after annealing at 1250 °C for 3.5 h in air, where substrates were put in the ceramic ZnO box to prevent desorption of Zn and O atoms from the surface and maintain the surface stoichiometry. The *m*-plane GaN layers were grown on the *m*-plane ZnO substrate by PLD at various temperatures ranging from room temperature to 700 °C. Since only the growth performed below 300 °C was by layer-by-layer mode, thick *m*-plane GaN layers were grown on low-temperature *m*-GaN buffer layers/lattice matched *m*-ZnO substrate at 700 °C. High quality *m*-GaN layers were confirmed by XRD, AFM and transmission electron microscopy (TEM) measurements.[34] From the TEM image, shown in Figure 9.5, *m*-plane GaN was almost coherent along the [0001] direction. Nonpolar *a*-plane GaN layers were also grown on lattice-matched *a*-plane ZnO at room temperature by PLD. Hydrothermally grown *a*-plane ZnO substrates ($10 \times 10 \times 0.5\,mm^3$; SPC Goodwill, Russia) were annealed at 1000 °C for 30 min in the PLD growth chamber with a background pressure of 2×10^{-10} Torr. The RHEED pattern changed from a broad spotty pattern to a bright streaky one, indicating an atomically flat surface. Similar to the *m*-plane GaN growth, room-temperature GaN buffer layers grown on *a*-ZnO substrates were used for high quality *a*-plane GaN layers.

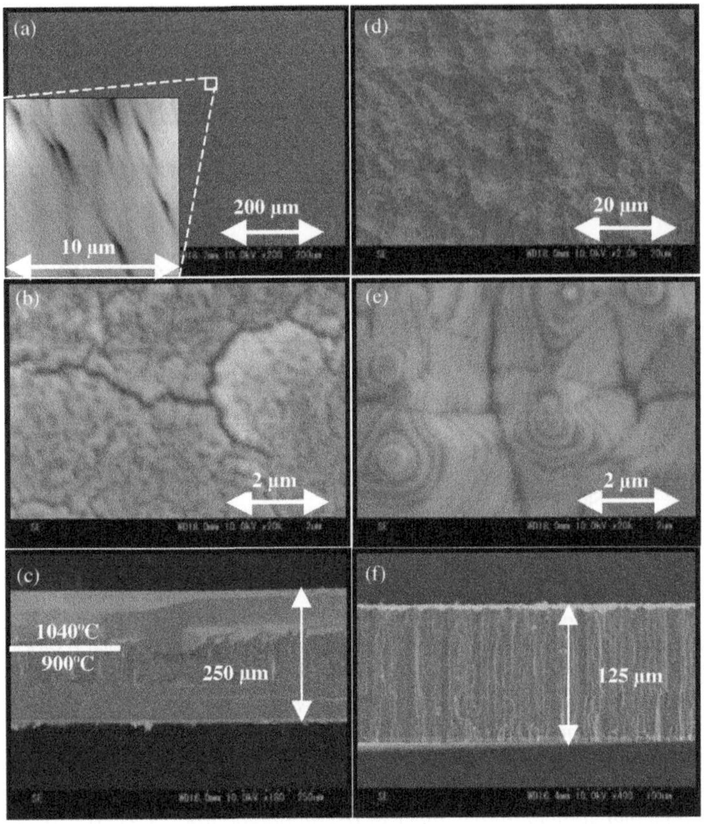

Figure 9.4 *(a–c) SEM images of the free-standing GaN layers grown by the two-step growth method by HVPE. (a) Top surface; (b) back-side surface; (c) side cut. (d–f) SEM images of the free-standing GaN layers grown at 900 °C. (d, e) Top surface; (f) side cut. (a, inset) AFM image of a 10 × 10 μm² area with surface roughness of 1.44 nm. Reprinted from S. W. Lee, et al., Strain-free GaN thick films grown on single crystalline ZnO buffer layer with in situ lift-off technique, Appl. Phys. Lett., 90, 061907. Copyright (2007) American Institute of Physics*

9.3 Compositional Analysis

Here we give one example of compositional analysis of ZnO films deposited by PLD. Rutherford backscattering spectroscopy (RBS) analysis revealed that the composition of the GaN films deposited at 600 °C on ZnO is stoichiometric when liquid Ga target was used in PLD growth. A typical RBS spectrum is shown in Figure 9.6. However, films deposited by a solid GaN target at a lower substrate temperature of 850 °C were nonstoichiometric because of insufficient N radicals during the growth of GaN, whereas for higher substrate temperatures, stoichiometric films were achieved.[7,36] Secondary ion mass spectroscopy (SIMS) analysis showed that oxygen is one of the impurities in the GaN layers and its concentration was much higher at the interface. Si and C were found throughout the entire depth of the samples and are most likely from the sapphire substrates.

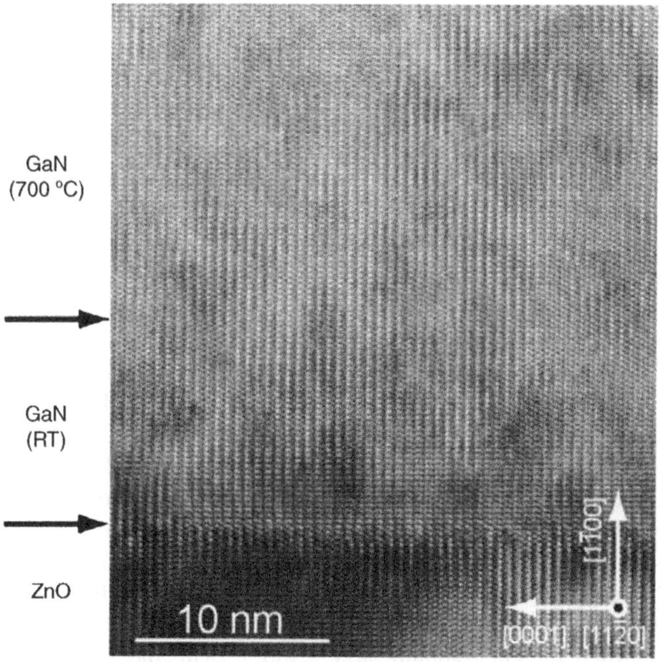

Figure 9.5 Cross-sectional TEM image of m-plane GaN (grown at 700 °C)/GaN (grown at room temperature)/m-ZnO substrate taken from the [11$\bar{2}$ 0] direction. Reprinted from A. Kobayashi, et al., Room temperature epitaxial growth of m-plane GaN on lattice-matched ZnO substrates, Appl. Phys. Lett., 90, 041908. Copyright (2007) with permissions from American Instituite of Physics

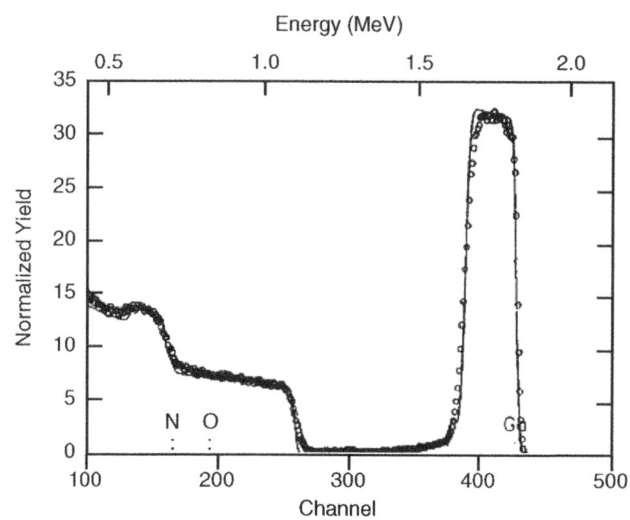

Figure 9.6 RBS spectrum of GaN thin films grown on ZnO/sapphire, He^{2+} beam energy is 2.3 MeV and the detector angle is 168°. The experimental curve is simulated with the theoretical curve to attain the composition of the films. Reprinted from R. F. Xia, et al., Growth of gallium nitride thin films by liquid-target pulsed laser deposition, J. Vac. Sci. Technol. A 15, 2207. Copyright (1997) with permission from American Vacuum Society

9.4 Structural Analysis

Various characterization methods have been used to analyze the structural properties of GaN and specific to the theme of this section, GaN on ZnO with a few examples on other substrates such as fused silica. XRD analysis showed that the GaN films grown directly on fused silica substrates are not only polycrystalline with different orientations, (1010), (0002) and (1011) but also are characterized with low intensities of diffraction peaks.[8] The XRD spectrum is dominated by the (0002) reflection when GaN films were deposited at a substrate temperature (T_s) of 600 °C; its significance is low for films grown at higher or lower temperatures. The FWHM of the rocking curve for the (0002) diffraction was 0.21, 0.20, and 0.52° for GaN/ZnO/Si (111), GaN/ZnO/sapphire (0001) and ZnO/sapphire, respectively.[36] The FWHM of the (0002) reflection for ZnO films on sapphire was 0.17° for a substrate temperature of 750 °C and O_2 pressure of 10^{-5}–10^{-4} Torr. The X-ray φ scans of ZnO films deposited at two deposition conditions, (a) $T_s = 500$ °C with a pulse repetition rate of 15 Hz and (b) $T_s = 750$ °C with a pulse repetition rate of 10 Hz, were recorded. The X-ray analysis showed that the substrate and the epitaxial layer are oriented with respect to each other by a 30° rotation of the unit cells. The in plane epitaxial and substrate relationship is ZnO[$10\bar{1}0$]||Al_2O_3[$11\bar{2}0$]. A similar fashion of orientation can be seen for AlN and GaN on sapphire, as shown in Figure 9.7. The FWHM of the rocking curve of the (0002) diffraction for GaN, AlN, ZnO growth on sapphire was less than 0.15°. The AlN films were deposited on ZnO/sapphire at T_s=950 °C with the same technique and its FWHM is 0.23°.[7] The (0002) diffraction peaks are clearly resolved for the structure of ZnO/GaN, showing high crystalline quality of the films, as shown in Figure 9.8.

Matsuoka *et al.*[22] observed that the etching rate of the O-face ZnO is faster than that of the Zn-face by about one order of magnitude. The (0001) Zn-face ZnO on sapphire substrates was used for the growth of $In_{0.22}Ga_{0.78}N$ layers. The single crystal ZnO layers were nitrided using ammonia at 76 Torr and 500 °C for 10 min. In the nitrided substrates, the RHEED showed a ring pattern, indicating that a polycrystalline film formed on the surface of the ZnO substrate. The XRD analysis revealed only the (0002) reflection of both ZnO and InGaN, and no other peaks could be observed (Figure 9.9). The RHEED pattern was streaky for the films grown on ZnO indicating single crystal nature whereas films deposited on sapphire were polycrystalline. The rough and smooth mirror-like surfaces were observed for InGaN films grown on ZnO and sapphire, respectively. The XRD line width of InGaN grown on ZnO is about 20% smaller than that of films grown on sapphire substrates. It is clear that the remarkable improvement in the crystalline quality of the films has been achieved. Ueda *et al.*[12] found that if pulsed laser deposited ZnO layer thickness is up to 50 nm, the FWHM for GaN grown on it is smaller by a factor of two than that of thicker ZnO films. However, for a 200 nm-thick ZnO buffer layer, the rocking curve was broader due to segregation of cracks in the subsequent GaN layer.

Wang *et al.*[10] reported that the FWHM of rocking curve of the (0002) diffraction for GaN/ZnO is 0.38°, which was wider, if ZnO buffer layer had not been used. The FWHM decreased from 0.45 to 0.22° with increasing ambient pressure from 10^{-5} to 1 Torr for GaN films grown on ZnO, as shown in Figure 9.10(a). Similarly, this value decreased with increasing substrate temperature, as shown in Figure 9.10(b). The GaN films deposited at >500, 700, and 800 °C under 0.1 Torr in N_2 ambient showed amorphous, polycrystalline,

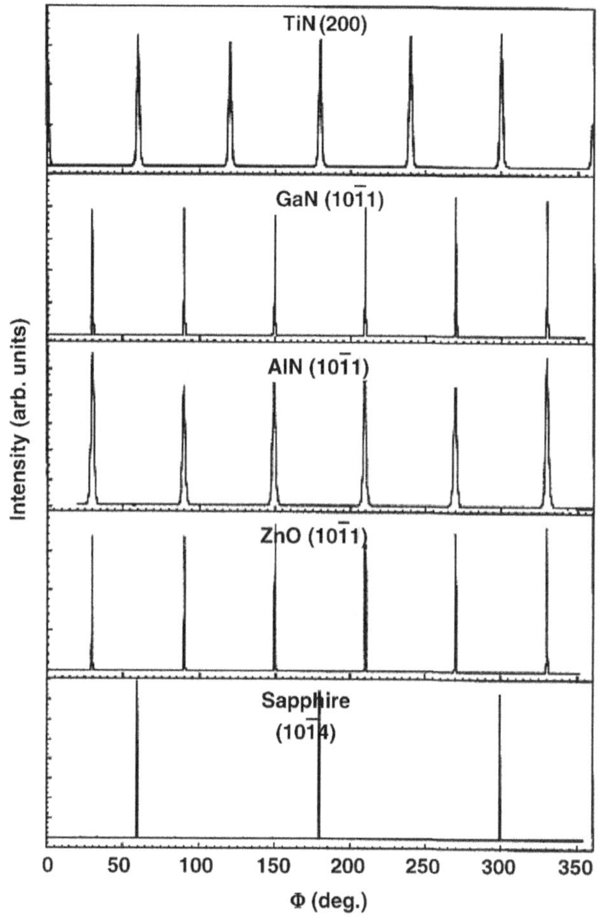

Figure 9.7 *X-ray φ scans of nitride compounds grown on sapphire substrate (0001). Reprinted from R.D. Vispute, et al., Advances in pulsed laser deposition of nitrides and their integration with oxides, Appl. Surface Sci. 127–129, 431. Copyright (1998) with permission from Elsevier*

and epitaxial nature, respectively. The decomposition of GaN occurred, if the deposition had been carried out above the substrate temperature of 800 °C. The RHEED analysis showed a streaky pattern, indicating high quality of the crystallinity. The RHEED patterns of ZnO and GaN grown on it are parallel to the [1$\bar{1}$02] direction, as shown in Figure 9.11(a). The GaN growth on chemical–mechanical polished ZnO substrates showed better than that on unpolished substrates by means of RHEED pattern studies, which showed weak and broken diffraction lines for the growth of GaN on unpolished ZnO substrates. A clear 2 × 2 RHEED reconstruction was observed for GaN layers deposited on both Zn- and O-face ZnO substrates, as shown in Figure 9.11(b), which indicated Ga polarity of GaN thin films.[16] Two RHEED patterns were recorded on ZnO along one azimuth and along another azimuth by rotating it 30°, which were parallel to the [$\bar{2}$110] and [1$\bar{1}$00] directions of ZnO, respectively (Figure 9.12). Almost similar patterns were observed for GaN grown

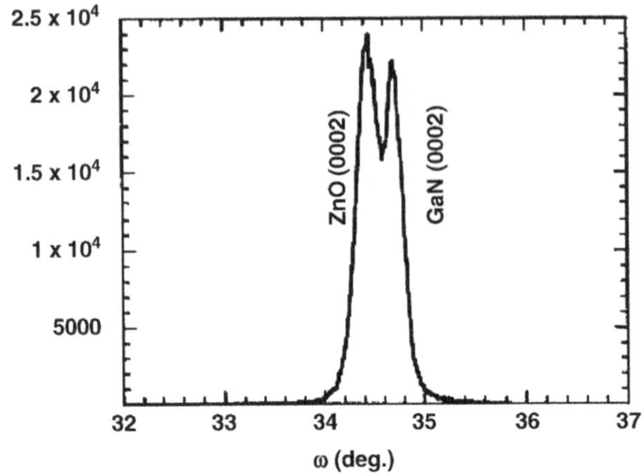

Figure 9.8 *X-ray rocking curve of GaN on sapphire with ZnO buffer layer. Reprinted from R.D. Vispute, et al., Advances in pulsed laser deposition of nitrides and their integration with oxides, Appl. Surface Sci. 127–129, 431. Copyright (1998) with permission from Elsevier*

on ZnO. Ueda *et al.*[12] observed that the RHEED showed a ring pattern indicating that a polycrystalline film had formed by a reaction between ammonia and the single crystal ZnO. The (0002) reflection was predominant when the GaN films were grown with ZnO buffer layer. An additional $(11\bar{2}1)$ reflection was observed along with (0002). In this sample sharp, rock-like three-dimensional structures were observed.

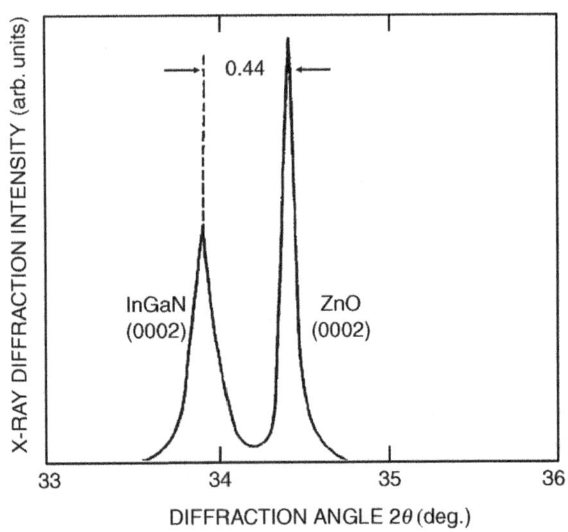

Figure 9.9 *X-ray rocking curve of InGaN/ZnO on sapphire substrate. Reprinted from T. Matsuoka, N. Yoshimoto, T. Sasaki and A. Katsui, Wide-gap semiconductor indium gallium nitride and indium gallium aluminum nitride grown by MOVPE, J. of Electro. Mater. 21, 157 (1992). Copyright (1992) with permission from Springer*

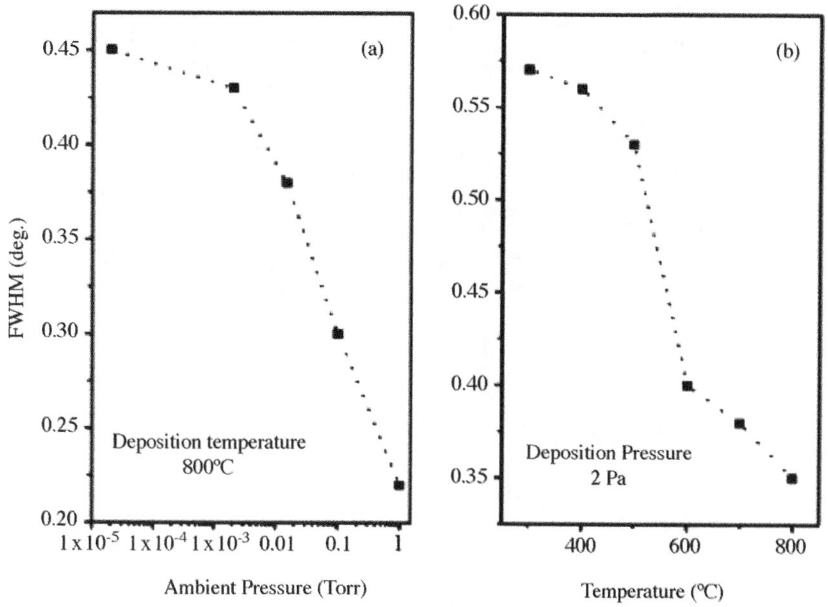

Figure 9.10 *Variation of FWHM of XRD rocking curve for GaN/ZnO grown on sapphire vs (a) ambient pressure and (b) deposition temperature. Reprinted from R. P. Wang, H. Muto, Y. Yamada, and T. Kusumori, Effect of ZnO buffer layer on the quality of GaN films deposited by pulsed laser ablation, Thin Solid Films, 411, 69. Copyright (2002) with permission from Elsevier*

9.5 Surface Studies

For a comprehensive analysis of the properties of GaN films grown on ZnO substrates, a whole slew of surface sensitive techniques can be applied. For example, scanning electron microscopy (SEM) analysis (Figure 9.13) showed that the GaN surface has large grain sizes, but is smooth within the grains when the GaN is deposited with ZnO buffer on sapphire substrates. Without ZnO buffer, however, the surface was found to be rough with many small domains.[12] In this study, the surface of the Zn-face of ZnO was noted to be smoother than that of the O-face. The rms roughness of the O-face surface is an order of magnitude higher than that of the Zn-face. Surprisingly, the GaN layers grown on the Zn-face of ZnO are rough compared with those grown on O-face ZnO substrates.[16] The GaN layers grown on ZnO substrates without any buffer layer at a substrate temperature above 760 °C easily peeled off.[16] Therefore, the low temperature buffer layer was added to prevent reaction between NH_3 and ZnO substrates. AFM images (Figure 9.14) show that the surface of the GaN layers is smooth for when the GaN buffer was used whereas it is rough for the AlN buffer case in part perhaps due to the higher lattice mismatch between AlN and ZnO compared with that of GaN and ZnO, but most likely due to the relatively low surface mobility of Al as compared with Ga. Yun *et al.*[37] studied the growth of GaN on different substrates without using GaN buffer layers. The GaN grown on the air annealed O-face ZnO substrates showed uniform and densely populated flower-like patterned surfaces with an rms value of 1.98 nm. The ZnO surface was also scanned by

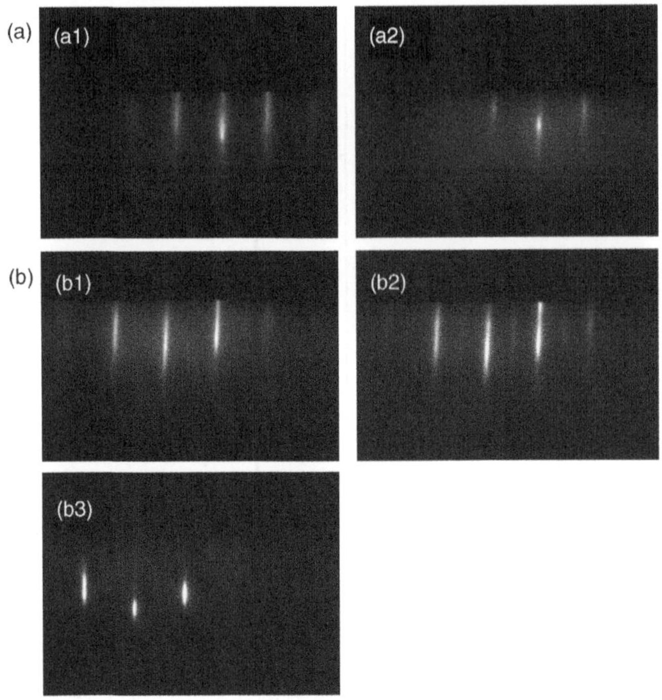

Figure 9.11 (a) RHEED patterns of O-face ZnO and GaN/ZnO: (a1) ZnO along [1$\bar{1}$02]; and (a2) GaN along [1$\bar{1}$02]. (b) RHEED pattern of GaN/ZnO: (b1) during growth (on Zn-face ZnO); (b2) cooling down to 350 °C (on Zn-face ZnO); and (b3) cooling down to 350 °C (on O-face ZnO)

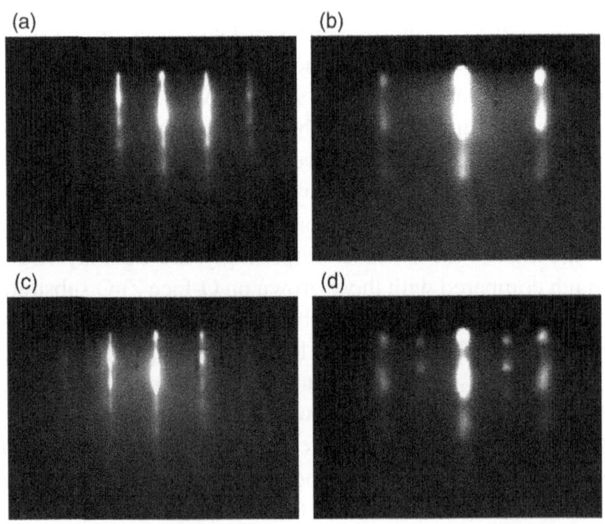

Figure 9.12 RHEED patterns of ZnO and GaN/ZnO: (a) ZnO along [$\bar{2}$110]; (b) ZnO along [1$\bar{1}$00]; (c) GaN along [$\bar{2}$110]; and (d) GaN along [1$\bar{1}$00]. Reprinted from R. P. Wang, H. Muto, Y. Yamada, and T. Kusumori, Effect of ZnO buffer layer on the quality of GaN films deposited by pulsed laser ablation, Thin Solid Films, 411, 69. Copyright (2002) with permission from Elsevier

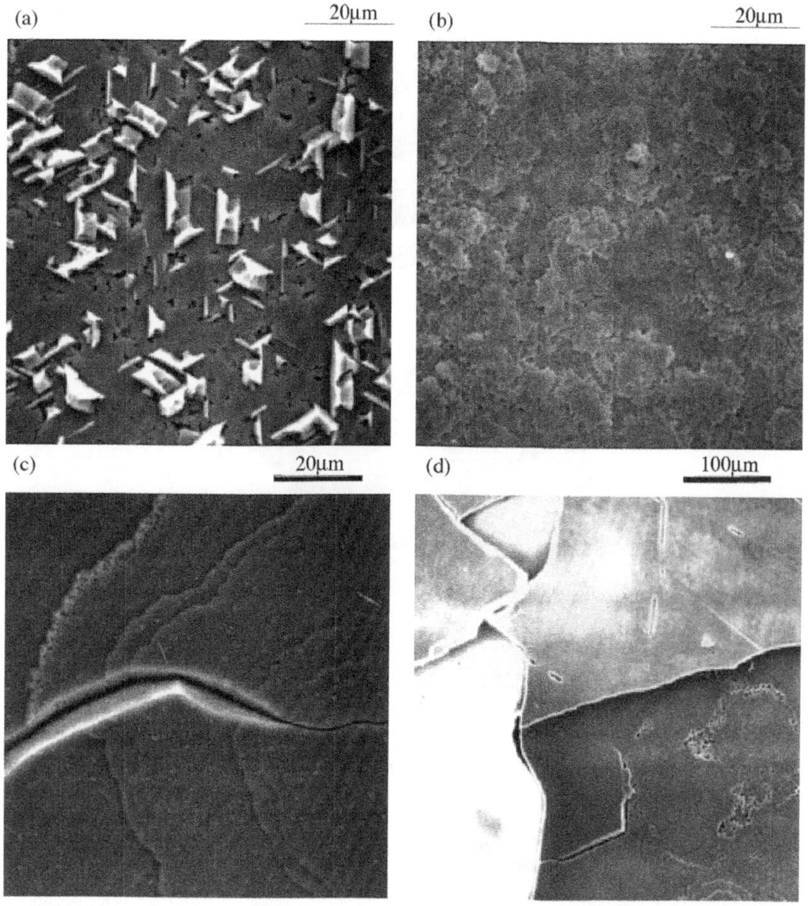

Figure 9.13 SEM images of (a) GaN/sapphire, (b) GaN/ZnO(25 nm)/sapphire, (c) GaN/ZnO (200 nm)/sapphire and (d) the same as (c) but at different location. Reprinted from T. Ueda, et al., Vapor phase epitaxy growth of GaN on pulsed laser deposited ZnO buffer layer , J. Cryst. Growth, 187, 340. Copyright (1998) with permission from Elsevier

AFM, prior to the deposition of GaN, which showed smooth and sparsely scattered small size granules with an rms value of 0.658 nm, as shown in Figure 9.15. TEM analysis showed that the density of extended defects is in the order of $10^8\,cm^{-2}$ on the surface of GaN, which is about three orders of magnitude lower than that in the buffer layer region.[16]

9.6 Optical Properties

9.6.1 Transmission Analysis

Optical measurements in general and transmission measurements in particular can be used to discern various transitions taking place in films. In the case of GaN on ZnO, the direct

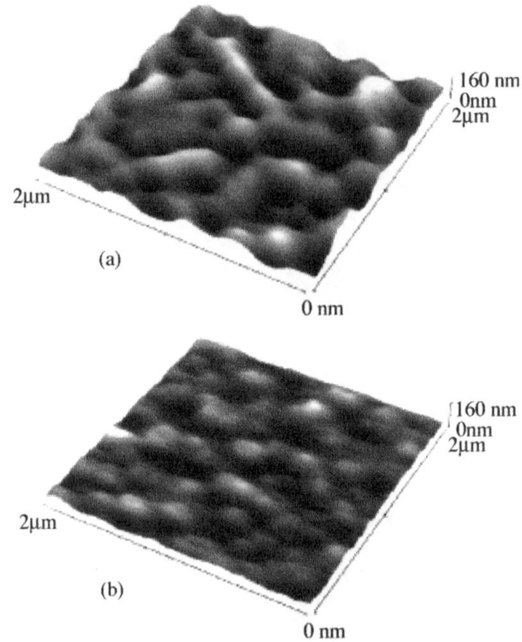

Figure 9.14 AFM images of (a) GaN grown on O-face ZnO, with AlN buffer and (b) GaN buffer. Reprinted from F. Hamdani, et al., Microstructure and optical properties of epitaxial GaN on ZnO (0001) grown by reactive molecular beam epitaxy, J. Appl. Phys., 83, 983. Copyright (1998) with permission from American Institute of Physics

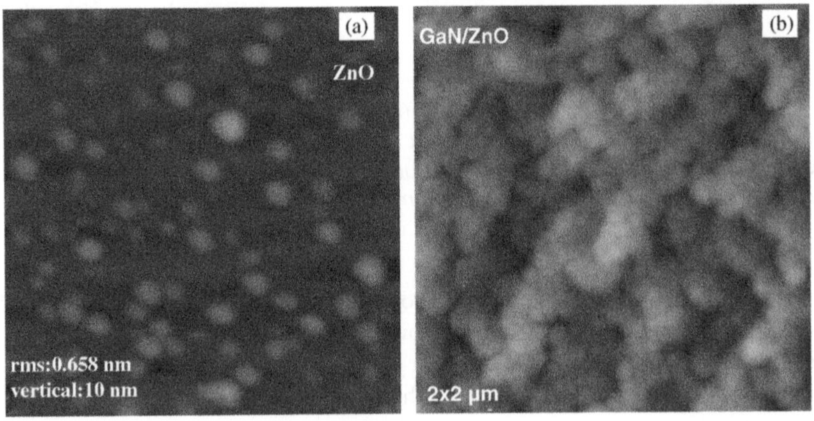

Figure 9.15 AFM images of (a) O-face ZnO and (b) GaN grown on O-face ZnO, without buffer layer. Reprinted from F. Yun, M. A. Reshchikov, L. He, T. King, D. Huang, H. Morkoç, J. Nause, G. Cantwell, H. P. Maruska, and C.W. Litton, Comparative analysis of MBE-grown GaN films on SiC, ZnO and $LiGaO_2$ substrates, in Defect and Impurity Engineered Semiconductors and Devices III, edited by S. Ashok, J. Chevallier, N.M. Johnson, B.L. Sopori and H. Okushi (Mater. Res. Soc. Symp. Proc. Volume 719, Warrendale, PA, 2002) F8.21. Copyright (2002) with permission from MRS

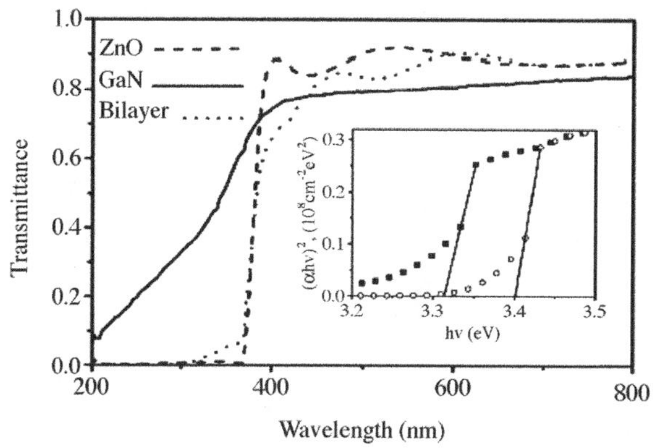

Figure 9.16 *Transmission spectra of ZnO, GaN and GaN/ZnO. The inset shows the band gaps of 3.3 and 3.4 eV related to ZnO and GaN, respectively. Reprinted from R. P. Wang, H. Muto, Y. Yamada, and T. Kusumori, Effect of ZnO buffer layer on the quality of GaN films deposited by pulsed laser ablation, Thin Solid Films, 411, 69. Copyright (2002) with permission from Elsevier*

optical band gaps of 3.3 and 3.4 eV for ZnO and GaN, respectively, were observed by optical transmission spectra.[7] A similar band gap value (3.45 eV) is reported for GaN grown on ZnO/fused silica substrates by PLD.[36]

Figure 9.16 shows sharp absorption edges for ZnO and GaN/ZnO except for GaN on sapphire, indicating the high quality of the layers.[10]

9.6.2 Cathodoluminescence Analysis

An electron beam can be used to create electron–hole pairs followed by observation of the recombination processes in semiconductors such as GaN and ZnO. Because very high excitation intensities can be obtained and the depth of the electron beam can be varied, this technique has been popular for analyzing GaN and ZnO. Sun *et al.*[38] studied defect levels in the energy gap, emission positions and their intensities as a function of distance from the GaN/ZnO interface in the samples, with different sheet densities (in the range of 6. 16.98.508 $\times 10^{14}$ cm$^{(-2)}$), grown by HVPE on ZnO buffered sapphire substrates by using cathodoluminescence (CL) spectroscopy. The near band edge, blue emission, and yellow luminescence peaks were observed at 3.4, 2.95 and 2.2 eV, respectively, in the CL spectra at room temperature (Figure 9.17). The intensities of the near-band-edge emission and yellow emission increased as the raster center was moved from the interface towards the surface of the GaN layer. Eventually, the dislocation density decreased with increasing distance from the GaN/sapphire interface (d_{int}) > 10 μm in sample 1 having a sheet interface concentration (n_s) of 6×10^{14} cm^{-2}. Therefore, it can be asserted that the dislocations are not the sources for yellow luminescence. However, they may decorate the complexes formed by Ga vacancies, which are believed to be the source of this luminescence.[39] The broad blue luminescence (BL) band pronounced at 2.9 eV for $d_{int} > 0.2$ μm might be due to Zn participation because the epitaxial layer was treated by ZnO buffer. The PL signature in GaN doped with Zn or contaminated by Zn from prior uses of Zn was studied by Reshchikov *et al.*[40] The assessment agrees well with the aforementioned BL emission in GaN grown on ZnO buffer layers or

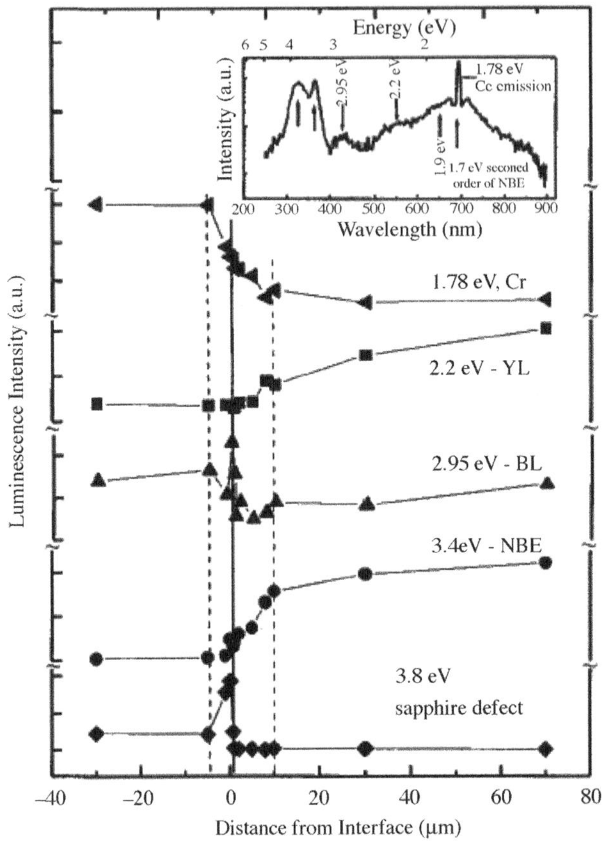

Figure 9.17 Cathodoluminescence intensity vs distance from the interface for different peaks; sapphire defect, near-band-edge emission (NBE), blue emission (BL), yellow luminescence (YL), chromium (Cr) from sapphire (sample 1, $n_s = 6 \times 10^{14}\,cm^{-2}$). Note that the scans are nor normalized to a constant value and the GaN/sapphire interface is referenced as $d_{int} = 0$. Reprinted from X. L. Sun, et al., Depth-dependent investigation of defects and impurity doping in GaN/sapphire using scanning electron microscopy and cathodoluminescence spectroscopy. J. Appl. Phys., 91, 6729. Copyright (2002) with permission from American Institute of Physics

substrates. Note that a similar in shape and position BL band is usually observed in GaN heavily doped with Mg. However, it has a different origin and is attributed to transitions from a deep donor to a shallow Mg acceptor.[41] For sample 2, $n_s = 1.6 \times 10^{15}\,cm^{-2}$, three spectra are noted at different distances from the GaN/sapphire interface ($d_{int} = 0.3$, 1 and 4 μm) at low temperatures, as shown in Figure 9.18. The peak at 3.517 eV above the band edge was attributed to the free electron concentration at the degenerate interfacial layer.[38] The intensity of the peak decreased for $d_{int} = 1$ μm and finally disappeared for $d_{int} = 4$ μm, i.e. as the depth increased, the strong neutral donor bound exciton (D^0X) emission segregated at 3.483 eV. A shoulder on the D^0X line at about 3.503 eV was assigned to the excited state of the free exciton.[38] The delineation of additional peaks, i.e. 3.41, 3.30, 3.28 and 3.19 eV, was observed as d_{int} increased. The peak at 3.28 eV was related to the donor–acceptor pair (DAP) recombination and its phonon replica is at 3.19 eV. The peak at

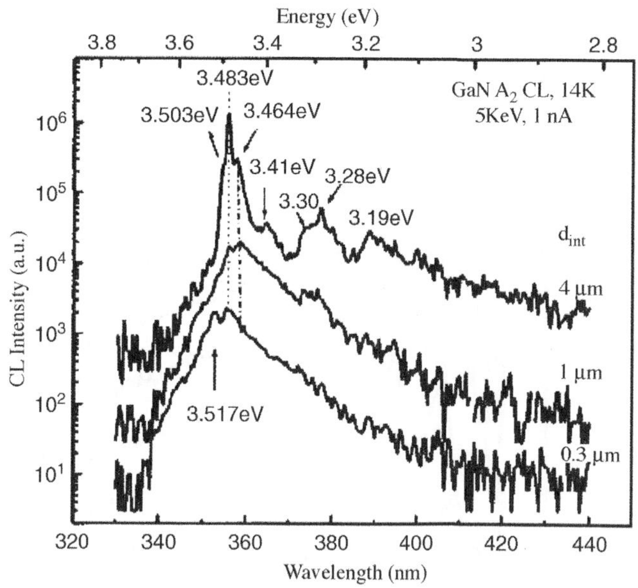

Figure 9.18 *CL spectra of sample 2 ($n_s = 1.6 \times 10^{15}\,cm^{-2}$) at different interface distances. Reprinted from X. L. Sun, et al., Depth-dependent investigation of defects and impurity doping in GaN/sapphire using scanning electron microscopy and cathodoluminescence spectroscopy. J. Appl. Phys., 91, 6729. Copyright (2002) with permission from American Institute of Physics*

3.41 eV frequently appears in GaN and is attributed by Xia *et al.*[36] to excitons bound to stacking faults. The peak at 3.30 eV could be due to either ZnO or GaN cubic phase mixed with the hexagonal phase.[38] Figure 9.19 shows the CL spectra of sample 3, $n_s = 9.8 \times 10^{15}\,cm^{-2}$.

In the case of sample 4, $n_s = 5 \times 10^{16}\,cm^{-2}$ and 17 µm thickness, the sapphire substrate was nitridated prior to GaN growth. In the low temperature CL spectra, D^0X at 3.483 eV was predominant as the depth increased from the interface and the other two known peaks at 3.4 and 3.3 eV are ascribed to excitons bound to stacking faults and or a cubic GaN domain. At the interface ($d_{int} \leq 5\,\mu m$), the 3.563 eV peak is due to band filling at the degenerate doping levels and the free-electron recombination.[38] The same feature has been observed in heavily doped ($10^{19}\,cm^{-2}$) samples.[42] The D^0X emission intensity decreased when CL spectra were recorded for columns or columnar regions, which might be due to grain boundaries, or high defect levels which can act as efficient nonradiative recombination centers.

Both samples 2 ($1.6 \times 10^{15}\,cm^{-2}$) and 3 ($9.8 \times 10^{15}\,cm^{-2}$) showed free excitons in the CL spectra which have been attributed to the quality of the samples. A weak residual acceptor level peak is seen in the luminescence spectra of these samples. This could be due to either Ga vacancy or C.[38] In sample 4, the near-band-edge emission quenches near the interface and grain boundaries where very high defect densities appear. A broad emission band at 3.56 eV is likely due to free electron recombination band showing high degenerate doping near the interface and grain boundaries (Figure 9.20).

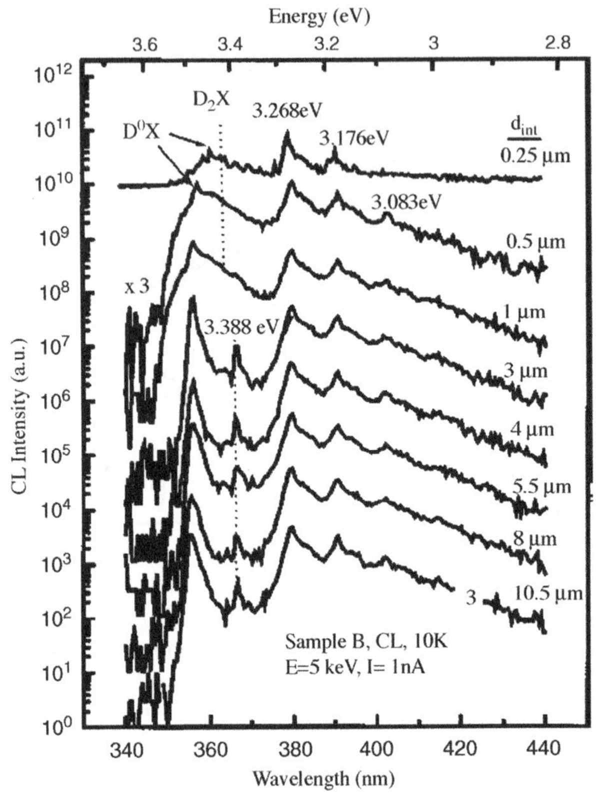

Figure 9.19 CL spectra of sample 3 ($n_s = 9.8 \times 10^{15}\,cm^{-2}$) at different interface distances. Reprinted from X. L. Sun, et al., Depth-dependent investigation of defects and impurity doping in GaN/sapphire using scanning electron microscopy and cathodoluminescence spectroscopy. J. Appl. Phys., 91, 6729. Copyright (2002) with permission from American Institute of Physics

9.6.3 Photoluminescence Analysis

Similar to the case of high energy electron induced electron–hole pair generation, above gap photon excitation can also be used to follow carrier recombination in an effort to determine the quality and optical processes taking place in the layers. Photoluminescence (PL) of three sets of bulk ZnO samples grown by Air Force Laboratories (Hanscom), Cermet, Inc., and what was then the Eagle Picher Company to use as substrates for GaN growth have been studied. All samples demonstrated good optical quality with very high quantum efficiency. Therefore, the quantum efficiency of the Cermet sample in the near-band-edge region exceeded 20%. The PL spectrum at 10 K of this sample is shown in Figure 9.21. The crystal quality of the sample was confirmed to be good by PL spectra showing low FWHM of 0.55 meV for the peak at 3.3597 eV. This peak is tentatively attributed to the exciton bound to neutral donor. Identification of the exciton structure of ZnO is quite controversial in the literature, so further studies are required. The main peak at ~3.36 eV was repeated three times on the low-energy tail of the exciton emission at energies which are multiples of the LO phonon energy (about 71 meV). Another sharp

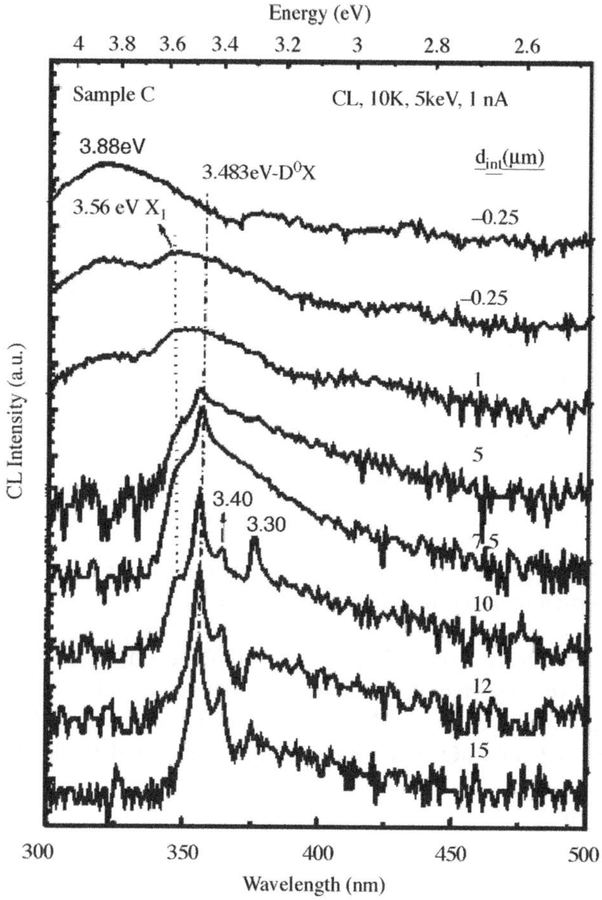

Figure 9.20 CL spectra of sample 4 ($n_s = 5 \times 10^{16}\,cm^{-2}$) at different interface distances. Reprinted from X. L. Sun, et al., Depth-dependent investigation of defects and impurity doping in GaN/sapphire using scanning electron microscopy and cathodoluminescence spectroscopy. J. Appl. Phys., 91, 6729. Copyright (2002) with permission from American Institute of Physics

peak was observed at 3.3206 eV with LO phonon replicas at 3.2505 eV and 3.180 eV, which is attributed to donor-acceptor transition.[43] As for the defect-related features, only the broad band with the maximum at about 2.4 eV (green band) is resolved. The low-temperature PL spectrum of the Cermet sample is nearly identical to that of the Hanscom sample. However, the FWHM of the 3.36 eV peak is lower approximately by a factor of three than that of the Hanscom sample. The broad defect related band was observed at about 2.1 eV instead of the green band. In the Eagle Picher sample, the PL spectra from two faces (Zn and O) were compared at room temperature, in an effort to identify the transitions. Now it is easy to analyze PL spectra of GaN deposited on ZnO, if any traces of ZnO or mixed phase of GaN and ZnO appeared. The low-temperature PL spectrum of GaN grown on ZnO compared with GaN grown on other substrates is shown in Figure 9.22. The FWHM of the excitonic peak was found to be about 13 meV for GaN/ZnO and a sharp

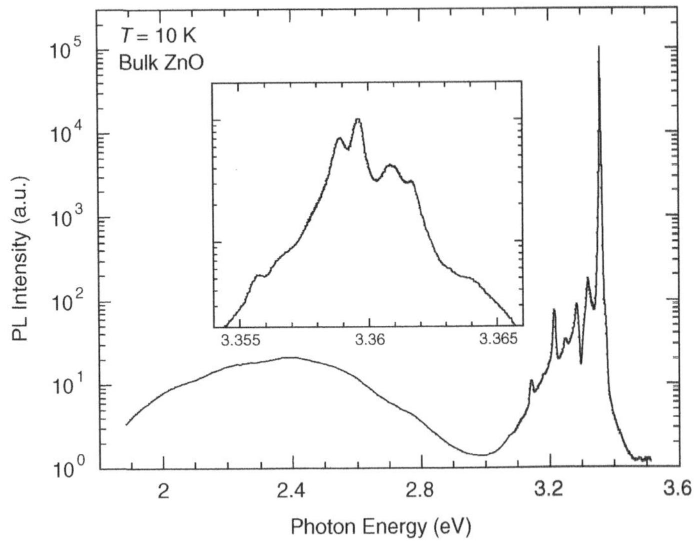

Figure 9.21 *PL spectra of ZnO substrate (produced by Cermet Inc.) at 10 K. Theinset shows the enlarged fine structure in the near-band-edge region*

peak at 3.358 eV is due to excitonic emission from the underlying ZnO substrate in PL spectra. Gu *et al.* reported that the high radiative efficiency and weaker yellow luminescence was observed in GaN/ZnO, as compared with other substrates.[18] The exciton peak of GaN/ZnO was very slightly blue-shifted while that for GaN/SiC

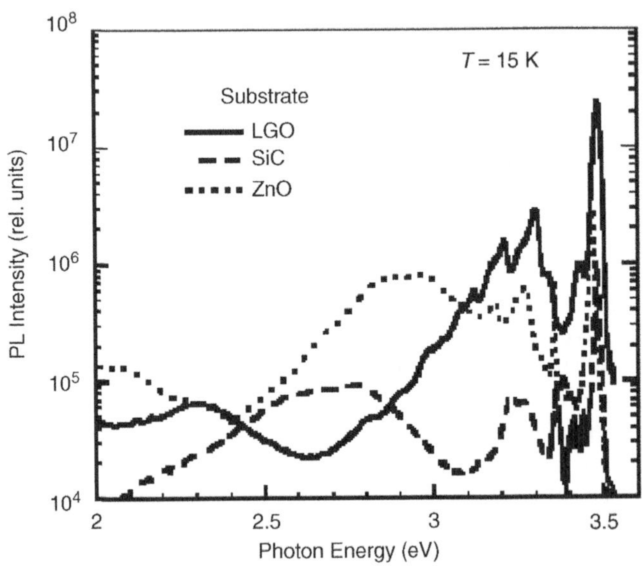

Figure 9.22 *Low temperature PL spectra of GaN grown without buffer on different substrates. Reprinted from F. Yun, et al., Mat. Res. Soc. Symp. Proc., 719, F8.21. Copyright (2002) with permission from MRS*

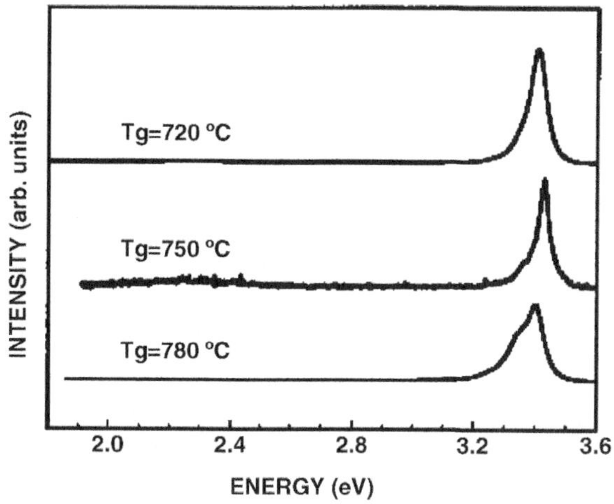

Figure 9.23 *Room temperature PL spectra of GaN grown without buffer on O-face ZnO at different growth temperatures. Reprinted from F. Hamdani, A. E. Botchkarev, H. Tang, W. Kim, and H. Morkoç, Effect of buffer layer and substrate surface polarity on the growth by molecular beam epitaxy of GaN on ZnO, Appl. Phys. Lett. 71, 3111. Copyright (1997) with permission from American Institute of Physics*

is red-shifted. It is a clear indication that the blue and red shifts are due to compressive and tensile strains of GaN/ZnO and GaN/SiC, respectively.

Hamdani *et al.*[21] demonstrated that the growth temperature is one of the important factors while depositing GaN on ZnO by using NH_3 in the MBE system. Room temperature PL spectra of GaN grown on the O-face ZnO substrates at different temperatures, i.e. 720, 750, and 780 °C, are shown in Figure 9.23. At room temperature the FWHM of the band edge emission increased from 36 to 60 meV with increasing growth temperature from 720 to 780 °C, respectively, due to increasing etching rate by NH_3 on the substrates with effect of temperature. The FWHM was found to be 8 meV at 4.2 K for the films grown at 720 °C. Moreover, any GaN growth was not observed at higher growth temperatures such as 800 °C. At low temperature, the intensity ratio of the free exciton peak and the donor bound exciton of GaN films grown on the O-face ZnO is two orders of magnitude higher than that of films on the Zn-face ZnO substrates. However, Gu *et al.* also achieved good quality GaN layers on Zn-face ZnO substrates.[18]

Figure 9.24 shows the low temperature PL and reflectivity spectra of the films grown on O-face ZnO substrates at growth temperature of 750 °C. A, B and C excitons at 3.476, 3.489 and 3.511 eV, respectively, and interference fringes due to multiple reflections of emission beam between the surface of GaN and GaN/ZnO interface are seen in the reflectivity spectrum. In the PL spectrum, the main peaks were attributed to free A exciton, a bound exciton and DAP.[21,44] Similarly, the PL and reflectivity spectra were recorded for GaN films grown on the Zn-face of ZnO substrates,[21] as shown in Figure 9.25. A and B excitons only pronounced at 3.481 and 3.496 eV, respectively, and the broadening of free exciton (A) in the reflectivity spectrum reflects the lower crystalline quality of the GaN films. Moreover, the donor bound exciton, DX, with a binding energy of 12 meV occurred instead of the free A exciton peak in the PL spectrum when comparing with Figure 9.24.

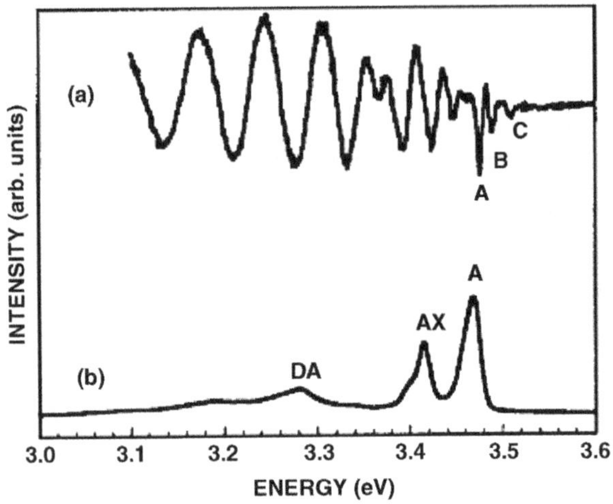

Figure 9.24 *Low temperature (a) differential reflectivity and (b) PL spectra of GaN grown without buffer on O-face ZnO. A, B and C, free excitons; AX, bound exciton; and DA, donor–acceptor pair. Reprinted from F. Hamdani, A. E. Botchkarev, H. Tang, W. Kim, and H. Morkoç, Effect of buffer layer and substrate surface polarity on the growth by molecular beam epitaxy of GaN on ZnO, Appl. Phys. Lett. 71, 3111. Copyright (1997) with permission from American Institute of Physics*

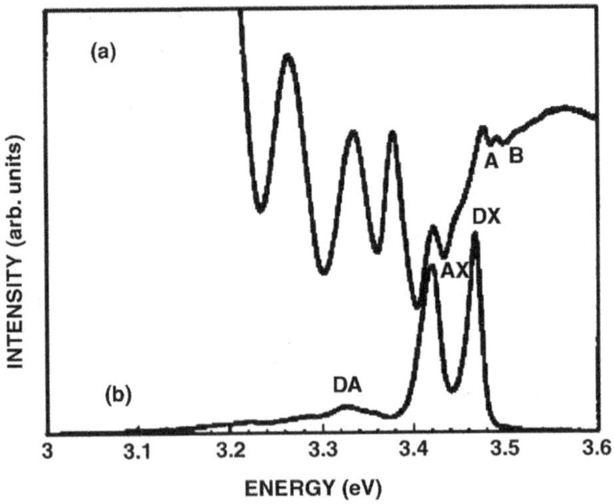

Figure 9.25 *Low temperature (a) differential reflectivity and (b) PL spectra of GaN grown without buffer on Zn-face ZnO. DX, bound exciton. Reprinted from F. Hamdani, A. E. Botchkarev, H. Tang, W. Kim, and H. Morkoç, Effect of buffer layer and substrate surface polarity on the growth by molecular beam epitaxy of GaN on ZnO, Appl. Phys. Lett. 71, 3111. Copyright (1997) with permission from American Institute of Physics*

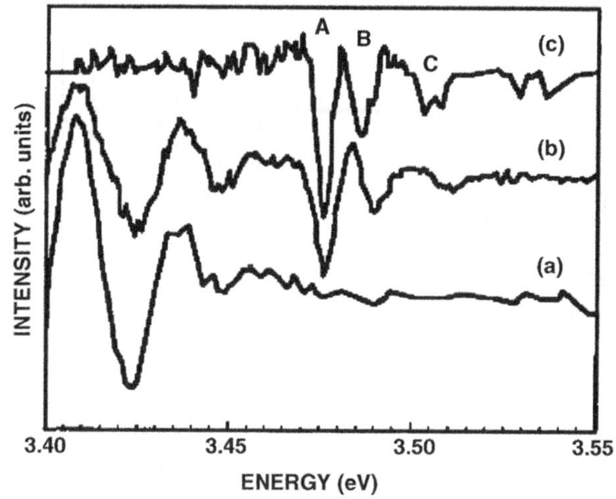

Figure 9.26 *Reflectivity derivative spectra of GaN grown on O-face ZnO with different buffer layers: (a) AlN; (b) GaN; and (c)* $In_{0.20}Ga_{0.80}N$. *The growth temperature of the buffer layers was 650 °C. Reprinted from F. Hamdani, A. E. Botchkarev, H. Tang, W. Kim, and H. Morkoç, Effect of buffer layer and substrate surface polarity on the growth by molecular beam epitaxy of GaN on ZnO, Appl. Phys. Lett. 71, 3111. Copyright (1997) with permission from American Institute of Physics*

An additional shift of the free exciton peak on about 5 meV indicates that the GaN films grown on the Zn-face ZnO substrates are under more compressive stress than the films on the O-face ZnO due to the difference in interface bonding between ZnO and GaN. The reflectivity derivative spectra of the GaN grown on the O-face ZnO with different buffer layers such as GaN, AlN and $In_{0.20}Ga_{0.80}N$ are given in Figure 9.26. A, B and C excitons clearly appeared and shifted to lower photon energies for GaN/ZnO with the $In_{0.20}Ga_{0.80}N$ buffer layer whereas in other cases no red shift was observed. This might be due to the strain relaxation between the $In_{0.20}Ga_{0.80}N$ buffer layer and GaN.[21] Figure 9.27 shows the reflectivity derivative spectra of ZnO and GaN/ZnO in which the ground and excited levels of free excitons labeled A_1, A_2, B_1, B_2, etc., occurred for GaN/ZnO, when $In_{0.20}Ga_{0.80}N$ was used as a buffer layer indicating better quality of the GaN epilayers.[21]

Hamdani *et al.*[16] found from PL spectra that the FWHM of the room temperature band edge emission was 80 and 36 meV for GaN/ZnO and ZnO substrate, respectively, and the observation of free exciton in the low-temperature PL spectrum is confirmation of the high quality of both materials. The intensity ratio of the free and bound excitons is higher for GaN/ZnO compared with ZnO in the low-temperature PL spectra, and thus it can be concluded that the quality of the epitaxial layers is higher, when ZnO is used as the substrates. The positions of A, B, C excitons in GaN grown on a ZnO substrate obtained from the reflectivity spectra slightly blue-shifted compared with the positions in bulk GaN. It indicates small compressive strain between these layers. However, the excitonic positions, in the case of InGaN buffer which is closely lattice matched to the GaN, shifted to lower energy levels compared with the positions of GaN grown directly on ZnO. On the other hand, the yellow luminescence did not appear at low temperatures, further proving the high quality of the samples, though at very low excitation levels it has been observable. The excitonic peaks were not observed in GaN

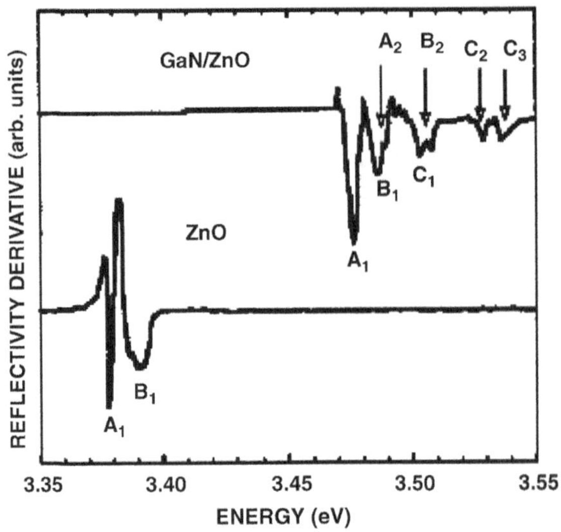

Figure 9.27 *Reflectivity derivative spectra of ZnO substrate and GaN grown on ZnO with thickness of 2 μm with InGaN buffer layer. Reprinted from F. Hamdani, et al., Microstructure and optical properties of epitaxial GaN on ZnO (0001) grown by reactive molecular beam epitaxy, J. Appl. Phys., 83, 983. Copyright (1998) with permission from American Institute of Physics*

grown on AlN buffer layers.[21] In PL spectra, A (3.375 eV), B (3.386 eV) free exciton and D^0X (3.364 eV) peaks were seen in GaN/ZnO whereas A, B and A2 exciton peaks with binding energy of 67 meV appeared in ZnO substrate.[44] The segregation of free exciton is a sign of low carrier concentration and high crystalline quality of the ZnO substrate. The properties of the grown GaN layers depend not only on the type of the buffer or substrate used for the growth but also on the polarity of the buffer or substrate. One can see the segregation of different peaks from GaN, if the Zn-face or the O-face ZnO substrates were used,[16] as shown in Figure 9.28.

In the PL spectra at 77 K from GaN grown on sapphire substrate with ZnO buffer layer, the BL band is seen.[12] Note that the band edge emission peak at 3.45 eV segregated in both types of samples with and without buffer layers. However, in the ZnO/GaN layers, a wide peak at 2.8 eV is due to the diffusion of Zn into GaN. Detchprohm *et al.*[13] studied the low temperature PL properties of GaN films grown on sputter deposited ZnO at different ambients such as Ar and O_2. The peaks at 3.486 and 3.479 eV are due to free and donor bound excitons, respectively, observed in both GaN samples. However, the FWHM (2.58 meV) of PL peak due to exciton bound to neutral donor for GaN grown in Ar is less than that of the GaN grown in O_2. It was concluded that the films grown using ZnO buffer showed high crystalline quality. The donor bound excitons and free excitons are observed in the PL spectra of GaN grown on sapphire with ZnO buffer layer. The very weak yellow luminescence at 2.2 eV was also present.[14]

In the Raman spectra, the coupled plasmon-phonon mode, in addition to two E_2 modes at 570 and 143 cm^{-1}, was a broad line at approximately 510–530 cm^{-1}.[12] The broad nature of this mode indicates that GaN films might have a higher carrier concentration. The

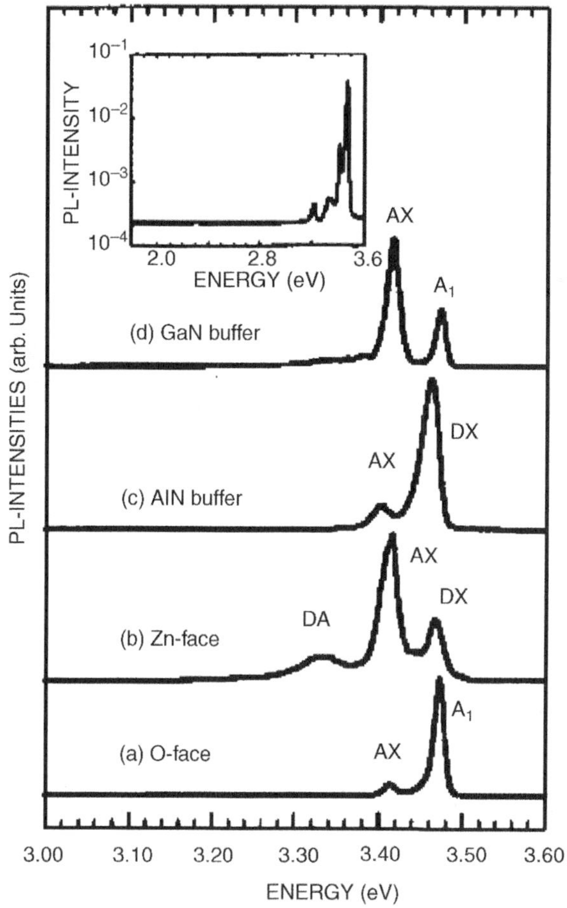

Figure 9.28 *PL spectra of GaN grown on (a) O-face ZnO, (b) Zn-face ZnO, (c) with AlN buffer on O-face ZnO and (d) with GaN buffer on O-face ZnO.* A_1*, free exciton; DX, donor bound exciton; and AX, acceptor bound exciton. The inset shows the spectrum over a wider energy range including red and green wavelengths. Reprinted from F. Hamdani, et al., Microstructure and optical properties of epitaxial GaN on ZnO (0001) grown by reactive molecular beam epitaxy, J. Appl. Phys., 83, 983. Copyright (1998) with permission from American Institute of Physics*

measured carrier concentration is in the order of 10^{19} cm^{-3}. Regarding the surface, terrace-like flat features were observed when growing GaN films on ZnO buffers.[12]

9.7 Electrical Properties

Perhaps the most sensitive characterization method, one which also bodes well for electronic devices, is electrical characterization, such as Hall measurements. In this vein, Detchprohm *et al.*[13] found that the growth of GaN on sputtered ZnO in O_2 is better than that of sputtered ZnO in Ar. The measured room temperature Hall mobility is in

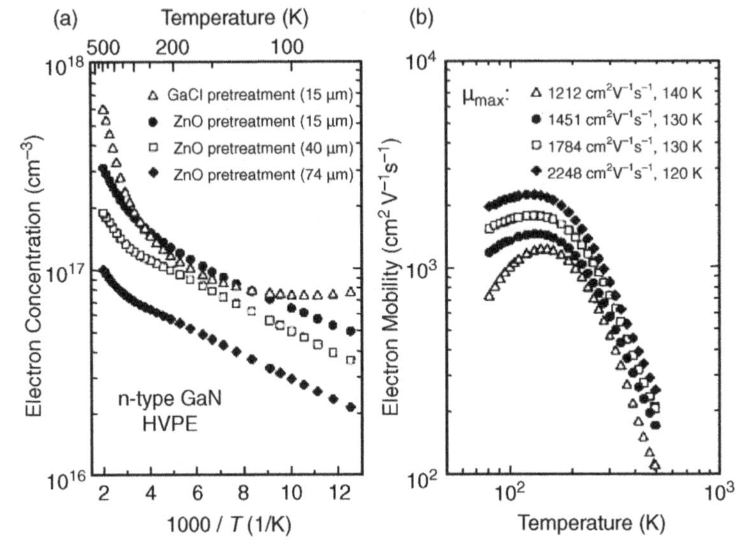

Figure 9.29 *(a) Variation of electron concentration vs inverse temperature and (b) electron mobility vs temperature for GaN layers grown by HVPE. Reprinted from R. J. Molnar, W. Götz, L. T. Romano, N. M. Johnson, Growth of gallium nitride by hydride vapor-phase epitaxy, J. Cryst. Growth, 178, 147. Copyright (1997) with permission from Elsevier*

the range of 420–520 $cm^2 V^{-1} s^{-1}$ for GaN films grown directly on sapphire. At room temperature, the carrier concentration of $9 \times 10^{15} - 4 \times 10^{16} cm^{-3}$ for GaN films grown on ZnO is two or three orders of magnitude lower than that of the films grown on sapphire. A relatively lower mobility value of 60 $cm^2 V^{-1} s^{-1}$ and a higher carrier concentration of $6 \times 10^{19} cm^{-3}$ for GaN films grown on ZnO by PLD are reported in the literature.[7] The carrier concentration and mobility of $In_{0.23}Ga_{0.77}N$ films grown on ZnO substrates at 800 °C are $1 \times 10^{18} cm^{-3}$ and 100 $cm^2 V^{-1} s^{-1}$, respectively.[22] The electron concentration and the Hall mobility are $7 \times 10^{16} cm^{-3}$ and 880 $cm^2 V^{-1} s^{-1}$ at 293 K and $4 \times 10^{16} cm^{-3}$ and 2248 $cm^2 V^{-1} s^{-1}$ at 120 K for 74 μm-thick ZnO pretreated GaN sample. A little variation in electrical properties was noted for over 2 inch sample, showing homogeneity of the sample grown by Molnar *et al.*[14] Electrical measurements, which were carried out on n-GaN grown by HVPE and treated by either GaCl, or ZnO, effectively showed thickness dependence. As depicted in Figure 9.29, the Hall mobility increased with increasing thickness of the GaN film/ZnO/sapphire. This is a clear indication that the defect concentration decreased with increasing thickness of the film ranging from 4.5 to 74 μm. A two-layer model was proposed by Look and Molnar,[45] and extended by Götz *et al.*,[46] to obtain more accurately the bulk mobility and electron concentration when a highly conductive interfacial layer is inserted. The effective areal density of free electrons $n_{s,eff}$ and the effective Hall mobility μ_{eff} are related to the free electron concentration $n(x)$ and electron mobility $\mu(x)$:

$$n_{s,eff}(d) = \left[\int_0^d n(x)\mu(x)dx \right]^2 \Big/ \int_0^d n(x)\mu(x)^2 dx \qquad (9.1)$$

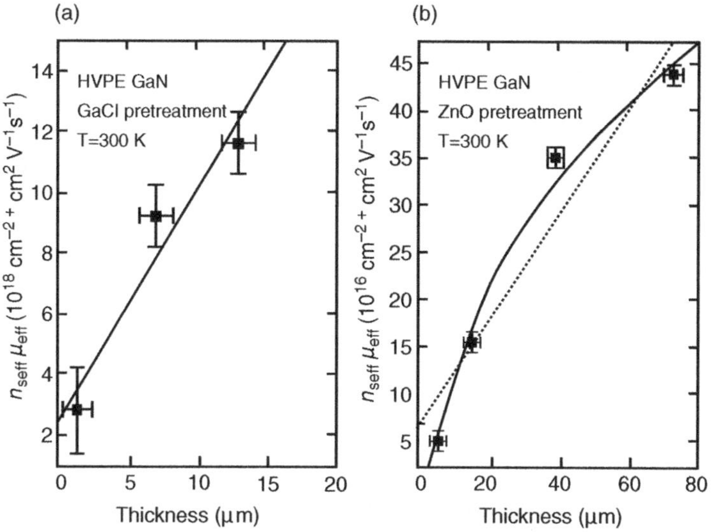

Figure 9.30 *Product of electron density ($n_{s,eff}$) and mobility (μ_{eff}) vs thickness of GaN layer for (a) GaCl-treated samples and (b) ZnO-treated samples. The solid and dotted lines are the linear fit to the experimental data. Reprinted from W. Götz, L. T. Romano, J. Walker, N. M. Johnson, and R. J. Molnar, Hall-effect analysis of GaN films grown by hydride vapor phase epitaxy, Appl. Phys. Lett., 72, 1214. Copyright (1998) with permission from American Institute of Physics*

and

$$\mu_{eff}(d) = \int_0^d n(x)\mu(x)^2 dx \Big/ \int_0^d n(x)\mu(x) dx \tag{9.2}$$

The final product can be derived from Equations (9.1) and (9.2):

$$n_{s,eff}\mu_{eff}(d) = \int_0^d n(x)\mu(x) dx \tag{9.3}$$

For layer 1: $0 < x \leq d_1$, $n(x) = n_1$, $\mu(x) = \mu_1$ and for layer 2: $d_1 < x \leq d_1 + d_2 = d$, $n(x) = n_2$, $\mu(x) = \mu_2$.

Equation (9.3) becomes ($d_1 \ll d_2 \sim d$)

$$n_{s,eff}\mu_{eff}(d) \approx n_1\mu_1 d_1 + n_2\mu_2 d_2 \tag{9.4}$$

Equation (9.4) predicts a linear dependence of $n_{s,eff}\mu_{eff}$ on film thickness (d). Experimentally, linearity was observed for the product of $n_{s,eff}\mu_{eff}$ vs thickness for GaCl-treated samples, whereas for the ZnO-treated samples, a slightly nonlinear behavior was observed, as shown in Figure 9.30. One could believe that there might be some disturbance at the interface, for which some arbitrary decay parameters are introduced in the equations:

$$n(x) = n_1 \exp(-x/\delta_n) + n_2 \tag{9.5}$$

$$\mu(x) = \mu_2 - \mu_1 \exp(-x/\delta_\mu) \tag{9.6}$$

where n_2 and μ_2 represent the electron concentration and mobility for a thick GaN film, respectively, and $n_1 + n_2$ and $\mu_2 - \mu_1$ are roughly related to electron concentration and mobility near the GaN/sapphire interface, respectively. The terms δ_n and δ_μ represent the increase and decrease of the electron concentration and mobility, respectively. From Equations (9.1), (9.2), (9.5) and (9.6):

$$n_{s,\text{eff}}(d) = A(d)^2/B(d) \quad \text{and} \quad \mu_{\text{eff}}(d) = B(d)/A(d) \tag{9.7}$$

where

$$A(d) = n_1\mu_1 d + n_1\mu_2\delta_\mu[\exp(-d/\delta_\mu)-1] - n_2\mu_1\delta_n[\exp(-d/\delta_\mu)-1]$$

and

$$B(d) = n_1{\mu_1}^2 d - 1/2 n_1\mu_2^2\delta_\mu[\exp(-2d/\delta_\mu)-1] + 2n_1\mu_1\mu_2\delta_\mu[\exp(-d/\delta_\mu)-1] \\ -n_2\mu_1^2\delta_n[\exp(-d/\delta_n)-1] \tag{9.8}$$

The theoretically determined electron concentration and mobility are: $n_2 = 3.5 \times 10^{16}\,\text{cm}^{-3}$ and $\mu_2 = 903\,\text{cm}^2\,\text{V}^{-1}\,\text{s}^{-1}$ for a thick film; and $n_1 + n_2 = 4.3 \times 10^{17}\,\text{cm}^{-3}$ and $\mu_2 - \mu_1 = 216\,\text{cm}^2\,\text{V}^{-1}\,\text{s}^{-1}$ for the interface between sapphire and GaN. The reported highest mobility value of GaN grown on sapphire with ZnO buffer is $2248\,\text{cm}^2\,\text{V}^{-1}\,\text{s}^{-1}$ at 120 K.[14] TEM analysis revealed that there is no interface between sapphire and GaN when ZnO is used as a buffer. These results solely support the electrical analysis.

9.8 GaN/ZnO Hybrid Devices

Because p-type ZnO had been relatively hard to attain, hybrid heterostructures in which the p-type material was of some other semiconductor have been explored. Furthermore, for some device features integration of GaN and ZnO technologies paves the way for enhanced performance and/or make certain device functions possible. To reiterate, semiconductor ZnO has a direct band gap of 3.3 eV with a wurtzite structure. In many respects ZnO competes with GaN for device applications. It is also clear that GaN has been shown to be capable of producing very high performance electronic and optical devices. Issues such as reliability, efficiency, etc., are being explored which is indicative of a success story. Power field effect transistors are capable of producing over 500 W of CW power in the communication band, LEDs have become very popular with current emphasis on efficiency while they are already more efficient than fluorescent bulbs, and blue lasers are already used in the Sony Play Station III for high definition viewing. The same, however, cannot necessarily be said about ZnO at this point. As already mentioned, the applications of ZnO overlap a good deal with those of GaN. On the electronic side, the relatively low mobility of ZnO compared with GaN and stronger electron–phonon coupling together with relatively low thermal conductivity are serious shortcomings for ZnO. However, transparent thin film transistors built in poly-ZnO appear to hold some potential, and so do the already established surface acoustic wave devices.

It still remains to be seen, however, as to how competitive ZnO would be with existing technologies. On the optical device front, ZnO needs to show high p-type conductivity along with heterojunctions for competitive devices to be built. Again the competition is GaN, which is well on its way to dominating the optical device development arena in the short wavelength end of the visible spectrum and UV. One advantage that ZnO has over GaN is the 60 meV exciton binding energy, compared with 25 meV for GaN. In addition, ZnO appears to be a more efficient light emitter compared with GaN. If lasers utilizing excitonic transitions were to be built, ZnO would have an advantage over GaN provided that p-type conductivity is obtained and other necessary processing capabilities are developed for ZnO. Further, electromechanical coupling of ZnO, particularly along the *c*-direction is higher than that of GaN, which could pave the way for applications such as surface acoustic wave devices. Further yet, if and when the highly hyped and touted nanostructures were to be of use, ZnO appears to be better suited for producing those structures. The worldwide shortage of In at a time of expanding demand for ITO seems to be opening the door for ZnO-based transparent oxides, i.e. AlZnO (or AZO for short), to be explored. If successful, this application area is huge.

In this section, we focus our attention on GaN/ZnO hybrid devices. As already mentioned, ZnO buffer layers and ZnO substrates have the potential to improve the quality of GaN-based emitters. ZnO itself is also a very promising material for UV-visible light emission. Hereafter, we review the recent progress on hybrid ZnO/GaN heterojunction LEDs and some other hybrid optical devices.

9.8.1 Hybrid ZnO/GaN Heterojunction LED

Alternative approaches which are based on hybrid heterojunctions combining p-doped GaN and n-doped ZnO have been realized. Alivov *et al.* demonstrated room temperature electroluminescence (EL) from n-ZnO:Ga/p-GaN:Mg (grown by MBE)/*c*-sapphire[47] and n-ZnO:Ga (grown by MOCVD)/p-AlGaN:Mg (grown by HVPE)/GaN (grown by HVPE)/6H-SiC[48] heterojunction LEDs, having emission wavelengths of 430 nm and 389 nm, respectively. In the former case, diode-like current–voltage (*I–V*) characteristics were observed under forward bias. The latter provided much better *I–V* characteristics with threshold voltage $\sim$3.2 V as shown in Figure 9.31 and low reverse leakage current $\sim 10^{-7}$ A at room temperature. This UV emission at 389 nm at room temperature when the diode is forward biased and operated up to 500 K was attributed to the radiative recombination in ZnO.

To improve optical and thermal characteristics of LEDs, UV LEDs based on p-n junction n-$Zn_{0.9}Mg_{0.1}O$/n-ZnO/p-$Al_{0.16}Ga_{0.84}N$/p-GaN triple heterostructures were grown by RF plasma-assisted MBE.[49,50] The charge distribution and band diagram of this structure are shown in Figure 9.32. The free carrier concentration in MgZnO and ZnO was assumed to be $\sim 3 \times 10^{17}$ cm^{-3}, with free hole concentrations in AlGaN and GaN assumed to be 1×10^{17} and 3×10^{17} cm^{-3}, respectively. The measured *I–V* characteristics shown in Figure 9.33(a) have rectifying characteristics with a turn-on voltage of $\sim$3.2 V. Strong UV emission at $\sim$390 nm assumed to originate from excitonic transition in ZnO was observed up to 650 K [Figure 9.33(b)].

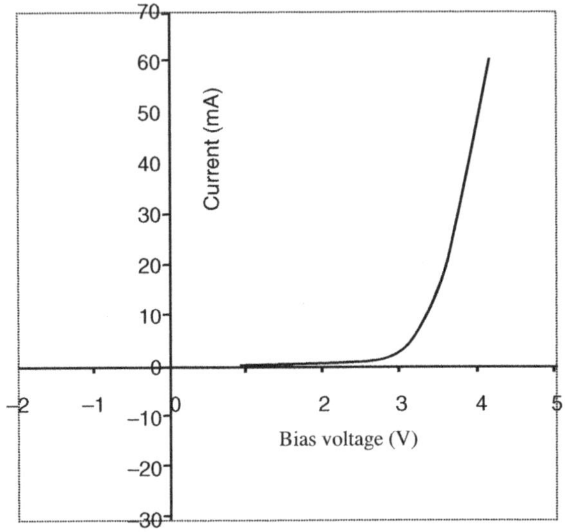

Figure 9.31 *Room temperature I–V characteristics of the n-ZnO/p-$Al_{0.12}Ga_{0.88}N$ structure. Reprinted from Y. I. Alivov, et al., Fabrication and characterization of n-ZnO/p-AlGaN heterojunction light-emitting diodes on 6H-SiC substrates, Appl. Phys. Lett., 83, 4719. Copyright (2003) with permission from American Institute of Physics*

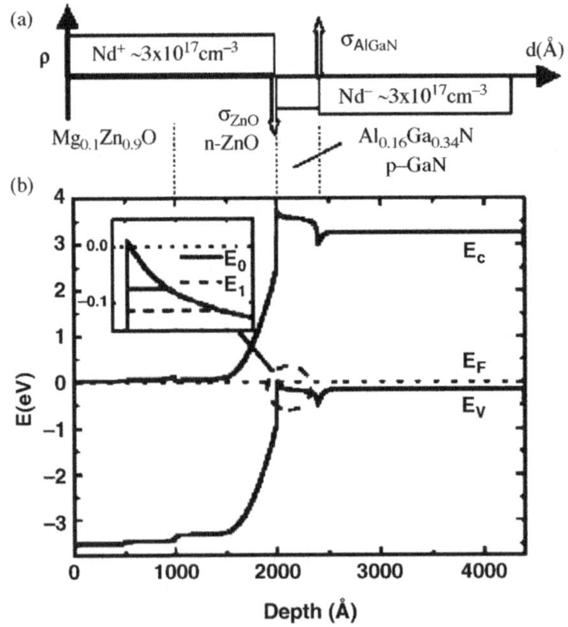

Figure 9.32 *(a) Charge distribution and (b) band diagram of the p-n junction MgZnO/ZnO/AlGaN/GaN heterostructure. The layer thicknesses are 100, 100, 40 and 200 nm, respectively. Reprinted from A. Osinsky, J. W. Dong, M. Z. Kauser, B. Hertog, A. M. Dabiran, P. P. Chow, S. J. Pearton, O. Lopatiuk, and L. Chernyak, MgZnO/AlGaN heterostructure light-emitting diodes, Appl. Phys. Lett., 85, 4272. Copyright (2004) with permission from American Institute of Physics*

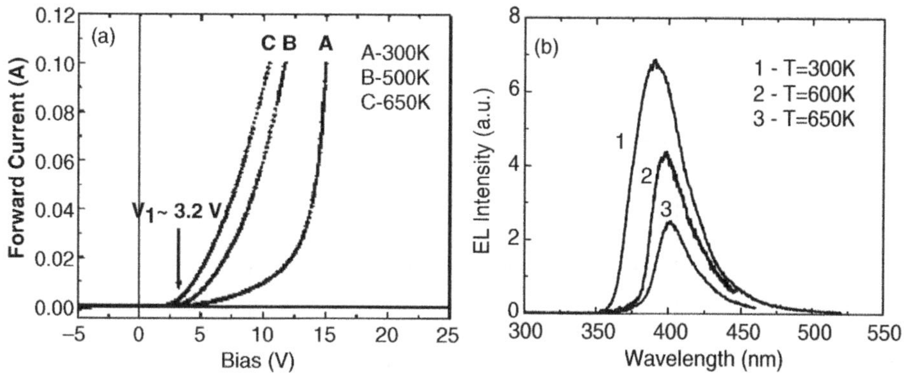

Figure 9.33 *(a) Forward bias I–V characteristics of triple heterostructure UV LED measured at different temperatures. Curves A, B, and C correspond to measurements at 300, 500 and 650 K, respectively. At 100 mA current, the series resistance is ~150 Ω and is ~105 Ω for 300–650 K. Reprinted from H. S. Yang, et al., Fabrication of Hybrid n-ZnMgO/n-ZnO/p-AlGaN/p-GaN Light-Emitting Diodes, Jpn. J. Appl. Phys., 44, 7296 Copyright (2005) with permission from the Japan Society of Applied Physics*

p-GaN/n-ZnO:In (p-n) and p-GaN/i-ZnO/n-ZnO:In (p-i-n) heterojunction LEDs were fabricated by the vapor cooling condensation system,[51] in which the material sources were put on a tungsten boat and heated. Then the sublimated material vapor gases were allowed to condense and deposit onto the substrate cooled by liquid nitrogen. The electron concentration and mobility of the deposited n-ZnO:In were $1.7 \times 10^{20}\,cm^{-3}$ and $3.7\,cm^2\,V^{-1}\,s^{-1}$, respectively, and for i-ZnO were $7.6 \times 10^{15}\,cm^{-3}$ and $2.17\,cm^2\,V^{-1}\,s^{-1}$. For the PL spectra, near-band-edge emission of 380 nm and broad band emission due to what was assumed to be the oxygen vacancies of 540 nm for n-ZnO:In and near-band-edge emission of 382 nm having FWHM of 13 nm for i-ZnO were observed. Figure 9.34 shows the schematic diagrams of the p-n and p-i-n heterojunction LED structures. As shown in Figure 9.35, the room temperature *I–V* characteristics of these LED structures show rectifying diode-like behavior. For the p-n heterojunction LEDs, the forward turn-on voltage and reverse breakdown voltage were 3 and –15 V, respectively, compared with 7 and –23 V for the p-i-n heterojunction LEDs. In the room temperature EL spectrum, the p-n heterojunction LEDs have a broad emission band at 432 nm, which is attributed to the transition from the conduction band to the acceptor level ascribed to the Mg-doped p-GaN when electrons are injected from the n-ZnO:In into the Mg-doped p-GaN. The EL emission peak of p-i-n heterojunction LEDs is at 385 nm, which corresponds to the PL emission wavelength of the i-ZnO.

High brightness UV-blue-green InGaN-based quantum well (QW) LEDs are now commercially available. However, the performance of InGaN-based green LEDs is still far from sufficient for use in solid-state lighting. This is because higher In composition and thick well width are needed to obtain the green emission and the internal field becomes large when the In composition and well width increases. InGaN with high In composition becomes unstable at high temperatures, which is required for high quality n-doped GaN growth. To overcome these challenges, hybrid n-ZnO/(InGaN/GaN) multi-quantum wells (MQWs)/p-GaN LEDs were achieved by Bayram *et al.*[52] The reason why n-ZnO layer by PLD is used is that ZnO layer can be grown at lower temperatures as compared with conventional MOCVD growth to protect the QWs. Note that the refractive index of ZnO at

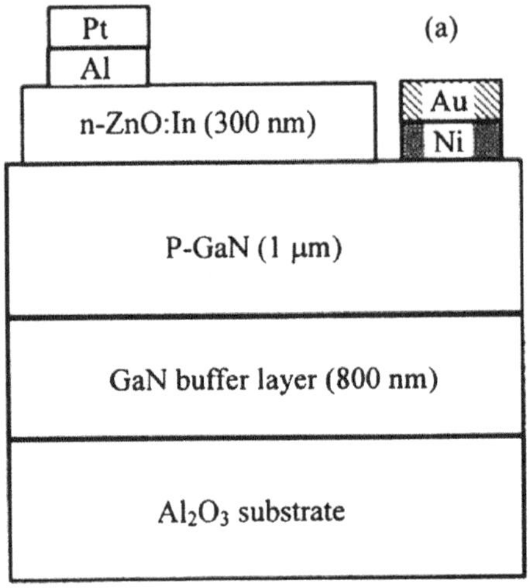

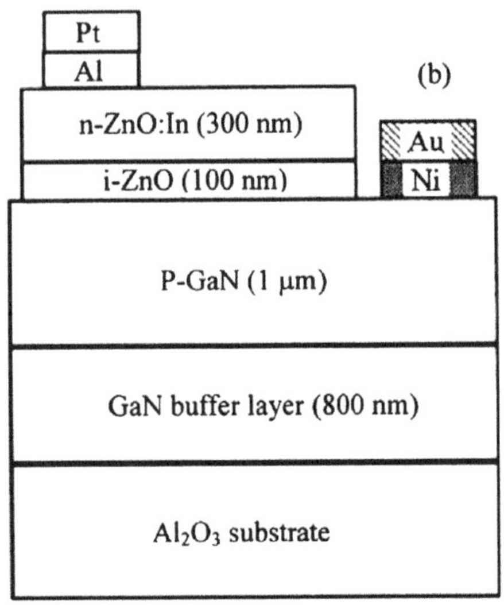

Figure 9.34 *The schematic diagrams of (a) p-GaN/n-ZnO:In and (b) p-GaN/i-ZnO/n-ZnO:In heterojunction LED structures. Reprinted from R. W. Chuang, R. X. Wu, L. W. Lai, and C. T. Lee, ZnO-on-GaN heterojunction light-emitting diode grown by vapor cooling condensation technique, Appl. Phys. Lett., 91, 231113. Copyright (2007) with permission from American Institute of Physics*

500 nm is ~2.0 compared with ~2.5 for GaN. In this manner, lower critical angle loss can be expected for light extraction through ZnO-capped LEDs. Hybrid n-ZnO/(InGaN/GaN) MQWs/p-GaN LEDs were grown by MOVCD and PLD. The device structure is illustrated in the inset of Figure 9.36. p-GaN was grown on a 600 nm-thick AlN buffer layer on sapphire by MOCVD. To activate the Mg dopant, thermal annealing was performed at 1000 °C for 30 s. Hole carrier concentration of p-GaN was determined to be 7.8×10^{17}

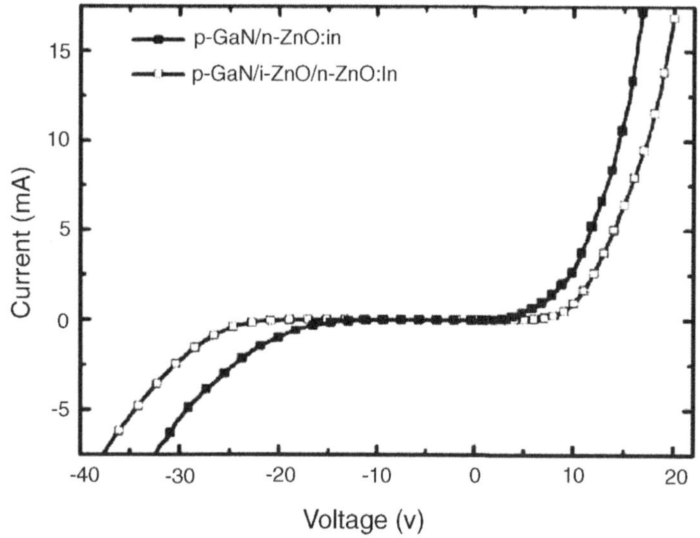

Figure 9.35 *I–V characteristics at room temperature of p-n and p-i-n heterojunction LEDs. Reprinted from R. W. Chuang, R. X. Wu, L. W. Lai, and C. T. Lee, ZnO-on-GaN heterojunction light-emitting diode grown by vapor cooling condensation technique, Appl. Phys. Lett., 91, 231113. Copyright (2007) with permission from American Institute of Physics*

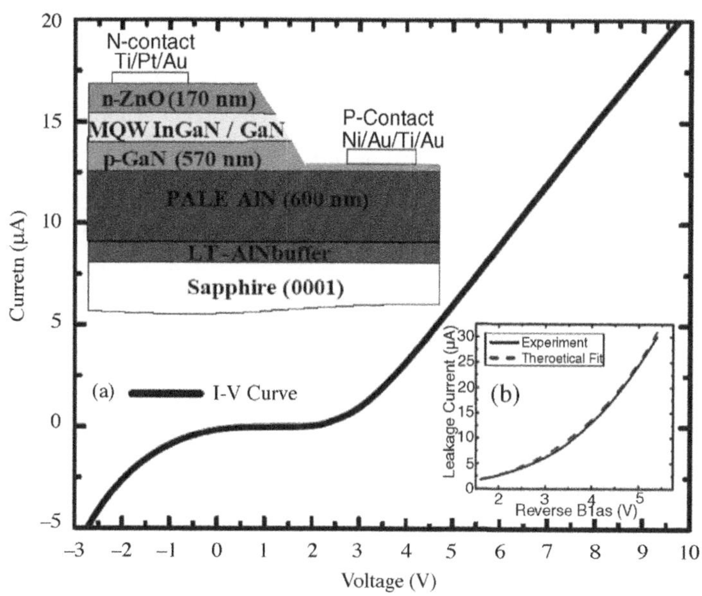

Figure 9.36 *(a) I–V curve of the fabricated LEDs. The device structure is illustrated in the inset. (b) Leakage current vs reverse-bias voltage: experimental and theoretical fits for the reverse voltage range from −1.6 to −5.4 V. Reprinted from C. Bayram, F. Hosseini Teherani, D. J. Rogers, and M. Razeghi, A hybrid green light-emitting diode comprised of n-ZnO/(InGaN/GaN) multi-quantum-wells/p-GaN, Appl. Phys. Lett., 93, 081111. Copyright (2008) with permission from American Institute of Physics*

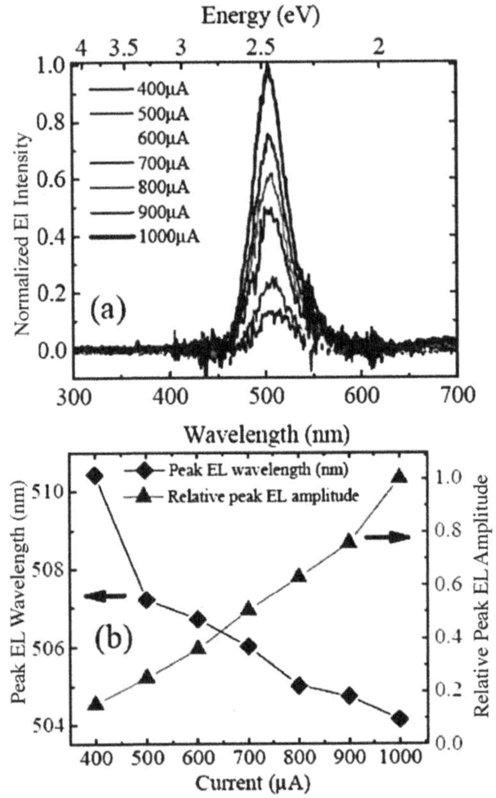

Figure 9.37 (a) Normalized EL intensity of the hybrid green LED at room temperature. (b) Peak EL wavelength and amplitude with respect to injection current. Reprinted from C. Bayram, F. Hosseini Teherani, D. J. Rogers, and M. Razeghi, A hybrid green light-emitting diode comprised of n-ZnO/(InGaN/GaN) multi-quantum-wells/p-GaN, Appl. Phys. Lett., 93, 081111. Copyright (2008) with permission from American Institute of Physics

cm^{-3} by Hall effect measurements. Five periods of InGaN/GaN quantum wells were grown in nitrogen ambient. Each period consisted of a 2 nm-thick InGaN quantum well with a 4 nm-thick GaN barrier. An n-ZnO layer was grown on top of the (InGaN/GaN) MQW/p-GaN by PLD to complete the LED structure. A typical *I–V* curve for a hybrid green LED is shown in Figure 9.36(a). The turn-on voltage was ~2.5 V and the calculated on-series resistance was 4.75 Ω. This high value is due to (1) high contact resistance because the meal contacts to the p-GaN were not annealed and (2) the closeness (~70 nm) of the p-contact to the GaN/AlN interface. As shown in Figure 9.36(b), a nonlinear increase in the leakage current with respect to the reverse-bias voltage was observed. The EL spectra under pulsed injection (duty cycle of 10% and frequency of 5 kHz), as shown in Figure 9.37, show linear dependence on the current density. A blue shift from 510 to 504 nm was observed as the current increased from 400 to 1000 μA and the FWHM decreased slightly from 194 to 179 meV. This is due to the band gap renormalization.[53]

9.8.2 ZnO-based Hybrid Microcavity

ZnO-based hybrid microcavities which consist of GaN-based distributed Bragg reflector (DBR), ZnO-based cavity, and dielectric DBR are studied for a new type of coherent optical source, namely polariton lasers. In the 1990s, the studies on semiconductor microcavities paved the way to the development of cavity polariton technology.[54] Cavity polaritons which are the elementary optical excitations in semiconductor microcavities may be viewed as a superposition of excitons and cavity photons. The major feature of cavity polariton technology centers on large and unique optical nonlinearities which would lead to a new class of optical devices such as polariton lasers exhibiting very low threshold and polariton parametric amplifiers with ultrafast response.[55] Among the wide-bandgap semiconductors, GaN and ZnO are promising candidates for low threshold polariton lasers operating at room temperature because of their large oscillator strengths and large exciton binding energies, particularly ZnO with its unmatched exciton binding energy of 60 meV in bulk.

To reiterate, ZnO is a wide-bandgap semiconductor having a large exciton binding energy (60 meV), much larger than that of GaN (23 meV), and a large oscillator strength, and therefore, is a potential candidate like GaN for the realization of room temperature polariton devices. The most adopted structure for the observation of polariton lasing is a model ZnO-based microcavity proposed by Zamfirescu *et al.*[56] The structure consists of a λ-thick ZnO cavity layer sandwiched between ZnO/$Zn_{0.7}Mg_{0.3}O$ DBRs having 14 and 15 pairs of $\lambda/4$ thick layers on the top and at the bottom, respectively. Amazingly, the vacuum Rabi splitting is as large as 120 meV in this model ZnO-based microcavity, suggesting room temperature polariton lasing to be possible. However, the ZnO technology is not so well developed as yet as compared with that of GaN. Instead of all ZnO-based microcavities, ZnO-based hybrid microcavities which consist of GaN-based DBR, ZnO-based cavity layer and dielectric DBR were grown by MOCVD, plasma-assisted MBE and plasma-enhanced CVD, respectively.[57] Figure 9.38 shows a cross-sectional image of a ZnO-based hybrid microcavity which consists of a λ-thick ZnO cavity layer sandwiched between a 29 pair $Al_{0.5}Ga_{0.5}N$/GaN bottom DBR and an 8 pair SiO_2/SiN_x top DBR. The bottom $Al_{0.5}Ga_{0.5}N$/GaN DBR was directly grown on a 200 nm-thick AlN buffer layer on a (0001) sapphire substrate by low-pressure MOCVD. The AlN buffer was chosen to avoid cracking due to the built-in strain caused by lattice mismatch. The Al composition in the AlGaN layer is nearly 50% as determined from XRD measurements, which also revealed clear interference fringes indicative of smooth interfaces between layers. The $Al_{0.5}Ga_{0.5}N$/GaN pair layer thickness was determined to be 77 nm from both XRD measurements and cross-sectional SEM images.

A λ-thick (optical thickness being $\sim$160 nm for $\lambda \approx 380$ nm in air) ZnO cavity layer was grown on the bottom $Al_{0.5}Ga_{0.5}N$/GaN DBR by plasma-assisted MBE. The top dielectric DBR which consisted of $\lambda/4$-thick SiO_2 and SiN_x layers was deposited on the ZnO half-microcavity by ultra-high vacuum remote plasma-enhanced chemical vapor deposition (UHV-RPECVD) to complete the microcavity structure. Angle-resolved measurements are conventional means to trace the cavity polariton modes in microcavities without changing the position or the temperature. By using the abovementioned ZnO microcavity samples, the angle-resolved PL spectra were measured at room temperature over the range 0–40° using 325 nm excitation light from a He-Cd laser. The results are shown in

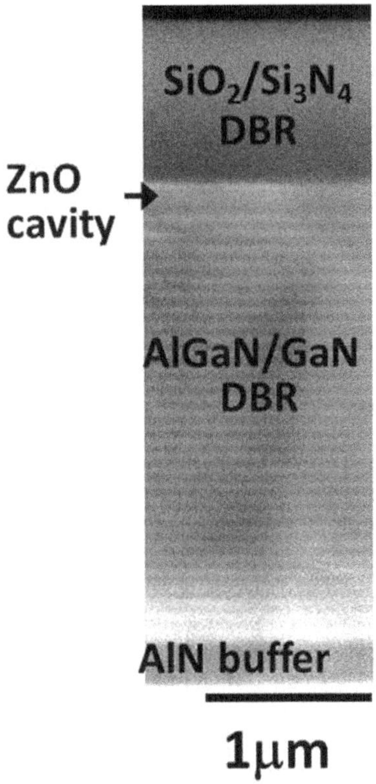

Figure 9.38 Cross-sectional SEM image of a ZnO-based hybrid microcavity. Reprinted from R. Shimada, J. Xie, V. Avrutin, Ü. Özgür, and H. Morkoç, Cavity polaritons in ZnO-based hybrid microcavities, Appl. Phys. Lett., 92, 011127. Copyright (2008) with permission from American Institute of Physics

Figure 9.39(a), where the dotted line indicates the uncoupled exciton mode while the solid lines are guides to the eye. As the angle increases, the lower polariton mode approaches the uncoupled exciton mode, while the upper polariton mode is dispersed from the exciton mode toward the cavity mode. These mode positions are plotted as a function of angle in Figure 9.39(b), indicating a typical anticrossing behavior between the cavity mode and exciton mode when the cavity mode crosses the exciton. Since the stopband width of the bottom DBR is narrow (~150 meV) due to relatively low refractive index contrast in semiconductor DBR layers, the upper polariton features are not clear at large angles, making it difficult to observe a clear anticrossing behavior. In addition, the relaxation processes at the lower polariton branch and thermalization issues due to the large vacuum Rabi splitting might be also responsible for poor resolution of the upper polariton branch. Yet, the anticrossing behavior is clearly seen in Figure 9.39(b) and confirms the strong coupling regime in ZnO-based hybrid microcavities. At the resonant condition of $\theta = 22°$, the vacuum Rabi splitting is estimated to be ~50 meV. This value is far below the 120 meV predicted by Zamfirescu *et al.*[56] The reason for such a large discrepancy is not clear at present but might be partly attributed to the experimental problems such as the

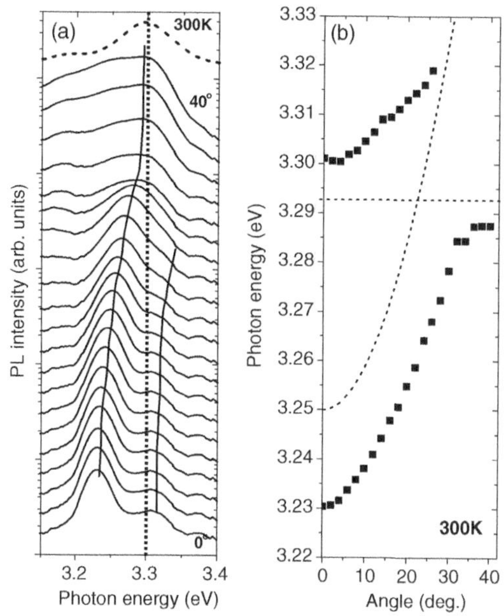

Figure 9.39 *(a) Angle-resolved PL spectra at room temperature in the range of 0 – 40° for a λ-thick ZnO hybrid microcavity. The dotted line is the exciton mode and the solid lines are guides to the eye. (b) Experimental dispersion curves for upper and lower cavity polariton modes. The dotted curves represent the uncoupled cavity and excition modes. Reprinted from R. Shimada, J. Xie, V. Avrutin, Ü. Özgür, and H. Morkoç, Cavity polaritons in ZnO-based hybrid microcavities, Appl. Phys. Lett., 92, 011127. Copyright (2008) with permission from American Institute of Physics*

inhomogeneous broadening in the ZnO cavity layer and the low Q-value in the microcavity. Polariton lasing depends on the formation of a Bose–Einstein condensation at the lower energy trap states in the lower polariton branch. High resolution spectroscopy of the lower polariton branch is imperative for the development of the polariton laser in the strong coupling regime.

9.9 Conclusions

In summary, we have reviewed the growth and characterization of GaN on ZnO substrates and on sapphire with ZnO buffer layer and related ZnO-based hybrid devices. The structural analysis revealed that GaN grown on ZnO showed good quality crystallinity. The reported CL measurements have given a clear picture about the quality of the layers and the effect of the interface between ZnO and GaN. In PL studies, the observation of A and B free exciton peaks bodes well for the quality of the GaN films grown on ZnO. By using a two-layer model, the exact concentration and mobility of the films can be obtained. Finally, ZnO-buffered GaN layers give us fruitful results rather than using just sapphire substrates for the growth of GaN by HVPE. Moreover, for high performance and reliable

GaN-based devices, ZnO buffer layer and ZnO substrates also play a major role in fabrication of free-standing GaN substrates and in the nonpolar GaN growth technique.

Furthermore, ZnO offers some potential in providing optoelectronic devices and encouraging progress has been made in the research phase. Despite this progress there is still a number of important issues that are in need of further investigation before this material can be transitioned to commercial use. The task is made more difficult by the highly successful GaN which competes for similar applications. However, there are some niche applications of ZnO which are not addressed by GaN which, if explored fully, might pave the way for some ZnO-based devices.

Acknowledgements

The work at VCU was supported by grants from BMDO through Cermet, ONR (Drs C. E. C. Wood and Y. S. Park), AFOSR (Drs G. L. Witt, D. K. Johnstone and T. Steiner) and NSF (Drs L. Hess and U. Varshney). They also acknowledge their colleagues, J. Nause at Cermet, Gene Cantwell at then Eagle Picher now at ZN technologies, Drs M. Alexander and M. Callahan at Air Force Research Labs-Hanscom Air Force Base, Dr D.C. Reynolds (retired) at Air Force Research Labs-Wright Patterson Air Force Base for discussions. The investigations of GaN layers grown on ZnO at VCU were carried out by Xing Gu, L. He and M. Reshchikov.

References

[1] H. Morkoç, *Nitride Semiconductors and Devices*, Springer Verlag Series in Materials Science, Spronger Verlag, New York, 1998.
[2] M. J. Manfra, K. W. Baldwin, A. M. Sergent, K. W. West, R. J. Molnar and J. Caissie, *Appl. Phys. Lett.* **85**, 5394 (2004).
[3] H. Morkoç, *Handbook of Nitride Semiconductors and Devices*, Wiley -VCH, Weinheim, 2008, Vol. 3, Ch. 1.
[4] H. Morkoç, *Handbook of Nitride Semiconductors and Devices*, Wiley -VCH, Weinheim, 2008, Vol. 1, Ch. 4.
[5] H. Morkoç, *Handbook of Nitride Semiconductors and Devices*, Wiley -VCH, *Weinheim*, 2008, Vol. 1, Ch. 3.
[6] R. D. Vispute, V. Talyansky, Z. Trajanovic, S. Choopun, M. Downes, R. P. Sharma, T. Venkatesan, M. C. Woods, R. T. Lareau, K. A. Jones and A. A. Illiadis, *Appl. Phys. Lett.* **70**, 2735 (1997); R. D. Vispute, V. Talyansky, R. P. Sharma, S. Choopun, M. Downes, T. Venkatesan, K. A. Jones, A. A. Iliadis, M. Asif Khan and J. W. Yang, *Appl. Phys. Lett.* **71**, 102 (1997).
[7] R. D. Vispute, V. Talyansky, R. P. Sharma, S. Choopun, M. Downes, T. Venkatesan, Y. X. Li, L. G. Salamanca-Riba, A. A. Iliadis, K. A. Jones and J. McGarrity, *Appl. Surf. Sci.* **127–129**, 431 (1998).
[8] R. F. Xiao, H. B. Liao, N. Cue, X. W. Sun and H. S. Kwok, *J. Appl. Phys.* **80**, 4226 (1996).
[9] X. W. Sun, R. F. Xiao and H. S. Kwok, *J. Appl. Phys.* **84**, 5776 (1998).
[10] R. P. Wang, H. Muto, Y. Yamada and T. Kusumori, *Thin Solid Films* **411**, 69 (2002).
[11] A. Kobatyashi, H. Fujioka, J. Ohta and M. Oshima, *Jpn. J. Appl. Phys.* **43**, L53 (2004).
[12] T. Ueda, T. F. Huang, S. Spruytte, H. Lee, M. Yuri, K. Itoh, T. Baba and J. S. Harris Jr, *J. Cryst. Growth* **187**, 340 (1998).

[13] T. Detchprohm, K. Hiramatsu, H. Amano and I. Akasaki, *Appl. Phys. Lett.* **61**, 2688 (1992).

[14] R. J. Molnar, W. Götz, L. T. Romano and N.M. Johnson, *J. Cryst. Growth* **178**, 147 (1997).

[15] S. Gu, R. Zhang, J. Sun, L. Zhang and T. F. Kuech, *Appl. Phys. Lett.* **76**, 3454 (2000).

[16] F. Hamdani, M. Yeadon, D. J. Smith, H. Tang, W. Kim, A. Salvaor, A. E. Botchkarev, J.M. Gibson, A.Y. Plolyakov, M. Skowronski and H. Morkoç, *J. Appl. Phys.* **83**, 983 (1998).

[17] X. Gu, F. Xiu, J. Li, D. Johnstone, H. Morkoç and J. Nause, *J. Mater. Sci.* **15**, 373 (2004).

[18] X. Gu, A. Teke, D. Johnstone, H. Morkoç, B. Nemeth and J. Nause, *Appl. Phys. Lett.* **84**, 2268 (2004).

[19] T. Ohgaki, S. Sugimura, N. Ohashi, I. Sakaguchi, T. Sekiguchi and H. Haneda, *J. Cryst. Growth* **275**, e1143 (2005).

[20] S. Heinze, A. Dadgar, F. Bertram, A. Krtschil, J. Bläsing, H. Witte, S. Tiefenau, T. Hempel, A. Diez, J. Christen and A. Krost, *Proc. SPIE* **6474**, 647406 (2007).

[21] F. Hamdani, A. E. Botchkarev, H. Tang, W. Kim and H. Morkoç, *Appl. Phys. Lett.* **71**, 3111 (1997).

[22] T. Matsuoka, N. Yoshimoto, T. Sasaki and A. Katsui, *J. of Electron. Mater.* **21**, 157 (1992).

[23] C. Sartel, S. Gautier, S. Ould Saad Hamady, N. Maloufi, J. Martin, A. Sirenko and A. Ougazzaden, *Superlatt. Microstruct.* **40**, 476 (2006).

[24] A. Ougazzaden, D. J. Rogers, F. Hosseini Teherani, T. Moudakir, S. Gautier, T. Aggerstam, S. Ould Saad, J. Martin, Z. Djebbour, O. Durand, G. Garry, A. Lusson, D. McGrouther and J. N. Chapman, *J. Cryst. Growth* **310**, 944 (2008).

[25] D. J. Rogers, F. Hosseini Teherani, A. Ougazzaden, S. Gautier, L. Divay, A. Lusson, O. Durand, F. Wyczisk, G. Garry, T. Monteiro, M. R. Correira, M. Peres, A. Neves, D. McGrouther, J. N. Chapman and M Razeghi, *Appl. Phys. Lett.* **91**, 071120 (2007).

[26] D. C. Park, S. Fujita and S. Fujita, *J. Mater. Sci. Lett.* **19**, 631 (2000).

[27] K. C. Kim, S. W. Kang, O. Kryliouk, T. J. Anderson, D. Craciun, V. Craciun, and R. K. Singh, *Mater. Res. Soc. Symp. Proc.*, **764**, 407 (2003).

[28] H. W. Kim and N. H. Kim, *Appl. Surf. Sci.* **236**, 192 (2004).

[29] T. Suzuki, H. J. Ko, A. Setiawan, J. J. Kim, K. Saitoh, M. Terauchi and T. Yao, *Mater. Sci. Semicond. Process.* **6**, 519 (2003).

[30] A. Kobayashi, Y. Kawaguchi, J. Ohta, H. Fujioka, K. Fujiwara and A. Ishii, *Appl. Phys. Lett.* **88**, 181907 (2006).

[31] K. S. A. Butcher, Afifuddin, P. P.-T. Chen, M. Godlewski, A. Szczerbakow, E. M. Goldys, T. L. Tansley and J. A. Freitas Jr, *J. Cryst. Growth* **246**, 237 (2002).

[32] S. W. Lee, T. Minegishi, W. H. Lee, H. Goto, H. J. Lee, S. J. Lee, Hyo-Jong Lee, J. S. Ha, T. Goto, T. Hanada, M. W. Cho and T. Yao, *Appl. Phys. Lett.* **90**, 061907 (2007).

[33] H. Kato, K. Miyamoto, M. Sano and T. Yao, *Appl. Phys. Lett.* **84**, 4562 (2004).

[34] A. Kobayashi, S. Kawano, Y. Kawaguchi, J. Ohta and H. Fujioka, *Appl. Phys. Lett.* **90**, 041908 (2007).

[35] A. Kobayashi, S. Kawano, K. Ueno, J. Ohta and H. Fujioka, H. Amanai, S. Nagao and H. Horie, *Appl. Phys. Lett.* **91**, 191905 (2007).

[36] R. F. Xia, X.W. Sun, Z. F. Li, H. S. Kwok, Q.Z. Liu and S.S. Lau, *J. Vac. Sci. Technol., A* **15**, 2207 (1997).

[37] F. Yun, M. A. Reshchikov, L. He, T. King, D. Huang, H. Morkoç, J. Nause, G. Cantwell, H. P. Maruska and C.W. Litton, *Mater. Res. Soc. Symp. Proc.* **719**, *F8.* 21 (2002).

[38] X. L. Sun, S. H. Goss, L. J. Brillson, D. C. Look and R. J. Molnar, *J. Appl. Phys.* **91**, 6729 (2002).

[39] J. Neugebauer and C. G. Van de Walle, *Appl. Phys. Lett.* **69**, 503 (1996).

[40] M. A. Reshchikov, H. Morkoç, R. J. Molnar, D. Tsvetkov and V. Dmitriev, *Mater. Res. Soc. Symp. Proc.* **743**, L11.1 (2003).

[41] M. A. Reshchikov, G.-C. Yi and B. W. Wessels, *Phys. Rev. B* **59**, 13176 (2000).

[42] B. Arnaudov, T. Paskova, E. M. Goldys, R. Yakimova, S. Evtimova, I. G. Ivanov, A. Henry and B. Monemar, *J. Appl. Phys.* **85**, 7888 (1999).

[43] D. C. Reynolds, D. C. Look, B. Jogai and H. Morkoç, *Solid State Commun.* **101**, 643 (1997).

[44] F. Hamdani, A. Botchkarev, W. Kim, H. Morkoç, M. Yeadon J. M Gibson, S. C. Y. Tsen, D. J. Smith, D. C. Reynolds, D. C. Look, K. Evans, C. W. Litton, W. C. Mitchel and P. Hemenger, *Appl. Phys. Lett.* **70**, 467 (1997).
[45] D. C. Look and R. J. Molnar, *Appl. Phys. Lett.* **70**, 3377 (1998).
[46] W. Götz, L. T. Romano, J. Walker, N. M. Johnson and R.J. Molnar, *Appl. Phys. Lett.* **72**, 1214 (1998).
[47] Y. I. Alivov, J. E. Van Nostrand, D. C. Look, M. V. Chukichev and B. M. Ataev, *Appl. Phys. Lett.* **83**, 2943 (2003).
[48] Y. I. Alivov, E. V. Kalinina, A. E. Cherenkov, D. C. Look, B. M. Ataev, A. K. Omaev, M. V. Chukichev and D. M. Bagnall, *Appl. Phys. Lett.* **83**, 4719 (2003).
[49] A. Osinsky, J. W. Dong, M. Z. Kauser, B. Hertog, A. M. Dabiran, P. P. Chow, S. J. Pearton, O. Lopatiuk and L. Chernyak, *Appl. Phys. Lett.* **85**, 4272 (2004).
[50] H. S. Yang, S. Y. Han, Y. W. Heo, K. H. Baik, D. P. Norton, S. J. Pearton, F. Ren, A. Osinsky, J. W. Dong, B. Hertog, A. M. Dabiran, P. P. Chow, L. Chernyak, T. Steiner, C. J. Kao and G. C. Chi, *Jpn. J. Appl. Phys.* **44**, 7296 (2005).
[51] R. W. Chuang, R. X. Wu, L. W. Lai and C. T. Lee, *Appl. Phys. Lett.* **91**, 231113 (2007).
[52] C. Bayram, F. Hosseini Teherani, D. J. Rogers and M. Razeghi, *Appl. Phys. Lett.* **93**, 081111 (2008).
[53] T. Takeuchi, C. Wetzel, S. Yamaguchi, H. Sakai, H. Amano, I. Akasaki, Y. Kaneko, S. Nakagawa, Y. Yamaoka and N. Yamada, *Appl. Phys. Lett.* **73**, 1691 (1998).
[54] C. Weisbuch, M. Nishioka, A. Ishikawa and Y. Arakawa, *Phys. Rev. Lett.* **69**, 3314 (1992).
[55] M. S. Skolnick, T. A. Fisher and D. M. Whittaker, *Semicond. Sci. Technol.* **13**, 645 (1998).
[56] M. Zamfirescu, A. Kavokin, B. Gil, G. Malpuech and M. Kaliteevski, *Phys. Rev. B* **65**, 161205 (2002).
[57] R. Shimada, J. Xie, V. Avrutin, Ü. Özgür and H. Morkoç, *Appl. Phys. Lett.* **92**, 011127 (2008).

10

Room Temperature Stimulated Emission and ZnO-Based Lasers

D.M. Bagnall
University of Southampton, Southampton, UK

10.1 Introduction

The fundamental emission properties of semiconductors under high excitation conditions are well described by Klingshirn.[1] In general, increasing the excitation levels of semiconductors results in increased interaction between excitonic species and carriers, and these interactions can lead to the observation of nonlinear spontaneous emission bands with quadratic dependence on excitation intensity. At low temperatures and intermediate density regimes, biexcitonic, exciton–exciton or exciton–carrier emissions can be expected, though small biexciton binding energies only allow biexcitonic emissions to be observed at very low temperature. At higher temperatures exciton ionization increases free carrier densities and exciton–carrier emissions are more likely to occur, then at very high excitation intensities exciton densities reach such levels that they overlap. Beyond this "Mott density", phase space filling and Coulomb interactions cause excitons to lose their individual character and an "electron–hole plasma" results.

Some of these spontaneous emission bands, under the right conditions, can provide a mechanism for stimulated emission and in the right circumstances, optical gain. Stimulated emission is identified by a strongly super-linear increase of the optical luminescence output as a function of excitation power above a certain threshold and a spectral narrowing of the emission. If these effects are accompanied by a suitable optical cavity then laser action or "lasing" can be confirmed by the occurrence of Fabry–Perot modes or spatially directed emission.[1]

Zinc Oxide Materials for Electronic and Optoelectronic Device Applications, First Edition.
Edited by Cole W. Litton, Donald C. Reynolds and Thomas C. Collins.

All of the effects outlined above have been observed in many semiconductor systems, however the nature of emission for given temperature ranges and excitation levels are strongly dependent upon both the fundamental material properties and the actual material quality of a given sample. For a long time ZnO was known to have rather desirable fundamental properties, with a wide band gap and a large exciton binding energy that ought to ensure violet and blue *excitonic* emissions at room temperature. However, deposition of suitable ZnO layers with low defect densities and low impurity concentrations has remained problematic. So, as with preceding developments of other semiconductor lasers the main aim of researchers wishing to realize low-threshold high-brightness blue and violet laser diodes based on ZnO was to first find techniques by which suitably perfect ZnO layers could be deposited. The second aim would be to optimize band-engineering by alloying with Al or Cd to allow the formation of double-heterostructures, quantum wells or superlattices and thereby enhance carrier densities and gain. The final requirement is for the realisation of p-type ZnO and thereby formation of diodes and carrier injection.

10.2 Emission Mechanisms

In the 1970s Klingshirn, Hvam and co-workers[2–9] carried out thorough studies of high excitation effects in ZnO. In perhaps the most complete work, Klingshirn[5] studied one- and two-photon excitation studies in the range 2–300 K. Emission bands were attributed to the interactions of free excitons with phonons ($E_X - m$LO), free excitons with free electrons ($E_{ex} - E_{el}$), excitons with excitons ($E_X - E_X$), bound excitons with phonons, and free excitons and bound excitons.

Above 150 K only three intrinsic high-excitation emission mechanisms are observed. During exciton–exciton scattering, one exciton decays to create a photon and scatters another exciton from the lowest ($n = 1$) excitation state into higher excitation states. Emissions due to these processes are labelled the "*P*" band, and individual emission lines are labelled P_n (where $n = 1, 2, 3, \ldots, \infty$ and represents the excitation state of the remaining exciton). The energy of the emitted photon (P_n) can be readily calculated according to:

$$P_{\mathrm{n}} = E_{\mathrm{ex}} - E_{\mathrm{b}}^{\mathrm{ex}}\left(1 - \frac{1}{n^2}\right) - \frac{3}{2}kT \qquad (10.1)$$

where E_{ex} is the free exciton emission energy, E_b^{ex} is the exciton binding energy and kT is the thermal energy.

A second important band is due to exciton–electron scattering.[7] A feature of this emission band is a strong shift to lower energy with increasing temperature. The emission energy for this process is approximated by:

$$h\omega_{\mathrm{max}}^{E_{\mathrm{ex}} - E_{\mathrm{el}}} = E_{\mathrm{ex}} - 7.74\,kT \qquad (10.2)$$

The "*N*" band, a final source of stimulated emission at room temperature, is due to electron–hole plasma recombination. Excitons begin to overlap at the Mott density ($\sim 10^{18}$ cm^{-3}). A distinctive feature of the *N* band is that, because of band-gap renormalization, it moves to lower energy with increasing excitation.[1]

Although much of the literature that reports stimulated emission and lasing in ZnO assigns emissions to either P_n or N bands, Klingshirn *et al.*[10–12] have recently argued that these assignments have been made too readily and often with relatively weak evidence. In particular it is felt that although excitonic processes are likely at low temperatures and low excitation densities, at room temperature and with high excitation and exciton densities close to the Mott density there are more likely mechanisms including those based upon exciton–phonon and exciton–electron processes.

10.3 Stimulated Emission

10.3.1 Bulk ZnO

Bulk ZnO, pumped by electron beam,[13–15] was shown to produce stimulated emission at cryogenic temperatures during the 1960s. Later studies with optical pumping also produced stimulated emission at cryogenic temperatures.[3,6–8,16] Throughout this time stimulated emission at room temperature had proved illusive. The first, brief, report of stimulated emission in ZnO crystals at room temperature was by Koch *et al.*[7] in 1978 and then Reynolds *et al.*[17] were the first to report optically pumped lasing in bulk ZnO as late as 1997.

10.3.2 Epitaxial Layers

Although ZnO thin films were regularly studied for application as a transparent conductor and as a piezoelectric material, up until 1996 relatively little attention had been paid to the ZnO system for its potential use in the fabrication of short wavelength laser diodes. However, a number of developments, particularly the remarkable success of the GaN-based system, caused a number of researchers to look at ZnO in more detail.

In 1996 Yu *et al.*[18] reported room temperature stimulated emission from ZnO epitaxial layers. The samples, produced at the Tokyo Institute of Technology, were hexagonal ZnO microcrystalites grown on sapphire substrates by a technique they called laser molecular beam epitaxy (laser-MBE). High excitation effects were studied using the frequency-tripled output (355 nm, 15 ps) of a Nd:YAG laser. Two spontaneous emission mechanisms and two stimulated emission mechanisms were identified at room temperature. Initially at low excitation a single broad spontaneous luminescence peak, attributed to free exciton recombination, was observed at 3.3 eV. At intermediate excitation levels the P_2 band, attributed to exciton–exciton scattering process in which one exciton is scattered into the $n = 2$ state, emerged 70 meV below the free exciton band in good agreement with Equation (10.1). Above a threshold of 24 kW cm^{-2} the much sharper "P" band was observed 100 meV below the free-exciton emission band. With increasing excitation the P band showed a superlinear increase in intensity, and was seen to suppress the spontaneous emission background, conditions highly indicative of stimulated emission. Beyond 50 kW cm^{-2} the appearance of a second narrow band (N), with its peek moving to lower energies with increasing excitation, was attributed to stimulated emission due to electron–hole plasma recombination. Figure 10.1 shows the development of the P and N lines with increasing excitation.[18] These remarkable results represented a considerable improvement on all earlier works and even now, there are few reports of lower threshold intensities for ZnO epilayers.

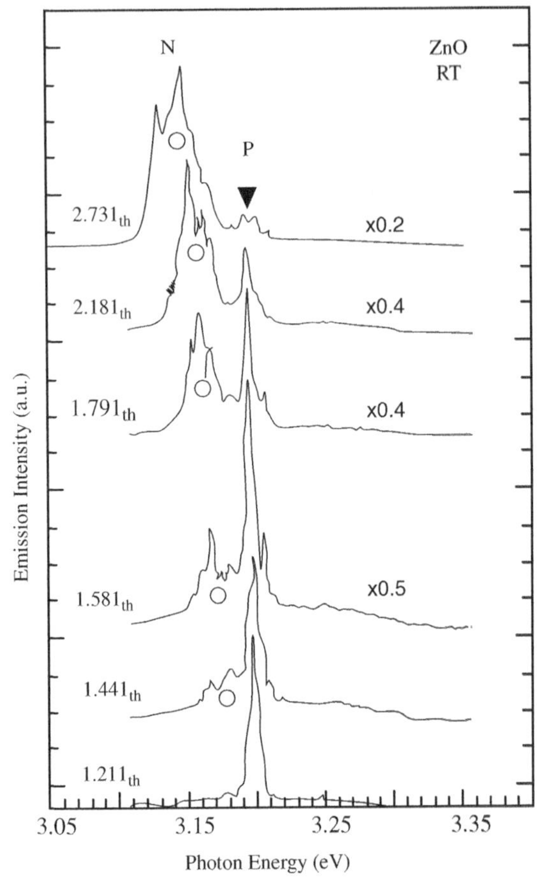

Figure 10.1 *Lasing spectra of the ZnO microcrystallite film pumped using the frequency-tripled output of a mode-locked Nd:YAG laser at various pumping intensities. Reprinted from Z. K. Tang, et al., Room-temperature ultraviolet laser emission from self-assembled ZnO micro-crystallite thin films, Appl. Phys. Lett. 72, 3270. Copyright (1998) with permission from American Institute of Physics*

The dramatic improvement in performance can be attributed to a number of features; most significantly, the excellent crystallinity and purity of the ZnO microcrystals produced by the laser-MBE process. High quality material leads to longer carrier lifetimes, reduced losses by nonradiatative processes, narrow spectral features and lowers absorption. Each of these effects reduces the threshold for stimulated emission and it is for these reasons research efforts towards improved optical devices in any semiconductor system are predominately based around the perfection of growth systems and processes. Perhaps the most important fundamental aspect was the excitonic nature of the stimulated emission. The exciton through its stability in ZnO even at room temperature and its narrow energy distribution provides for low thresholds. In addition the exciton–exciton scattering process which puts the emission peak 100 meV below the fundamental absorption edge is of additional benefit as self-absorption is significantly reduced.

In a later work on the same samples,[19] the stimulated emission threshold is shown to be a strong function of crystallite size with the lowest threshold occurring at a film thickness of 55 nm. So, it was shown for the first time in the case of ZnO that the spatial confinement of excitons can have the expected beneficial effect of driving down threshold values.

Shortly after the first report of stimulated emission at room temperature, the first room temperature ZnO lasers were reported, by workers at Tohoku University[20] and at the Tokyo Institute of Technology.[19,21,22]

Workers at Tohoku University had been studying the growth of ZnO by plasma-enhanced MBE on sapphire substrates.[23,24] Under relatively high excitation by the frequency-tripled Nd:YAG laser, regions of single crystal epilayers had been found to exhibit stimulated emission and, after cavities were formed by cleaving, some localized regions were seen to lase. The stimulated emission and lasing thresholds were found to be in the range 200–240 kW cm^{-2}. In further studies at higher temperatures stimulated emission due to exciton–exciton scattering and electron–hole plasma were clearly resolved (Figure 10.2).[25] At room temperature the threshold for stimulated emission at the "P"

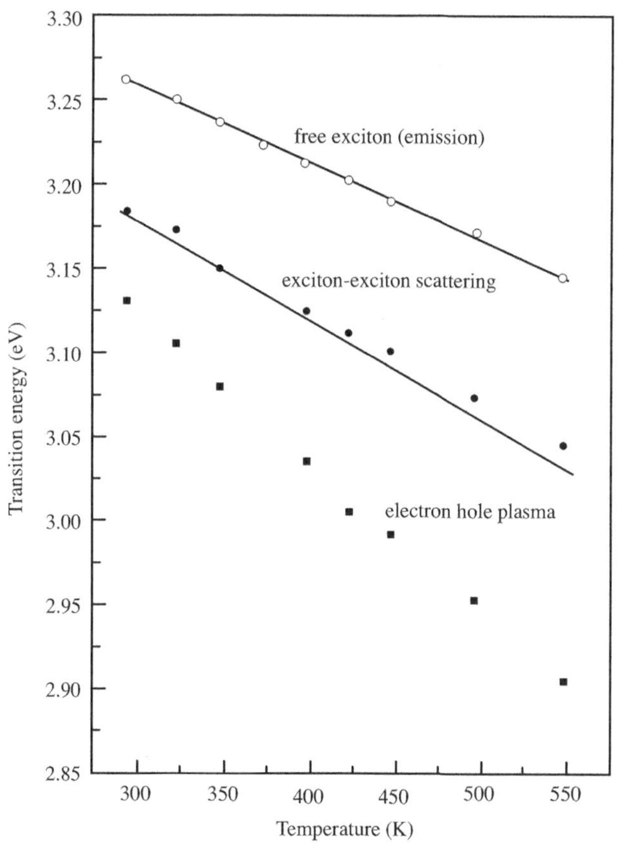

Figure 10.2 *Temperature dependence of peak energy positions. Reprinted from D.M. Bagnall, et al., High temperature excitonic stimulated emission from ZnO epitaxial layers, Appl. Phys. Lett. 73, 1038. Copyright (1998) with permission from American Institute of Physics*

band was 400 kW cm^{-2} while the "N" band had a threshold of 800 kW cm^{-2}. Even at 550 K both features, with peaks at 3.05 eV and 2.9 eV, were still observed with thresholds of 1.2 and 1.9 MW cm^{-2}, respectively.

In further works on ZnO produced by the Tokyo Institute of Technology[26–29] and Tohoku University,[30,31] further information on lasing and stimulated emission have been described. Two publications in particular have provided detailed analysis of gain spectra produced by the variable stripe length method of Shaklee.[32]

Yu *et al.*[19] describe the lasing and gain features of the microcrystalline samples of the Tokyo Institiute of Technology. Gain spectra are plotted for three excitation fluencies; at 2.1 μJ cm^{-2} the peak in the gain spectrum of 50 cm^{-1} is at 3.19 eV, at 3.1 μJ cm^{-2} the peak is 165 cm^{-1} at 3.175 eV and at 3.4 μJ cm^{-2} it is 220 cm^{-1} at 3.17 eV. A maximum gain of 280 cm^{-1} is found at a fluence of 3.8 μJ cm^{-2} whereupon further increases in excitation lead to free carrier densities close to the Mott transition and, as a consequence, electron–hole plasma formation. The maximum values of gain are in reasonably good agreement with the values of 320 cm^{-1} (at 3.0 μJ cm^{-2}) and 300 cm^{-1} determined by the same authors for the same samples by analysis of lasing spectra.[29]

In the second work specifically dedicated to studies of gain spectra, Chen *et al.* [30] have studied ZnO produced by plasma-MBE at Tohoku University. Once again, values are determined by the stripe length method at room temperature. At 180 kW cm^{-2} a peak gain of about 40 cm^{-1} at 3.17 eV is attributed to exciton–exciton scattering, at 220 kW cm^{-2} the mechanism produces symmetric excitonic gain spectra with a maximum of 177 cm^{-1} with a red-shift of 6 meV (Figure 10.3). At 300 kW cm^{-2} a gain value in excess of 550 cm^{-1} is reached at 3.125 eV but this gain is attributed to the electron–hole plasma recombination mechanism.

10.3.3 Quantum wells and Superlattices

The formation of quantum well structures within a semiconductor matrix can allow tuning of emission wavelengths, the realization of enhanced exciton binding energies, enhanced carrier confinement and thereby enhanced gain. Alloying to provide material such as ZnMgO with a larger band gap can be used to surround a ZnO quantum well, while alloying to provide a material such as ZnCdO with narrower band gap can provide quantum wells within a ZnO matrix.

10.3.4 ZnMgO/ZnO Structures

The ZnMgO alloy system allows for band-gap engineering from 3.37 eV (ZnO) to around 4 eV ($Zn_{0.67}Mg_{0.33}O$) at room temperature[33] and photoluminescence studies of ZnO quantum wells within ZnMgO and superlattices have shown tuning of room temperature emission wavelengths from 3.36 to around 3.87 eV.[34,35]

Relatively few researchers have described stimulated emission from ZnO-based quantum well samples. This lack of success reflects the difficulties associated with the growth of these structures, in particular avoiding the precipitation of cubic MgO and the maintenance of excellent structural quality. Success in this regard seems to be enhanced by growth on lattice matched $ScAlMgO_4$ substrates rather than sapphire, although some recent reports indicate increasing success with growth on sapphire.[36–38]

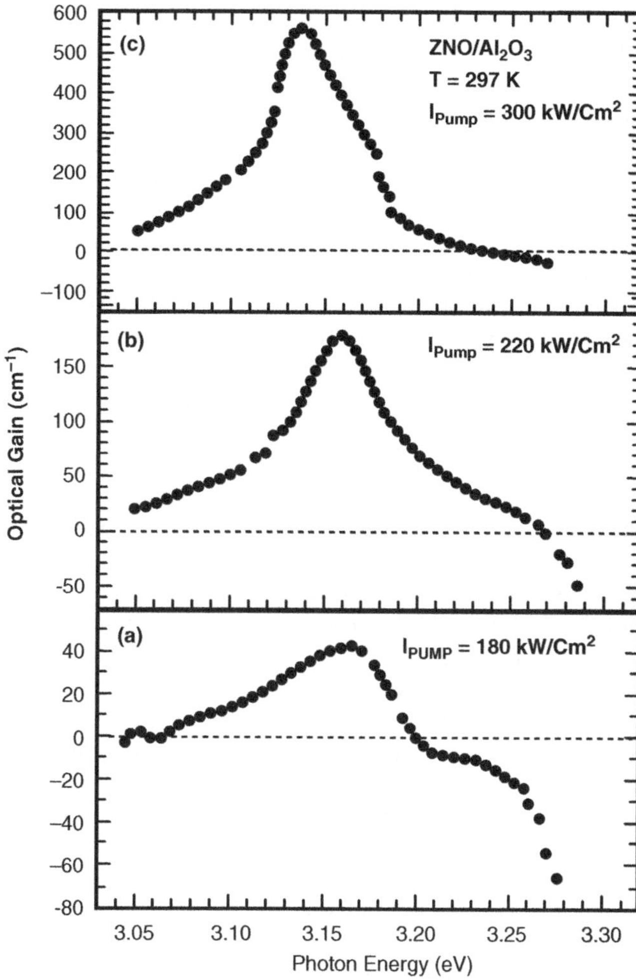

Figure 10.3 *Optical gain spectrum of a ZnO epilayer at excitation density of (a) 180 kW cm^{-2}, (b) 220 kW cm^{-2} and (c) 300 kW cm^{-2} at room temperature. Reprinted from Y. Chen, N.T. Tuan, Y. Segawa, H. Ko, S. Hong and T. Yao, Stimulated emission and optical gain in ZnO epilayers grown by plasma-assisted molecular-beam epitaxy with buffers, Appl. Phys. Lett. 78, 1469. Copyright (2001) with permission from American Institute of Physics*

Workers at the Tokyo Institute of Technology and their collaborators have performed the most detailed studies of ZnO-based superlattice systems.[36–39]

H. D. Sun *et al.*[40] provided one of the earliest detailed studies of stimulated emission from a range of ZnO/$Zn_{1-x}Mg_xO$ multiple quantum well (MQW) samples grown on $ScAlMgO_4$ substrates by laser-MBE. The magnesium content of the barrier layers was kept at $x = 0.12$ to provide a barrier of around 0.2 eV. Barrier thickness was kept at 5 nm, while the quantum wells ranged from 0.69 to 4.65 nm. Optical excitation was carried out using an excimer laser and edge emission was detected. At low excitation intensities a spontaneous emission peak due to localized excitons is observed but as excitation levels were increased this peak was found to saturate before a stimulated emission peak due to exciton–exciton scattering emerged. Exciton–exciton scattering was found to be the main stimulated emission mechanism for all temperatures from 5 K through to room temperature. Although at higher temperatures, and higher excitation levels, stimulated emission due to electron–hole plasma recombination was simultaneously observed, this would probably be from localized regions of the material. Exciton binding energies were readily deduced and found to be enhanced for all quantum well samples, increasing with decreasing well width to a maximum value of 86 meV in the case of 1.75 nm wide quantum wells. Thresholds for stimulated emission at 5 and 300 K were determined at around 220 and 480 kW cm^{-2}, respectively, values an order of magnitude greater than those observed for the best ZnO epilayers, though the authors suggest that shorter excitation pulses and subtraction of reflected and transmitted light might reduce these values.

Ohtomo *et al.*[36] have produced 10-period superlattices of $ZnO/Zn_{0.88}Mg_{0.12}O$ and $ZnO/Zn_{0.74}Mg_{0.26}O$ on $ScAlMgO_4$ substrates using combinatorial laser-MBE. ZnO quantum well thicknesses were varied from 0.7 to 4.7 nm while barrier layers were kept at 5 nm. Excitation was carried out using a frequency-tripled Nd:YAG laser with 15 ps pulses, and edge emission was studied in the temperature range 294–380 K. At room temperature all of the $ZnO/Zn_{0.88}Mg_{0.12}O$ superlattices were found to exhibit stimulated emission at energies in the range 3.2–3.4 eV with thresholds below 22 kW cm^{-2}. The lowest threshold of 11 kW cm^{-2}, reported for the sample with 4.7 nm wells, compares favorably with typically optical pumping thresholds reported for high quality ZnSe- and GaN-based systems. The stimulated emission mechanism is tentatively assigned to exciton–exciton scattering. Clear threshold characteristics are provided for temperatures up to 377 K. and the characteristic temperature is determined at 87 K.

In other work of note, J. W. Sun *et al.*[41] have observed room temperature stimulated emission at 3.33 eV and excitation intensities of around 200 kW cm^{-2}, from $ZnO/Mg_{0.2}Zn_{0.8}O$ MQW structures grown by plasma-assisted MBE on sapphire. The emission mechanism is assigned to exciton–exciton scattering and the enhanced exciton binding energy calculated to be 122 meV in good agreement with other similar work.[38,39]

10.3.5 ZnO/ZnCdO Structures

Early studies of $Zn_yCd_{1-y}O$ alloy films were carried out by the Tokyo Institute of Technology,[42–44] where it was found that a band gap of 3.0 eV at room temperature could be achieved with $y=0.07$, when layers were grown on $ScAlMgO_4$ (0001) substrates by pulsed laser deposition.[44] More recent work by Sadofev *et al.*[45] has shown that this range can be extended to as far as 2.15 eV and emission in the yellow spectral range at Cd compositions as high as $y=0.32$. Strong band-gap-related emission was reported for both epilayers and quantum wells grown using MBE at remarkably low growth temperatures.

The success of this MBE approach has recently yielded the first evidence of room temperature stimulated emission and lasing in ZnCdO quantum well samples.[46] Samples containing multiple $Zn_yCd_{1-y}O$ quantum wells with $y=0.1$ and widths of 3.2 nm were found to lase at 5 K with a threshold of 60 kW cm^{-2} and with the emission peak at 2.77 eV.

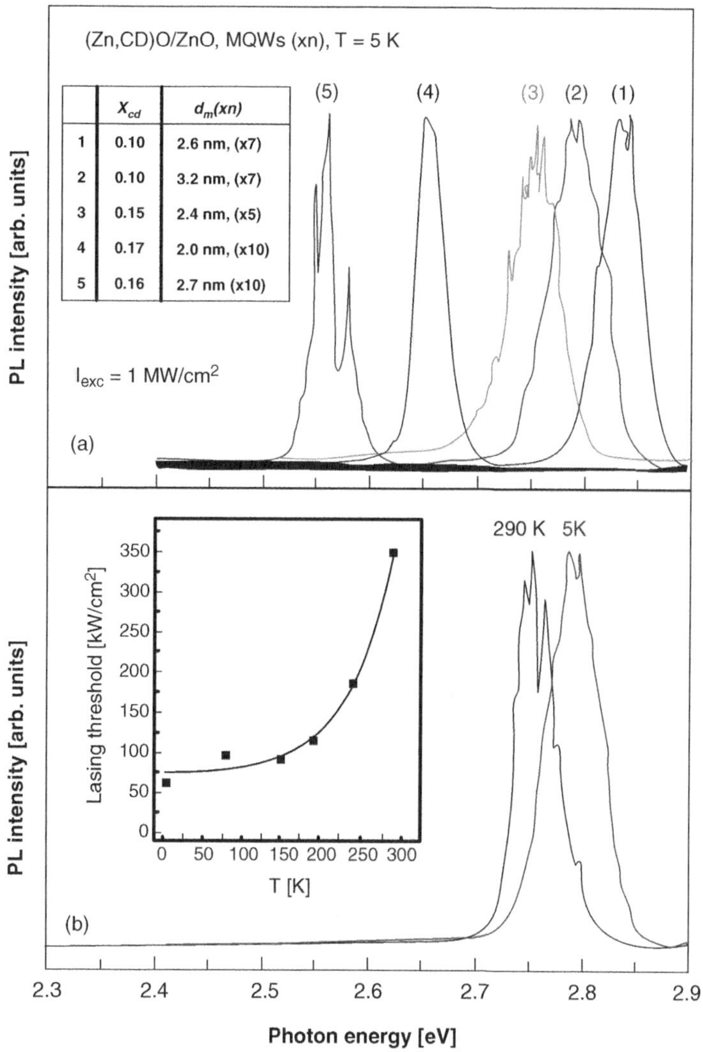

Figure 10.4 *Laser emission spectra from ZnCdO MQWs. (a) Variation of the laser emission with structure design; 1–4 are on sapphire substrates, 5 is on ZnO substrate. (b) Change of laser emission from low to room temperature for structure 2. Inset: temperature dependence of the threshold.*[46]

At room temperature laser emission at 2.75 eV was found at a threshold of 350 kW cm^{-2} for the same sample. Other samples with $y = 0.16$ and well width 2.7 nm were found to lase with emission peaks as low as 2.53 eV corresponding to emission wavelengths around 490 nm [Figure 10.4]. The lasing thresholds are found to be relatively robust to temperature increases, a feature indicative of gain mechanisms related to the recombination of carriers localized within the quantum wells. These are encouraging developments for a ZnO-based technology, with emission extending well into the green, there is potential advantage over GaN-based systems, though this work is at an early stage.

10.4 Zinc Oxide Lasers

10.4.1 Introduction

According to modern convention demonstrations of "laser action" or "lasing" requires coherence, and this is conventionally achieved by placing gain media inside an optical cavity. This cavity will then favor the growth of particular frequencies and phases. Thus selective amplification is a result of positive feedback for electromagnetic waves that form a standing pattern in the cavity. In general, to prove lasing it is necessary to provide evidence of the formation of Fabry–Perot modes.

Mode spacing ($\Delta\lambda$) is given by:

$$\Delta\lambda = \frac{\lambda_0^2}{2L(n - \lambda_0 \mathrm{d}n/\mathrm{d}\lambda)}$$

where L is the cavity length and n is the index of refraction at λ_0 (the wavelength of one of the modes).

Meanwhile, to a first approximation, the gain threshold (G_{th}) is given by:

$$G_{\mathrm{th}} = \frac{\alpha}{\beta} + \frac{1}{2\beta L} \ln \frac{1}{R_1 R_2}$$

where α is the loss per unit length, β is the gain and R_1 and R_2 are the mirror (or facet) reflectivities.

To observe modes directly it is desirable to have a short cavity length as this will increase the mode separation but small cavity lengths require higher threshold intensities. High mirror reflectivities are very desirable as this reduces thresholds. Resolution of laser modes from cavities prepared by simple cleaving is difficult, as the process tends to produce poor quality mirrors with low reflectivities, and relatively large cavity lengths and this combined with excitation pulse variations, and spatial variations in the sample often produces broad Gaussian and Lorentzian lineshapes when laser spectra are generated by boxcar integration or multiple charge-coupled device (CCD) acquisitions. To obtain spectra with clear mode structure it is necessary to resolve emissions spatially and temporally as well as spectrally.[47] Thus, excitation stripes should be narrow to avoid the effect of spatial variations along the laser bar and the emission from a single excitation pulse should be collected and dispersed into a high resolution monochromator/CCD system.

The first room temperature ZnO laser was reported in 1997 by the Tohoku University group.[20] Samples were grown by plasma-enhanced MBE on sapphire substrates. To produce laser cavities as-grown samples are cleaved to produce "bars" with cavity lengths in the range 300–1000 μm. By moving an intense excitation stripe, generated by a frequency tripled Nd:YAG laser, along the bar high intensity regions emitting speckled light were located. When these regions were studied in detail a threshold for stimulated emission was determined at 240 kW cm^{-2} and lasing was proven by the presence of modulations in the spectra taken from single excitation pulses. As far as we are aware there is only one other instance where a ZnO cavity has been fabricated; however, one of the most fascinating features of ZnO is its almost unique tendency to form natural laser cavities. "Microstructural", "powder" and "nanowire" ZnO lasers have all been reported within the last few years.

10.4.2 Microstructural Lasers

There have now been many reports of ZnO laser cavities formed as a result of microstructural features of thin films. The Tokyo Institute of technology and their collaborators[19,22,28,29] found regularly spaced sharp lines in lasing spectra reminiscent of cavity modes (Figure 10.5.). This was surprising since there had been no attempt to form a laser cavity. Furthermore, the mode spacing and therefore cavity length was found to

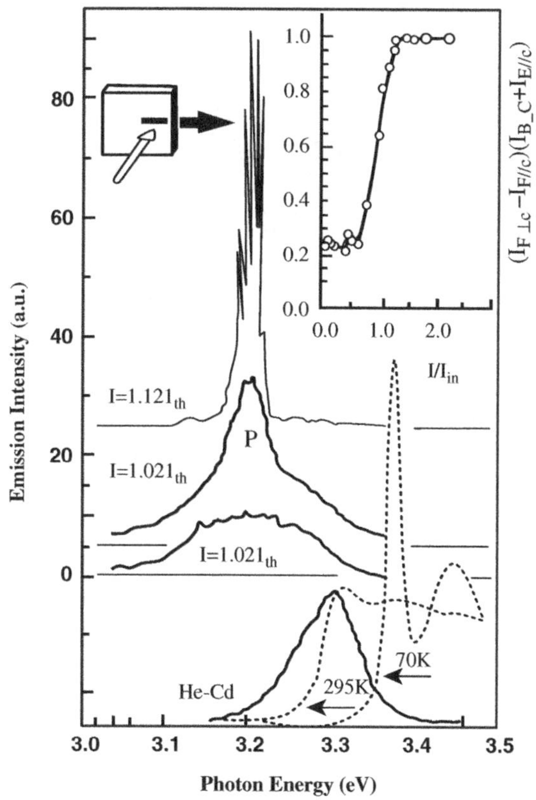

Figure 10.5 The lower trace shows the absorption spectrum (dotted curve) and photoluminescence spectrum, measured at 70 K and room temperature, respectively. The upper traces show spontaneous and stimulated emission spectra under pumping intensities of $0.2I_{th}$, $1.02I_{th}$ and $1.12I_{th}$ provided by the frequency-tripled and mode-locked Nd:YAG laser (355 nm, 15 ps). The inset on the right-hand side shows the degree of polarization of the UV emissions plotted as a function of the pumping intensity. Reprinted from Z. K. Tang, et al., Room-temperature ultraviolet laser emission from self-assembled ZnO microcrystallite thin films, Appl. Phys. Lett. 72, 3270. Copyright (1998) with permission from American Institute of Physics

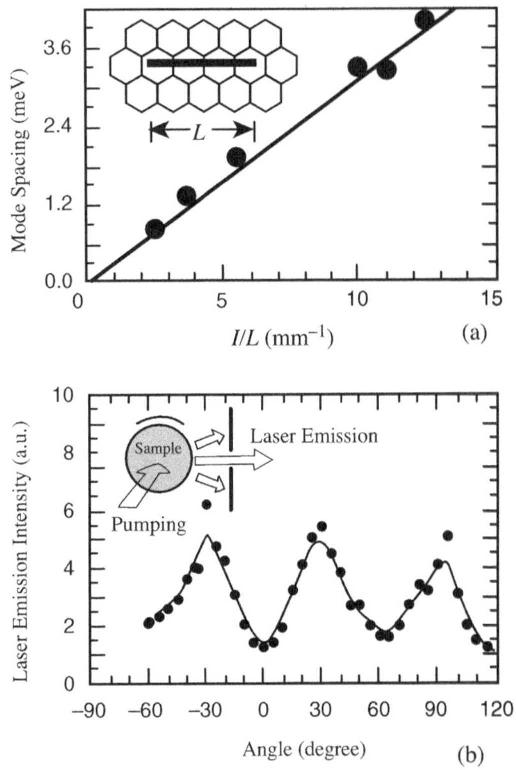

Figure 10.6 *(a) The mode spacing (solid circles) plotted as a function of the stripe length L of the pumping laser beam. The solid line is calculated for a Fabry–Pérot resonant cavity with length L. A model of natural Fabry–Pérot microcavity formed by the hexagonal microcrystallites is schematically drawn in the inset. (b) The laser emission intensity plotted as a function of rotating angle. The experimental set-up is shown in the inset. The maximum and minimum of the laser emission intensity repeats every 60°. Reprinted from Z. K. Tang, et al., Room-temperature ultraviolet laser emission from self-assembled ZnO microcrystallite thin films, Appl. Phys. Lett. 72, 3270. Copyright (1998) with permission from American Institute of Physics*

change, in steps, with the length of the excitation stripe (Figure 10.6.). The authors concluded that the laser cavity was formed by the {1100} facets of parallel hexagonal ZnO microcrystallites, where many hexagons are covered by the excitation stripe. The necessary mirrors were described as a result of the combination of the change in refractive index due to excitation and the grain boundaries.

Workers at Northwestern University have reported laser action from nonepitaxial ZnO (Figure 10.7).[48] Polycrystalline films were grown on amorphous fused silica substrates by laser ablation. They found laser cavities to "self-form" as a result of strong optical scattering in the films. The scattering mean free path is estimated to be of the order of the emission wavelength of ZnO and it is suggested that emitted light could return via a "random" or complex path to the scatterer from which it was scattered and thus form a

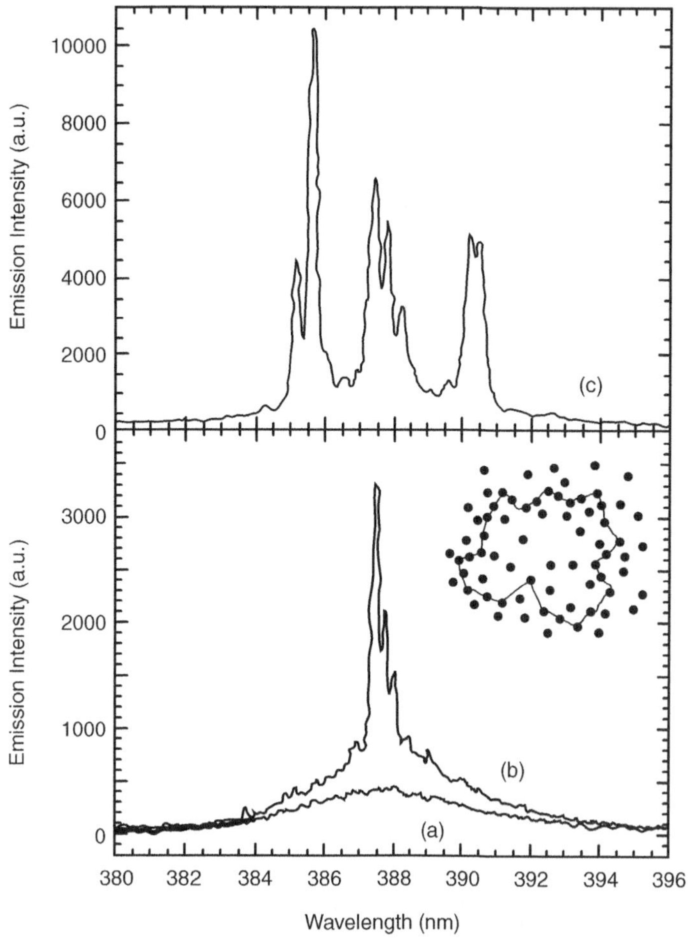

Figure 10.7 *Emission spectra from a ZnO "powder laser" when the excitation area is (a) 2700 μm², (b) 3800 μm² and (c) 4500 μm². The excitation intensity is 400 kW cm⁻². The inset is a schematic diagram showing the formation of a closed-loop path for light through multiple optical scattering in a random medium. Reprinted from H. Cao, Y. G. Zhao, H. C. Ong, S. T. Ho, J. Y. Dai, J. Y. Wu, and R. P. H. Chang, Ultraviolet lasing in resonators formed by scattering in semiconductor polycrystalline films, Appl. Phys. Lett., 73, 3656. Copyright (1998) with permission from American Institute of Physics*

closed-loop path (Figure 10.7, inset). A number of subsequent publications by Northwestern and collaborators have examined[49–51] and modelled[52] random cavity formation. Models are suggested in which regions of net gain, regions of loss and scattering centers can lead to the formation of cavities.

Other groups have also reported lasing as a result of microstructural cavities. Mitra and Thareja[53] have observed lasing from ZnO pellets and polycrystalline thin films formed by pulsed laser deposition. Cho *et al.*[54] have observed lasing from polycrystalline films deposited by the oxidation of metallic zinc.

What is clear from all these studies is that grain boundaries, features normally avoided, in ZnO can play a significant role in the formation of laser cavities. Spatially resolved electron energy loss spectroscopy (EELS) has revealed significant changes in refractive index to form reflective layers of around 10 nm in the vicinity of grain boundaries.[55] This change is speculated to be due to the formation of space-charge regions at the grain boundaries.

10.4.3 Powder Lasers

ZnO has also provided the first demonstration of laser action in a powder.[56] ZnO nanoparticles were produced by the precipitation reaction of zinc acetate dihydride and diethylene glycol. The nanocrystals of around 50 nm agglomerate to form micrometer-sized clusters (Figure 10.8). Excitation with the 4th harmonic of a Nd:YAG laser (266 nm) above a threshold of around 0.3 nJ at room temperature results in a very sharp peak (FWHM = 12 nm) in the emission spectrum at around 380 nm (Figure 10.9). Similar to the lasing observed in polycrystalline ZnO films, optical cavities are formed by multiple scattering and wave interference within the disordered medium. Whether these lasers are simply complex cavities or whether the Anderson localization plays a part is still a matter for debate.

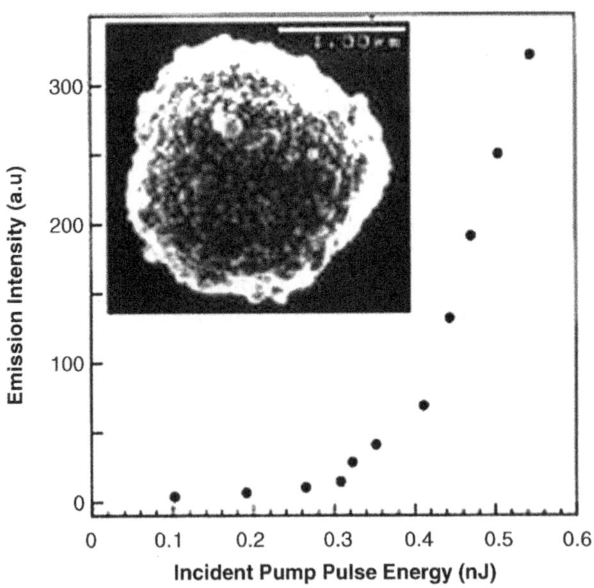

Figure 10.8 *Spectrally integrated intensity of emission from the ZnO cluster vs the incident pump pulse energy. The inset is the SEM image of the ZnO cluster. Reprinted from H. Cao, J. Y. Xu, E.W. Seelig, and R.P.H. Chang, Microlaser made of disordered media, Appl. Phys. Lett., 76, 2997. Copyright (2000) with permission from American Institute of Physics*

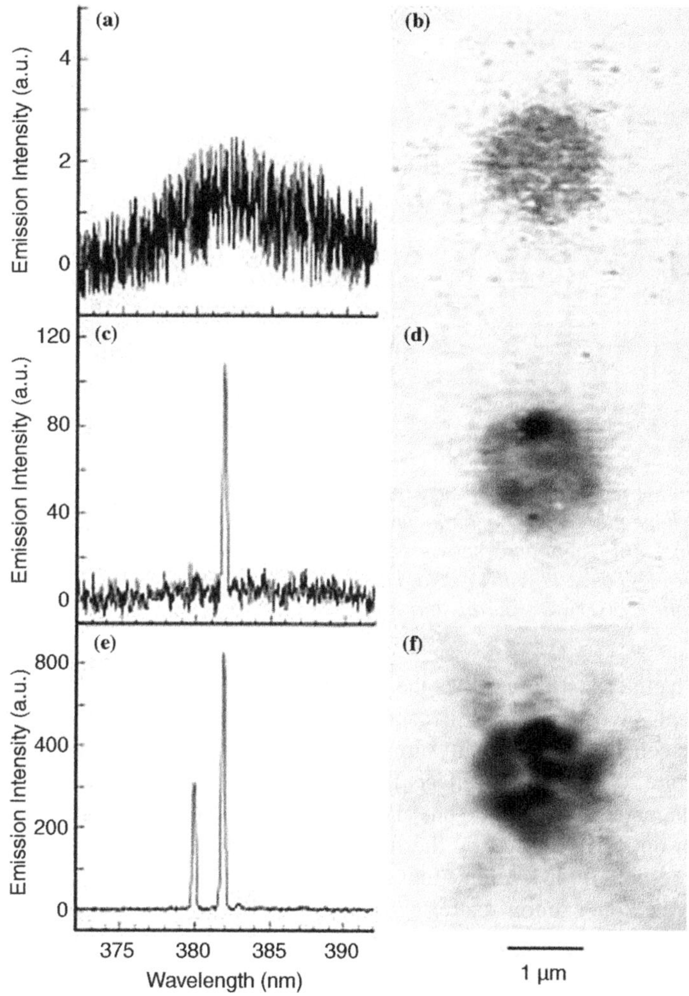

Figure 10.9 *(a, c, e) Emission spectra from the ZnO cluster shown in Figure10.8. (b, d, f) The corresponding spatial distributions of emission intensity in the cluster. The incident pump pulse energy is 0.26 nJ for (a) and (b), 0.35 nJ for (c) and (d), and 0.50 nJ for (e) and (f). Reprinted from H. Cao, J. Y. Xu, E.W. Seelig, and R.P.H. Chang, Microlaser made of disordered media, Appl. Phys. Lett., 76, 2997. Copyright (2000) with permission from American Institute of Physics*

10.4.4 Nanowire Lasers

Perhaps one of the most intriguing developments over the last few years has been the demonstration of room temperature ZnO nanowire "nanolasers".[57,58] Physical vapor deposition techniques can be used to fabricate arrays of ZnO nanowires via the vapor–liquid–solid (VLS) process (Figure 10.10).[57] Sapphire substrates are coated with a thin

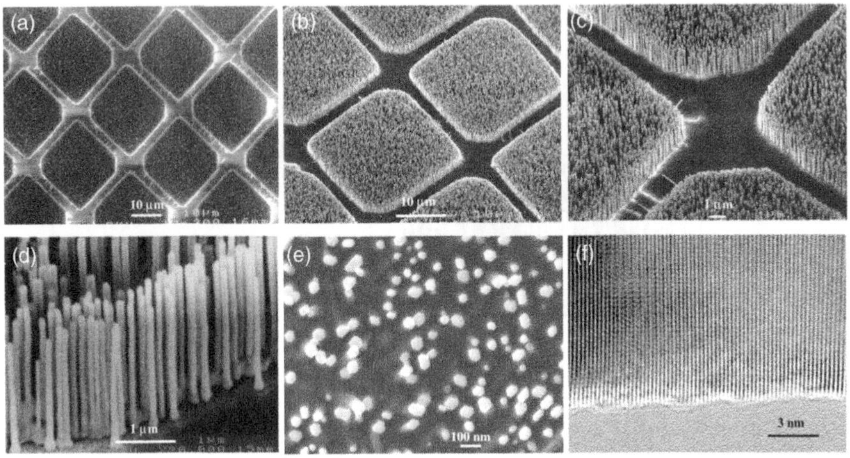

Figure 10.10 (a–e) SEM images of ZnO nanowire arrays grown on sapphire substrates. A top view of the well-faceted hexagonal nanowire tips is shown in (e). (f) High-resolution TEM image of an individual ZnO nanowire showing its <0001> growth direction. Reprinted from M. Huang, S. Mao, H. Feick, H. Yan, Y. Wu, H. Kind, E. Weber, R. Russo, P. Yang, Science, 292, 1897. Copyright (2001) with permission from AAAS

film of gold, as the substrate reaches the growth temperature the gold forms nanoclusters and subsequent ZnO wire growth is catalyzed by the gold. Very well faceted hexagonal wires with excellent crystal quality are commonly produced. The distribution of wire diameters, from 20 to 150 nm, reflects the distribution of gold clusters, though 95% of wires have diameters between 70 nm and 100 nm. Lengths are readily controlled up to 10 μm. When optically excited by the frequency quadrupled output of a Nd:YAG laser (266 nm) a threshold of 40 kW cm^{-2} and clear mode structure in emission spectra indicate lasing due to exciton–exciton scattering [Figure 10.11(a,b)]. The length of the nanowire forms the laser cavity length and the sapphire/ZnO and ZnO/air interfaces at either end of the wires provide the mirrors [Figure 10.11(c)].

The low threshold for laser action in ZnO can be attributed to the single mode nature of the lasers (fortuitously wires with diameters in the range 80–120 nm are single mode waveguides for UV light), the giant oscillator strength effect and the high crystal quality (which also manifests itself in long luminescence lifetimes). A lot of nanotechnological ingenuity will be required to turn these nanolasers into practicable devices but the prospects for integration into "system-on-a-chip", use as optical probes or pixelated displays are enticing. In addition to offering a glimpse of a nanotechnological future the work of the Berkeley group[57] also independently verifies that stimulated emission and lasing can be achieved in ZnO crystals with thresholds as low as 40 kW cm^{-2}.[18]

10.4.5 ZnO Laser Diodes

Both the fabrication of p-type ZnO and the reproducibility of laser-quality ZnO remain seriously challenging and, as yet, only one group has reported on the fabrication of ZnO-based laser diodes.

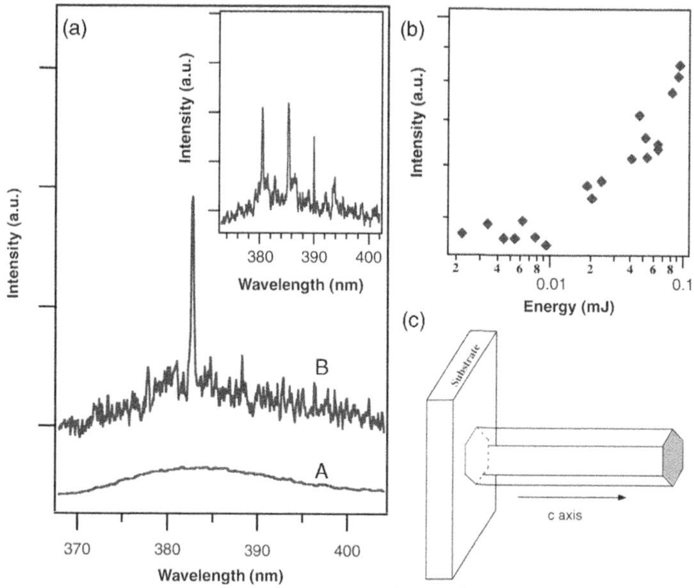

Figure 10.11 (a) Emission spectra from nanowire arrays below (A) and above (B and inset) the lasing threshold. The pump power for these spectra is 20, 100 and 150 kW cm^{-2}, respectively. The spectra are offset for easy comparison. (b) Integrated emission intensity from nanowires as a function of optical pumping energy intensity. (c) Schematic illustration of a nanowire as a resonance cavity with two naturally faceted hexagonal end faces acting as reflecting mirrors. Reprinted from M. Huang, S. Mao, H. Feick, H. Yan, Y. Wu, H. Kind, E. Weber, R. Russo, P. Yang, Science, 292, 1897. Copyright (2001) with permission from AAAS

Ryu *et al.*[59] fabricated arsenic-doped p-type ZnO/BeZnO and gallium-doped n-type BeZnO/ZnO heterojunction structures with undoped ZnO/BeZnO MQW active layers. This work represents a significant milestone with stimulated emission reported with a threshold of 420 A cm^{-2} and emission at 3.21 eV. The authors suggest exceptionally high exciton binding energies (263 meV) in the MQW active layers, exciton–exciton scattering as the main emission mechanism and the existence of Fabry–Perot modes. However, Klingshirn *et al.*[12] have suggested alternative explanations for the experimental evidence presented.

It is hoped that in the near future the fabrication of more devices with more detailed analysis will yield a greater understanding and a concerted effort toward reproducible, efficient and lasting ZnO-based lasers, though it is clear that this is a challenging goal.

10.5 Conclusions

Although ZnO epitaxial growth technologies are still relatively immature the best reported room temperature "figures of merit" for ZnO-based structures, 11 kW cm^{-2} thresholds, 300 cm^{-1} gain and 87 K temperature coefficients, compare favorably with those of ZnSe-

and GaN-based structures under similar conditions. In the context of emission characteristics there is no fundamental reason why ZnO-based devices should not provide a competitive alternative to GaN-based devices in some if not all circumstances. The main issues remain reproducibility of "laser quality" and p-type material as part of complete laser structures. There is still plenty of room for progress in this regard as the total research investment in ZnO laser technology is still a tiny fraction of the total investment put in to ZnSe and GaN technologies. The prevalence of the exciton–exciton gain mechanism at room temperature, the relatively low growth temperatures, the availability of ZnO substrates, excellent transport properties and the possibility of micro-laser self-formation are all factors that provide justification for continued work. However, the ZnMgCdO system does not seem as flexible in terms of band-engineering as the GaAlInN system and issues such as device lifetimes and stability have yet to be addressed.

In this chapter we have only considered the stimulated emission and lasing with relevance to traditional band-to-band short-wavelength devices. Two other types of device deserve brief consideration. As yet, no experimental or theoretical work has been published considering the potential of ZnO for use in infrared quantum cascade lasers. With current solubility limits the band offsets in the ZnO/ZnMgO system may not allow for devices at the communications wavelengths but with high electron densities and excellent transport characteristics ZnO ought to have high power applications in the 2–5 μm wavelength range. A second type of device has recently received theoretical attention. Zamfirescu *et al.*[60] have suggested ZnO as the semiconductor most suitable for the realization of room temperature "polariton lasers". The polariton laser is a device based on the strong coupling of light with excitons in a semiconductor microcavity. This revolutionary laser requires no population inversion to achieve optical amplification, instead, coherent light amplification is based on the mechanism of the Bose condensation of exciton polaritons. Zamfirescu *et al.* found record values for longitudinal transverse splitting of the exciton resonances in ZnO, values two orders of magnitude larger than those of GaAs. A ZnO-based microcavity structure is proposed that would have high gain and thresholds as low as $3\,kW\,cm^{-2}$ at room temperature.

References

[1] C. Klingshirn, *Semiconductor Optics*, Springer-Verlag, Heidelberg, 1997.
[2] J.M. Hvam, *Solid State Commun.* **12**, 95 (1973).
[3] J.M. Hvam, *Phys. Status Solid B* **63**, 511 (1974).
[4] C. Klingshirn, E. Ostertag and R. Levy, *Solid State Commun.* **15**, 883 (1974).
[5] C. Klingshirn, *Phys. Status Solid B* **71**, 547 (1975).
[6] J.M. Hvam, *Solid State Commun.* **26**, 987 (1978).
[7] S.W. Koch, H. Haug, G. Schmieder, W. Bohnert and C. Klingshirn, *Phys. Status Solid B* **89**, 431 (1978).
[8] J.M. Hvam, *Phys. Status Solid B* **93**, 581 (1979).
[9] J.M. Hvam, G. Blattner, M. Reuscher and K. Klingshirn, *Phys. Status Solid B* **118**, 179 (1983).
[10] C. Klinshirn, J. Fallert, O. Gogolin, M. Wissinger, R. Hauschild, M. Hauser, H. Kalt and H. Zhou, *J. Lumin.* **128**, 792 (2008).
[11] C. Klinshirn, R. Hauschild, J. Fallert and H. Kalt, *Phys. Rev. B* **75**, 115203 (2007).
[12] C. Klinshirn, J. Fallert, H. Zhou and H. Kalt, *Appl. Phys. Lett.* **91**, 126101 (2007).
[13] F.H. Nicoll, *Appl. Phys. Lett.* **9**, 13 (1966).

[14] J.R. Packard, D.A. Campbell and W.C. Tait, *J. Appl. Phys.* **38**, 5255 (1967).
[15] S. Iwai and S. Namba, *Appl. Phys. Lett.* **16**, 354 (1970).
[16] T. Goto and D.W. Langer, *J. Appl. Phys.* **42**, 5066 (1971).
[17] D.C. Reynolds, D.C. Lock, B. Jogai and H. Morkoc, *Solid Stat. Commun.* **101**, 643 (1997).
[18] P. Yu, Z. K. Tang, G. K. L. Wong, M. Kawasaki, A. Ohtomo, H. Koinuma and Y. Segawa, in *23rd International Conference on the Physics of Semiconductors*, edited by M. Scheffler and R. Zimmermann, World Scientific, Singapore, 1966, p. 1453.
[19] P. Yu, Z. K. Tang, G. K. L. Wong, M. Kawasaki, A. Ohtomo, H. Koinuma and Y. Segawa, *J. Cryst. Growth* **184–185**, 601 (1998).
[20] D.M. Bagnall, Y.F. Chen, Z.Q. Zhu, T. Yao, S. Koyama, M.Y. Shen and T. Goto, *Appl. Phys. Lett.* **70**, 2230 (1997).
[21] A. Ohtomo, M. Kawasaki, T. Koida, K. Masubuchi, H. Koinuma, Y. Sakurai, Y. Yoshida, T. Yasuda and Y. Segawa, *Appl. Phys. Lett.* **72**, 2466 (1998).
[22] Z. K. Tang, P. Yu, G. K. L. Wong, M. Kawasaki, A. Ohtomo, H. Koinuma and Y. Segawa, *Appl. Phys. Lett.* **72**, 3270 (1998).
[23] Y. Chen, D.M. Bagnall, Z. Zhu, T. Sekiuchi, K. Park, K. Hiraga, T. Yao, S. Koyama, M.Y. Shen and T. Goto, *J. Cryst. Growth* **181**, 165 (1997).
[24] Y. Chen, D.M. Bagnall, H. Koh, K. Park, K. Hiraga, Z. Zhu and T. Yao, *J. Appl. Phys.* **84**, 3912 (1998).
[25] D.M. Bagnall, Y.F. Chen, Z.Q. Zhu, T. Yao, M.Y. Shen and T. Goto, *Appl. Phys. Lett.* **73**, 1038 (1998).
[26] Z. K. Tang, P. Yu, G. K. L. Wong, M. Kawasaki, A. Ohtomo, H. Koinuma and Y. Segawa, *Solid State Commun.* **103**, 459 (1997).
[27] Y. Segawa, A. Ohtomo, M. Kawasaki, H. Koinuma, Z.K. Tang, P. Yu and G.K.L. Wong, *Phys. Status Solidi B* **202**, 669 (1997).
[28] M. Kawasaki, A. Ohtomo, I. Ohkubo, H. Koinuma, Z.K. Tang, P. Yu, G.K.L. Wong, B.P. Zhang and Y. Segawa, *Mater. Sci. Eng., B* **56**, 239 (1998).
[29] Z. K. Tang, P. Yu, G. K. L. Wong, M. Kawasaki, A. Ohtomo, H. Koinuma and Y. Segawa, *Nonlinear Optics* **18**, 355 (1997).
[30] Y. Chen, N.T. Tuan, Y. Segawa, H. Ko, S. Hong and T. Yao, *Appl. Phys. Lett.* **78**, 1469 (2001).
[31] D.M. Bagnall, Y.F. Chen, M.Y. Shen, Z.Q. Zhu, T. Goto and T. Yao, *J. Cryst. Growth* **184–185**, 605 (1998).
[32] K.L. Shaklee, R.E. Nahory and R.F. Leheny, *J. Lumin.* **7**, 284 (1973).
[33] A. Ohtomo, M. Kawasaki, T. Koida, K. Musubuchi, H. Koinuma, Y. Sakurai, Y. Yoshida, T. Yasuda and Y. Segawa, *Appl. Phys. Lett.* **72**, 2456 (1998).
[34] A. Ohtomo, M. Kawasaki, Y. Sakurai, I. Ohkubo, R. Shiroki, Y. Yoshida, T. Yasuda, Y. Segawa, and H. Koinuma, *Mater. Sci. Eng., B* **56**, 263 (1998).
[35] T. Makino, N.T. Tuan, H.D. Sun, C.H. Chia, Y. Segawa, M. Kawasaki, A. Ohtomo, K. Tamura, T. Suemoto, H. Akiyama, M. Baba, S. Saito, T. Tomita and H. Koinuma, *Appl. Phys. Lett.* **78**, 1979 (2001).
[36] A. Ohtomo, K. Tamura, M. Kawasaki, T. Makino, Y. Segawa, Z.K. Tang, G.K.L. Wong, Y. Matsumoto and H. Koinuma, *Appl. Phys. Lett.* **77**, 2204 (2000).
[37] H.D. Sun, T. Makino, Y. Segawa, M. Kawasaki, A. Ohtomo, K. Tamura and H. Koinuma, *J. Appl. Phys.* **91**, 1993 (2002).
[38] Y. Segawa, H.D. Sun, T. Makino, M. Kawasaki and H. Koinuma, *Phys. Status Solidi* **192**, 14 (2002).
[39] T. Makino, Y. Segawa, M. Kawasaki and H. Koinuma, *Semicond. Sci. Technol.* **20**, S78 (2005).
[40] H.D. Sun, T. Makino, N.T. Tuan, Y. Segawa, Z. K. Tang, G. K. L. Wong, M. Kawasaki, A. Ohtomo, K. Tamura and H. Koinuma, *Appl. Phys. Lett.* **77**, 4250 (2000).
[41] J.W. Sun, Y.M. Lu, Y.C. Liu, D.Z. Shen, Z.Z. Zhang, B.H. Li, J.Y. Zhang, B. Yao, D.X. Zhao and X.W. Fan, *J. Phys. D: Appl. Phys.* **40**, 6541 (2007).
[42] A. Ohtomo, K. Tamura, K. Saikusa, T. Takahashi, T. Makino, Y. Segawa, H. Koinuma and M. Kawasaki, *Appl. Phys. Lett.* **75**, 2635 (1999).
[43] T. Makino, N.T. Tuan, Y. Segawa, C.H. Chia, M. Kawasaki, A. Ohtomo, K. Tamura and H. Koinuma, *Appl. Phys. Lett.* **77**, *1632*, (2000).

[44] T. Makino, Y. Segawa, M. Kawasaki, A. Ohtomo, R. Shiroki, K. Tamura, T. Yasuda and H. Koinuma, *Appl. Phys. Lett.* **78**, 1237 (2001).

[45] S. Sadofev, S. Blumstengel, J. Cui, J. Puls, S. Rogaschwski, P. Schafer and F. Henneberger, *Appl. Phys. Lett.* **89**, 201907 (2006).

[46] S. Sadofev, S. Kalusniak, J. Puls, P. Schafer, S. Blumstengel and F. Henneberger, *Appl. Phys. Lett.* **91**, 231103 (2007).

[47] D. M. Bagnall, A. Ralston, K.P. O'Donnell, P.C. Smith, P. Wright and B. Cockayne, *Appl. Phys. Lett.* **66**, 1455 (1995).

[48] H. Cao, Y. G. Zhao, H. C. Ong, S. T. Ho, J. Y. Dai, J. Y. Wu and R. P. H. Chang, *Appl. Phys. Lett.* **73**, 3656 (1998).

[49] H. Cao, Y. G. Zhao, H. C. Ong and R. P. H. Chang, *Phys. Rev. B* **59**, 15107 (1999).

[50] H. Cao, Y. G. Zhao, X. Liu, E. W. Seelig and R. P. H. Chang, *Appl. Phys. Lett.* **75**, 1213 (2000).

[51] H. Cao, J. Y. Xu, S.-H. Chang, S. T. Ho, E. W. Seelig, X. Liu and R. P. H. Chang, *Phys. Rev. Lett.* **84**, 5584 (2000).

[52] A. L. Burin, M. A. Ratner, H. Cao and R.P.H. Chang, *Phys. Rev. Lett.* **87**, 5503 (2001).

[53] A. Mitra and R.K. Thareja, *J. Appl. Phys.* **89**, 2025 (2001).

[54] S. Cho, J. Ma, Y. Kim, Y. Sun, G.K.L. Wong and J. B. Ketterson, *Appl. Phys. Lett.* **75**, 2761 (1999).

[55] H.C. Ong, J.Y. Dai, A.S.K. Li, G.T. Du, R.P.H. Chang and S.T. Ho, *J. Appl. Phys.* **90**, 1663 (2001).

[56] H. Cao, J. Y. Xu, E.W. Seelig and R.P.H. Chang, *Appl. Phys. Lett.* **76**, 2997 (2000).

[57] M. Huang, S. Mao, H. Feick, H. Yan, Y. Wu, H. Kind, E. Weber, R. Russo and P. Yang, *Science* **292** 1897 (2001).

[58] M. Huang, Y. Wu, H. Feick, N. Tran, E. Weber and P. Yang, *Adv. Mater.* **13**, 113 (2001).

[59] Y.R. Ryu, J.A. Lubguban, T.S. Lee, H.W. White, T.S. Yeong, C.J. Youn and B.J. Kim, *Appl. Phys. Lett.* **90**, 131115 (2007).

[60] M. Zamfirescu, A. Kavokin, B. Gil, G. Malpuech and M. Kaliteevski, *Phys. Rev. B* **65**, 161205 (R) (2002).

11

ZnO-Based Ultraviolet Detectors

*Jian Zhong and Yicheng Lu**
Department of Electrical and Computer Engineering,
Rutgers University, Piscataway, NJ, USA
*(*Email address: ylu@rci.rutgers.edu)*

11.1 Introduction

The ultraviolet (UV) region covers the spectral wavelength of 400–10 nm. The UV radiation is commonly divided into three regions:[1,2]

- UV-A (400–320 nm);
- UV-B (320–280 nm);
- UV-C (280–10 nm).

Another widely used division of the UV wavelengths is:

- near-UV (400–300 nm);
- middle-UV (300–200 nm);
- far-UV (200–100 nm);
- extreme-UV (100–10 nm).

In UV semiconductor photodetectors, electron–hole pairs are generated when incident photons at UV wavelengths are absorbed in the semiconductor. The photogenerated electrons and holes are separated by the electric field due to the presence of built-in potential or external applied voltage. A photocurrent is then produced and collected proportional to the incident UV radiation power.

ZnO is a wide-band-gap semiconductor that is currently undergoing a renaissance because of its many promising properties. The earliest work on ZnO can be dated back to

Zinc Oxide Materials for Electronic and Optoelectronic Device Applications, First Edition.
Edited by Cole W. Litton, Donald C. Reynolds and Thomas C. Collins.

the 1930s, particularly optical characterizations in the 1950s and 1960s.[3–5] ZnO has the following advantages:[6] (i) a 60-meV exciton binding energy for more efficient optical emission and detection (much higher than the 25 meV of GaN and 4 meV of GaAs), which is 2.4 times the thermal energy at room temperature; (ii) a commercially available native substrate; (iii) radiation resistance (one order higher than that of GaN), which is particularly useful for detectors operating in hostile environments;[7] (iv) low cost for the basic materials; (v) wet-chemical processing is feasible, which eases the fabrication process and cuts the production cost; and (vi) high-quality epitaxial films and nanostructures can be grown at a low temperature ($\sim$400 °C), hundreds of degrees lower than GaN. A high saturation velocity has been predicted to be $\sim 3 \times 10^7$ cm s^{-1},[8] potentially leading to fast electronic and optoelectronic devices.

The UV photoreponses in ZnO films were first reported by Mollwo[9] and Miller[10] in the early 1950s. Since then, photoconduction from various types of ZnO materials has been extensively studied, including single crystal,[11,12] polycrystalline films,[13–18] epitaxial films[19–21] and nanostructures.[22–26] Recently, in the research and development of p-type doping,[27,28] ZnO has demonstrated potentially superior properties for optoelectronic devices, including light-emitting diodes (LEDs) and UV photodetectors.

By alloying ZnO with MgO to form the ternary compound $Mg_xZn_{1-x}O$, the energy band gap of ZnO can be increased from 3.3 to 7.8 eV.[29–31] Ryu *et al.* proposed $Be_xZn_{1-x}O$ as a good candidate for modulating the band gap to the deep UV region.[32] Since BeO has the same wurtzite structure of ZnO, the band gap of $Be_xZn_{1-x}O$ can be tuned from 3.37 to 10.6 eV without phase segregation. The quaternary alloy, $Zn_{1-x-y}Be_xMg_yO$, continuously tunes the band gap from 3.7 to 4.9 eV.[33] The band gap engineering for ZnO opens up for the broad application areas, such as solar blind and UV-sensitive space detection, *in situ* combustion monitoring, and biodetection.

Several good reviews on semiconductor UV photodetectors have been published,[1,2] which cover the basic theory of photoconductive and photovoltaic detectors built on other wide-band-gap semiconductors. Table 11.1 summarizes the basic material parameters of semiconductors used for UV detection.

ZnO UV photodetectors are categorized in the following device configurations:

1. photoconductive detector;
2. Schottky barrier photodetector;
3. metal–insulator–metal (MIS) structure;
4. p-n junction and p-i-n photodiode;
5. field-effect phototransistor, including thin-film transistor (TFT);
6. integrated surface acoustic wave (SAW) and photoconductive wireless UV detectors (UV-SAW);[35]

Among these, p-n photodiodes have the advantages of fast photoresponse speed and low dark current. However, because of the difficulties in obtaining high quality and reliable p-type materials, ZnO homojunction-based photodetectors show poor performances at the current stage. In this chapter, we will first review photoconductivity mechanism of ZnO for UV detection. Various types of ZnO-based UV photodetectors will be discussed, including photoconductive-type, Schottky-type, UV-SAW, and $Mg_xZn_{1-x}O$ photodetectors, etc. Finally, ZnO one-dimensional (1D) nanostructure-based UV detectors are reviewed and compared with the thin-film devices.

Table 11.1 *Basic material parameters of semiconductors used for UV detection*[2,34]

	Si	GaAs	GaP	3C-SiC	4H-SiC	6H-SiC	Diamond	GaN	AlN	ZnS	ZnSe	CdS	ZnO
E_g (eV)	1.12	1.43 direct	2.27	2.39	3.2	2.86	5.5	3.39 direct	6.2 direct	3.6 direct	2.82 direct	2.50 direct	3.37 direct
Thermal conductivity ($W\,cm^{-1}\,K^{-1}$)	1.5	0.5	1.1	3.2	4.9	4.9	20	1.3	3.19		(0.18)		
Melting point (*K*)	1683	1513	1740	2830	2830		3773				1100	1750	1975
Electron saturation velocity ($10^7\,cm\,s^{-1}$)	1	2	1.25		2	2	2.7	2.5	1.4				3
Mobility ($cm^2\,V^{-1}\,s^{-1}$)													
Electrons	1400	8500	350	1000	950	400	2200	1000	135	165		340	205
Holes	600	400	100	50	120	75	1600	30	14	5		340	180
Dielectric constant	11.8	12.5	11.1	9.7	9.7	9.7	5.5	8.9	8.5	9.6			9.1
Breakdown field ($10^5\,V\,cm^{-1}$)	3	6	10	20	20	24	100	26	20				

11.2 Photoconductivity in ZnO

It has been found that the photoresponse of ZnO consists of two components: a fast response and a slow response. The fast response results from the reversible solid-state process, such as intrinsic interband or excitonic transition, while the slow one is governed by the surface-related oxygen adsorption and photodesorption process, or by the bulk defect related recombination process.[11,15,36–39]

In polycrystalline ZnO films, the slow photoreponse component is found to be the dominant process.[15,38,40] The slow process in the photoresponse of polycrystalline ZnO has been mainly attributed to a two-step process, in which free carriers interact with surface states and grain boundaries:

i. Under the dark condition, oxygen adsorbs at the surface states (e.g. oxygen deficiency sites) to create a negatively charged ion by capturing a free electron from n-type ZnO:

$$O_2(g) + e^- \rightarrow O_2^-(ad) \tag{11.1}$$

Thus, a depletion region and band bending are created near the surface, resulting in a decrease of the film conductivity.

ii. Under UV irradiation, forced by the electrical field in the surface depletion region, the photogenerated holes move toward the surface to neutralize the negative oxygen ion. This process is called the photodesorption of O_2:

$$h^+ + O_2^-(ad) \rightarrow O_2(g) \tag{11.2}$$

When the free carriers generated by optical absorption are spatially separated, the film conductivity is increased.

This model has been supported by photoconductivity decay measurements of ZnO, where the decay is slower in vacuum but faster in oxygen.[15,36,37,41] A hole-capture model in polycrystalline ZnO films was proposed by using photo-Hall and transient photoconductivity measurements.[36–38] The transient photoconductivity study shows that the hole-capture process is of nonthermal nature; the excitation rate of photoconductivity does not depend on temperature but depends on ambient oxygen partial pressure. The photoconductivity relaxation is attributed to electron tunneling to the surface oxygen states. The slow photoconductivity process associated with the surface trap states has been also observed in the epitaxial ZnO films.

In general, the fast photoresponse process in ZnO is related to the solid-state process such as intrinsic interband photoconductivity, which is a general photoconduction process in semiconductor materials. When the energy of incident photons is higher than the energy band gap, electrons are excited from the valence band to the conduction band. Free electrons and holes are generated:

$$h\nu \rightarrow h^+ + e^- \tag{11.3}$$

Thus, the film conductivity is increased. This process is characterized by the cut-off wavelength of the photoresponse corresponding to the energy band gap.

Other photoconduction mechanisms, including exciton photoconductivity and phonon-assisted photoconductivity, have been also connected with the fast photoresponse process in ZnO. An exciton is a hydrogen-like two-particle system composed of an electron and a hole formed by coulomb interaction. Since the exciton is typically quasi-charge-neutral, it does not directly contribute to the photoconductivity. However, exciton photoconductivity has been observed in II–VI semiconductors including ZnO,[42,43] which may be due to ionization of the excitons. The mechanisms of the exciton ionization could be: (i) ionization under the electrical field; (ii) exciton–phonon or exciton–impurity interaction; and (iii) the Auger process. A phonon-assisted photoconductivity is caused by interaction between photogenerated carriers and phonons. Oscillations in photoconductivity of single crystal ZnO have been observed in the infrared (IR) region. By studying the oscillation energies in the surface photoconductivity spectra, Lüth attributed the oscillations to the bulk longitudinal optical (LO) phonons and surface phonons.[44]

In order to make a high speed UV photodetector, it is critical to use the high quality ZnO films in which the slow photoresponse process is suppressed by reducing the trap density and background carrier concentration, and the fast process is dominant.

Effects of ambient gas. Zinc oxide shows a strong chemisorption behavior at the surface, which has been widely used for various chemical/gas sensing. The conductivity and photoconductivity of ZnO films are strongly dependent on ambient gas conditions.[15,45–47] An oxidizing atmosphere, such as CO_2 and water vapor, effectively accelerates the decay of photoconductivity in ZnO films.[15] It is suggested that oxygen in water vapor may diffuse into the interior part of ZnO films from the surface through pores inside. The oxygen adsorption modifies the electrical conductivity and photoresponse time of the ZnO films.

Effects of oxygen plasma treatment and surface passivation. Enhancement of UV photoresponse properties of a ZnO photoconductor is obtained by use of an oxygen plasma treatment. Liu and Kim reported that an oxygen plasma treatment dramatically improved the UV detection properties of ZnO metal–semiconductor–metal (MSM) photoconductors grown by radio frequency (RF)-magnetron sputtering,[48] through reducing the decay time constant and increasing the on/off ratio of photocurrent. The as-grown ultrathin (~20 nm) ZnO film was exposed to the oxygen plasma for 30 min using an inductively coupled plasma reactive ion etching (ICP-RIE) system with ICP power of 125 W and RIE power of 5 W, oxygen pressure of 1.5×10^{-2} Torr and a flow rate of 30 sccm. Figure 11.1 shows the current–voltage (*I–V*) characteristics of (a) an as-grown ZnO MSM photoconductor and (b) an oxygen-plasma-treated ZnO MSM photoconductor. The dark leakage current of ZnO photoconductor is reduced from 5 mA to 400 pA at 3 V after the oxygen plasma treatment. The normalized photocurrent transient is measured in air and in vacuum (10^{-6} Torr) ambient and compared for the as-grown and oxygen-plasma-treated ZnO MSM photoconductor, respectively, shown in Figure 11.2 (a) and (b).

The as-grown ZnO photoconductor exhibits a slow photocurrent decay profile with a time constant in an order of tens of minutes, which is strongly dependent on the oxygen ambient pressure [Figure 11.2(a)]. The slow photocurrent decay process indicates that a slow charge-transfer process due to the oxygen chemisorption occurs, which is slower in vacuum than in air. After the oxygen plasma treatment, the ZnO photoconductor becomes insensitive to the ambient with a dramatically reduced decay time constant (below 10 ms)

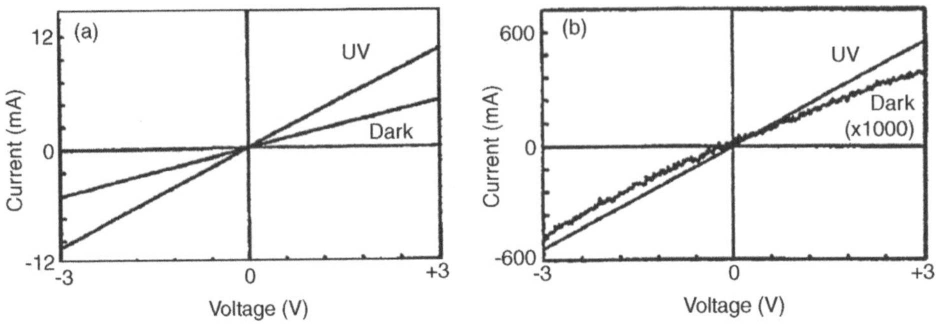

Figure 11.1 *I–V characteristics of ZnO photoconductors with (a) as-grown ZnO film and (b) oxygen-plasma-treated ZnO film. Reprinted from M. Liu and H. K. Kim, Ultraviolet detection with ultrathin ZnO epitaxial films treated with oxygen plasma, Appl. Phys. Lett. 84, 173. Copyright (2004) with permission from American Institute of Physics*

[Figure 11.2(b)]. It is suggested that the plasma treatment suppresses the chemisorption sites (primarily the oxygen deficiency sites) on the ZnO surface and the oxygen vacancies in bulk ZnO, leading to reduction of the chemisorption of oxygen through the surface and decrease of free carrier concentration in the bulk for the dark current. In the plasma treatment process, highly diffusive atomic oxygen may effectively fill the chemisorption sites on the ZnO surface, while positive oxygen ions in the plasma may neutralize the negatively charged oxygen molecular ions adsorbed on the ZnO surface.

Surface passivation effectively suppresses the slow photoresponse decay in the ZnO photoconductor. One of the surface passivation schemes is to deposit a SiO_2 layer after wet chemical etching of bare ZnO surface in 30% H_2O_2 for 15 min.[49] However, the quantum efficiency is also reduced.

Effects of trap levels. The slow UV photoresponse process of ZnO photodetector is controlled by the trap levels on the surface and in the bulk ZnO. Sharma *et al.* reported a very slow decay and rise time of photocurrent (~800 ms) in polycrystalline ZnO films

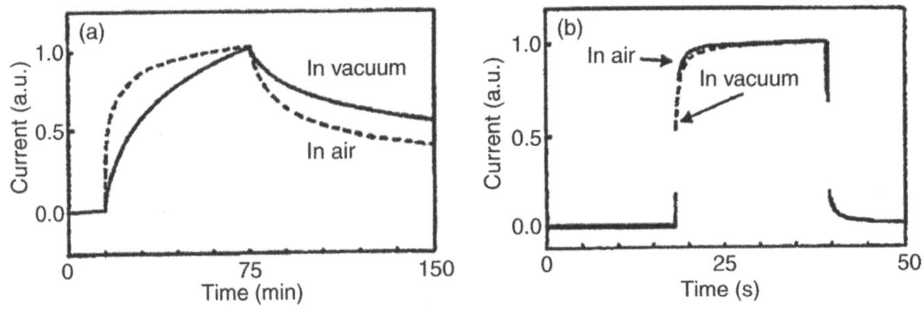

Figure 11.2 *Photocurrent transients of ZnO photoconductors with (a) as-grown ZnO film measured in air (dashed curve) and in vacuum (10^{-6} Torr) (solid curve) ambient and (b) oxygen-plasma-treated ZnO film measured in air (solid curve) and in vacuum (10^{-6} Torr) (dashed curve) ambient. Reprinted from M. Liu and H. K. Kim, Ultraviolet detection with ultrathin ZnO epitaxial films treated with oxygen plasma, Appl. Phys. Lett. 84, 173. Copyright (2004) with permission from American Institute of Physics*

grown by unbalanced RF-magnetron sputtering. They attributed the observed slow time response to the trap levels in the band gap.[50]

Oh *et al.* found the trap centers in ZnO layers with different Zn/O flux ratios by admittance spectroscopy.[51] The ZnO layers exhibited three electron-trap centers ET1 (33 meV), ET2 (0.14 eV) and ET3 (65 meV), near the conduction-band edge, where ET1 and ET2 are ascribed to singly and doubly ionized Zn interstitials, respectively. The trap densities of ET1, ET2, and ET3 were estimated to be $3 \times 10^{15}\,\text{cm}^{-3}$, $2 \times 10^{16}\,\text{cm}^{-3}$ and $3 \times 10^{15}\,\text{cm}^{-3}$, respectively. Since the ZnO layers used were grown under Zn-rich growth conditions, it is suggested that the large photoresponse time observed in ZnO layers can be ascribed to Zn-interstitial type defects.[52]

Epitaxial ZnO thin films grown by pulsed laser deposition (PLD) were investigated by deep level transient spectroscopy (DLTS) and by Fourier transform infrared photocurrent (FTIR-PC) spectroscopy.[53] The IR photocurrent of undoped ZnO films shows transitions between 100 meV and 500 meV from deep defect states to the band edges, including intrinsic deep defects E1 at ~110 meV and E3 at ~320 meV. Cordaro and Shim reported trap levels T1 at ~170 meV and T2 at ~330 meV below the conduction band edges, and related them to the second ionization of zinc interstitial and the oxygen vacancy, respectively.[54] There are reports of trap levels L1 (close to T1) and L2 at ~170 and ~260 meV and E1, E3 (close to T2) and E4 at ~120, ~290 and ~570 meV, respectively.[55,56] However, the chemical origins of many trap states related to the photoconduction process in ZnO are still unclear.

Doping effects. Doping and co-doping with multiple elements (N, Te, P, Cu, etc.) are used to improve the photoresponse of n-type ZnO films.[19,57–59] Photosensitivity is defined as a ratio of the photocurrent to the dark current:

$$\text{Photosensitivity} = \frac{I_{\text{photo}}}{I_{\text{dark}}} \approx 1 + \frac{\Delta n}{n_0} \tag{11.4}$$

where n_0 is the background carrier concentration and Δn is the photogenerated excess carrier density. When the equilibrium carrier concentration is suppressed, the photosensitivity is increased. The approach to use the doping and co-doping methods is to reduce the background carrier concentration of n-type ZnO films by incorporating compensation dopants of group V elements.

Ultraviolet photoresponse of ZnO film deposited by PLD was enhanced by phosphorous doping and post-annealing.[57] Phosphorous introduced a deep level in the energy band gap, heavily compensating the defect-related shallow donors in the intrinsically n-type ZnO films. On the other hand, the incorporated acceptor atoms such as N and P may compensate the oxygen vacancies at the ZnO surface, reducing the photoresponse time.

Porter *et al.* proposed to co-dope ZnO films with nitrogen and tellurium during PLD to improve the photoconductivity.[58] The isoelectronic impurity Te is chosen as a co-dopant due to the anion-rich nature of ZnTe, in contrast to the cation-rich nature of ZnO. After N and Te co-doping, the sample's dark resistivity increases from 10^{-2}–10^{-1} to 10^{3}–$10^{5}\,\Omega\,\text{cm}$. A peak photoconductivity was achieved at a Te concentration of $10^{20}\,\text{cm}^{-3}$, presumably limited by the solid solubility of Te in ZnO.

Effects of surface polarity. Surface polarity is known to strongly affect the photoconductivity of ZnO. In this study, ZnO single crystal samples were prepared by a hydrothermal method.[60] After chemical–mechanical polishing, the ZnO substrates were annealed at 850 °C for 2 h. One side of the c-plane ZnO substrates has an O-face, and the

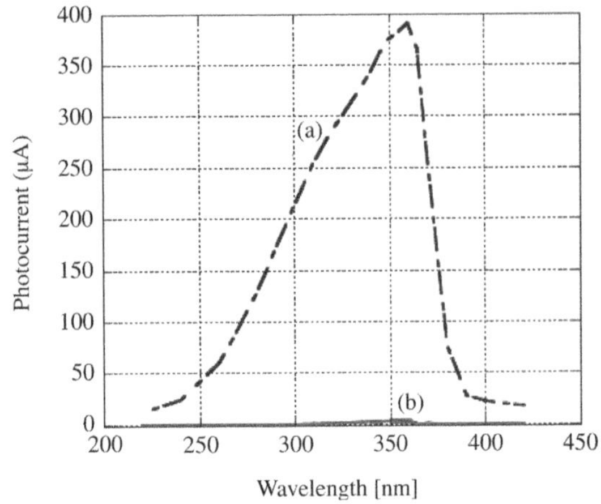

Figure 11.3 *Spectral response of the photocurrent of c-plane ZnO single crystals: (a) O-face; and (b) Zn-face. Reprinted from F. Masuoka, et al. Phys. Stat. Sol. (c) 3, 1238 (2006). Copyright Wiley-VCH Verlag GmbH & Co. KGaA. Reproduced with permission*

other side has a Zn-face. The spectral responses of photocurrent for different faces of the c-plane ZnO single crystals were measured under UV irradiation of $\sim 10\,\mathrm{mW\,cm^{-2}}$. Figure 11.3 shows the spectral response of the O-face and the Zn-face of a c-plane ZnO single crystal. Photocurrent of the O-face is about two orders of magnitude larger than that of the Zn-face. It is also noted that the photoresponse of the O-face ZnO in the shorter wavelength of 250–200 nm is stronger than that of the Zn-face ZnO.

The different behavior for the photoresponsivity of O- and Zn-face ZnO single crystals is explained by an energy band diagram model illustrated in Figure 11.4. On the O-face, the surface-terminated oxygen atoms act as the acceptor-like surface states. They are deep

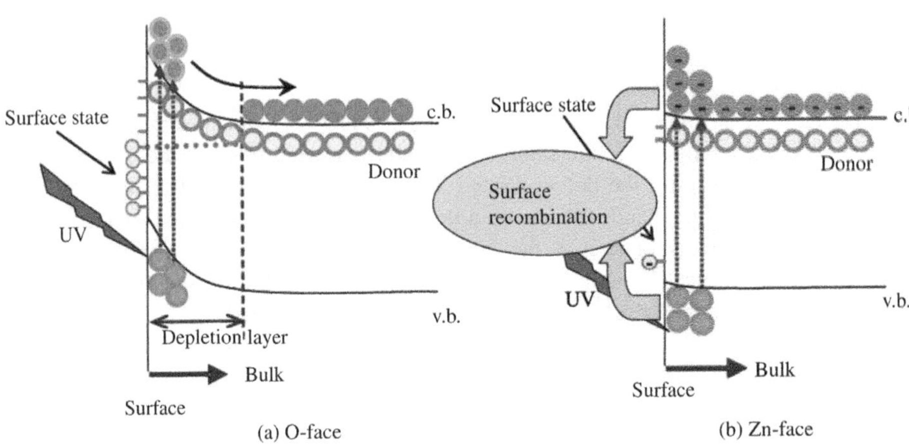

Figure 11.4 *Energy band diagram model for (a) O-face and (b) Zn-face of a c-plane ZnO single crystal. Reprinted from F. Masuoka, et al. Phys. Stat. Sol. (c) 3, 1238 (2006). Copyright Wiley-VCH Verlag GmbH & Co. KGaA. Reproduced with permission*

levels in the energy gap. The depletion layer is extended from the surface into the bulk due to generation of electron trapping in these surface states. The extended built-in electric field effectively separates electrons from holes, leading to increased photoexcited carrier lifetime for a higher photoresponsivity. On the other hand, on the Zn-face, the built-in electric field only exists near the surface; therefore the surface recombination is fast. Particularly, at shorter wavelength than the absorption band edge, the surface absorption and recombination dominates for the Zn-face ZnO.

11.2.1 Persistent Photoconductivity

Persistent photoconductivity (PPC) has been observed in various types of ZnO, including polycrystalline,[38,40] epilayer,[61,62] and nanowires (NWs).[63,64] PPC is closely related to the slow photoconduction process. It has been reported that polycrystalline ZnO samples show a very slow relaxation (over a time interval of hours or days) of conductivity after illumination.[36–38,40] PPC is associated with the defects in semiconductor materials, such as neutral donor (DX) centers in III–V alloys. Under the light illumination, these DX centers can transfer between a nonconducting defect localized state (DLS) in the band gap and a conducting DLS resonant with the conduction band, which is a delocalized hydrogenic-like state,[65] giving rise to PPC in the semiconductors.

A slow photoresponse component dominates the photoresponse of ZnO photodetectors when PPC is present. In a PPC process, photogenerated holes are usually captured by the hole traps, resulting in a reduced electron–hole recombination rate and an increased lifetime of nonequilibrium electrons. Therefore, a large photoconductive gain is expected to accompany PPC. The presence of PPC dramatically affects the performance of ZnO-based UV photodetectors by reducing the UV/visible rejection ratio and decreasing the photoresponse speed.

The influence of grain boundaries on photoconductivity of polycrystalline ZnO was studied by Zhang.[40] The photoresponse is related to the changes of barrier heights of grain boundaries when the films are irradiated with UV light. It is suggested that the hole traps contributing to the persistent photoconductivity are located at grain boundaries. Several defect related photoconduction mechanisms have been proposed. Tomm *et al.* observed that room temperature photocurrent peak is significantly below the room temperature band edge in the ZnO films grown by laser-assisted molecular beam epitaxy (MBE).[42] It is suggested that the photoconduction process is defect related. However, the nature of the defects is unclear.

Zhang *et al.* proposed a two-hole capture model through the oxygen vacancy (V_O) to explain the slow photoconductivity decay.[66] The energy levels of the intrinsic defects in ZnO were calculated by the local density approximation (LDA) method. They found that the +2 charge ionized oxygen vacancy had a lower energy than a charge neutral V_O due to a large lattice distortion formed after capturing two holes. Therefore, a charge neutral oxygen vacancy can capture two holes. The reduction of delocalized holes causes a decrease of electron–hole recombination rate, which is responsible for the fast photoconduction decay. A fast phonon-assistant conductivity relaxation is suppressed because of the deep energy level from the hole traps and large lattice distortion. As a result, a slow process dominates.

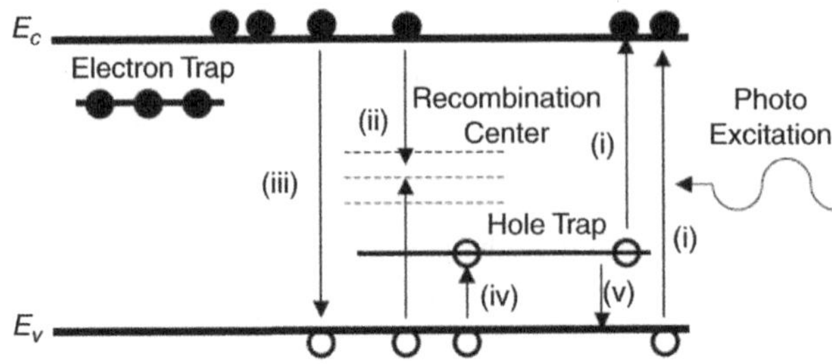

Figure 11.5 *Schematic illustration of a proposed model for persistent photoconductivity in ZnO involving hole traps and defect levels: (i) photogeneration of electron–hole pair; (ii) recombination through recombination centers in the mid gap; (iii) band-to-band recombination; (iv) hole capture process; and (v) hole emission process through hole traps. Adapted from K. Moazzami, et al., Sub-bandgap photoconductivity in ZnO epilayers and extraction of trap density spectra, Semicond. Sci. Technol, 21, 717 (2006)*

Moazzami *et al.* studied the sub-band-gap photoconductivity in n-type ZnO films grown by plasma-assisted MBE on c-plane sapphire.[49] The carrier trapping dynamics were investigated by time-resolved photoluminescence (PL) and photoconductive transient under optical excitation wavelengths of 400–700 nm. Figure 11.5 illustrates a proposed model for PPC in ZnO involving hole traps and defect levels.

Under an optical excitation, electron–hole pairs are generated, either between the defect states and the band edge, or between the conduction band (CB) and the valence band (VB). The PPC could be attributed to the capture of holes in the hole trap states, followed by hole emission and carrier recombination. A deep hole trap may act as a negative charge when it is filled with a hole, similar to the hole trap observed in GaN. The excess hole captured by the hole traps are thermally emitted into the VB, and recombine with the electrons in the CB. A rate equation model describes the persistent photoconductive process in n-type ZnO:[49]

$$\frac{\mathrm{d}n}{\mathrm{d}t} = \phi\sigma(N_\mathrm{T}-N)-\frac{p}{\tau_\mathrm{rec}} \tag{11.5}$$

$$\frac{\mathrm{d}p}{\mathrm{d}t} = Nse^{-\frac{E_\mathrm{p}}{kT}}-\frac{p}{\tau_\mathrm{cap}}-\frac{p}{\tau_\mathrm{rec}} \tag{11.6}$$

$$\frac{\mathrm{d}N}{\mathrm{d}t} = \phi\sigma(N_\mathrm{T}-N)-Nse^{-\frac{E_\mathrm{p}}{kT}}+\frac{p}{\tau_\mathrm{cap}} \tag{11.7}$$

where n and p are the electron and hole density in the CB and VB, respectively, N_T is the total density of hole traps, N is the number of filled hole traps, ϕ is the incident photon flux, σ is the photon capture cross-section of the hole traps, τ_rec is the band edge carrier recombination lifetime, τ_cap is the hole capture lifetime, E_p is the energy separation between the hole trap state and VB and s is the hole escape frequency.

In order to derive the trap density spectrum, assuming the hole capture lifetime is much larger than the band edge carrier recombination lifetime, the relationship between hole traps and conduction band electrons based on the rate equations is derived from Equations (11.5)–(11.7):

$$\int N_{\mathrm{T}}(E_i)\exp(-se^{-\frac{E_i}{kT}}t)\mathrm{d}E = \Delta n(t) \tag{11.8}$$

where i represents the ith hole trap state.

Lany and Zunger proposed that oxygen vacancies act as a source of PPC in n-type ZnO through the first-principle electronic structure calculations.[67] They suggested that the neutral oxygen vacancy V_{O}^{0} is a localized ground state in the energy band gap; however, the ionized 2+ excited state V_{O}^{2+} becomes an excited state resonant with the conduction band. In this model, the light induced charge state transition follows:

$$V_{\mathrm{O}}^{0} \rightarrow V_{\mathrm{O}}^{2+} + 2e \tag{11.9}$$

which leads to the configuration change for the oxygen vacancy from the nonconducting ground to the conducting excited state. To return to the ground state, the metastable state needs to be thermally activated across an energy barrier while an electron must be captured by the DLS from the CB, as a result, n-type PPC occurs.

Reemts and Kittel suggested that PPC in highly porous ZnO films could be attributed to a lattice relaxation process of surface states, when electrons are photoexcited into the surface states located in the energy band gap of ZnO.[68] A stretched exponential law was used to fit the photoresponse decay.

11.2.2 Negative Photoconductivity

Negative photoconductivity has been observed in ZnO NW assemblies [such as composites of ZnO-AAM (anodic alumina membrane) and ZnO-CNT (carbon nanotubes)].[69,70] The potential applications include optical logic circuits and high-resolution image sensors for integrated nanoscale optoelectronics.

Sen *et al.* studied the negative photoconduction in a planar two-dimensional network consisting of ZnO particles (with diameters of 0.1–100 μm) and single-walled carbon nanotubes (ZnO-CNT).[69] Under UV light illumination (340 nm, ~10 mW cm^{-2}), the conductivity of ZnO-CNT decreased by 3% compared with the dark value. An intensity-dependent photocurrent was obtained on ZnO-CNT devices and fitted to a double exponential decay with characteristic time constants of ~450 and 75 s, respectively. The negative photoconduction is attributed to the interface-mediated charging/discharging effect involving oxygen chemisorption/desorption on the surface of ZnO particles that form a heterojunction with CNTs. The exponential photocurrent decay is explained by the surface charge stated modified band energy diagram at the ZnO/CNT interface, approximately follows:

$$\begin{aligned} &\Delta n_{\mathrm{s}} \sim -I \\ &p \sim e^{E_{\mathrm{V}}-E_{\mathrm{F}}/kT} \\ &E_{\mathrm{V}} \sim -I \end{aligned} \tag{11.10}$$

where Δn_s is the surface charge density, I is the light intensity, E_V is the position of the valence band, and p is the effective hole concentration.

Transition from negative photoconductivity to positive photoconductivity in ZnO NWs was realized through thermally annealing AAM as the NW growth template.[70] High density vertical ZnO NW arrays were grown in AAM by laser-assisted chemical vapor deposition (CVD) [Figure 11.6(a)]. Before the CVD growth, Sn catalyst was deposited by electrochemical deposition into the bottom of the AAM channels. As shown in Figure 11.6(b), negative photoconductivity is observed in the ZnO NW. It is expected that the anion impurities introduce electron trapping in the inner alumina layer. Since ZnO NW and the inner AAM channel forms a core–shell heterostructure, electrons can tunnel into these trapping states above the conduction band upon photoexcitation, lowering the conduction electron concentration in the NW. When an AAM was thermally annealed at 500 °C for 13 h in pure argon gas, positive photoconductivity is observed [Figure 11.6(c)]. This is attributed to thermal decomposition of the acid anion and thus reduction of the impurity levels in the alumina.

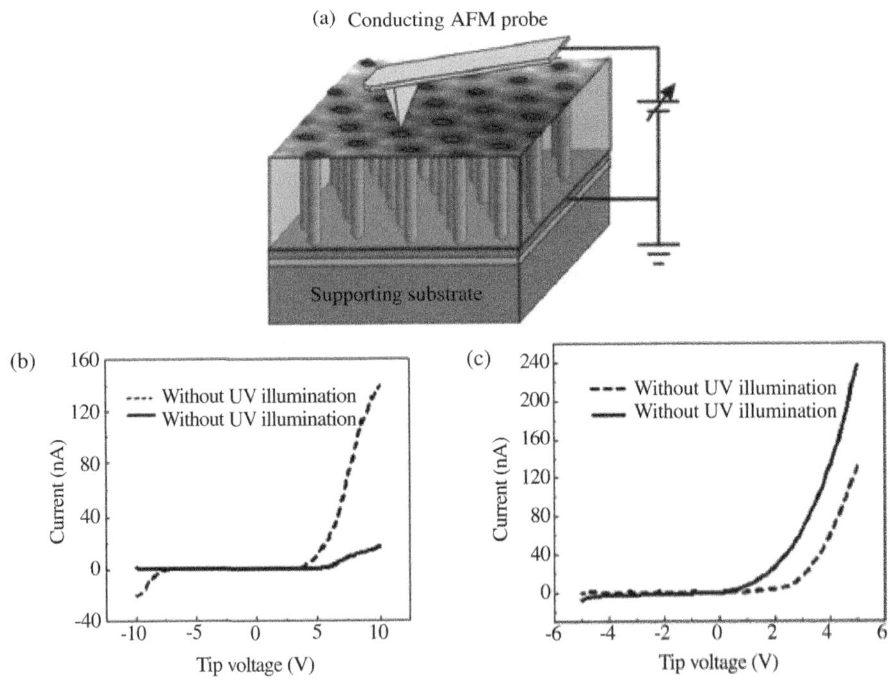

Figure 11.6 (a) Schematic illustration of an atomic force microscopy (AFM) probe to measure individual ZnO NWs in a vertical NW array. (b) I–V curves measured from a single ZnO NW in an unannealed AAM, showing negative photoconductivity. (c) I–V curves measured from a single ZnO NW in an annealed AAM, showing positive photoconductivity. Reprinted from Z. Fan, et al., Electrical and photoconductive properties of vertical ZnO nanowires in high density arrays, Appl. Phys. Lett. 89, 213110. Copyright (2006) with permission from American Institute of Physics

11.3 ZnO Film-Based UV Photodetectors

11.3.1 Photoconductive UV Detector

The photoconductive photodetector is the simplest device, acting as a photo-sensitive resistor. When an incident photon has energy higher than the band gap energy of ZnO, an electron–hole pair is produced. The electron and hole are separated under the electric field. The photogenerated current is collected by the electrodes of the samples under the proper bias. The photoconductive UV detector exhibits an internal photoelectric gain. However, the photoconductive type of detector has the limitation of the gain-bandwidth. The ZnO film thickness should be optimized so that the ZnO layer can absorb sufficient incident photons without producing a large noise current due to the ZnO intrinsic conductivity. The circular or interdigital (IDT) metal pattern is deposited on top of the ZnO surface for the electrode. Figure 11.7 shows the schematic diagram of a ZnO IDT MSM photoconductive photodetector on a sapphire substrate; Figure 11.7(a) is the top view and (b) the cross-sectional view.

Photoconductivity in polycrystalline ZnO films has been reported extensively.[14,16–18,36] Polycrystalline ZnO films were deposited by RF sputtering,[17,40,71] ion-beam sputtering,[16] and solution-based synthesis,[14,36] etc. While exhibiting large photoresponse, the photodetector made of polycrystalline films suffers from slow speed and large noise due to the dark conductivity.

Liu *et al.* reported the first epitaxial a-plane ZnO film-based UV photoconductivity detector.[19] In comparison with its polycrystalline counterpart, the epitaxial ZnO-based detector significantly improves the photoresponse time. The photodetector is made of a 1 μm thick a-plane ZnO epitaxial layer grown on r-plane sapphire using metal organic chemical vapor deposition (MOCVD). Diethylzinc (DEZn) and oxygen were used as the Zn metal organic source and oxidizer, and NH_3 was used as an *in situ* dopant for MOCVD growth. The typical growth conditions were the following: chamber pressure of 50 Torr, 10–15 lpm of N_2 (carrier gas), 1000 sccm of O_2, DEZn bubbler temperature of 10 °C, 100 sccm Ar flow through DEZn bubbler, growth temperature in the range of 350–600 °C. Growth rates of 1–2 $\mu m\, h^{-1}$ were obtained. Unintentionally doped ZnO generally exhibits n-type behavior because of the presence of various defects such as oxygen vacancies, zinc interstitials and impurities. In order to reduce the background electron concentration, the compensation doping process, such as *in situ* nitrogen doping, was carried out by introducing NH_3 during MOCVD growth of ZnO.

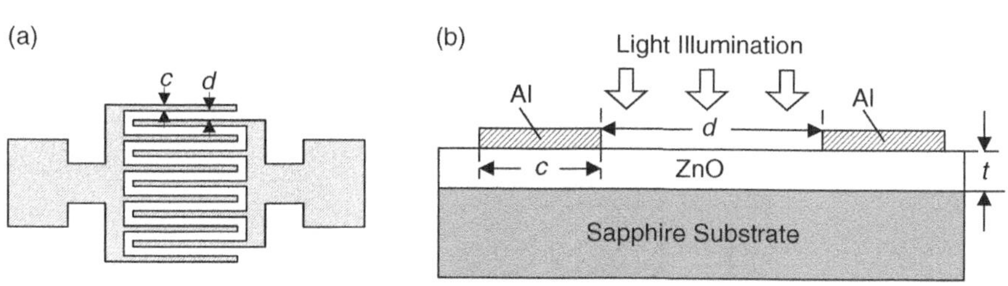

Figure 11.7 *Schematic drawing of a ZnO IDT MSM photoconductive photodetector on a sapphire substrate. (a) Top view and (b) cross-sectional view, where c and d are the metal electrode finger width and interelectrode spacing, respectively, and t is the thickness of the ZnO film*

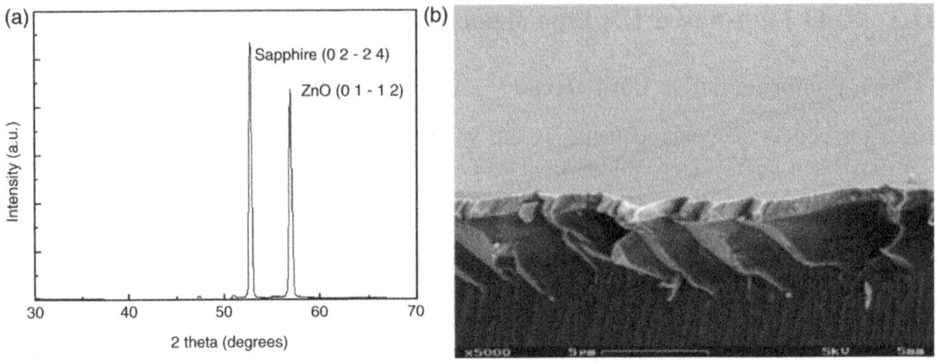

__Figure 11.8__ (a) X-ray θ–2θ scan of ZnO on r-plane sapphire and (b) SEM image of the surface morphology of ZnO films grown in the presence of NH_3. Reprinted from Y. Liu, et al., J. Electron. Mater, 29, 69. Copyright (2000) with permission from TMS

The ZnO films deposited on $(01\bar{1}2)$ Al_2O_3 with their $(11\bar{2}0)$ a-plane parallel to the surface, as confirmed by the X-ray diffraction (XRD) θ–2θ scans [Figure 11.8(a)]. The in-plane orientation of the films was analyzed by X-ray ϕ-scans, and the epitaxial relationship was determined to be $(11\bar{2}0)$ ZnO//$(01\bar{1}2)$ Al_2O_3 and [0001] ZnO//$[0\,\bar{1}\,1\,1]$ Al_2O_3.[72] The lattice mismatch parallel to and perpendicular to the [0001] direction of ZnO is 1.53 and 18.3%, respectively. The a-plane ZnO films grown on r-sapphire are dense with a smooth surface, as seen in the SEM image shown in Figure 11.8(b). A cross-sectional high resolution transmission electron microscopy (HR-TEM) lattice image of the interface between ZnO and r-sapphire is observed to be atomically sharp and semicoherent, as shown in Figure 11.9.

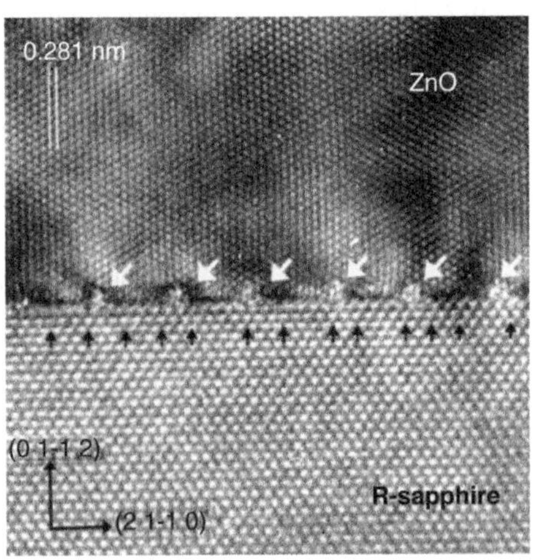

__Figure 11.9__ HR-TEM image of the interface between ZnO and r-sapphire. The view is along the c-axis of ZnO. Highly strained regions with a repeat distance twice that of the misfit dislocations are observed. Reprinted from C. R. Gorla, et al., J. Appl. Phys. 85, 2595. Copyright (1999) with permission from American Institute of Physics

UV photodetectors were fabricated based on MSM circular structures. The device material is made up of 1-μm-thick N-doped ZnO grown on r-sapphire. Al (200 nm thick) was used as the contact metal. A circular pattern was used for contact electrodes. The spacing of the two electrodes varied from 2 to 16 μm. e-beam evaporation, photolithography and wet chemical etching techniques were used to generate the metal electrode patterns. The spectral photoresponse of the MSM photodetectors was measured using a 75 W Xe lamp UV light source, an Oriel monochromator for monochromic light power output, and a chopper to generate the light pulse with a frequency ranging from 1 to 1000 Hz. A Newport 1830-C optical power meter was employed to calibrate the light beam intensity. *I–V* characteristics were measured using a Tektronics Semiconductor Work-Bench 372.

The measured dark and photoilluminated *I–V* characteristics are shown in Figure 11.10 (a). The dark current for a bias voltage of 5 V was only 450 nA. The reason for relatively low dark current is that the resistivity of the ZnO film was high. From the *I–V* measurement data, we calculated the resistivity of the ZnO film to be $3.1 \times 10^5\ \Omega$ cm. The wavelength of 365 nm was used for the photoilluminated *I–V* measurements. The light power was 6.4 nW. The linear *I–V* relationships under both forward and reverse bias exhibit ohmic metal—semiconductor contacts. The detector operates in the photoconductive mode.

The gain of the photoconductor can be expressed as the following:

$$G_{ph} = \frac{\tau_p}{t_{tr}}\left(1 + \frac{\mu_p}{\mu_n}\right) \tag{11.11}$$

where τ_p is the effective recombination time for excess minority carriers, t_{tr} is the transit time of electrons in the device, μ_n and μ_p are the electron and hole mobilities, respectively. τ_p can be much larger than t_{tr} if the device is properly designed.

For the case of constant mobility when the bias V_b is small, the photoconductor gain $G_{ph} \propto V_b$, the electron transit time can be expressed as:

$$t_{tr} \propto \frac{d}{\mu_n E} \propto \frac{d^2}{\mu_n V_b} \tag{11.12}$$

where E is the applied electric field at low bias voltage V_b.

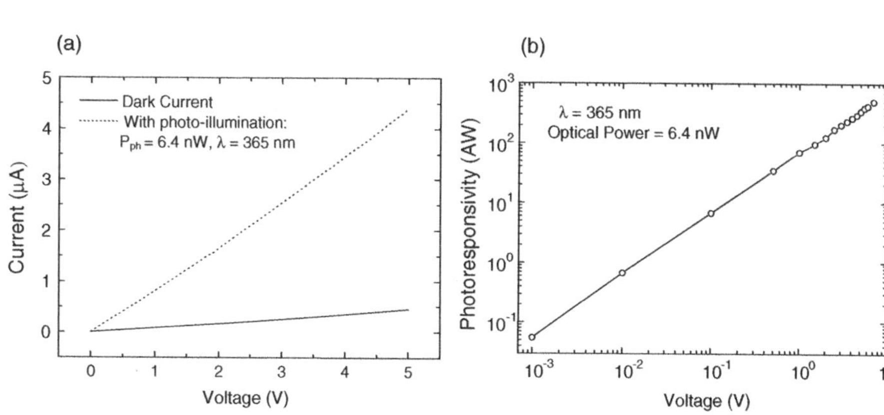

Figure 11.10 *(a) Dark (solid line) and photoilluminated (dashed line) I–V characteristics from a ZnO MSM photoconductive photodetector and (b) photoresponsivity vs bias voltage of a ZnO MSM photoconductive photodetector. Reprinted from Y. Liu, et al., J. Electron. Mater, 29, 69 (2000)*

As shown in Figure 11.10(b), the photoconductors show a linear photoresponse vs bias voltage from 1 mV to 5 V. Such a linear relationship indicates that the device operates in the constant mobility range. For $d = 10\,\mu m$, $\lambda = 365\,nm$, G_{ph}/V_b is approximately $120\,V^{-1}$. Assuming $\mu_n \sim 100\,cm^2\,V^{-1}\,s^{-1} \gg \mu_p$, the minority carrier lifetime (τ_p) can be estimated as:

$$\tau_p \approx \frac{G_{ph} d^2}{\mu_n V_b} \approx 1.2 \times 10^{-6}\,s \tag{11.13}$$

Figure 11.11 shows the spectral response of a MSM photoconductive detector. A sharp cut-off near 373 nm was observed. The photoresponse drops by more than two orders of magnitude across the cut-off wavelength within 15 nm of the band edge. These results are consistent with the transmission measurements. There is a long absorption tail below the band gap. The impurity levels within the band gap might be the source of the tail. Photoresponsivity of about $400\,A\,W^{-1}$ is obtained under 5 V bias.

We have also measured the photoresponse speed of the detector. The optical excitation source was a Ti:sapphire regenerative amplifier-pumped optical parametric amplifier, which produces visible ultra short pulses at a 175 kHz repetition rate. These pulses were compressed to less than 100 fs and frequency doubled in b-barium borate (BBO) to obtain UV pulses tunable between 300 nm and 375 nm. Neutral density filters are used to control the optical power on the detector. Shown in Figure 11.12 is the transient photoresponse of the ZnO detector under 5 V bias. The rise time is about 1 μs, while the fall time is about 1.5 μs.

The fast photoresponse of the photoconductors can be attributed to the improvement of the crystalline quality of the ZnO films, which reduces the defect-induced recombination. Simultaneously, the oxygen adsorption process is suppressed by reducing the electron carrier concentration (by N doping) and by greatly reducing the grain boundaries in the high-quality epitaxial ZnO film.

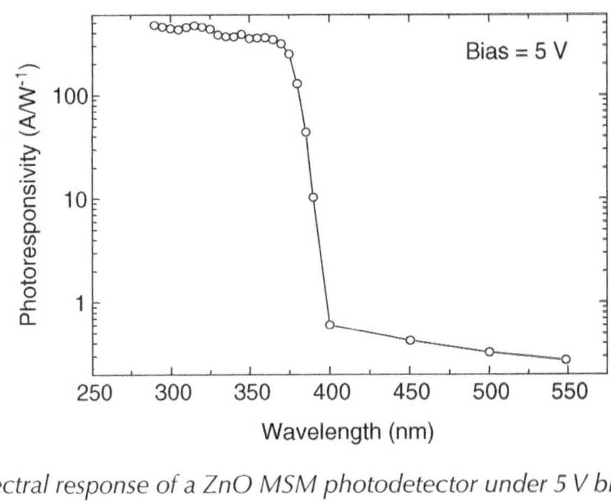

Figure 11.11 *Spectral response of a ZnO MSM photodetector under 5 V bias. Reprinted from Y. Liu, et al., J. Electron. Mater, 29, 69 (2000)*

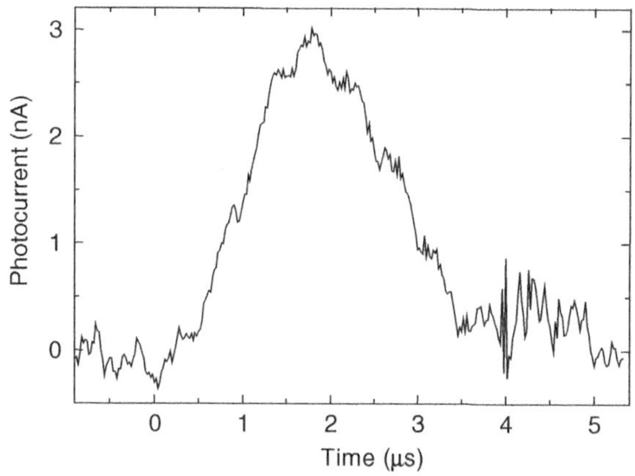

Figure 11.12 *Photocurrent transient of a ZnO MSM photodetector. Reprinted from Y. Liu, et al., J. Electron. Mater, 29, 69 (2000)*

11.3.2 Schottky Barrier UV Photodetectors

The Schottky barrier photodetector is among the simplest of the unipolar photovoltaic devices. Under optical illumination, the nonequilibrium minority carriers generated in the bulk of the semiconductor are collected by the built-in electrical field of the Schottky barrier formed at the semiconductor surface. The advantages of Schottky photodiodes are high speed, low noise performance, and ease of fabrication. Fabricius *et al.* made Schottky barrier type UV sensitive photodiodes using Au on a thin sputtered layer of polycrystalline ZnO.[13] These photodiodes exhibited slow operating speed (rise time ~20 μs and decay time ~30 μs) and low quantum efficiency (1%) due to the large amount of recombination centers in the polycrystalline ZnO layers.

The first Schottky barrier photodetector based on epitaxial ZnO film was reported by Liang *et al.* at Rutgers.[20] In this work, the Schottky barrier UV photodetector is made of nonpolar $(11\bar{2}0)$ a-plane ZnO epitaxial films grown by MOCVD. As described in Section 11.3.1, the epitaxial relationships between a-plane ZnO and r-plane sapphire substrate were determined to be $(11\bar{2}0)$ ZnO//$(01\bar{1}2)$ Al_2O_3 and (0001) ZnO//$(0\bar{1}11)$ Al_2O_3. Therefore, the *c*-axis of ZnO lies in the surface plane $(11\bar{2}0)$ of the ZnO film. In this nonpolar a-plane, absence of net surface charges (due to the presence of equal numbers of both Zn as well as O atoms on the surface) and a net zero dipole moment normal to the surface prevent divergence of surface energy.[73–75] As a result, Schottky contacts formed on a-plane ZnO surfaces would be more stable as compared with the polar c-plane surface. Furthermore, in this a-ZnO/r-Al_2O_3 system, the lattice mismatch between ZnO and the substrate along the *c*-axis is 1.53%, while along the direction perpendicular to the *c*-axis is 18.3%. The overall lattice mismatch is less than that of the polar (0001) ZnO film on c-plane sapphire system.

a-plane ZnO epitaxial films were grown on r-plane $(01\bar{1}2)$ sapphire substrates in a low-pressure MOCVD reactor. DEZn and O_2 were used as the source materials. The reactants were introduced in the reactor by two separate injectors to reduce gas phase reaction. The DEZn flow rate was 50–100 sccm and the O_2 flow rate was 1000–2000 sccm. The growth temperature range was from 380 to 420 °C. The typical growth rate was 1–2 μm h^{-1}. Because unintentionally doped ZnO film prepared by MOCVD generally exhibits n-type conduction due to oxygen vacancies or Zn interstitials, NH_3 (500 sccm) was used as a nitrogen compensation doping source to reduce the electron concentration. A 500 nm ZnO epilayer was grown with carrier concentration of about 10^{16} cm^{-3}.

UV photodetectors were designed and fabricated using both circular and IDT MSM structures. For the circular structure, the outer diameter is 320 μm, the inner diameter is 150 μm, and the gap is 10 μm. For the IDT structure, the fingers are 10 μm wide and 180 μm long, with a 10 μm gap. Aluminum and silver were used to form ohmic and Schottky contacts, respectively. The formation of Schottky contacts is greatly influenced by chemical reactions at the metal and semiconductor interface and diffusion of the metal into the semiconductor. This effect is much more pronounced for II–VI oxides as compared with III–V nitrides. In the case of ZnO, the less reactive metal Ag is used to form Schottky contacts for the UV photodetectors. The Ag-ZnO-Al circular structures were used for Schottky diode studies, while the Ag-ZnO-Ag structures served as Schottky-type MSM photodetectors. For comparison, the Al-ZnO-Al structures were also fabricated as MSM photoconductive detectors. e-beam evaporation was utilized for metal deposition. The thickness of Al and Ag layers was 2000 Å. For the Schottky contact, a 500 Å thick Au layer was finally deposited on the top of the Ag layer, to act as an oxide-resistant layer. It is well known that in fabrication of Schottky contacts, the surface states, contaminants and defects of the surface layer significantly affect the barrier height and leakage current. Oxygen plasma was used to clean the surface before and after the metallization process in order to minimize such effects.

Photoresponse measurements were performed using an Oriel optical system. The photoresponse speed of the detector was also measured. The optical excitation source was the 337.1 nm line of a N_2 pulse laser, with a pulse width of 2 ns at a repetition rate of 40 Hz. A Xe arc lamp and monochromator combination provided the light source. The signal from the detector was monitored by a lock-in amplifier. Neutral density filters were used to control the optical power on the detector. Optical energy on the detector was about 10 nJ per pulse. The signal from a 50 Ω load resistor was recorded by a digital scope with a time resolution better than 1 ns. The bias voltage was 9 V.

Shown in Figure 11.13 are the *I–V* characteristics of ZnO MSM circular devices, including both Schottky and ohmic types. The linear *I–V* relationship from the Al-ZnO-Al structure clearly indicates the ohmic behavior of the Al on n-type ZnO contact. On the other hand, the rectified *I–V* relationship from the Ag-ZnO-Al confirms Schottky junction formation between Ag and n-type ZnO. In a Schottky diode, the general *I–V* characteristics are represented by:

$$J = J_o\left[\exp\left(\frac{qV}{\eta kT}\right) - 1\right], \tag{11.14}$$

$$J_o = A^* T^2 \exp(-\Phi_B/kT), \tag{11.15}$$

where J_o is the saturation current density, η is the ideality factor, k is Boltzmann's constant, T is the absolute temperature, A^* is the effective Richardson coefficient, and Φ_B is the barrier height. The Schottky contact area is 1.77×10^{-4} cm^2. The values of $\eta = 1.50$ and

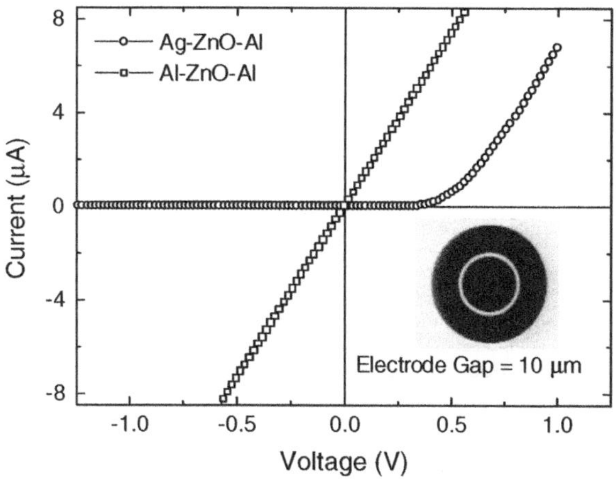

Figure 11.13 I–V characteristics of ZnO diodes with circular pattern. The electrode gap is 10 μm. Reprinted from S. Liang, et al., ZnO Schottky ultraviolet photodetectors, J. Cryst. Growth, 225, 110. Copyright (2001) with permission from Elsevier

$J_o = 2.38 \times 10^{-8}\,A\,cm^{-2}$ were obtained from curve fitting. The barrier height Φ_B is estimated to be about 0.84 eV by using $A^* \sim 32\,A\,cm^{-2}\,K^{-2}$ ($A^* = 4\pi q m^* k^2 h^{-3}$, where $m^* \sim 0.27 m_0$). At a reverse bias of 1 V, the leakage current (~0.1 nA) of the Ag-ZnO-Al device is about five orders of magnitude smaller than that of the Al-ZnO-Al device (~10 μA).

The Ag-ZnO-Ag MSM structure with IDT configuration was used to evaluate the UV detector performance. Figure 11.14 shows the dark and photoilluminated *I–V* characteristics of a Schottky detector. The wavelength and power of the illuminated light are 368 nm and 0.1 μW, respectively. The leakage current of the photodetector is about 1 nA at a bias of 5 V. The breakdown starts at about 8 V. The early and soft breakdown is mainly due to

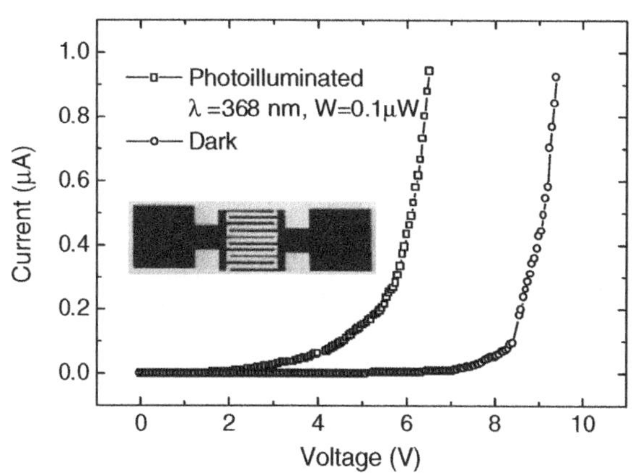

Figure 11.14 I–V characteristics of a ZnO Schottky photodetector with an IDT structure. The inset shows a SEM picture of the top view of the device. Reprinted from S. Liang, et al., ZnO Schottky ultraviolet photodetectors, J. Cryst. Growth, 225, 110. Copyright (2001) with permission from Elsevier

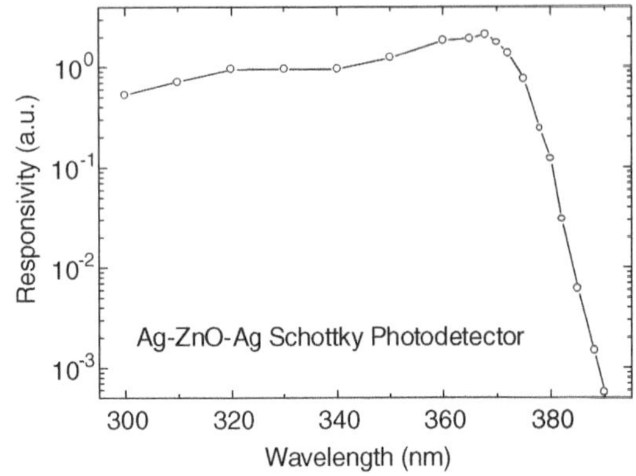

Figure 11.15 *Spectral response of a ZnO Schottky photodetector with an IDT structure. Reprinted from S. Liang, et al., ZnO Schottky ultraviolet photodetectors, J. Cryst. Growth, 225, 110. Copyright (2001) with permission from Elsevier*

the nonuniform field distribution, especially at the corners of the finger electrodes in the IDT patterns. The low-frequency photoresponsivity is about 1.5 $A W^{-1}$, which corresponds to a quantum efficiency of 2.5. We believe that the photoconductive effect that occurs at the high field contributes to the large gain (>1).

The spectral photoresponse of a IDT Schottky device is shown in Figure 11.15. The photoresponsivity drops more than three orders of magnitude from 370 to 390 nm. The sharp cut-off at wavelength of 370 nm agrees with the ZnO energy band gap of 3.35 eV. The responsivity decreases at the shorter wavelength range due to decrease of the penetrating depth of the light, resulting in an increase of the surface recombination.

Figure 11.16 shows photocurrent as a function of time from an IDT ZnO Schottky detector. The photoresponse has a fast component, which rises within 12 ns and falls to 66% of its peak value within 50 ns, as shown in the inset of Figure 11.16. After that a slow

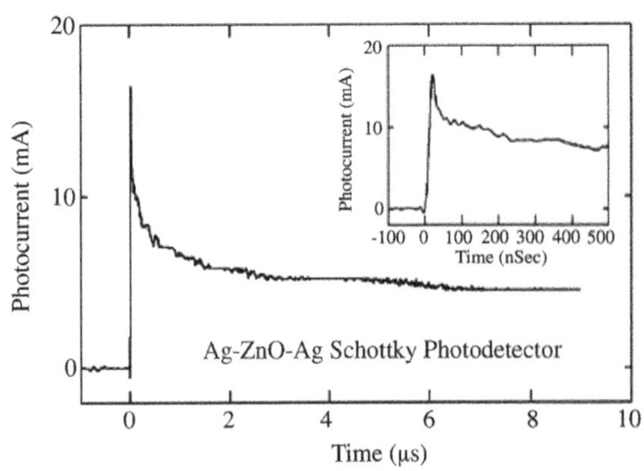

Figure 11.16 *Transient photocurrent of a ZnO Schottky photodetector with an IDT structure under a nitrogen gas laser pulsed excitation. The inset is a magnified response pulse. Reprinted from S. Liang, et al., ZnO Schottky ultraviolet photodetectors, J. Cryst. Growth, 225, 110. Copyright (2001) with permission from Elsevier*

Table 11.2 *Comparison of main device parameters of ZnO Schottky-type MSM photodetectors with Ag, Pd, and Ni metal contacts*[76–79]

Metal	Photoresponsivity ($A\,W^{-1}$ at 370 nm)	Quantum efficiency	Noise equivalent power (W)[a]	Normalized detectivity ($Hz^{1/2}\,W^{-1}$)
Ag	0.066	17.3%	6.8×10^{-13}	1.04×10^{12}
Pd	0.051	11.4%	1.13×10^{-12}	6.25×10^{11}
Ni	0.09	23.8%	6.4×10^{-12}	1.1×10^{11}

[a]Measured at a given bandwidth of 100 Hz and an applied bias of 1 V.

process follows lasting about 5 ms (not shown in the figure). The fast response in a MSM detector is usually related to the transit time of the photogenerated carriers, while the slow response in a ZnO MSM detector is usually attributed to the oxygen adsorption at the surface and grain boundaries.[15,40] The high quality epitaxial ZnO films were used for both Schottky and photoconductive detectors. Such a slow process was not observed in the photoconductive devices with the same material quality. Therefore, this slow process was attributed to the trapping and emission of photogenerated carriers in the ZnO surface of the Schottky detector. It should be noted that oxygen plasma was used to treat the ZnO surface only when making the Schottky detectors.

Table 11.2 summarizes the main device parameters for ZnO Schottky-type MSM photodetectors with various metal contact electrodes reported by Young *et al.*[76–79] As shown in Table 11.2, the UV responsivity, quantum efficiency, and noise spectra were studied and compared in the ZnO Schottky-type MSM UV detectors with various metal contact schemes. ZnO epitaxial films were grown on the c-plane sapphire substrate using plasma-assisted MBE. Room temperature Hall measurements showed the carrier concentration and mobility of as-grown ZnO films were $1.71 \times 10^{16}\,cm^{-3}$ and $26.4\,cm^2\,V^{-1}\,s^{-1}$, respectively. Schottky-type MSM photodetectors were made of IDT Schottky contacts deposited on top of a ZnO active layer. To form Schottky contact to ZnO, metals with large work functions are chosen, including Ag, Pd, Ni, and Ru. Depending on the ZnO-based material quality and surface conditions, the reported data vary from different research groups.

11.3.3 Integrated Surface Acoustic Wave and Photoconductive Wireless UV Detectors

11.3.3.1 The Principle of UV SAW Devices

Most UV photodetectors use either voltage or current as the output. In order to construct zero-power and wireless UV detector for applications such as distributed sensor network, UV-SAW photodetectors have been proposed based on acoustoelectric interaction.[80–82] In the UV-SAW photodetectors, the UV sensitive semiconductor layer is integrated with the SAW device built on the piezoelectric material. The incident light will be absorbed by the semiconductor and generate electron–hole pairs. These free carriers in the semiconductor layer will interact with the electric field accompanying the propagating SAW in the piezoelectric material, resulting in increase of insertion loss. Furthermore, a velocity reduction will occur due to the piezoelectric stiffening, resulting in a phase shift and time delay across the SAW device. Both the attenuation and phase changes correlated with the incident light intensity and wavelength can be used for the optical detection. The UV-SAW photodetectors offer advantages of low power consumption (due to the passive nature) and RF readout (for wireless communication).

Several UV-SAW photodetectors have been demonstrated with various piezoelectric and UV sensing layer integration. GaN-based UV SAW devices were demonstrated by using MOCVD grown GaN on c-plane sapphire substrates.[80,81] In these devices, GaN, which is a weakly piezoelectric wide-band-gap semiconductor, is used for both SAW excitation and UV sensing. A GaN-based SAW UV detector was reported to cause a 60 kHz oscillation frequency shift at 221.3 MHz when used as a delay element in the feedback path of an oscillator. The estimated electromechanical coupling coefficient K^2 is ~0.055% and the achieved relative oscillation frequency shift is 271 ppm. The maximum response was observed at the optical wavelength of 365 nm, which corresponds to the photon energy equal to the GaN band gap of 3.4 eV.[80] In an alternative configuration, a SAW launched from an input IDT transducer is used to sweep electron–hole pairs generated in the sensing area to an MSM photoconductive detector out of the sensing region (Figure 11.17).[81] The responsivity of this MSM photodetector is a function of the frequency and power of the SAW. Based on the acoustoelectric interaction, the coupling coefficient K^2 is the key factor to influence the change of velocity and attenuation. A higher K^2 is desirable for a larger acoustoelectric effect and hence a higher sensitivity.

In the case of a thin semiconducting layer system placed on the piezoelectric crystal's surface, the mobile carriers are confined to a conductive layer of thickness d, which is assumed to be much smaller than the wavelength of the SAW. The thin semiconducting layer can be a single layer semiconductor film under optical illumination. The longitudinal electric field of the SAW can only be screened at the piezoelectric surface, i.e. $z = 0$, where z is along the growth direction of the semiconductor thin film. The conductivity modulation in a thin sheet layer results in an effective modification of the relaxation frequency ω_c, which is a function of the SAW wave vector k; $k = 2\pi/\lambda$ for the homogeneous semiconductor.

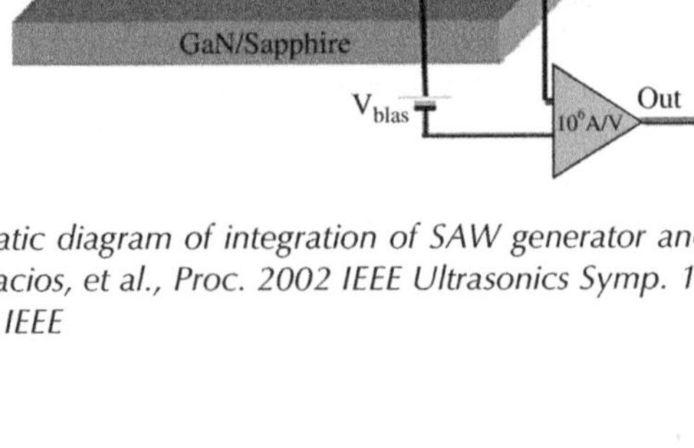

Figure 11.17 Schematic diagram of integration of SAW generator and a UV photodetector. Reprinted from T. Palacios, et al., Proc. 2002 IEEE Ultrasonics Symp. 1, 55. Copyright (2002) with permission from IEEE

Because the electric field in the homogeneous bulk semiconductor decreases exponentially with the height (e^{-kz}), a layer of thickness $1/k$ effectively carries this current. For the thin semiconducting layer, the available thickness to carry the current is defined as d. Thus the effective resistance and the time constant become $1/kd$ times larger:

$$\tau = \frac{\varepsilon_1 + \varepsilon_2}{\sigma \cdot kd} \tag{11.16}$$

and the conductivity relaxation frequency ω_c is:

$$\omega_c = \frac{kd \cdot \sigma}{\varepsilon_1 + \varepsilon_2} = \frac{\sigma_d k}{\varepsilon_1 + \varepsilon_2} \tag{11.17}$$

where $\sigma_d = \sigma \cdot d$ is the sheet conductivity of the thin film. Thus, the ratio ω_c/ω becomes frequency independent:

$$\frac{\omega_c}{\omega} = \frac{\frac{\sigma_d k}{\varepsilon_1 + \varepsilon_2}}{\omega} = \frac{\sigma_d}{\frac{\omega}{k}(\varepsilon_1 + \varepsilon_2)} = \frac{\sigma_d}{v_0(\varepsilon_1 + \varepsilon_2)} = \frac{\sigma_d}{\sigma_M} \tag{11.18}$$

where $\sigma_M = v_0(\varepsilon_1 + \varepsilon_2)$ is the conductivity at which the maximum loss occurs. Now Equation (11.18) becomes after separation of the real part and the imaginary part:

$$\frac{\Delta v}{v_0} = \frac{v - v_0}{v_0} = \frac{K^2}{2} \frac{1}{1 + (\sigma_d/\sigma_M)^2} \tag{11.19}$$

$$\Gamma = \frac{K^2}{2} k \frac{(\sigma_d/\sigma_M)}{1 + (\sigma_d/\sigma_M)^2} \tag{11.20}$$

The relationships of Equations (11.19) and (11.20) are plotted in Figure 11.18. Shown in Figure 11.18(a) is the relative change $\Delta v/v_0 = (v - v_0)/v_0$ of the sound velocity. Figure 11.18(b)

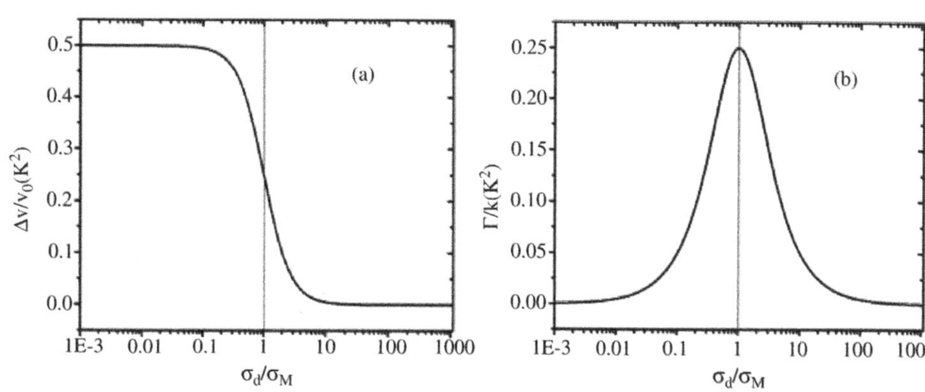

***Figure 11.18** Change in sound velocity $\Delta v/v_0$ (a) and attenuation coefficient per unit wave vector k (b) in units of effective coupling coefficient K^2 as a function of the ratio σ_d/σ_M. The thin conducting layer is assumed to be located on top of the piezoelectric surface. Reprinted with permission from J. Zhu*

is the attenuation Γ per unit length, as a function of the sheet conductivity of the thin semiconducting layer.

11.3.3.2 ZnO-Based Multilayer UV-SAW Photodetectors

Based on the acoustoelectric interaction, the device is designed by integrating a UV-sensitive semiconducting ZnO layer with piezoelectric ZnO layer on r-Al_2O_3 substrates. The detailed bottom-up layer structure consists of a piezoelectric a-ZnO layer on r-Al_2O_3 substrate for SAW generation, a $Mg_{0.2}Zn_{0.8}O$ buffer layer, and a semiconducting ZnO sensing layer [Figure 11.19(a)]. These three layers are all grown by MOCVD. A thick piezoelectric ZnO film (~2.0 μm) is first grown on r-Al_2O_3 substrates. As ZnO has a lower acoustic velocity than r-Al_2O_3, this layered system has dispersive acoustic velocity, with higher order wave modes excited at higher film thickness–frequency (hf) products. Based on the SAW simulation, the thickness is determined to ensure the excitation of the first higher order Rayleigh wave mode, the Sezawa wave mode, in the test devices below 1.0 GHz. The Sezawa wave mode is chosen as it has higher acoustic velocity, v_{SAW}, and maximum effective piezoelectric coupling, K_{eff}^2, than the base Rayleigh wave mode and the Love wave mode. The as-grown, unintentionally doped MOCVD ZnO shows n-type conductivity, with a carrier concentration of about $\sim 10^{17}\,cm^{-3}$. To compensate the excess carriers in ZnO film and ensure efficient SAW excitation, the piezoelectric ZnO layer is doped with Li to increase its resistivity to above $10^7\,\Omega cm$. Then, a 50 nm thick $Mg_{0.2}Zn_{0.8}O$ buffer layer is used to isolate the semiconducting ZnO layer from the piezoelectric ZnO layer and as a diffusion barrier for Li. The semiconducting ZnO layer's thickness, h_{mesa}, varies from 200 to 400 nm to analyze the thickness dependence of the SAW and optical responses. After the formation of the semiconducting ZnO sensing mesa by wet chemical etching, the Al IDT transducers are deposited and patterned on top of the piezoelectric ZnO surface. The top view of a prototype ZnO UV SAW device is shown in Figure 11.19(b).

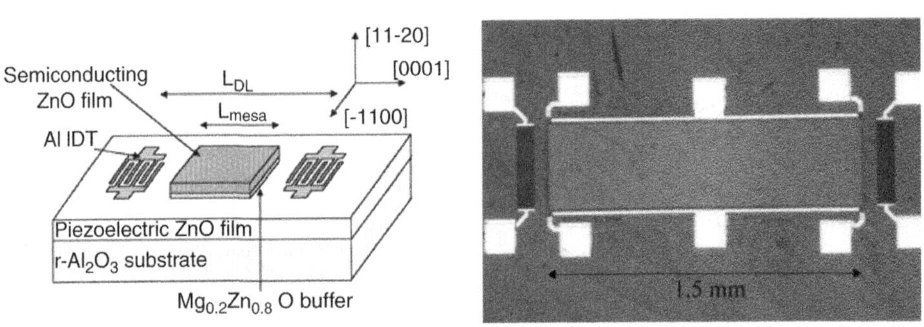

Figure 11.19 *(a) Schematic structure of a ZnO-based multilayer UV-SAW device on r-Al_2O_3 substrate. Reprinted from N. W. Emanetoglu, et al., Surface acoustic wave ultraviolet photodetectors using epitaxial ZnO multilayers grown on r-plane sapphire, Appl. Phys. Lett. 85, 3702 (2004) (b) Top view of the UV SAW device under an optical microscope. Reprinted with permission from J. Zhu*

For the horizontal device structure, there are three main parameters: the SAW wavelength λ, the delay line length L, and the UV sensing mesa length L_{mesa}. The SAW wavelength λ determines the operating frequency of the UV SAW device and is determined by the periodicity of the IDT electrodes. Three wavelengths, 8, 12 and 16 μm, are chosen based on the photolithography constraints and the piezoelectric film thickness. The aperture of the IDTs is chosen to be 380 μm. The ZnO UV SAW devices are comprised of two sets, with delay lengths, L, of 1.2 mm and 1.7 mm, and corresponding mesa lengths, L_{mesa}, of 1.0 mm and 1.5 mm, respectively. Each set contains SAW devices with 8, 12 and 16 μm acoustic wavelength. The length of the sensing mesa L_{mesa} is chosen based on ensuring sufficient phase shift and insertion loss change under UV illumination. The devices are aligned parallel to the ZnO c-axis, which is in the surface plane of the $(11\bar{2}0)$ ZnO film, to generate the Rayleigh type wave modes. In the IDT region, the base Rayleigh wave mode has an estimated maximum coupling coefficient of 1.9% with a velocity of 2768 m s^{-1}, while the maximum coupling coefficient for the Sezawa wave mode is estimated to be 6% with an acoustic velocity of 5658 m s^{-1}, leading to a larger photoresponse.

The ZnO SAW UV devices were tested using an HP 8753D network analyzer and Cascade Microtech probes for the RF response (the transmission parameter S_{21}). The ZnO UV SAW device was first measured under the dark condition. A 75 W Xe lamp with a monochromator system was used as the optical excitation source. Multiple UV filters were used to adjust the power density of the incident light. Two optical illumination procedures were used, illuminating the whole device surface and only illuminating the semiconducting ZnO mesa through a shadow mask, as shown in Figure 11.20. The light wavelengths were varied from 630 to 300 nm. The light power density was varied from 810 μW cm^{-2} to

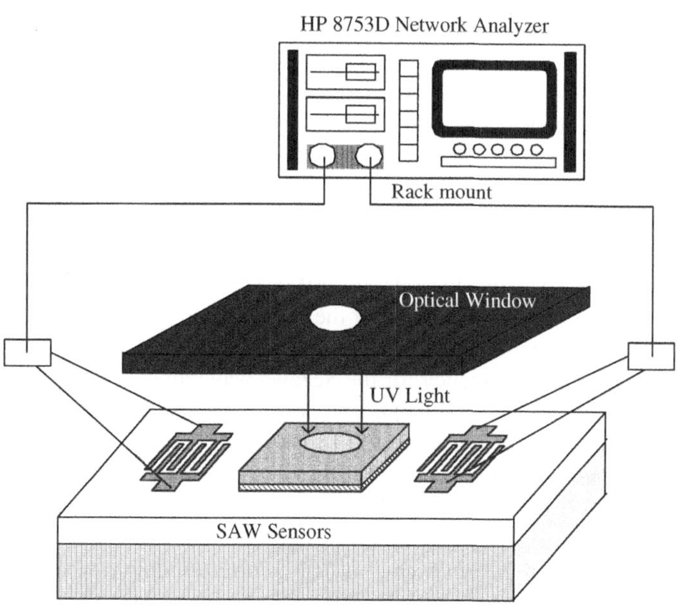

Figure 11.20 *Schematic of the UV-SAW testing set-up. Reprinted with permission from J. Zhu*

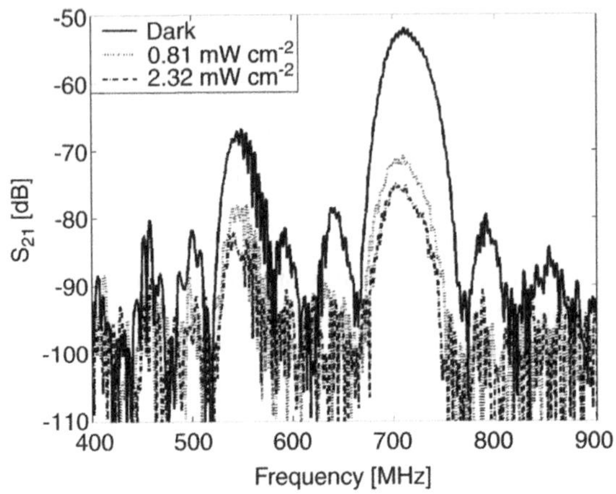

Figure 11.21 *Frequency response of the SAW UV detector under 365 nm light illumination, under dark, 0.81 mW cm*$^{-2}$ *and 2.32 mW cm*$^{-2}$ *conditions. Reprinted from N. W. Emanetoglu, et al., Surface acoustic wave ultraviolet photodetectors using epitaxial ZnO multilayers grown on r-plane sapphire, Appl. Phys. Lett. 85, 3702 (2004)*

2.32 mW cm^{-2} at $\lambda_{\text{light}} = 365$ nm. *I–V* measurements were used to determine the change in the resistance of the mesa active layer as a function of the light wavelength and power, then correlated with the SAW photoresponse.

Figure 11.21 shows the SAW response of a UV SAW device with $\lambda_{\text{SAW}} = 8\,\mu$m, $L = 1$ mm, and $h_{\text{mesa}} = 300$ nm, under dark and two different light power conditions ($\lambda_{\text{light}} = 365$ nm), while using a shadow mask to limit the illuminated area to the mesa. The power density of the incident light was adjusted by applying UV filters to the light beam. The base Rayleigh wave mode response is at 545.9 MHz, with an insertion loss of–67.8 dB. In comparison, the Sezawa wave mode response is at 711.3 MHz, corresponding to $v_{\text{SAW}} = 5690$ m s^{-1}, with an insertion loss of -52.5 dB. The Sezawa wave insertion loss increases to -75.3 dB when the light power is 2.32 mW cm^{-2}, while the Rayleigh wave insertion loss is $-$ 83.3 dB. In comparison, the UV SAW detector with a 200-nm-thick active region had an insertion loss of -33.8 dB for the Sezawa wave mode, due to its lower conductance. The insertion loss can also be improved by using better designed IDTs with narrower bandwidth and unidirectional transducers. When the whole device surface is illuminated, a shift in center frequency, up to 11 MHz, along with additional phase shift and insertion loss, was observed. This is due to the generation of carriers in the Li-doped piezoelectric ZnO layer under the IDTs, whose high resistivity ($> 10^7\,\Omega$ cm) is insufficient to totally suppress carrier generation.

The phase shift at center frequency for the Sezawa wave mode with respect to the dark condition, as a function of light wavelength and power level, is shown in Figure 11.22. The inset shows the transmission spectrum of ZnO epilayer grown on r-Al_2O_3 for comparison. The UV SAW device response closely follows the absorption spectrum. For light wavelengths above 400 nm, the phase shift is small. The phase shift increases rapidly as the light wavelength approaches the band edge at about 372 nm. The phase shift at 365 nm for a light power of 2.32 mW cm^{-2} is 107°, corresponding to a frequency shift of 1.36 MHz in an oscillator circuit, calculated for the standard oscillator circuit configuration with the SAW in the feedback path:

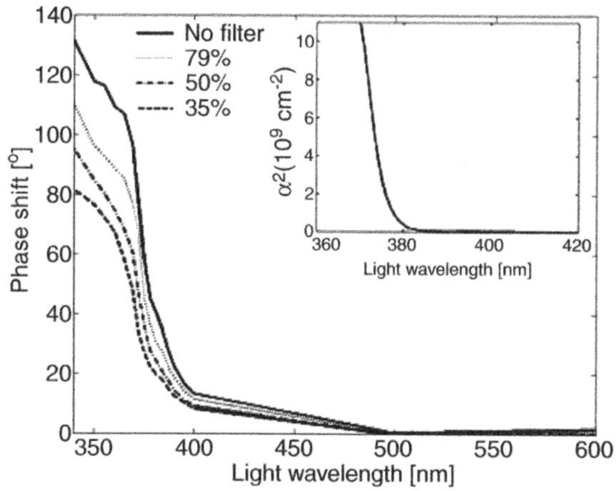

Figure 11.22 *Phase shift vs light wavelength for light power of 2.32, 1.83, 1.18 and 0.81 mW cm^{-2}. The responses have been normalized for constant power with respect to 365 nm. The inset shows the absorption spectrum of a ZnO thin film on r-Al_2O_3. Reprinted from N. W. Emanetoglu, et al., Surface acoustic wave ultraviolet photodetectors using epitaxial ZnO multilayers grown on r-plane sapphire, Appl. Phys. Lett. 85, 3702 (2004)*

$$\Delta f = \frac{1}{\tau}\frac{\Delta\phi}{360^{\circ}} = \frac{1}{\frac{L_{mesa}}{v_{SAW,mesa}} + \frac{L_{IDT}}{v_{SAW,IDT}}}\frac{\Delta\phi}{360^{\circ}} \tag{11.21}$$

where τ is the delay time across the device, $v_{SAW,mesa}$ the acoustic velocity in the mesa region, $v_{SAW,IDT}$ the acoustic velocity in the IDT region, L_{IDT} the delay path length in the IDT region, and $\Delta\phi$ the phase shift across the device. The Sezawa wave velocity in the mesa region, $v_{SAW,mesa}$, is estimated to be 5430 m s^{-1}. This 1.36 MHz frequency shift corresponds to a 0.19% relative shift for a light power of 2.32 mW cm^{-2}, which compares favorably with the 0.46% relative frequency shift at 40 mW cm^{-2} reported for a ZnO/$LiNbO_3$ SAW UV detector.[82] The estimated effective coupling, k_{eff}^2, at hf = 1710 is 1.6%, calculated using the transfer matrix method. It is estimated that the maximum effective coupling that could be obtained for this structure is 3.2% at hf = 2820. Thus, device performance can be improved by operating at a higher frequency using a smaller SAW wavelength.

Figure 11.23 shows the differential insertion loss for the Sezawa wave mode with respect to the dark condition, as a function of light wavelength and power density. The data have been normalized for constant power on the device. As the light wave approaches the band edge, at about 372 nm, the differential insertion loss rapidly increases, similar to the phase shift changes. At 365 nm, the differential insertion loss for the Sezawa mode under 2.32 mW cm^{-2} illumination is 22.8 dB.

11.3.3.3 Hybrid ZnO/$LiNbO_3$ SAW UV Photodetector

UV photodetectors have been developed by combining various SAW devices, such as SAW filter,[82] SAW oscillator[84] and SAW delay line,[85] with a ZnO photoconductor. In the ZnO/$LiNbO_3$ hybrid photodetector, semiconducting ZnO is used as a photosensitive layer and bulk $LiNbO_3$ is used to excite the SAW. A hybrid ZnO/$LiNbO_3$ SAW UV photodetector has piezoelectric coupling.[82] The $LiNbO_3$ substrate was used for SAW excitation, while the

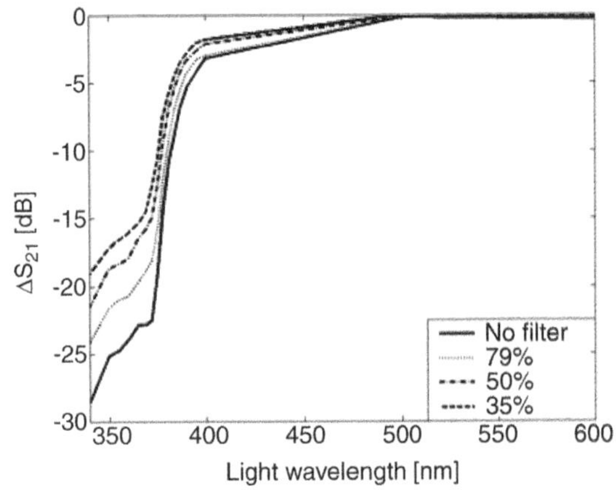

Figure 11.23 *Differential insertion loss vs light wavelength for light power of 2.32, 1.83, 1.18 and 0.81 mW cm*$^{-2}$*. The responses have been normalized for constant power with respect to 365 nm. Reprinted with permission from J. Zhu*

unbalanced RF-magnetron sputtering deposited ZnO film was used as the photosensitive layer. This device was reported to exhibit a frequency shift of 170 kHz at 37 MHz, for a UV light intensity of 40 mW cm^{-2}. The wavelength of the UV light is 365 nm.

In order to enhance the UV sensitivity, a ZnO/$LiNbO_3$ SAW oscillator configuration was used.[84] An Amplifier oscillator circuit was built up to improve the UV sensitivity. The UV sensitivity was evaluated by both of the oscillation voltage amplitude and frequency shift due to the acoustoelectron interaction described in Section 11.3.3.1. For the amplitude measurement, the UV sensitivity depended on the power supply V_{cc} of the oscillator circuit and an oscillation quench was observed with a higher UV light intensity. Kumar *et al.*[84] demonstrated a ZnO/$LiNbO_3$ UV detector by sputtering a thin layer of ZnO film on a commercially available 128° Y-X $LiNbO_3$ SAW filter. The thickness of ZnO was optimized to 71 nm to reduce the mass loading effect. The ZnO/$LiNbO_3$ UV-SAW photodetector was connected into the positive feedback loop of a high frequency amplifier to form an oscillator. By control of the power supply voltage, the UV detector exhibits a UV sensitivity of 450 nW cm^{-2} and a voltage responsivity of ~24 kV W^{-1}. For the frequency evaluation, there was ~28 kHz oscillation frequency ($f_o = 35.4$ MHz) down shift when UV illumination (~34 μW cm^{-2}) was chopped after 5 s.

An UV array detector has been developed by introducing a $LiNbO_3$-based slanted finger interdigital transducer (SFIT) and multiple ZnO optically active areas deposited by RF sputtering.[86] A SFIT can be treated as a combination of numerous subfilters, each with a uniform finger width and with a distinct center frequency. Figure 11.24 shows a schematic structure of a SFIT SAW UV array detector. The ZnO-based optically active areas are designed perpendicular to the SAW propagation direction between two ports of the SFIT electrode. Therefore, the SAW response introduced by each active area is distinguishable from the operating frequency. The magnitude of the corresponding frequency is related to the UV light intensity due to the acoustoelectric interaction between the SAW and photogenerated carriers in the ZnO photoconductive overlayer. When the size of an optical active area is reduced, the spatial distribution of the light field can be detected and recorded.

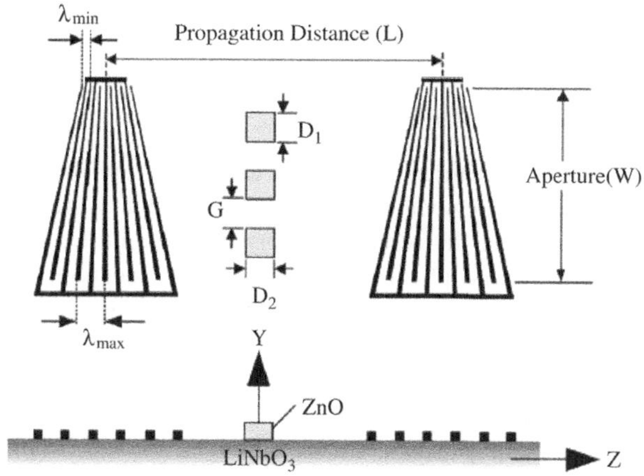

Figure 11.24 Schematic diagram of a hybrid $ZnO/LiNbO_3$ SAW UV array photodetector. Three ZnO active areas are shown by the squares in the top view. Reprinted from C. Ma, T. Huang, and J. Yu, Application of slanted finger interdigital transducer surface acoustic wave devices to ultraviolet array photodetectors, J. Appl. Phys., 104, 033528 (2008)

To determine the light field distribution, the response curves from each individual active area are first constructed to form the standard curves. Under UV illumination, the SAW insertion loss responses were then interpolated from the standard curves to deduce the light power intensity. The frequency shift is insignificant due to the photoelectric effect in the SFIT SAW device. From the insertion loss evaluation, a UV photosensitivity of 65 μW cm^{-2} at 380 nm was achieved with a minimum optically active area of 0.04 mm^2. The corresponding insertion loss difference is −58.33 dB. Figure 11.25 shows the

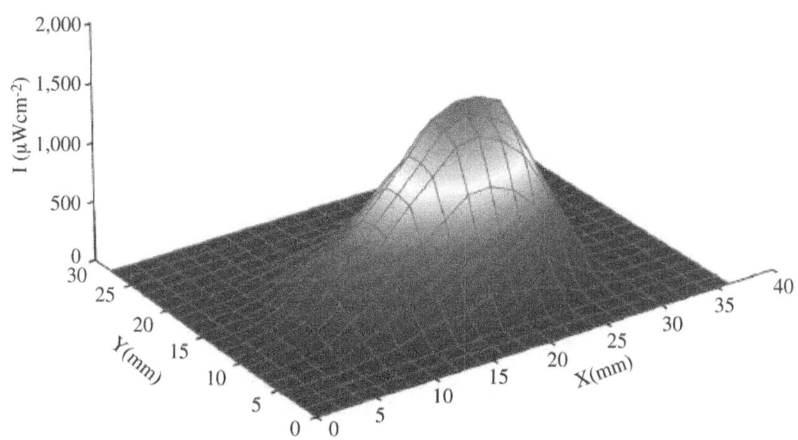

Figure 11.25 Measured light field distribution using a hybrid $ZnO/LiNbO_3$ SAW UV array photodetector. Reprinted from C. Ma, T. Huang, and J. Yu, Application of slanted finger interdigital transducer surface acoustic wave devices to ultraviolet array photodetectors, J. Appl. Phys., 104, 033528. Copyright (2008) with permission from American Institute of Physics

measured light field distribution at a UV wavelength of 380 nm from an SFIT SAW device with three ZnO active areas (Figure 11.24).

11.3.4 Photodetectors Using ZnO TFT

A field-effect phototransistor is a three terminal device in which photodetection occurs due to photogenerated carriers in the semiconductor channel upon optical excitation and the presence of an electric field at the semiconductor and the gate oxide interface for the carrier extraction. ZnO TFT is an optically transparent field-effect transistor primarily used for display technology. Figure 11.26 shows the schematic diagram of a bottom-gate ZnO-based TFT optical detector structure.

The ZnO channel layer is used as a photoconductor, where photogenerated carriers are collected by source and drain electrodes. The gate electrode is separated from the active channel layer through a layer of gate dielectric (e. g. SiO_2). By control of the voltage applied to the gate electrode, the charge density in the channel and the source-to-drain conductance are changed through formation of a space charge layer in the channel. For the n-channel ZnO TFT device, a positive gate voltage increases the electron density in the channel, while a negative gate voltage decreases the electron density. As a result, a photoconductive gain in the phototransistor is modulated by varying the gate voltage. The TFT-based phototransistors offer the advantages of high photoconductive gain, a large current transfer ratio, and versatility in designing the optically controlled circuits (e.g. photoinverter, photodetector, and opto-isolator, etc.). The photocurrent in the TFT is proportional to the gate voltage.

ZnO TFT-based optical detectors were reported by Bae *et al.*[87,88] ZnO TFTs were fabricated using the RF-magnetron sputtering technique at room temperature or at 300 °C for the ZnO channel layer. RF sputtering is chosen due to its capability of low temperature deposition, low cost, and convenience in fabrication. A wide range of materials were selected for the source and drain electrodes of the ZnO TFTs. These include Al, Au, Ti and transparent oxides (NiO_x), etc. p-Si and glass are used as the substrates. On Si substrates, thermally grown SiO_2 provides high quality gate dielectric. A poly-4-vinylphenol (PVP) polymer gate dielectric was also demonstrated.[89] The optical response of TFTs shows

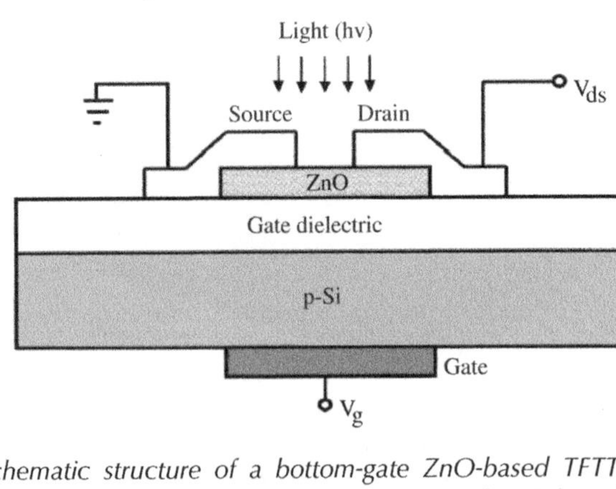

Figure 11.26 *Schematic structure of a bottom-gate ZnO-based TFTTFT optical detector. Adapted from H. S. Bae, et al., Photodetecting properties of ZnO-based thin-film transistors, Appl. Phys. Lett. 83, 5313. Copyright (2003) with permission from American Insitute of Physics*

high gain in UV wavelength (340 nm), as well as in blue (450 nm) and green (540 nm) wavelengths. The dynamic UV response of TFTs was measured to be ~300 ms. All indicate that the elimination of mid gap states in the sputtered ZnO channel is critically important for improving the photodetection properties of TFTs. The UV photosensitivity is dramatically enhanced when the TFT is operated under the depletion mode with a negative gate bias instead of the accumulation mode with a positive gate bias. This is due to reduction of dark current in the ZnO channel under the depletion mode.

11.3.5 $Mg_xZn_{1-x}O$ UV Photodetector

There is keen interest in the solar blind UV photodetectors, which find broad applications in space engineering, flame detection, and biotechnology. ZnO can alloy with MgO to form the ternary compound $Mg_xZn_{1-x}O$ to extend the energy band gap, and therefore the detection spectrum into the shorter wavelength region. The direct energy band gap of wurtzite-structured $Mg_xZn_{1-x}O$ can be tuned from 3.3 ($x=0$) to 4.0 eV ($x=0.34$), corresponding to a cut-off wavelength from ~375 to ~305 nm. Further increase of Mg composition would extend the indirect band gap of cubic-structured $Mg_xZn_{1-x}O$ up to 7.8 eV. Difficulties for the growth of high quality $Mg_xZn_{1-x}O$ films are partly from the fact that MgO (cubic) and ZnO (hexagonal) have different crystal structures. According to the phase diagram, the solid solubility of Mg in hexagonal ZnO is less than 4%. However, by using nonequilibrium growth methods, high quality wurtzite $Mg_xZn_{1-x}O$ films with up to 34% Mg incorporation have been achieved without phase segregation.

The energy band gap of $Mg_xZn_{1-x}O$ follows Vegard's law:

$$E_g(Mg_xZn_{1-x}O) = (1-x)E_g(ZnO) + xE_g(MgO) \tag{11.22}$$

Figure 11.27 shows the energy band gap of $Mg_xZn_{1-x}O$ as a function of Mg composition. It is shown that the $Mg_xZn_{1-x}O$ films have wurtzite and cubic crystal structures for

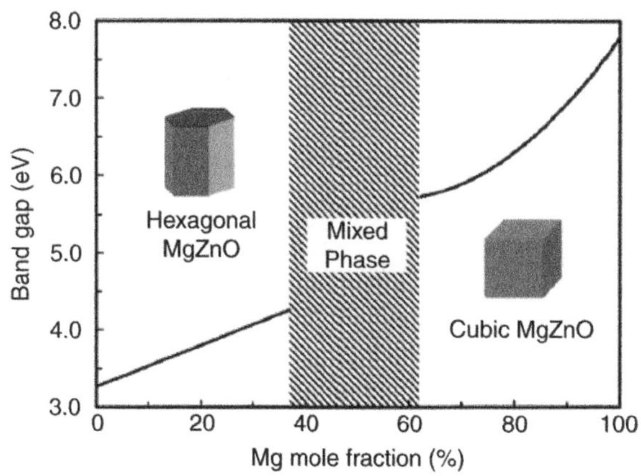

Figure 11.27 *Band gap energy of $Mg_xZn_{1-x}O$ films as a function of Mg composition. Reprinted from W. Yang, et al., Compositionally-tuned epitaxial cubic $Mg_xZn_{1-x}O$ on Si(100) for deep ultraviolet photodetectors Appl. Phys. Lett. 82, 3424. Copyright (2003) with permission from American Institute of Physics*

Mg composition <37% and >62%, respectively. For Mg composition between 37% and 62%, a mixed hexagonal and cubic phase occurs.

Yang *et al.* reported the first $Mg_xZn_{1-x}O$ MSM photoconductive detector grown by PLD on c-plane sapphire substrates.[91] Shown in Figure 11.28(a) is an optical microscope image of a $Mg_{0.34}Zn_{0.66}O$ UV detector with a size of $250 \times 1000\,\mu m^2$. The IDT metal electrodes are 250 μm long, 5 μm wide, and have an interelectrode spacing of 5 μm. A ~150 nm thick Cr/Au bilayer was patterned as the metal contacts by conventional photolithography and ion milling. To achieve ohmic contact, a thin layer (~3 nm) of chromium was used. A monochromator (150 W xenon lamp, 1200 lines mm^{-1} grating) and a nitrogen gas laser (337.1 nm, pulse duration ≤4 ns) were used as the excitation source to characterize the $Mg_{0.34}Zn_{0.66}O$ detector. Figure 11.28(b) shows the linear relationship of the *I–V* curves for both dark current and photocurrent. At 5 V bias, the dark current and photocurrent under UV illumination (308 nm, 0.1 μW) are ~40 nA and 124 μA, respectively, indicating a responsivity of ~1200 A W^{-1}. The visible rejection (*R* 308 nm/*R* 400 nm) is more than four orders of magnitude.

The spectra response of a $Mg_{0.34}Zn_{0.66}O$ UV detector under illumination is plotted in Figure 11.29(a). The peak response is at 308 nm. The −3 dB cut-off wavelength is 317 nm. The inset of Figure 11.29(a) is the responsivity as a function of bias voltage. A linear relationship was obtained between 0.5 V and 5 V, indicating no carrier mobility saturation or sweep-out effect up to the applied bias. Figure 11.29(b) shows the temporal response of an $Mg_{0.34}Zn_{0.66}O$ UV detector with 3 V bias and 50 Ω load. The 10–90% rise and fall time are 8 ns and 1.4 μs, respectively. The signal drops to zero at ~30 μs.

A prototype deep UV detector based on cubic-phase $Mg_xZn_{1-x}O$ thin films was demonstrated for solar-blind detection.[90] To reduce the lattice mismatch between the film and the substrate, a thin $SrTiO_3$ buffer layer was used to grow $Mg_xZn_{1-x}O$ ($x > 0.62$) on Si (100) substrates by PLD. The epitaxial relationship was established as $Mg_xZn_{1-x}O$ (100)//$SrTiO_3$(100)//Si(100) and $Mg_xZn_{1-x}O$[100]//$SrTiO_3$[100]//Si[100]. The $Mg_xZn_{1-x}O$

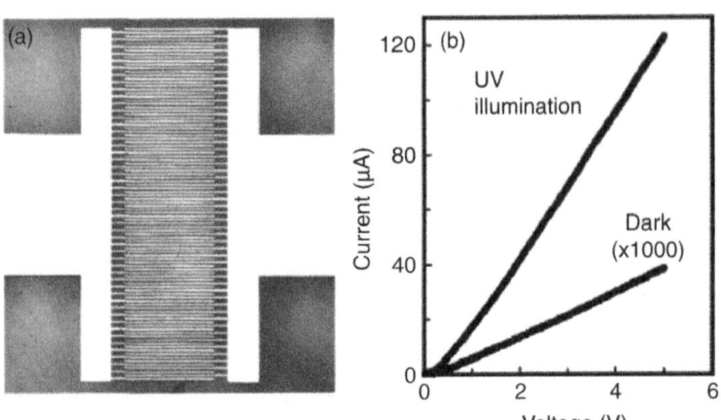

Figure 11.28 *(a) Optical microscope image of a $Mg_{0.34}Zn_{0.66}O$ MSM UV detector. The Cr/Au fingers are 250 μm long, 5 μm wide with an interfinger spacing 5 μm. (b) I–V characteristics of dark current and photocurrent under 308 nm, 0.1 μW UV light illumination. Reprinted from W. Yang, et al., Ultraviolet photoconductive detector based on epitaxial $Mg_{0.34}Zn_{0.66}O$ thin films, Appl. Phys. Lett. 78, 2787. Copyright (2001) with permission from American Institute of Physics*

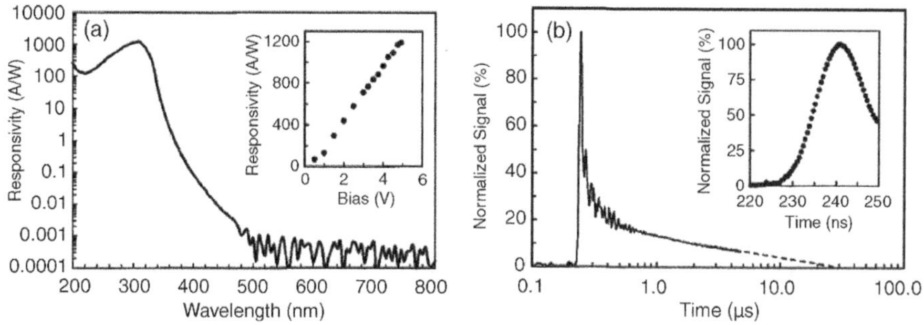

Figure 11.29 *(a) Spectral response of a $Mg_{0.34}Zn_{0.66}O$ detector at 5 V bias. The inset shows the responsivity as a function of bias voltage under 308 nm, 0.1 μW UV light illumination. (b) Temporal response of a $Mg_{0.34}Zn_{0.66}O$ detector excited by nitrogen gas laser pulses. The inset shows the enlarged pulse response. Reprinted from W. Yang, et al., Ultraviolet photoconductive detector based on epitaxial $Mg_{0.34}Zn_{0.66}O$ thin films, Appl. Phys. Lett. 78, 2787. Copyright (2001) with permission from American Institute of Physics*

photodetector with MSM structure was fabricated. The IDT metal electrodes (250 μm long, 5 μm wide, with 5 μm spacing) were patterned from a 250 nm gold film, followed by rapid thermal annealing at 400 °C. A peak photoresponsivity was achieved at 225 nm.

Liu *et al.* reported a $Mg_xZn_{1-x}O$ p-n homojunction diode grown on c-plane sapphire by plasma-assisted MBE.[92] The spectra response of a $Mg_{0.24}Zn_{0.76}O$ p-n photodiode under the reverse bias of 0 and 6 V is plotted in Figure 11.30(a). The left inset of Figure 11.30(a) shows the schematic diagram of the $Mg_xZn_{1-x}O$ p-n photodiode structure. The cut-off wavelength is ~345 nm, corresponding to 24% Mg incorporation. The visible rejection ratio (*R* 325 nm/*R* 400 nm) achieved is four orders of magnitude under 6 V reverse bias. The peak responsivity at 325 nm is 4×10^{-4} A W^{-1} at 9 V reverse bias. A linear relationship

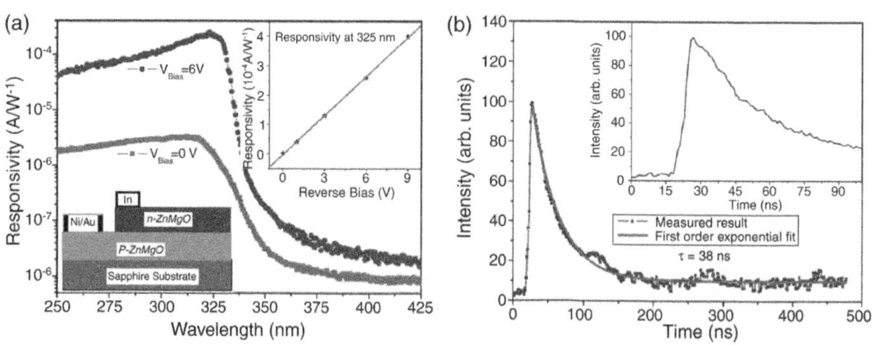

Figure 11.30 *(a) Spectral response of a $Mg_{0.24}Zn_{0.76}O$ p-n photodiode under the reverse bias of 0 and 6 V. The left inset of shows the schematic diagram of the $Mg_{0.24}Zn_{0.76}O$ p-n photodiode. The right inset shows the responsivity at 325 nm vs reverse bias voltage. (b) Transient response of the $Mg_{0.24}Zn_{0.76}O$ p-n photodiode excited by a 266 nm Nd:YAG laser with a 50 Ω load. The inset shows the enlarged pulse response. Reprinted from K. W. Liu, et al., $Zn_{0.76}Mg_{0.24}O$ homojunction photodiode for ultraviolet detection, Appl. Phys. Lett. 91, 201106 (2007)*

is obtained between 0 V and 9 V, indicating no carrier mobility saturation up to 9 V. Figure 11.30(b) shows a transient response of the $Mg_{0.24}Zn_{0.76}O$ p-n photodiode under the excitation of 266 nm from a Nd:YAG laser with a 50 Ω load. The photodiode has fast photoresponse with a rise time of 10 ns and fall time of 150 ns. The thermal limited detectivity was calculated as 1.8×10^{10} cm Hz$^{1/2}$ W^{-1} at 325 nm with a noise equivalent power of 8.4×10^{-12} W Hz$^{-1/2}$ at room temperature.

11.4 ZnO NW UV Photodetectors

One-dimensional ZnO nanostructures have attracted increasing attention due to their promising optical and electrical properties. The electron–hole interaction will have orders of magnitude enhancement in a nanostructure, due to the dramatically increased electronic density of states near the van Hove singularity. Bulk ZnO has a small exciton Bohr radius (1.8–2.3 nm).[93,94] The quantum confinement effect in ZnO NWs is observable at the scale of an exciton Bohr radius. On the other hand, the giant oscillator strength effect occurs in ZnO NWs with diameters larger than the bulk exciton Bohr radius but smaller than the optical wavelength,[95,96] making ZnO NWs suitable for high sensitivity UV detection.

Single crystalline ZnO NWs haven been synthesized by various techniques, such as MOCVD,[97–99] CVD,[100,101] chemical vapor transport and condensation (CVTC),[102,103] catalyst-assisted MBE,[104] template-assisted growth[105] and solution-based synthesis.[106] ZnO NWs have the same lattice constants and crystal structure of bulk, confirmed by XRD and TEM data.[94,107] Many bulk properties are still preserved in ZnO NWs. In comparison with its bulk counterpart, ZnO nanostructures possess certain significant characteristics for UV detection, including: (i) high surface-to-volume ratio and large density of surface trap states (primarily oxygen-related hole traps) greatly increase the photogenerated carrier lifetime and modify the effective carrier mobility; and (ii) reduced dimension decreases the carrier transit time in the active area of the nanoscale device. As a result, a large photoconductive gain is expected. A photoconductive gain as high as $\sim 10^8$ was reported in a ZnO single NW UV photodetector,[101] which is promising for single-photon detection.

In general, there are two types of device configurations of ZnO NW photodetectors: (i) a vertical structure prepared by self-assembled growth or template-assisted growth; and (ii) a horizontal structure, in which ZnO nanostructure photodetectors are fabricated using the "pick-and-place" manipulation of randomly dispersed ZnO NWs. In the latter, ZnO NWs are usually scratched from the growth substrate, then sonicated in an organic liquid drop and dispersed onto a template. After picking up an appropriate single NW, metal contacts are deposited on both ends of the NW using photo- or nanolithography. Most horizontal devices are randomly located on the template surface, and then the photoconduction properties of ZnO NWs are characterized and studied.

11.4.1 Photoconductive Gain in a ZnO NW

11.4.1.1 Theoretical Background

A simplified approach to evaluate the carrier transport and photoconduction in an individual ZnO NW is based on classical principles governing the carrier generation and

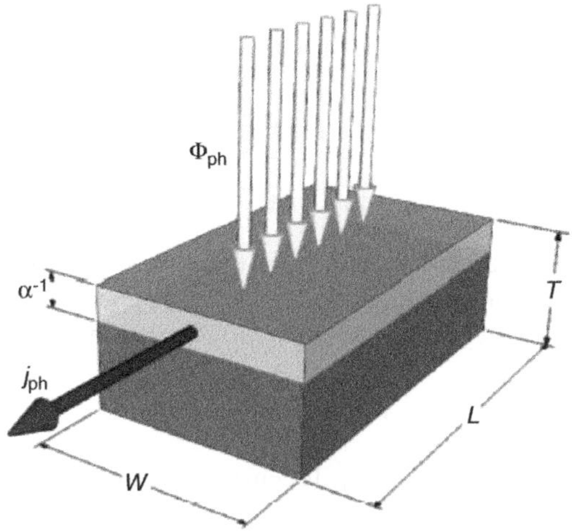

Figure 11.31 *Schematic of a NW structure with an arbitrary volume of length L, width W and thickness T, under a photon flux* Φ_{ph}. *Free carriers are optically generated within an absorption depth* α^{-1}, *leading to a photocurrent density* j_{ph}. *Reprinted from J. D. Prades, et al., Toward a Systematic Understanding of Photodetectors Based on Individual Metal Oxide Nanowires, J. Phys. Chem. C, 112, 14639 (2008)*

transport in semiconductors, providing that many bulk properties are still kept in the ZnO NW. The key factors for modeling a ZnO NW device are all surface-related, including surface state modified carrier mobility and carrier lifetime. A single ZnO NW photodetector was modeled by Prades *et al.*,[108] as an arbitrary volume of length L, width W, and thickness T, which is illustrated in Figure 11.31.

The current density in the NW follows the classical rule:

$$j_{ph} = q\Delta n_{ph} v \tag{11.23}$$

where q is the elemental charge, Δn_{ph} is the photogenerated carrier density and v is the velocity of the carriers. By assuming a constant carrier generation profile within the optical absorption depth α^{-1}, Δn_{ph} is expressed by:

$$\Delta n_{ph} = \frac{\eta F}{V_{ph}} \tau(F) = \frac{\eta F}{\alpha^{-1} WL} \tau(F) \tag{11.24}$$

where η is the quantum efficiency, F is the photon absorption rate, V_{ph} is the photogeneration volume (Figure 11.31), α is the optical absorption coefficient and τ is the carrier lifetime, which is a function of photon absorption rate F. The continuity equation gives:

$$\frac{\partial \Delta n_{ph}}{\partial t} = g_{ph} - \frac{\Delta n_{ph}}{\tau(F)} \tag{11.25}$$

where g_{ph} is the carrier photogeneration rate.

Case 1: Surface Modified Carrier Mobility. If the photogenerated excess carrier density is independent of the carrier lifetime, the time dependent photocurrent can be derived from Equation (11.25):

$$i_{\text{ph}}(t) = I_{\text{ph}}(1-e^{-t/\tau_{\text{r}}}) \tag{11.26}$$

$$i_{\text{ph}}(t) = I_{\text{ph}}e^{-t/\tau_{\text{f}}} \tag{11.27}$$

where I_{ph} is the steady-state photocurrent, τ_{r} is the photocurrent rise time and τ_{f} is the photocurrent fall time. When an external electric field is applied along the length direction, the carrier drift velocity is related to the applied voltage V by:

$$v = \mu^{*}E = \frac{\mu^{*}V}{L} \tag{11.28}$$

where μ^{*} is the effective carrier mobility. It consists of contributions from the bulk (μ_{B}) and the surface (μ_{S}):

$$\frac{1}{\mu^{*}} = \frac{1}{\mu_{\text{B}}} + \frac{1}{\mu_{\text{S}}} \tag{11.29}$$

The photocurrent in the NW is derived as:

$$I_{\text{ph}} = j_{\text{ph}}(\alpha^{-1}W) = q\frac{W}{L}\beta\eta\tau\mu^{*}V\Phi_{\text{ph}} \tag{11.30}$$

where β is the fraction of photons not reflected by the surface and Φ_{ph} is the incident photon flux. The photoconductive gain G_{ph} is defined as the ratio between the number of electrons collected and the number of photons absorbed per unit time:

$$G_{\text{ph}} = \frac{I_{\text{ph}}}{qF} \approx \frac{1}{L^{2}}\eta\tau\mu^{*}V \tag{11.31}$$

Case 2: Surface Modified Carrier Lifetime. In the second case, carrier lifetime is treated as a function of photon absorption rate due to the presence of high density surface trap states. Figure 11.32 illustrates the photoconduction process in a single ZnO NW photodetector. Similar to the oxygen adsorption–photodesorption process described in Section 11.2, the hole-trapping mechanism through the surface states (such as dangling bonds at the surface) governs the photoconduction in ZnO NWs. Under illumination with photon energy higher than the energy band gap, electron–hole pairs are generated [Figure 11.32(a)]. Holes migrate to the surface along the potential field provided by the band bending [Figure 11.32(b)] and discharge of the negatively charged adsorbed oxygen ions, and are trapped at the surface states [Figure 11.32(c)]. The unpaired electrons are either collected at the anode or recombine with holes generated when oxygen is readsorbed at the surface.

Soci *et al.* developed a model to describe the internal photoconductive gain in semiconductor NWs, where the high density of surface trap states enhances the carrier lifetime, leading to photoconductive gain.[101] When surface hole traps are filled by the photogenerated holes upon illumination, the depletion region near the surface becomes narrow and the band bending is flattened. This increases the free hole concentration and the probability of electon–hole recombination. The carrier lifetime is a function of photon

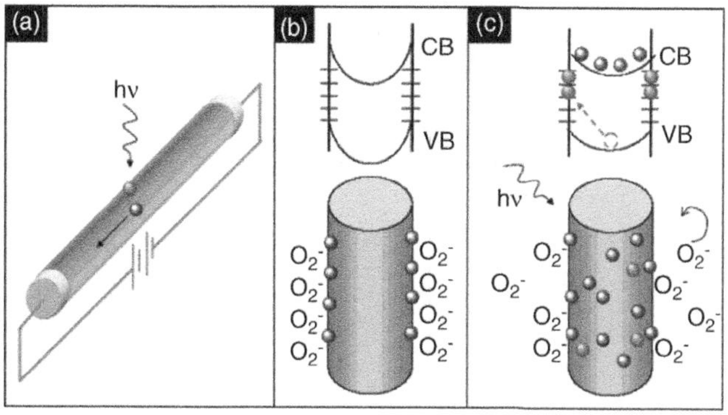

Figure 11.32 (a) Photoconduction in a single NW photodetector. (b, c) Surface trapping and photoconduction mechanism in a ZnO NW: (b) in the dark; and (c) under optical excitation. Reprinted from C. Soci, et al., ZnO Nanowire UV Photodetectors with High Internal Gain, Nano Lett 7, 1003. Copyright (2007) with permission from American Chemical Society

absorption rate:

$$\tau(F) = \tau^0 \frac{1}{1 + (F/F_0)^n} \tag{11.32}$$

where τ^0 is the carrier lifetime at low excitation and F_0 is the photon absorption rate when trap saturation occurs. Assuming the photoabsorption occurs in the entire ZnO NW, from Equations (11.24) and (11.32), the photocurrent in the NW is given by:

$$I_{\mathrm{ph}} = q \Delta n_{\mathrm{ph}} v A = q\eta\left(\frac{\tau^0}{\tau_{\mathrm{t}}}\right)\frac{F}{1 + (F/F_0)^n} \tag{11.33}$$

where v is the carrier drift velocity and τ_{t} is carrier transit time. The photoconductive gain G_{ph} is derived from Equation (11.33):

$$G_{\mathrm{ph}} = \frac{I_{\mathrm{ph}}}{qF} = \left(\frac{\tau^0}{\tau_{\mathrm{t}}}\right)\frac{\eta}{1 + (F/F_0)^n} \tag{11.34}$$

The gain–bandwidth (3 dB bandwidth) product of the NW photodetector is obtained by:

$$G_{\mathrm{ph}}B = \left(\frac{1}{2\pi\tau_{\mathrm{t}}}\right)\frac{1}{1 + (F/F_0)^n} \tag{11.35}$$

11.4.1.2 Experimental Results

Large photoconductivity of a single ZnO NW was first reported by Kind *et al.* in 2002.[22] Single crystalline ZnO NWs with diameters ranging from 50 to 300 nm were dispersed on prefabricated gold electrodes. Electrical resistivities without and with UV light irradiation were measured in a four-terminal configuration. The conductivity of ZnO NW under UV irradiation increases by four to six orders of magnitude compared with the dark current with a response time in the order of seconds (Figure 11.33). The photoresponse has a cut-off wavelength of ~370 nm. The slow process can be suppressed by reducing the trap

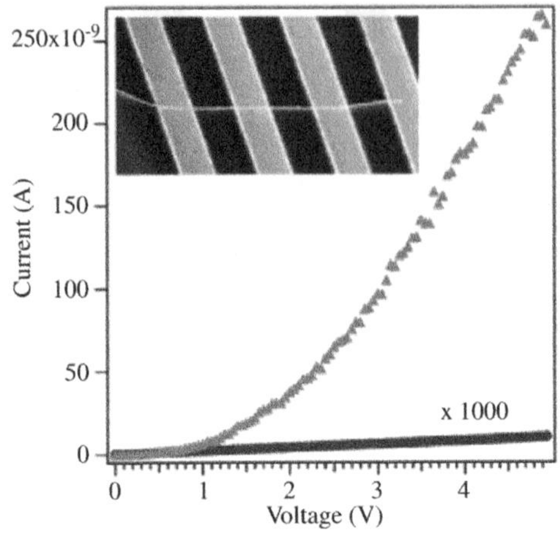

Figure 11.33 *I–V characteristics of dark current and photocurrent of a single ZnO NW under 365 nm, 0.3 mW cm^{-2} UV light illumination. The inset is a field emission scanning electron microscopy (FESEM) image of a ZnO NW with 60 nm diameter bridging four Au electrodes. Reprinted from H. Kind, H. Yan, et al., Adv. Mater, 14, 158. Copyright (2002) with permission from John Wiley and Sons*

density and background carrier concentration. ZnO NWs also show a reversible switching behavior between dark conductivity and photoconductivity when the UV lamp is turned on and off. This suggests that ZnO NWs are good candidates for optoelectronic switches.

The photoconduction properties of ZnO NW photodetectors are controlled by the geometrical size of the NWs and electrodes, ambient gas, and surface coatings. The electron mobility in ZnO varies depending on the diameter of the NWs. For example, mobility values of 2–30 cm^2 V^{-1} s^{-1} were reported for NWs with radii < 100 nm,[109–111] and ~200 cm^2 V^{-1} s^{-1} in thicker NWs.[101] By surface passivation of the ZnO NWs, electron mobility up to 1000 cm^2 V^{-1} s^{-1} was reported.[111–114]

Photoconduction was investigated in the ZnO NWs which were grown by CVD, with diameters of 150–300 nm and lengths of 10–14 μm.[101] After the NWs were transferred onto a SiO_2/Si substrate, Ti/Au electrodes were patterned on top of these NWs using photolithography for photocurrent detection. Figure 11.34(a) shows the photocurrent spectra of an array of ZnO NWs. The photoresponsivity of ZnO NWs increases by about two orders of magnitude after keeping the sample under vacuum (10^{-4} Torr) for about 20 min. This is consistent with the increased carrier lifetime caused by a reduced oxygen readsorption rate in vacuum. Figure 11.34(b) plots the photoconductive gain from Equation (11.34). A high internal photoconductive gain of 2×10^8 is obtained, corresponding to a gain–bandwidth product of 6 GHz. The intrinsic material parameters of ZnO NWs are also estimated based on the experimental results. An intrinsic carrier concentration as low as ~10^{13} cm^{-3} and high electron mobility of ~270 cm^2 V^{-1} s^{-1} are achieved. These material data obtained from the ZnO NW are much better than ZnO thin films, indicating a low bulk defect density in the NW.

Surface functionalization and passivation are used to dramatically improve the photoresponsivity of ZnO nanostructures.[114,115] It was found that a single ZnO nanobelt coating with a thin layer (~20 nm) of plasma polymerized acrylonitrile (PP-AN) showed

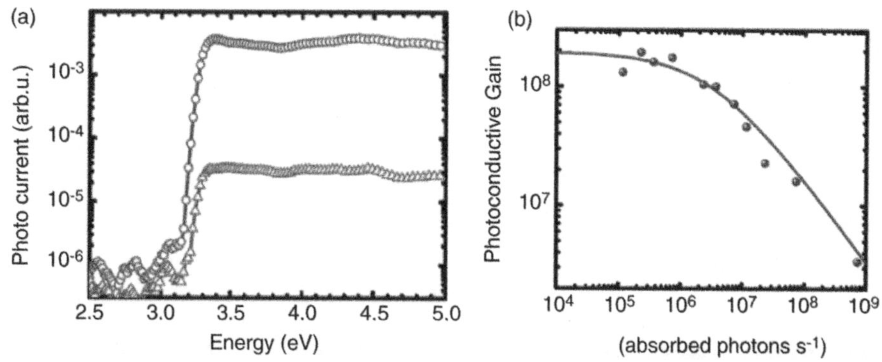

Figure 11.34 *(a) Spectral response of an array of ZnO NWs in air (triangles) and in vacuum (10^{-4} Torr) (circles), at an applied bias of 0.3V and a modulation frequency of 160 Hz. (b) Plot of photoconductive gain vs photon absorption rate. Reprinted from C. Soci, et al., ZnO Nanowire UV Photodetectors with High Internal Gain, Nano Lett 7, 1003. Copyright (2007) with permission from American Chemical Society*

up to 750 times higher UV photoresponse ($\lambda = 365$ nm) than that of a bare ZnO nanobelt.[114] The mechanism for this giant photoconductivity is attributed to the efficient exciton dissociation under UV excitation, which is caused by enhanced electron hopping transfer from the ZnO VB to the PP-AN then transit to the ZnO CB. In addition, various schemes of surface passivation were conducted with Si_3N_4/SiO_2,[111] polymide,[112] poly (methyl methacrylate) (PMMA)[113] and polyacrylonitrile coating.[114] To remove surface contamination, oxygen plasma cleaning is typically applied to the sample prior to the spin coating of polymers or CVD deposition of SiO_2.

11.4.2 Noise Characteristics of ZnO NW UV Photodetector

Noise equivalent power (NEP) and normalized detectivity (D^*) are the main parameters to characterize the signal-to-noise performance of a photodetector. NEP is the optical input power for which the signal-to-noise ratio is equal to one. D^* is the reciprocal of the NEP, which is normalized for bandwidth and the photodetector area.

Noise characteristics of vertically well-aligned ZnO NW UV photodetectors were reported.[116] To fabricate the vertical-structured device, undoped ZnO NWs were first grown on a ZnO:Ga/glass template by a self-catalyzed vapor–liquid–solid method at 600 °C. To prevent the metal particles from falling onto the bottom and shorting the device, a thin spin-on-glass (SOG) film was spin coated onto the top of ZnO NWs followed by thermal annealing. Au was used as the top electrode, with a diameter of 200 μm. Figure 11.35(a) shows the FESEM image of as-grown ZnO NWs on a ZnO:Ga/glass template, and Figure 11.35(b) is a top view of ZnO NWs coated with SOG.

Figure 11.36(a) shows the spectral response of a vertical ZnO NW photodetector. The peak photoresponse occurs at ~365 nm, with a UV to visible rejection ratio of more than three orders of magnitude (R 365 nm/R 425 nm). The ZnO NW photodetector has a maximum quantum efficiency of 12.6%. Figure 11.36(b) shows the measured low-frequency noise power spectra under various applied biases. In a ZnO NW photodetector,

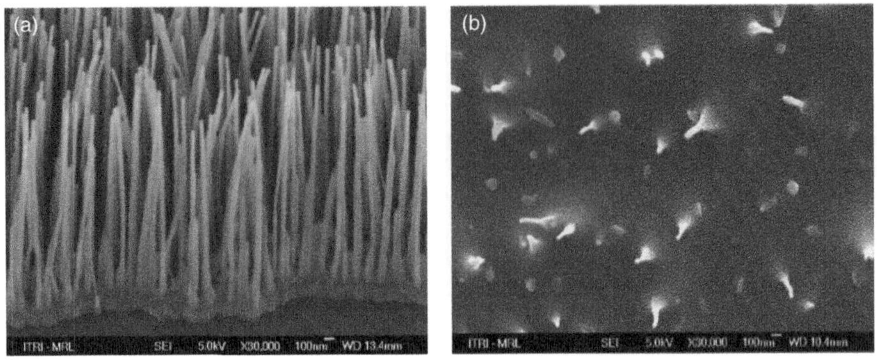

Figure 11.35 FESEM images of (a) the cross-sectional view of as-grown ZnO NWs on a ZnO: Ga/glass template and (b) the top view of ZnO NWs coated with SOG. Reprinted from C. Lu, S. Chang, S. Chang, Y. Chiou, C. Kuo, H. Chang, C. Hsu, and I. Chen, IEEE sensors J. 7, 1020. Copyright (2007) IEEE

at low frequency ($\leq$100 Hz), the noise power spectra are dominated by $1/f$ noise. Above 100 Hz, the noise decreases as $\sim 1/f^2$, possibly due to interfacial traps in the SOG film.

The total noise current power is estimated by:

$$\langle i_n \rangle^2 = \int_0^B S_n(f)\mathrm{d}f = \int_0^1 S_n(1)\mathrm{d}f + \int_1^B S_n(f) = S_0[\ln(B) + 1] \tag{11.36}$$

where $S_n(f)$ is the spectral density of the noise power and B is the photodetector bandwidth. We assume that $S_n(f) = S_n(1\,\mathrm{Hz})$ when $f < 1$ Hz. NEP and D^* are then given by:

$$\mathrm{NEP} = \frac{\sqrt{\langle i_n \rangle^2}}{R} \tag{11.37}$$

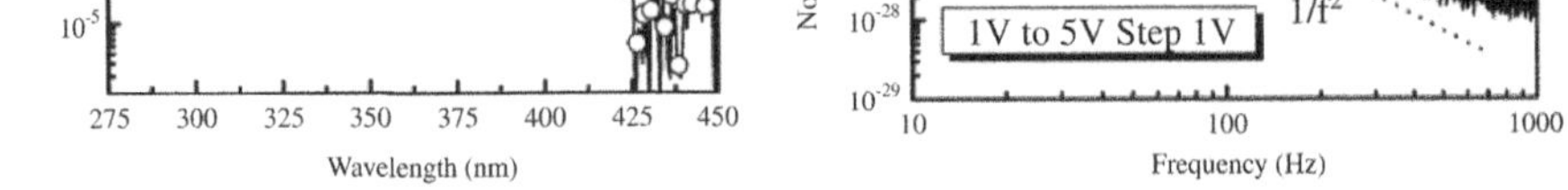

Figure 11.36 (a) Spectral response of ZnO NW photodetector at an applied bias of 1 V. (b) Measured low-frequency noise power spectra under various applied biases. Reprinted from C. Lu, S. Chang, S. Chang, Y. Chiou, C. Kuo, H. Chang, C. Hsu, and I. Chen, IEEE sensors J. 7, 1020. Copyright (2007) IEEE

$$D^* = \frac{\sqrt{A}\sqrt{B}}{\text{NEP}} \tag{11.38}$$

where R is the responsivity of the photodetector, A is the area of the photodetector and B is the bandwidth. For a given bandwidth of 100 Hz at 1 V bias, the NEP and D^* of the NW photodetector are 5.73×10^{-11} W and 6.17×10^{9} cm Hz$^{1/2}$ W^{-1}, respectively. The NEP of the NW photodetector is more than one order of magnitude larger than that of the planar-structured ZnO MSM UV photodetectors (see Table 11.2). This is presumably attributed to the much larger surface area and larger carrier trapping and detrapping effects of the NWs.

11.5 Conclusions

This chapter reviews the progress of ZnO UV photodetectors. ZnO-based materials and nanostructures possess great promise for UV detection technologies. Various types of ZnO-based UV photodetectors have been demonstrated. The ZnO photoconductive MSM photodetectors show a large photoresponsivity and fast photoresponse time due to the excellent material quality of the ZnO epitaxial films. The ZnO epitaxial film-based Schottky barrier photodetector has been demonstrated, showing a faster photoresponse component of tens of nanoseconds. Through band gap engineering, $Mg_xZn_{1-x}O$ photodetectors extend the cut-off wavelength to the deep UV region, which is important for solar-blind applications. ZnO can be made multifunctional through proper doping. The semiconducting ZnO has been integrated with piezoelectric ZnO to form a multilayer structure where the acousto-optic interaction is used for the novel UV detection technology, such as UV-SAW detectors. Unlike the traditional photodetectors which have the output in current or voltage, the ZnO UV-SAW photodetector has its output in the frequency domain, suitable for a wireless sensor and low power operation. Recently, ZnO nanostructures have been emerging as one of most widely used functional nanomaterials. The single crystalline 1D ZnO nanostructures can be grown on various substrates, including glass, GaN, Si, metals, and flexible substrates at low temperature. ZnO NW-based photodetectors demonstrate the very large internal gain, which is promising for unique applications such as single-photon detection.

Currently, the development of ZnO p-n junction- based UV detectors is still in the early stage due to the constraints of lack of high quality and reliable p-doping technology. With successful development of ZnO p-type materials and multiple quantum well technology, ZnO-based devices will become one of the major UV detection technologies and find broad applications.

Acknowledgements

This work was supported by NSF, AFOSR, and NJ Commission of Science and Technology. One of the authors (YL) acknowledges the technical assistance from his PhD students and postdoctors, S. Liang, Y. Liu, H. Sheng, N. W. Emanetoglu, Y. M. Chen, J. Zhu, Y. Chen, C. R. Gorla and S. Muthukumar.

References

[1] M. Razeghi and A. Rogalski, *J. Appl. Phys.* **79**, 7433 (1996).
[2] E. Monroy, F. Omnès and F. Calle, *Semicond. Sci. Technol.* **18**, R33 (2003).
[3] C. W. Bunn, *Proc. Phys. Soc., London* **47**, 835 (1935).
[4] E. Mollwo, *Z. Angew. Phys.* **6**, 257 (1954).
[5] D. C. Reynolds and T. C. Collins, *Phys. Rev.* **185**, 1099 (1969).
[6] D. C. Look, *Mater. Sci. Eng., B* **80**, 381 (2001).
[7] P. Schreiber, T. Tang, G. Smith, T. Pickenpaugh, P. Gehred and C. Litton, *Proc. SPIE* **3629**, 230 (1999).
[8] J. D. Albrecht, P. P. Ruden, S. Limpijumnong, W. R. L. Lambrecht and K. F. Brennan, *J. Appl. Phys.* **86**, 6864 (1999).
[9] E. Mollwo, in *Proceedings of the Photoconductivity Conference*, edited by R. G. Breckenridge, John Wiley & Sons. Ltd, New York, 1954, p. 509.
[10] P. H. Miller, in *Proceedings of the Photoconductivity Conference*, edited by R. G. Breckenridge, John Wiley & Sons, Ltd, New York, 1954, p. 287.
[11] R. J. Collins and D. G. Thomas, *Phys. Rev.* **112**, 388 (1958).
[12] G. Heiland, *J. Phys. Chem. Solids*, **22**, 227 (1961).
[13] H. Fabricius, T. Skettrup and P. Bisgaard, *Appl. Opt.* **25**, 2764 (1986).
[14] A. M. Fernandez and P. J. Sebastian, *J. Phys. D: Appl. Phys.* **26**, 2001 (1993).
[15] Y. Takahashi, M. Kanamori, A. Kondoh, H. Minoura and Y. Ohya, *Jpn. J. Appl. Phys.* **33**, 6611 (1994).
[16] D. H. Zhang and D. E. Brodie, *Thin Solid Films* **238**, 95 (1994).
[17] D. H. Zhang and D. E. Brodie, *Thin Solid Films* **251**, 151 (1994).
[18] A. E. Jimenez-Gonzailez and P. K. Nair, *Semicond. Sci. Technol.* **10**, 1277 (1995).
[19] Y. Liu, C. R. Gorla, S. Liang, N. Emanetoglu, Y. Lu, H. Shen and M. Wraback, *J. Electron. Mater.* **29**, 69 (2000).
[20] S. Liang, H. Sheng, Y. Liu, Z. Huo, Y. Lu and H. Shen, *J. Cryst. Growth* **225**, 110 (2001).
[21] D. C. Oh, T. Suzuki, T. Hanada, T. Yao, H. Makino and H. J. Ko, *J. Vac. Sci. Technol., B* **24**, 1595 (2006).
[22] H. Kind, H. Yan, B. Messer, M. Law and P. Yang, *Adv. Mater.*, **14**, 158 (2002).
[23] K. Keem, H. Kim, G. Kim, J. S. Lee, B. Min, K. Cho, M. Sung and S. Kim, *Appl. Phys. Lett.* **84**, 4376 (2004).
[24] Y. W. Heo, B. S. Kang, L. C. Tien, D. P. Norton, F. Ren, J. R. Laroche and S. J. Pearton, *Appl. Phys. A* **80**, 497 (2005).
[25] C. S. Lao, M. C. Park, Q. Kuang, Y. L. Deng, A. K. Sood, D. L. Polla and Z. L. Wang, *J. Am. Chem. Soc.* **129**, 12096 (2007).
[26] Y. Jin, J. Wang, B. Sun, J. C. Blakesley and N. C. Greenham, *Nano Lett* **8**, 1653 (2008).
[27] A. Tsukazaki, A. Ohtomo, T. Onuma, M. Ohtani, T. Makino, M. Sumiya, K. Ohtani, S. F. Chichibu, S. Fuke, Y. Segawa, H. Ohno, H. Koinuma and M. Kawasaki, *Nat. Mater.* **4**, 1 (2005).
[28] Y. Ryu, T. Lee, J. A. Lubguban, H. W. White, B. Kim and C. Youn, *Appl. Phys. Lett.* **88**, 241108 (2006).
[29] A. Ohtomo, M. Kawasaki, T. Koida, K. Masubuchi, Y. Sakurai, Y. Yoshida, T. Yasuda, Y. Segawa and H. Koinuma, *Appl. Phys. Lett.* **72**, 2466 (1998).
[30] S. Muthukumar, J. Zhong, Y. Chen, T. Siegrist and Y. Lu, *Appl. Phys. Lett.* **82**, 742 (2003).
[31] S. Choopun, R. D. Visoute, W. Yang, R. P. Sharma, T. Venkatesan and H. Shen, *Appl. Phys. Lett.* **80**, 1529 (2002).
[32] Y. R. Ryu, T. S. Lee, J. A. Lubguban, A. B. Corman, H. W. White, J. H. Leem, M. S. Han, Y. S. Park, C. J. Youn and W. J. Kim, *Appl. Phys. Lett.* **88**, 052103 (2006).
[33] C. Yang, X. M. Li, Y. F. Gu, W. D. Yu, X. D. Gao and Y. W. Zhang, *Appl. Phys. Lett.* **93**, 1121114 (2008).
[34] Y. Lu and J. Zhong, in *Semiconductor Nanostructures for Optoelectronic Applications*, edited by S. Todd, Artech House Inc., Norwood, MA, 2004, Ch. 6.

[35] N. W. Emanetoglu, J. Zhu, Y. Chen, J. Zhong, Y. Chen and Y. Lu, *Appl. Phys. Lett.* **85**, 3702 (2004).
[36] S. A. Studenikin, N. Golego and M. Cocivera, *J. Appl. Phys.* **87**, 2413 (2000).
[37] S. A. Studenikin and M. Cocivera, *J. Appl. Phys.* **91**, 5060 (2002).
[38] S. A. Studenikin, N. Golego and M. Cocivera, *J. Appl. Phys.* **83**, 2104 (1998).
[39] P. Sharma, A. Mansingh and K. Sreenivas, *Appl. Phys. Lett.* **80**, 553 (2002).
[40] D. H. Zhang, *J. Phys. D: Appl. Phys.* **28**, 1273 (1995).
[41] H. V. Hove and A. Luyckx, *Solid State Commun.* **4**, 603 (1966).
[42] J. W. Tomm, B. Ullrich, X. G. Qiu, Y. Segawa, A. Ohtomo, M. Kawasaki and H. Koinuma, *J. Appl. Phys.* **87**, 1844 (2000).
[43] N. V. Joshi, *Photoconductivity: Art, Science, and Technology*, Marcel Dekker, New York, 1990.
[44] H. Lüth, *Phys. Rev. Lett.* **29**, 1377 (1972).
[45] Y. Shapira, R. B. McQuistan and D. Lichtman, *Phys. Rev. B* **15**, 2163 (1977).
[46] G. Heiland, *Z. Physik* **138**, 459 (1954).
[47] H. J. Krusemeyer and D. G. Thomas, *J. Phys. Chem. Solids* **4**, 78 (1958).
[48] M. Liu and H. K. Kim, *Appl. Phys. Lett.* **84**, 173 (2004).
[49] K. Moazzami, T. E. Murphy, J. D. Phillips, M. C.-K. Cheung and A. N. Cartwright, *Semicond. Sci. Technol.* **21**, 717 (2006).
[50] P. Sharma, K. Sreenivas and K. V. Rao, *J. Appl. Phys.* **93**, 3963 (2003).
[51] D. C. Oh, T. Suzuki, J. J. Kim, H. Makino, T. Hanada, M. W. Cho and T. Yao, *Appl. Phys. Lett.* **86**, 032909 (2005).
[52] D. C. Oh, T. Suzuki, H. Makino, T. Hanada, H. J. Ko and T. Yao, *Phys. Status Solidi C* **3**, 946 (2006).
[53] H. Frenzel, H. v. Wenckstern, A. Weber, H. Schmidt, G. Biehne, H. Hochmuth, M. Lorenz and M. Grundmann, *Phys. Rev.* **76**, 035214 (2007).
[54] J. F. Cordaro and Y. Shim, *J. Appl. Phys.* **60**, 4186 (1986).
[55] A. Rohatgi, S. K. Pang, T. K. Gupta and W. D. Straub, *J. Appl. Phys.* **63**, 5375 (1988).
[56] F. D. Auret, S. A. Goodman, M. Hayes, M. J. Legodi, H. A. van Laarhoven and D. C. Look, *Appl. Phys. Lett.* **79**, 3074 (2001).
[57] Y. W. Heo, K. Ip, S. J. Pearton and D. P. Norton, *Phys. Status Solidi A* **201**, 1500 (2004).
[58] H. L. Porter, A. L. Cai, J. F. Muth and J. Narayan, *Appl. Phys. Lett.* **86**, 211918 (2005).
[59] N. Kouklin, *Adv. Mater.* **20**, 2190 (2008).
[60] F. Masuoka, K. Ooba, H. Sasaki, H. Endo, S. Chiba, K. Maeda, H. Yoneyama, I. Niikura and Y. Kashiwaba, *Phys. Status Solidi C* **3**, 1238 (2006).
[61] B. Claflin, D. C. Look, S. J. Park and G. Cantwell, *J. Cryst. Growth* **287**, 16 (2006).
[62] A. Y. Polyakov, N. B. Smirnov, A. V. Govorkov, A. I. Belogorokhov, E. A. Kozhukhova, A. V. Markov, A. Osinsky, J. W. Dong and S. J. Pearton, *Appl. Phys. Lett.* **90**, 132103 (2007).
[63] S. Ahn, H. J. Ji, K. Kim, G. T. Kim, C. H. Bae, S. M. Park, Y. Kim and J. S. Ha, *Appl. Phys. Lett.* **90**, 153106 (2007).
[64] A. Bera and D. Basak, *Appl. Phys. Lett.* **93**, 053102 (2008).
[65] P. M. Mooney, *J. Appl. Phys.* **67**, R1 (1990).
[66] S. B. Zhang, S. H. Wei and A. Zunger, *Phys. Rev. B* **63**, 075205 (2001).
[67] S. Lany and A. Zunger, *Phys. Rev. B* **72**, 035215 (2005).
[68] J. Reemts and A. Kittel, *J. Appl. Phys.* **101**, 013709 (2007).
[69] S. Sen, D. Chowdhary and N. A. Kouklin, *Appl. Phys. Lett.* **91**, 093125 (2007).
[70] Z. Fan, D. Dutta, C. Chien, H. Chen, E. C. Brown, P. Chang and J. G. Lu, *Appl. Phys. Lett.* **89**, 213110 (2006).
[71] H. Y. Kim, J. H. Kim, M. O. Park and S. Im, *Thin Solid Films* **398–399** 93 (2001).
[72] C. R. Gorla, N. W. Emanetoglu, S. Liang, W. E. Mayo, Y. Lu, M. Wraback and H. Shen, *J. Appl. Phys.* **85**, 2595 (1999).
[73] J. G. Fripiat, A. A. Lucas, J. M. Andre and E. G. Derouane, *Chem. Phys.* **21**, 101 (1977).
[74] P. W. Tasker, *J. Phys. C* **12**, 4977 (1979).
[75] H. Sheng, S. Muthukumar, N. W. Emanetoglu and Y. Lu, *Appl. Phys. Lett.* **80**, 2132 (2002).

[76] S. J. Young, L. W. Ji, R. W. Chuang, S. J. Chang and X. L. Du, *Semicond. Sci. Technol.* **21**, 1507 (2006).
[77] S. J. Young, L. W. Ji, S. J. Chang and Y. K. Su, *J. Cryst. Growth* **293**, 43 (2006).
[78] S. J. Young, L. W. Ji, T. H. Fang, S. J. Chang, Y. K. Su and X. L. Du, *Acta Mater.* **55**, 329 (2007).
[79] S. J. Young, L. W. Ji, S. J. Chang, S. H. Liang, K. T. Lam, T. H. Fang, K. J. Chen, X. L. Du and Q. K. Xue, *Sens. Actuators, A* **141**, 225 (2008).
[80] D. Ciplys, R. Rimeika, M.S. Shur, S. Rumyantsev, R. Gaska, A. Sereika, J. Yang and M.A. Khan, *Appl. Phys. Lett.* **80**, 2020 (2002).
[81] T. Palacios, F. Calle, J. Grajal, E. Monroy, M. Eickhoff, O. Ambacher and F. Omnes, *Proc. 2002 IEEE Ultrasonics Symp.* **1**, 55 (2002).
[82] P. Sharma and K. Sreenivas, *Appl. Phys. Lett.* **83**, 3617 (2003).
[83] J. Zhu, PhD thesis, Rutgers University, 2007.
[84] S. Kumar, P. Sharma and K. Sreenivas, *Semicond. Sci. Technol.* **20**, L27 (2005).
[85] T. Huang and C. Ma, *Jpn. J. Appl. Phys.* **47**, 6507 (2008).
[86] C. Ma, T. Huang and J. Yu, *J. Appl. Phys.* **104**, 033528 (2008).
[87] H. S. Bae, M. H. Yoon, J. H. Kim and S. Im, *Appl. Phys. Lett.* **83**, 5313 (2003).
[88] H. S. Bae, M. H. Yoon, J. H. Kim and S. Im, *J. Appl. Phys.* **97**, 076104 (2005).
[89] K. Lee, K. Kim, J. Choi, M. S. Oh, D. K. Hwang, S. Jang, E. Kim and S. Im, *J. Phys. D: Appl. Phys.* **41**, 135102 (2008).
[90] W. Yang, S. S. Hullavarad, B. Nagaraj, I. Takeuchi, R. P. Sharma, T. Venkatesan, R. D. Vispute and H. Shen, *Appl. Phys. Lett.* **82**, 3424 (2003).
[91] W. Yang, R. D. Vispute, S. Choopun, R. P. Sharma, T. Venkatesan and H. Shen, *Appl. Phys. Lett.* **78**, 2787 (2001).
[92] K. W. Liu, D. Z. Shen, C. X. Shan, J. Y. Zhang, B. Yao, D. X. Zhao, Y. M. Lu and X. W. Fan, *Appl. Phys. Lett.* **91**, 201106 (2007).
[93] Y. B. Li, Y. Bando, T. Sato and K. Kurashima, *Appl. Phys. Lett.* **81**, 144 (2002).
[94] Z. W. Pan, Z. R. Dai and Z. W. Wang, *Science* **291**, 1947 (2001).
[95] J. Park, H. Choi, K. Siebein and R. K. Singh, *J. Cryst. Growth* **258**, 342 (2003).
[96] W. Heo, V. Varadarajan, M. Kaufman, K. Kim, D. P. Norton, F. Ren and P. H. Fleming, *Appl. Phys. Lett.* **81**, 3046 (2002).
[97] W. I. Park, D. H. Kim, S. W. Jung and G. Yi, *Appl. Phys. Lett.* **80**, 4232 (2002).
[98] S. Muthukumar, H. Sheng, J. Zhong, Z. Zhang, N. W. Emanaetoglu and Y. Lu, *IEEE Trans. Nanotechnol.* **2**, 50 (2003).
[99] J. Zhong, S. Muthukumar, Y. Chen, Y. Lu, H. M. Ng, W. Jiang and E. L. Garfunkel, *Appl. Phys. Lett.* **83**, 2003.
[100] J. Wu and S. Liu, *J. Phys. Chem. B* **106**, 9546 (2002).
[101] C. Soci, A. Zhang, B. Xiang, S. A. Dayeh, D. P. R. Aplin, J. Park, X. Y. Bao, Y. H. Lo and D. Wang, *Nano Lett* **7**, 1003 (2007).
[102] Y. B. Li, Y. Bando, T. Sato and K. Kurashima, *Appl. Phys. Lett.* **81**, 144 (2002).
[103] M. H. Huang, Y. Wu, H. Feick, N. Tran, E. Weber and P. Yang, *Adv. Mater.* **13**, 113 (2001).
[104] Y. W. Heo, V. Varadarajan, M. Kaufman, K. Kim, D. P. Norton, F. Ren and P. H. Fleming, *Appl. Phys. Lett.* **81**, 3046 (2002).
[105] Y. Li, G. W. Meng, L. D. Zhang and F. Phillipp, *Appl. Phys. Lett.* **76**, 2011 (2000).
[106] L. Vayssieres, *Adv. Mater.* **15**, 464 (2003).
[107] V. A. L. Roy, A. B. Djurisic, W. K. Chan, J. Gao, H. F. Lui and C. Surya, *Appl. Phys. Lett.* **83**, 141 (2003).
[108] J. D. Prades, R. Jimenez-Diaz, F. Hernandez-Ramirez, L. Fernandez-Romero, T. Andreu, A. Cirera, A. Romano-Rodriguez, A. Cornet, J. R. Morante, S. Barth and S. Mathur, *J. Phys. Chem. C* **112**, 14639 (2008).
[109] P. Chang, C. J. Chien, D. Stichtenoth, C. Ronning and J. G. Lu, *Appl. Phys. Lett.* **90**, 113101 (2007).
[110] Z. Y. Fan, D. W. Wang, P. C. Chang, W. Y. Tseng, W. A. Chiou and J. G. Lu, *Appl. Phys. Lett.* **85**, 5923 (2004).

[111] P. Chang, Z. Fan, C. Chien, D. Stichtenoth, C. Ronning and J. C. Lu, *Appl. Phys. Lett.* **89**, 133113 (2006).

[112] W. I. Park, J. S. Kim, G. Yi, M. H. Bae and H. Lee, *Appl. Phys. Lett.* **85**, 5052 (2004).

[113] W. Hong, B. Kimm, T. Kim, G. Jo, S. Song, S. Kwon, A. Yoon, E. A. Stach and T. Lee, *Colloids Surf., A* **313–314**, 378 (2008)

[114] J. H. He, Y. H. Lin, M. E. McConney, V. V. Tsukruk, Z. L. Wang and G. Bao, *J. Appl. Phys.* **102**, 084303 (2007).

[115] C. Lao, Y. Li, C. P. Wong and Z. L. Wang, *Nano Lett* **7**, 1323 (2007).

[116] C. Lu, S. Chang, S. Chang, Y. Chiou, C. Kuo, H. Chang, C. Hsu and I. Chen, *IEEE Sens. J.* **7**, 1020 (2007).

12

Room-Temperature Stimulated Emission from ZnO Multiple Quantum Wells Grown on Lattice-Matched Substrates

Takayuki Makino[1], Yusaburo Segawa[2], Masashi Kawasaki[3] and Hideomi Koinuma[4]
[1]WPI Advanced Institute for Materials Research, Tohoku University, Sendai, Japan
[2]Cross-Correlated Materials Research Group, Advanced Science Institute, RIKEN, Wako, Japan
[3]WPI Advanced Institute for Materials Research, Tohoku University, Sendai, Japan Cross-Correlated Materials Research Group, Advanced Science Institute, RIKEN, Wako, Japan, and CREST, Japan Science and Technology Agency, Tokyo, Japan
[4]Graduate School of Frontier Science, The University of Tokyo, Kashiwa, Chiba, Japan

12.1 Introduction

Recent demonstration of zinc oxide (ZnO) heterostructures opens up significant opportunities for fabrication of electronic and photonic nanodevices. In particular, semiconductor quantum structures with well-defined interfaces are the main components for nanoscale resonant tunneling devices, field effect transistors, and light-emitting devices (LEDs). There have been many studies recently on the properties of wide-gap semiconductors due to the industrial demand for short-wavelength LEDs or laser diodes. ZnO is a wide-gap semiconductor with a direct gap around 3.4 eV (i.e. in the near UV in the electromagnetic spectrum) and crystallizes preferentially in the hexagonal wurtzite-type structures.

Zinc Oxide Materials for Electronic and Optoelectronic Device Applications, First Edition.
Edited by Cole W. Litton, Donald C. Reynolds and Thomas C. Collins.

However, ZnO has some advantages over gallium nitrides[1,2] among which are the availability of fairly high-quality bulk single crystals and a large exciton binding energy (60 meV). It has also much simpler crystal growth technology, resulting in a potentially lower cost for the devices. High quality ZnO films can be grown at relatively low temperatures. The large exciton binding energy of 60 meV allows us to observe an intense near-band-edge emission at room and higher temperatures, because this value is 2.4 times greater than that of the room temperature (RT) thermal energy (approximately 25 meV). One crucial problem should be overcome for the real application of devices, that is, the growth of p-type conductivity ZnO layers. Despite all the progress that has been made, it remains the most extensive topic in current research, and most of the efforts are directed just to solving this problem.

In all commercially available semiconductor lasers including GaN lasers, a recombination of electrons and holes is used as the mechanism of laser action. In such cases, the threshold carrier density required to accomplish the inversion distribution of a population for an electron–hole system is one or two orders of magnitude higher than the Mott transition density. If an exciton-related recombination is used as the mechanism of laser action, the resultant threshold value is expected to be two or three orders of magnitude lower with higher quantum efficiency. Thus, the advantage of ZnO is that excitons can exist stably even at RT and even under a high-density condition.

In this chapter, we present a review of optical properties of ZnO multiple quantum well (MQW) structures grown on lattice-matched substrates based on the studies of the luminescence of quantum wells (QWs) both under weak and strong excitation. We also compare the properties of the QW deposited on sapphire. Studies on sapphire substrates revealed that the ZnO epilayers are adapted for optoelectronic applications in several aspects including an optically pumped laser oscillation of excitonic origin even at RT[3,4] and a band gap engineering tunable by preparing the (Zn,Mg)O and (Zn,Cd)O solid solutions.[5–7] In addition, growth of ZnO/(Mg,Zn)O (MQWs) were successfully performed.[8–11]

However, the abovementioned studies also revealed unsatisfactory properties of ZnO epilayers and superlattices. These cannot be avoided as long as lattice-mismatched sapphire substrates are used. In addition, the fact that neither luminescence (photoluminescence) nor a stimulated emission could be observed at RT in quantum wells grown on sapphire substrates is also a problem. These problems have been completely resolved by using lattice-matched substrates. We have adopted hexagonal $ScAlMgO_4$ (0001) with lattice constants of $a = 3.246$ Å and $c = 25.195$ Å.[12] This crystal has an in-plane lattice mismatch as small as 0.09%. $ScAlMgO_4$ can be regarded as a natural superlattice composed of alternating stacking layers of wurtzite-type $MgAlO_x$ and rocksalt (111)-ScO_y layers. The structure of a crystal grown by this method and a possible hetero-interface with ZnO are schematically shown in Figure 1 of Ohtomo *et al.*[13]

Although there is now a fairly good understanding of the basic properties of ZnO epilayers, it is only recently that the basic properties of its QWs have been studied in detail. The combination of ZnO and (Zn,Mg)O has been employed in a study by another research group.[14–21] We also conducted studies on this interesting and unexplored material, part of which are introduced in this chapter. In this chapter, we describe the optical properties of ZnO/(Zn,Mg)O MQWs grown on lattice-matched $ScAlMgO_4$ (SCAM) substrates.

Not only linear properties but also nonlinear optical properties are presented. The properties of the strongly excited electron–hole system in direct band gap material have been investigated for several decades. We studied the luminescence of ZnO MQWs under strong excitation. The bulk or thin film forms of this material have been studied previously by a variety of spectroscopic techniques. Compared with these, strong excitation effects have been only recently performed for the ZnO QW systems.

This chapter is organized as follows. The experimental procedures are briefly described in Section 12.2. In Sections 12.3–12.7, quantum confinement effects of excitons, the well-width dependence of the exciton–phonon coupling constants, and possible mechanism of a stimulated emission of ZnO MQWs are described. The summarizing remarks are given in Section 12.8.

12.2 Experimental Details

Samples of MQWs (10 periods) were grown by the method of laser molecular beam epitaxy (MBE). A QW is defined as stacks alternately deposited by using two kinds of very thin semiconductor layers (wells and barriers) that have different band gap energies. ZnO was used as a well layer material and a ZnMgO solid solution, the band gap of which is larger than that of ZnO, was used as a barrier layer. The magnesium concentration dependence of the band gap energy is given elsewhere.[22] It should be noted that the in-plane lattice mismatch between ZnO and these alloys is very small. The Mg concentration was set to 0.12 or 0.27, because the barrier height could be changed by a change in its concentration. Eighteen samples with different Mg concentrations and well widths (L_w), 6.9–46.5 Å, were prepared in order to estimate the L_w dependence of their optical properties. Barrier layer thickness was fixed at 50 Å. The samples were grown by the "combinatorial" method, the concept of which has been explained elsewhere.[23–27] The study has been performed using time-resolved luminescence spectroscopy at 5 K. Detection was achieved using a Hamamatsu streak camera with a time resolution of approximately 20 ps. The third harmonics (266 nm in wavelength) from a titan–sapphire picosecond pulsed laser was used for excitation.

12.3 Quantum Confinement Effect of Excitons in QWs

As mentioned in Section 12.1, the MQWs grown on sapphire substrates have several drawbacks due to the formation of rough interfaces caused by the use of lattice-mismatched substrates. One of these drawbacks is that a controllability of layer thickness is not sufficient for the quantum confinement effect to be elicited especially in the case of $L_w \leq 15$ Å. Moreover, we could not observe an exciton emission at RT. In other words, the efficiency of luminescence is not sufficiently high enough to enable observation of the exciton emission at RT.[8, 9] From a viewpoint of an optoelectrical device to be operable at RT, problems must be overcome. The use of lattice-matched substrates opens up the possibility of a solution to these drawbacks. We descibe how the properties can be improved by using SCAM substrates.[28] Bragg diffraction peaks and clear intensity oscillations due to Laue patterns corresponding to the layer thickness have been observed in the X-ray diffraction (XRD) patterns of the MQWs studied. This suggests a high crystallinity and a high degree of thickness homogeneity.

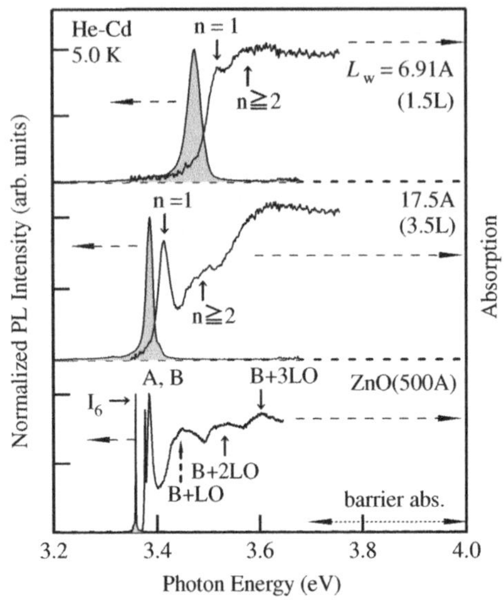

Figure 12.1 *PL and absorption spectra obtained from ZnO/$Mg_{0.12}Zn_{0.88}O$ MQWs measured at 5 K for L_w of 17.5 and 6.9 Å. The absorption energy of the barrier layers is shown by a horizontal arrow. Spectra obtained from a 500-Å-thick ZnO film are also shown. "B + LO, B + 2LO, and B + 3LO" correspond to exciton–phonon complex transitions, "$n = 1$" shows the lowest excitonic absorption of the well layers, and "$n \geq 2$" means the excited states of the exciton or higher interband transitions. Reprinted from Makino, T., et al., Room-temperature luminescence of excitons in ZnO/(Mg,Zn)Omulti-quantum wells on lattice-matched substrates, App. Phys. Lett., 77,975. Copyright (2000) with permission from American Institute of Physics*

Furthermore, observation of the atomic force microscopy (AFM) images revealed that the surface of an MQW is composed of well-defined atomically flat terraces and steps corresponding to the charge neutral unit of ZnO. In other words, the interface roughness in the heterostructure cannot be larger than 0.26 nm. It can be said that ZnO and MgZnO alloy layers grow in a two-dimensional growth mode on this substrate, resulting in the formation of a sharp hetero-interface between them.

We show, in Figure 12.1, luminescence (PL) and absorption spectra in ZnO/ $Mg_{0.12}Zn_{0.88}O$ MQWs on SCAM substrates measured at 5 K with well widths (L_w) of 17.5 and 6.9 Å. Also shown for comparison in the bottom panel of the figure is PL and absorption spectra in a 500-Å-thick ZnO epilayer on the lattice-matched SCAM substrate. Both the PL and absorption peaks shifted towards the higher energy side as L_w decreased. This shift resulted from a quantum confinement effect. For reference, we note that the exciton Bohr radius is ≈18 Å.[29] The absorption peaks ($n = 1$) arise from the lowest excitonic states of well layers. The peak energies of PL were constantly located on the lower energy side of those of absorption peaks.

The well width dependences of the peak energies of PL (closed circles) and absorption (open squares) are shown in Figure 12.2(b) and (c), respectively.[28] The lowest transition energy of excitons (open triangles) formed with confined electrons and holes was calculated using a model of one-dimensional, finite periodic square-well potential proposed by Gol'dman and Krivchnokov.[30] The exciton binding energy (60 meV)[31]

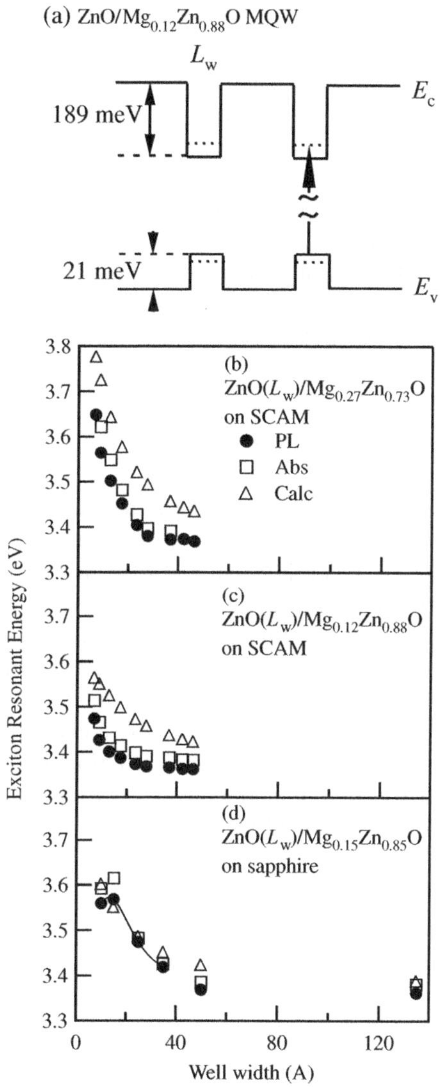

Figure 12.2 *(a) Diagram of conduction and valence bands between barrier and well layers in a ZnO/$Mg_{0.12}Zn_{0.88}O$ MQW.[9] The upward arrow shows the lowest interband transition. (b) Peak energies of PL (circles) and absorption (squares) are plotted against L_w in [ZnO/$Mg_{0.27}Zn_{0.73}O]_{10}$ on SCAM substrates. Results of calculation (triangles) of the interband transition energy that include the excitonic effect are also shown. (c) Similar except that the Mg content was ≈12%. (d) Similar except that the substrate was sapphire. The Mg content was ≈15%. Energies of PL excitation spectra (squares) are plotted instead of those of absorption spectra, due to the presence of 100-nm-thick ZnO buffer layers. Note that the peak energies of PL excitation spectra coincide with those of the absorption spectra. The curve is a visual guide.[28] Reprinted from Makino, T., et al., Room-temperature luminescence of excitons in ZnO/(Mg,Zn)Omulti-quantum wells on lattice-matched substrates, App. Phys. Lett., 77,975. Copyright (2000) with permission from American Institute of Physics*

is assumed to be independent of L_w here. The optical transition process on a ZnO/$Mg_{0.12}Zn_{0.88}O$ MQW is shown in Figure 12.2(a). This tendency of the L_w dependence of the exciton transition energy was consistent with the calculation results. As reported by Coli and Bajaj,[32] incorporation of the effects of exciton–phonon interaction and the quantum confinement in calculation, leads to values of the excitonic transitions that are in good agreement with our experimental data. Figure 12.2(d) shows the corresponding peak energy plot for MQWs grown on sapphire substrates. As shown in Figure 12.2(d), both peak energies have a maximum at L_w of 15 Å when sapphire substrates were employed. This indicates a critical thickness that prevents quantum confinement with respect to the exciton energy. It is speculated that the existence of the critical thickness corresponds to the poor controllability of layer thickness due to the lattice mismatching.

12.4 Exciton–Phonon Interaction in QWs

Despite the viable technological applications and remarkable achievements of ZnO, the fundamental properties, such as electron–phonon interaction are less understood. The intrinsic properties cannot be controlled but must be understood since they play an essential role in the device operations. The coupling constant between excitons and phonons in ZnO QWs (MQWs) has not been evaluated quantitatively so far. We quantify the coupling constant by estimating the temperature dependence of the absorption spectra. Figure 12.3 shows the temperature dependence of the full width at half-maximum (FWHM) of the excitonic absorption peaks for a ZnO epilayer [Figure 12.3(a)] and for a typical MQW sample with a QW width of 17.5 Å [Figure 12.3(b)]. The solid line represents the fitted results based on Equation (12.1). The temperature dependence of the FWHM can be approximately described by:[33]

$$\Gamma(T) = \Gamma_0 + \gamma_{\text{ph}} T + \Gamma_{\text{LO}}/[\exp(\hbar\omega_{\text{LO}}/k_{\text{B}}T) - 1] \qquad (12.1)$$

where Γ_0, $\hbar\omega_{LO}$ (72 meV), γ_{ph}, Γ_{LO} and k_B are the inhomogeneous linewidth at temperature (T) of 0 K, longitudinal optical (LO)-phonon energy, strengths of the exciton-acoustic-phonon and the exciton-LO-phonon couplings and the Bolzmann constant, respectively. It was experimentally found that $\hbar\omega_{LO}$ of the MQWs is not different from the bulk value.

Figure 12.4 (closed circles, left axis) shows the values of Γ_{LO} obtained for the epilayer and its well width (L_w) dependence obtained for ZnO/$Mg_{0.12}Zn_{0.88}O$ MQWs. The values of Γ_{LO} of the MQWs are smaller than those for the epilayers and monotonically decrease with decrease in L_w. Here we try to explain this result by the enhancement of exciton binding energy induced by the quantum confinement effect. Figure 12.4 (open circles, right axis) shows the L_w dependence of exciton binding energy. This dependence was determined by studying spectra of stimulated emission. As is well known, the major process that contributes to broadening of the exciton linewidth is scattering of 1S excitons into the free-electron–hole continuum or into excited excitonic states by absorbing LO phonons. If exciton binding energy exceeds $\hbar\omega_{LO}$ (72 meV), dissociation efficiency into the continuum states is greatly suppressed compared with the case of exciton binding energy smaller than $\hbar\omega_{LO}$. In such case, Γ_{LO} is reduced. Indeed, exciton binding energy

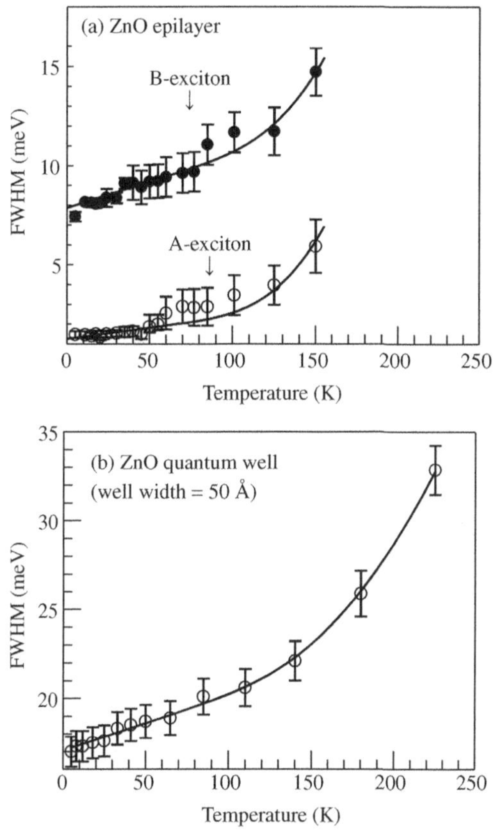

Figure 12.3 *(a) Width (full width at half-maximum, FWHM, circles) of A- and B-exciton absorption bands plotted as a function of temperature. Open circles are data of the A-excitons and the closed circles are data of the B-excitons. The solid curves represent the fitting results. (b) Similar plot for the MQW with Mg concentration of 0.12 and L_w of 46.5 Å.*[34,35] *Reprinted from Sun, H. D., et al., Temperature dependence of the exciton linewidth in ZnO/(Mg,Zn)O multi-quantum wells grown on lattice matched substrates. Appl. Phys. Lett. 78, 2464. Copyright (2001) with permission from American Institute of Physics*

exceeds $\hbar\omega_{LO}$ in the case of MQWs. A similar effect has also been observed in other QW systems.[37] A schematic explanation is described in the literature.[35,38]

12.5 The Localization Mechanism of the Exciton in a QW

It was found that the excitonic luminescence in the ZnO MQWs under investigation is due to radiative recombination from excitons localized by the potentials formed by the fluctuations of L_w and barrier height. Our spectral assignments are based on (1) the well width dependence of Stokes shift (difference between the energies of absorption and luminescence bands), (2) the temperature dependence of PL spectra, and (3) the spectral distribution (luminescence energy dependence) of decay time constants of

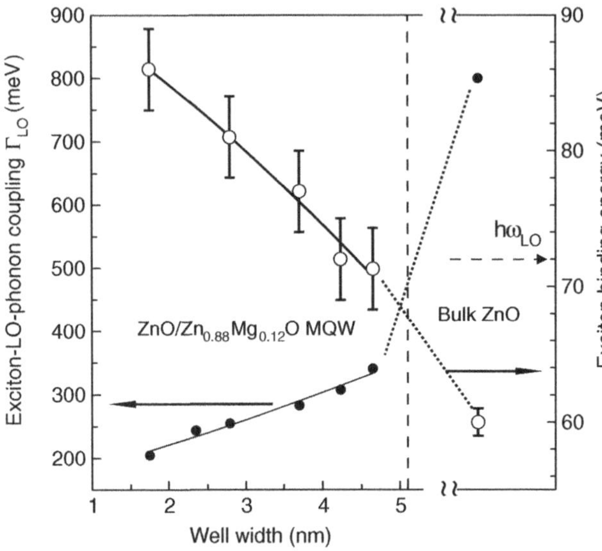

Figure 12.4 *Strengths of coupling between excitons and LO phonons Γ_{LO} (closed circles) and exciton binding energies (open circles) in bulk ZnO and MQWs with different L_ws.*[36] *The solid curve is a visual guide*

luminescence.[28,35,39,40] A typical example of the spectral distribution of decay time constants is shown in Figure 12.5(b). Here, the temperature dependence of the PL spectrum in a QW in the case of magnesium composition of 0.27 is described in detail.

Figure 12.5(a) shows the temperature dependence of PL (solid line) and that of absorption (broken line) spectra in ZnO(17.5 Å)/$Mg_{0.27}Zn_{0.73}O$ MQWs over a temperature range of 5–300 K. It should be noted that spectra obtained at temperatures between 95 K and 200 K had two peaks, both of which originated from a recombination of localized excitons. The separation of these peaks was 12–20 meV. Figure 12.5(b) shows PL decay time as a function of monitored photon energy at 5 K in the same MQW. The solid curve is the results of theoretical calculation based on the model of the excitonic localization.[28,41] Figure 12.6(a) summarizes peak energies of the PL spectra (E_{PL}^{pk}) (open circles and triangles) and the excitonic absorption energy (open squares) as functions of temperatures. It should be noted that the higher PL peak position does not coincide with that of absorption spectra even at temperatures near room temperature. We also examined, for comparison, the temperature dependence of PL peak energy in an MQW having a lower barrier height: a ZnO/$Mg_{0.12}Zn_{0.88}O$ MQW with a well width of 27.9 Å. Figure 12.6(b) shows the peak energies of PL (circles) and absorption (squares) spectra in this sample. In this case, the energies of luminescence and absorption are the almost the same as each other at temperatures near room temperature. Two kinds of MQWs having different barrier heights showed significantly different temperature dependences of PL spectra.

Following a temperature rise, the PL energy of ZnO(17.5 Å)/$Mg_{0.27}Zn_{0.73}O$ MQWs exhibited low energy shifts between 5 K and 50 K, blue-shifts between 50 K and 200 K, and again shifts to a low energy side at temperatures higher than 200 K. Furthermore, at temperatures between 95 K and 200 K, the spectra had two peaks, both of which originated

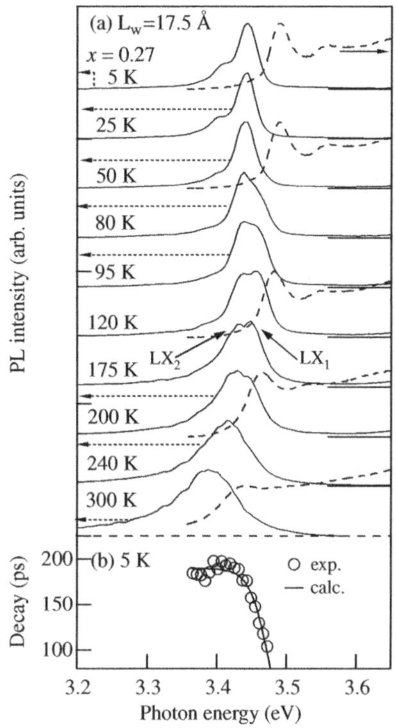

Figure 12.5 *(a) PL (solid line) and absorption (broken line) spectra in a ZnO (17.5 Å)/ $Mg_{0.27}Zn_{0.73}O$ MQW over the temperature range of 5–300 K. All of the spectra have been normalized and shifted in the vertical direction for clarity. (b) PL decay times as a function of monitored photon energy at 5 K in the same MQW. The solid curve is results of the theoretical calculation based on the model of the excitonic localization. Reprinted from Makino, T., et al., Temperature dependence of near ultraviolet photoluminescence in ZnO/(Mg,Zn)O multi-quantum wells, Appl. Phys. Lett. 78, 1979. Copyright (2001) with permission from American Institute of Physics*

from a recombination of localized excitons. The absorption peak energies both in ZnO epilayers and in MQWs are monotonically decreasing functions of temperature as was revealed in previous studies.[34,35] This can be explained in terms of narrowing of the fundamental energy gap induced by a temperature rise.

In general, when a dominant PL peak is assigned to a radiative recombination of localized excitons, its peak energy blue-shifts with increase in temperature in a range of low temperatures and red-shifts at higher temperatures. The E_{PL}^{pk} blue-shifts and continuously connects to that of free excitons due to thermal activation of localized excitons. The E_{PL}^{pk} of the free-excitonic emission is a monotonically decreasing function of temperature due to the band gap shrinkage. The temperature dependence shown in Figure 12.5(a) is, however, different from the abovementioned typical behavior. The temperature dependence of the recombination mechanism for localized excitons is thought to be closely related to the temperature variation in the decay time constant of their PL. Thus temperature dependence of PL decay times (τ_{PL}) in an MQW with a L_w of 27.9 Å is shown in Figure 12.7(c). The τ_{PL}

values exhibit nonmonotonical behavior with respect to temperature; the value increased in a low temperature range and decreased above a certain critical temperature.

The temperature dependence of the recombination mechanism for localized excitons can be explained[42] as follows: (i) For 5 K < T < 50 K, the relatively long relaxation time of excitons gives the excitons more opportunity to relax down into lower energy tail states caused by the inhomogeneous potential fluctuations before recombining. This is because radiative recombination processes dominate in this temperature range. This behavior produces a red-shift in the peak energy position with increasing temperature. (ii) For 50 K < T < 95 K, the exciton lifetimes decrease with increasing temperature. Thus, these excitons recombine before reaching the lower energy tail states. This behavior enhances a broadening of the higher-energy side emission and leads to a blue-shift in the peak energy. (iii) For 95 K < T < 200 K, further enhancement of high-energy emission components produces a new peak, as seen in Figure 12.6(a) (triangles). (iv) At temperatures above 200 K, since the excitons are less affected by the temperature-induced rapid change in their lifetimes and since the relaxation rate of the excitons increased due to an increase in the

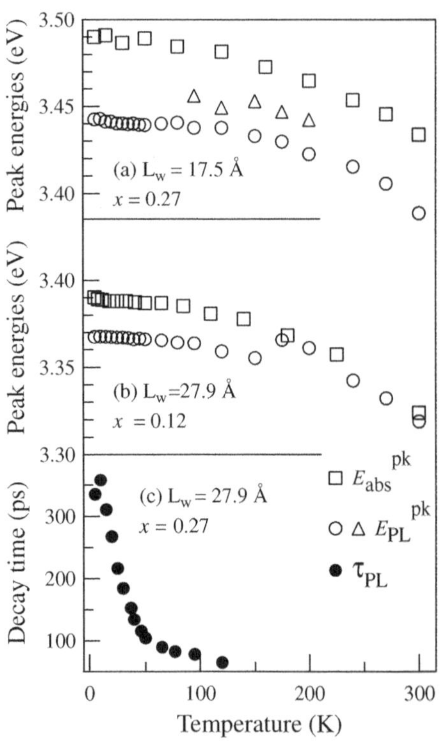

Figure 12.6 *PL (open circles and triangles) and absorption (open squares) peak positions as a function of temperature in ZnO(17.5 Å)/$Mg_{0.27}Zn_{0.73}O$ (a) and ZnO(27.9 Å)/$Mg_{0.12}Zn_{0.88}O$ (b) MQWs. (c) Temperature dependence of PL decay times, τ_{PL}, in ZnO(27.9 Å)/$Mg_{0.27}Zn_{0.73}O$ MQWs at temperatures of 5–120 K.[28] Reprinted from Makino, T., et al., Temperature dependence of near ultraviolet photoluminescence in ZnO/(Mg,Zn)O multi-quantum wells, Appl. Phys. Lett. 78, 1979. Copyright (2001) with permission from American Institute of Physics*

phonon population, blue-shift behavior becomes less pronounced. Since the energy of blue-shift is less than the temperature-induced band gap shrinkage, the peak position again exhibits a red-shift behavior. As mentioned above, the features of excitonic spontaneous emission in the well layers are sensitively affected by the dynamics of recombination of localized exciton states, which significantly vary with temperature. Readers should refer to the literature for details.[39]

12.6 Time-Resolved Luminescence in ZnO QWs

Time-resolved PL (TRPL) is a nondestructive and powerful technique that has been commonly used for the optical characterization of direct-gap semiconductors. One can measure by this time-domain spectroscopy the exciton recombination time, an important parameter related to material quality and device performance. The exciton lifetimes will vary with crystal quality, becoming longer as the quality improves or localization of the excitons become deeper.

Reynolds *et al.*[43] measured the recombination lifetime of the free excitons at 2 K in a single-crystal ZnO grown by the hydrothermal method.

In addition to the time-integrated luminescence, time-resolved luminescence of ZnO MQW were measured in order to investigate the exciton dynamics in this hetero-structure. Chia *et al.*[44] presented results of TRPL measurements for the ZnO QWs. Figure 12.7 shows

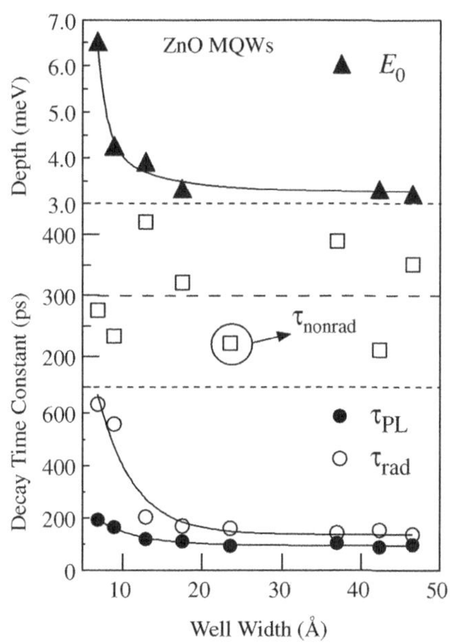

Figure 12.7 *Well-width dependences of localization depth (E_0, closed triangles) and recombination times, τ_{PL} (closed circles), τ_{rad} (open circles), and τ_{nonrad} (open squares). The solid curves are drawn as a guide to the eye. Reprinted from Makino, T., et al., Temperature dependence of near ultraviolet photoluminescence in ZnO/(Mg,Zn)O multi-quantum wells, Appl. Phys. Lett. 78, 1979. Copyright (2001) with permission from American Institute of Physics*

the TRPL data for the QWs as a function of well layer thickness. The time-resolved luminescence data fitted well with a single exponential decay curve function with a several hundreds of picoseconds decay component. Evaluated PL decay time is a monotonically decreasing function of the emission energy. In other words, decay time constants in the low energy side are longer than those in the higher counnterparts. This suggests that not only radiative recombination but also the transfer process to the tail state should contribute to the decay of the localized excitons in the QW system. Figure 12.7 shows the decay time constant of localized excitons (τ_{PL}, closed circles) and E_0 (closed triangles) as a function of the well width. When the well width becomes thin, it is found that both the τ_{PL} and E_0 become longer. The dependence of the radiative (τ_{rad}) and nonradiative (τ_{nonrad}) recombination times on the well layer thickness was deduced. Combined analysis of temperature dependences of PL decay times and of spectrally integrated PL intensity was carried out. Because the radiative recombination process is relatively dominant at a low temperature, the values of the nonradiative process have relatively large error bars. Nevertheless, it can be said that, as a general trend, the nonradiative recombination time is independent of the well width. The well width dependence of radiative recombination time can be explained as being due to the thermal release effect from the localized to delocalized states of excitons. It could be noticed that the width dependences of the localization depth of excitons and the radiative recombination times are similar with respect to each other. If the localization effect is absent in QWs, radiative recombination time is a linearly increasing function of temperature in the entire temperature range. Because of the localization effect, radiative recombination time becomes stagnant at low temperatures. It is necessary to systematically estimate the dependence of radiative recombination time on temprature for QWs having various well width in order to clarify the radiative recombination mechanism in ZnO MQWs.[40]

On the other hand, we discuss here the well width dependence of nonradiative recombination time. Normally, the nonradiative recombination becomes shorter in the MQWs in the case of thinner wells because of the degraded film qualities and carrier leakage outside the well region. However, we have not observed this tendency in this work. This may suggest that an efficient carrier confinement inside the well region could be realized in the QWs studied here. Some dispersion in the data is related to certain inaccuracy in the determination of the well width and magnesium concentration in the barrier layer, which is not taken into account.

12.7 Stimulated Emission in MQWs

This section is devoted to high excitation effects, stimulated emission and lasing, covering the range over QWs. As shown in Figure 12.8, the energies of spontaneous PL and absorption peaks are almost the same at RT.[28] The spontaneous emission spectrum was obtained under the condition of excitation using a 5 mW helium cadmium laser operated in the continuous-wave mode, while the stimulated emission spectrum was obtained under the condition of high-power excitation using a frequency-tripled mode-locked Nd:YAG laser (355 nm, 10 Hz, 15 ps) operated in the pulsed mode. The power of excitation was varied, as is described later. The agreement between the spontaneous emission and absorption peaks is an indication of the well-regulated hetero-interfaces as well as the

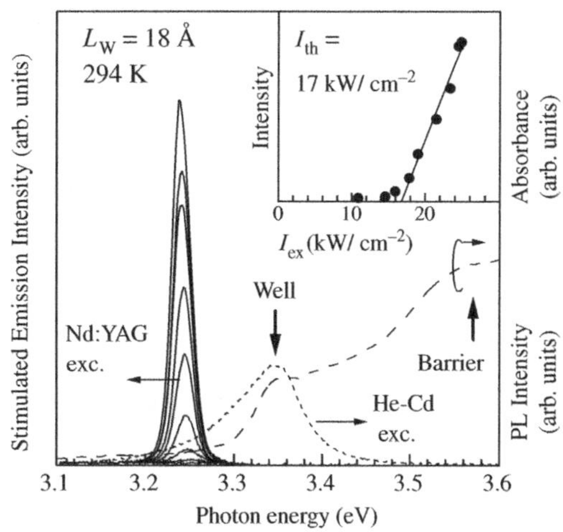

Figure 12.8 *Excitation intensity (I_{ex}) dependence of the stimulated emission spectra obtained from a ZnO/$Mg_{0.12}Zn_{0.88}O$ SL (L_w = 1.8 nm) under the condition of pulsed excitation at RT. Spontaneous PL (dotted line) under the condition of continuous-wave excitation and absorption (broken line) spectra are also shown. Inset shows the integrated intensity of the stimulated emission peak as a function of I_{ex}. Threshold intensity (I_{th}) is 17 kW cm^{-2}.[45] Ohtomo, A., et al., Room-temperature stimulated emission of excitons in ZnO/(Mg,Zn)O superlattices, Appl. Phys. Lett, 77, 2004. Copyright (2000) with permission from American Institute of Physics*

small compositional fluctuations in the barrier layers (well-depth fluctuations). Such RT PL in the MQWs grown on the sapphire substrates was not found.

We performed high-power excitation experiments to determine the characteristics of stimulated emission in ZnO MQWs in the optical pumping mode. Figure 12.8 shows stimulated emission spectra of MQWs with $x = 0.12$ and L_w = 17.5 Å that were obtained at RT. Strong and sharp emission peaks were observed at 3.24 eV above a very low threshold (I_{th} = 17 kW cm^{-2}), and their integrated intensities rapidly increased as the excitation intensity (I_{ex}) increased, as can be seen in the inset. Although such stimulated emission was not observed even at 4.2 K in the case using a sapphire substrate, RT stimulated emission in MQWs studied here was observed. This is one of the significant improvements achieved by applying lattice-matching conditions to a substrate.

The L_w dependences of peak energies of the absorption and stimulated emissions are summarized in Figure 12.9. Similar dependence of the threshold is also shown in Figure 12.9(c). The stimulated emission energy is 100 meV lower than that of the absorption peak. The lowest threshold value was 11 kW cm^{-2} in the case of L_w of 47 Å.

We tested the high-temperature operation of the stimulated emission from the viewpoint of possible applications to devices. Figure 12.10 shows the temperature dependence of the I_{stim}–I_{ex} curves of a MQW with $x = 0.26$ and L_w = 4.2 nm in the temperature range of 294–377 K. Here, I_{stim} is the intensity of the stimulated emission and I_{ex} is the excitation intensity. The threshold of the stimulated emission (I_{th}) increased gradually with increasing temperature. The inset shows the temperature dependence of I_{th} on a logarithmic scale. Characteristic temperature, which is an index of stability of threshold

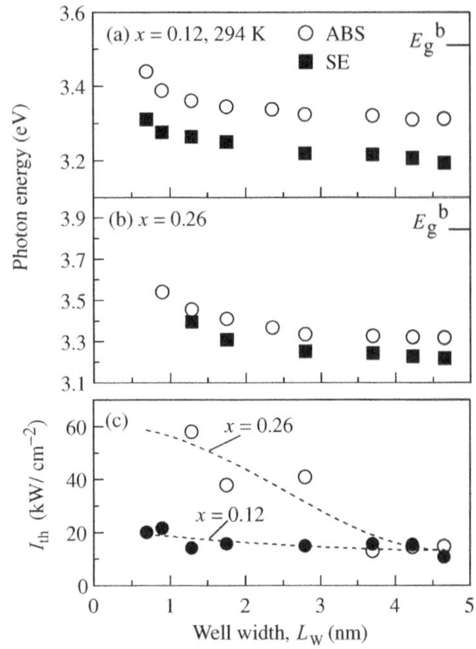

Figure 12.9 *Optical transition energies of sub-band absorption (open circles) and stimulated emission (closed squares) as a function of well layer thickness (L_W) for ZnO/$Mg_x Zn_{1-x}$ O MQWs with x = 0.12 (a) and x = 0.26 (b). Band gap energy of the barrier layers (E_g^b) is also shown. (c) L_W dependence of the stimulated emission threshold (I_{th}) in MQWs with x = 0.12 (closed circles) and x = 0.26 (open circles). Stimulated emission did not occur for the x = 0.26 films with L_W below 1 nm since the excitation energy is lower than the absorption energy.[45] Ohtomo, A., et al., Room-temperature stimulated emission of excitons in ZnO/(Mg,Zn)O superlattices, Appl. Phys. Lett, 77, 2004. Copyright (2000) with permission from American Institute of Physics*

characteristics with respect to temperature rise, was estimated to be 87 K.[45] This was significantly higher than that of a 55-nm-thick ZnO/sapphire (67 K),[46] which showed excitonic laser action with a threshold of 24 kW cm^{-2}. We speculate that this kind of improvement can be explained by the enhanced binding energy of excitons due to the quantum confinement effect.[36]

In order to clarify the mechanism of stimulated emission of these MQWs, the temperature dependence of the stimulated emission spectrum in the temperature range from low temperature to RT was estimated. Figure 12.11 shows the temperature dependence of the peak energy of the stimulated emission band in the QWs (Mg concentration of 0.12, L_Ws of 37.0 Å and 17.5 Å). For comparison, the same plot for thin films of ZnO grown on sapphire substrates is shown in Figure 12.11(a). It is already clear that the RT stimulated emission band in these thin films is what is called the P-line, which one of the typical phenomena of high-density exciton effects. Inelastic scattering between excitons (Auger-like process) gives rise to the appearance of this stimulated emission band. One of the two excitons participating in the collisional events is ionized, whereas the other is recombined radiatively after collision. At sufficiently low temperatures, peak

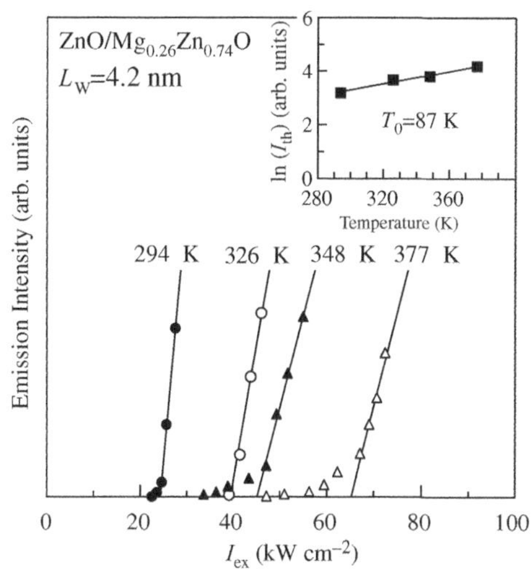

Figure 12.10 *Temperature dependence of emission intensity as a function of excitation intensity (I_{ex}) in a ZnO/$Mg_{0.26}Zn_{0.74}O$ SL ($L_w = 4.2$ nm). Inset shows the threshold intensity (I_{th}) as a function of temperature on a logarithmic scale.[45] Ohtomo, A., et al., Room-temperature stimulated emission of excitons in ZnO/(Mg,Zn)O superlattices, Appl. Phys. Lett, 77, 2004. Copyright (2000) with permission from American Institute of Physics*

energy of the relevant stimulated emission band is lower than the resonance energy of an exciton, the energy difference of which is equal to the exciton binding energy. The temperature dependence of the peak energy difference in QWs shows the same behavior as that of ZnO, as can be clearly seen in Figure 12.11. Thus, as a result of careful comparison with that of a ZnO thin film, it became clear that the mechanism of this stimulated emission is inelastic scattering processes between the excitons. Therefore, the L_w dependence of exciton binding energies can be experimentally determined from analysis of the energy position of the P_T line. As shown in Figure 12.4 (right axis), when L_w decreased, the exciton binding energy increased up to about 90 meV and exceeded $\hbar\omega_{LO}$ (72 meV). This enhancing effect is due to the quantum confinement effect. It has been revealed that these estimated exciton binding energies approximately agree with the theoretical values.[32,38] If QW structures were used, the phonon scattering process and thermo-broadening effect of excitonic linewidth can be controlled. These are favorable from the viewpoint of application.

Combinatorial concept-aided techniques adopted in the growth of our samples suppressed the variations in crystal growth conditions and hence the undesired uncertainty in the deduced spectroscopic results. In a related review article, the benefit of the combinatorial technique is described in detail.[23] For example, the well width dependence of the radiative and nonradiative recombination times of localized excitons was estimated by TRPL spectroscopy.[40,44] Well width dependence of *biexciton* binding energy was also estimated.[47] In addition, optical properties of (Cd,Zn)O/(Mg,Zn)O MQWs have not been explored so far except in our work.[40] These structures are advantageous from the viewpoint of almost perfect in-plane lattice-matching.[48,49]

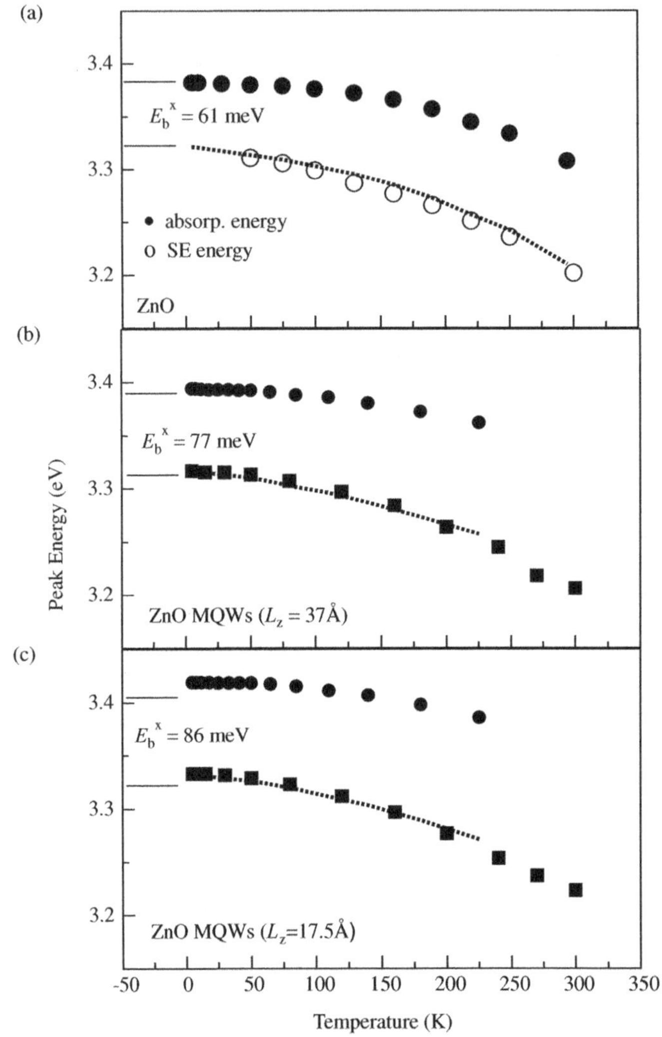

Figure 12.11 *Temperature dependence of peak energies of the P-bands (open circles and closed squares) and the free-exciton absorphon energies (closed circles) in a ZnO epitaxial layer (a) and in ZnO/$Zn_{0.88}Mg_{0.12}O$ MQWs (b), (c).*[36] *Sun, H. D.; Makino, T., et al., Stimulated emission induced by the inelastic exciton-excitonscattering in ZnO/(Mg,Zn)O multi-quantum wells, Appl. Phys. Lett. 77, 4250. Copyright (2000) with permission from American Insitute of Physics*

12.8 Summary

We present, in this chapter, a review of optical properties of ZnO MQWs grown using laser mocular beam epitaxy on the lattice-matched SCAM substrate. Compared with the cases of ZnO films and crystals, it is only recently that the basic optical properties of QWs have begun to be actively investigated. The strong stimulated emission was observed in ZnO/ZnMgO MQWs. The stimulated emission induced by exciton–exciton scattering occurred throughout the range of temperatures from 5 K to RT. At a certain excitation density, radiative recombination processes induced by both exciton–exciton scattering (P band) and electron–hole plasma (EHP) contribute to the stimulated emission at higher temperatures, but the excitation threshold of the *P* band is much lower than that of EHP emission. The demonstration of stimulated emission with excitonic origin at RT contributes to the possible of realization of a low-threshold UV or violet diode laser composed of ZnO-based MQWs. The exciton binding energies of the samples were also deduced as a function of the well layer thickness. The enhancement of exciton binding energy is due to the quantum confinement effect. This is desirable from the viewpoint of the stability of excitons at higher temperatures. ZnO is very favorable as compared with other wide-band-gap semiconductors in that this material lends itself nicely to the production of nanostructures from which functional devices have already been fabricated.

Moreover, a p-n junction is essential for the production of a current injection laser. A ZnO crystal usually shows n-type conductivity and it has not been possible to produce a p-type layer with low resistance. Such a feature of wide-band-gap semiconductors is called unipolarity in carrier doping. In this chapter, we have not discussed this problem.

Such high-quality ZnO QWs deposited on lattice-matched substrates are expected to have many applications for UV optoelectronics devices. Various applications of ZnO, such as its use in transparent electric-conduction films or a surface acoustic-wave device, are well-known. New functions of this substance have been found through research on high-quality single-crystalline thin films grown by the use of laser-assisted molecular beam epitaxy.

Acknowledgements

The authors thank their co-workers of the collaboration, Dr A. Ohtomo, Dr K. Tamura, and Professor Y. Matsumoto, for the elaborated scietific work. We would like to express our gratitude to the abovementioned researchers as well as to Dr Ngyuen Tien Tuan, Professor H. D. Sun, and Professor C. H. Chia, Institute of Physical and Chemical Research, Japan, for their expert assistance in all aspects of our research. Thanks for stimulating discussions also goes to colleagues at other institutes, namely to Professor S. Chichibu and Professor Y. Takagi. Financial support is acknowledged from JFE 21st Century Foundation, the Asahi Glass Foundation, and the inter-university cooperative program of the IMR, Tohoku University, Japan.

References

[1] H. Sakai, T. Koide, H. Suzuki, M. Yamaguchi, S. Yamasaki, M. Koike, H. Amano and I. Akasaki, *Jpn. J. Appl. Phys., Part 2* **34**, L1429 (1995).

[2] S. Nakamura and G. Fasol, *The Blue Laser Diode - GaN based Light Emitters and Lasers*, Springer-Verlag, Heidelberg, 1997.

[3] Z. K. Tang, G. K. L. Wong, P. Yu, M. Kawasaki, A. Ohtomo, H. Koinuma and Y. Segawa, *Appl. Phys. Lett.* **72**, 3270 (1998).

[4] P. Yu, Z. K. Tang, G. K. L. Wong, M. Kawasaki, A. Ohtomo, H. Koinuma and Y. Segawa, *Solid State Commun.* **103**, 459 (1997).

[5] A. Ohtomo, M. Kawasaki, T. Koida, K. Masubuchi, H. Koinuma, Y. Sakurai, Y. Yoshida, T. Yasuda, and Y. Segawa, *Appl. Phys. Lett.* **72**, 2466 (1998).

[6] M. Kawasaki, A. Ohtomo, R. Shiroki, I. Ohkubo, H. Kimura, G. Isoya, T. Yasuda, Y. Segawa and H. Koinuma, in *Extended Abstracts of the 1998 International Conference on Solid State Devices and Materials*, Business Center of the Academic Society Japan, Hiroshima, Japan, 1998, p. 356.

[7] T. Makino, Y. Segawa, M. Kawasaki, A. Ohtomo, R. Shiroki, K. Tamura, T. Yasuda, and H. Koinuma, *Appl. Phys. Lett.* **78**, 1237 (2001).

[8] A. Ohtomo, M. Kawasaki, Y. Sakurai, I. Ohkubo, R. Shiroki, Y. Yoshida, Y. Sakurai, T. Yasuda Y. Segawa and H. Koinuma, *Mater. Sci. Eng. B* **56**, 263 (1998).

[9] A. Ohtomo, M. Kawasaki, I. Ohkubo, H. Koinuma, T. Yasuda and Y. Segawa, *Appl. Phys. Lett.* **75**, 980 (1999).

[10] A. Ohtomo, R. Shiroki, I. Ohkubo, H. Koinuma and M. Kawasaki, *Appl. Phys. Lett.* **75**, 4088 (1999).

[11] T. Makino, Y. Segawa, A. Ohtomo, K. Tamura, T. Yasuda, M. Kawasaki and H. Koinuma, *Appl. Phys. Lett.* **79**, 1282 (2001).

[12] N. Kimizuka and T. Mohri, *J. Solid State Chem.* **78**, 98 (1989).

[13] A. Ohtomo, K. Tamura, K. Saikusa, T. Takahashi, T. Makino, Y. Segawa, H. Koinuma and M. Kawasaki, *Appl. Phys. Lett.* **75**, 2635 (1999).

[14] Several efforts to grow not MQWs but epilayers are currently being carried out using MgO or $MgAl_2O_4(111)$ substrates by means of plasma-assisted MBE, whose misfits are smaller than that of sapphire; Y. Chen, S.-K. Hong, H.-J. Ko, M. Nakajima, T. Yao and Y. Segawa, *Appl. Phys. Lett.* **76**, 245 (2000).

[15] T. Gruber, C. Kirchner, R. Kling, F. Reuss and A. Waag, *Appl. Phys. Lett.* **84**, 5359 (2004).

[16] B. Zhang, N. Binh, K. Wakatsuki, C. Liu, Y. Segawa and N. Usami, *Appl. Phys. Lett.* **86**, 032105 (2005).

[17] S. Sadofev, S. Blumstengel, J. Cui, J. Puls, S. Rogaschewski, P. Schäfer Y. Sadofyev and F. Henneberger, *Appl. Phys. Lett.* **87**, 091903 (2005).

[18] S. Krishnamoorthy, A. Iliadis, A. Inumpudi, S. Choopun, R. Vispute and T. Venkatesan, *Solid State Electron.* **46**, 1633 (2002).

[19] P. Misra, T. Sharma, S. Porwal and L. Kukreja, *Appl. Phys. Lett.* **89**, (2006).

[20] T. Yatsui, J. Lim, M. Ohtsu, S. An and G. Yi, *Appl. Phys. Lett.* **85**, 727 (2004).

[21] T. Bretagnon, P. Lefebvre, T. Guillet, T. Taliercio, B. Gil and C. Morhain, *Appl. Phys. Lett.* **90**, 201912 (2007).

[22] Y. Matsumoto, M. Murakami, Z. W. Jin, A. Ohtomo, M. Lippmaa, M. Kawasaki and H. Koinuma, *Jpn. J. Appl. Phys.* **38**, **L603** (1999).

[23] H. Koinuma, H. N. Aiyer and Y. Matsumoto, *Sci. Tech. Adv. Mater.* **1**, 1 (2000).

[24] A. Ohtomo, T. Makino, K. Tamura, Y. Matsumoto, Y. Segawa, Z. K. Tang, G. K. L. Wang, H. Koinuma and M. Kawasaki, *Proc. SPIE* 3941, 70 (2000).

[25] T. Makino, G. Isoya, Y. Segawa, C. H. Chia, T. Yasuda, M. Kawasaki, A. Ohtomo, K. Tamura, Y. Matsumoto and H. Koinuma,in *Proceedings of the First International Conference on Combinatorial and Composition Spread Techniques in Material and Device Development, San Jose*, edited by G. Jabbour, SPIE, Bellingham, 2000, Vol. 3941, p. 28.

[26] M. Ohtani, T. Fukumura, M. Kawasaki, K. Omote, T. Kikuchi, J. Harada, A. Ohtomo, M. Lippmaa, T. Ohnishi, D. Komiyama, R. Takahashi, Y. Matsumoto and H. Koinuma, *Appl. Phys. Lett.* **79**, 3594 (2001).

[27] M. Ohtani, T. Fukumura, M. Kawasaki, K. Omote, T. Kikuchi, J. Harada and H. Koinuma, *Appl. Phys. Lett.* **80**, 2066 (2002).

[28] T. Makino, N. T. Tuan, H. D. Sun, C. H. Chia, Y. Segawa, M. Kawasaki, A. Ohtomo, K. Tamura and H. Koinuma, *Appl. Phys. Lett.* **77**, 975 (2000).

[29] E. Mollwo, *Semiconductors: Physics of II–VI and I–VII Compounds, Semimagnetic Semiconductors*, Landolt–Börnstein New Series, edited by O. Madelung, M. Schulz and H. Weiss, Springer, Berlin, 1982, Vol. 17, p. 35.

[30] I. I. Gol'dman and V. Krivchenokov, *Problems of Quantum Mechanics*, Addison-Wesley, Reading, MA, 1961, p. 60.
[31] K. Hümmer, *Phys. Status Solidi B* **56**, 249 (1973).
[32] G. Coli and K. K. Bajaj, *Appl. Phys. Lett.* **78**, 2861 (2001).
[33] P. Lautenschlager, M. Garriga, S. Logothetidis and M. Cardona, *Phys. Rev. B* **35**, 9174 (1987).
[34] T. Makino, C. H. Chia, N. T. Tuan, Y. Segawa, M. Kawasaki, A. Ohtomo, K. Tamura and H. Koinuma, *Appl. Phys. Lett.* **76**, 3549 (2000).
[35] H. D. Sun, T. Makino, N. T. Tuan, Y. Segawa, M. Kawasaki, A. Ohtomo, K. Tamura and H. Koinuma, *Appl. Phys. Lett.* **78**, 2464 (2001).
[36] H. D. Sun, T. Makino, N. T. Tuan, Y. Segawa, Z. K. Tang, G. K. L. Wong, M. Kawasaki, A. Ohtomo, K. Tamura and H. Koinuma, *Appl. Phys. Lett.* **77**, 4250 (2000).
[37] N. T. Pelekanos, J. Ding, M. Hagerott, A. V. Nurmikko, H. Luo, N. Samarth and J. K. Furdyna, *Phys. Rev. B* **45**, 6037 (1992).
[38] H. D. Sun, T. Makino, Y. Segawa, M. Kawasaki, A. Ohtomo, K. Tamura and H. Koinuma, *J. Appl. Phys.* **91**, 1993 (2002).
[39] T. Makino, N. T. Tuan, H. D. Sun, C. H. Chia, Y. Segawa, M. Kawasaki, A. Ohtomo, K. Tamura, M. Baba, H. Akiyama, T. Suemoto, S. Saito, T. Tomita and H. Koinuma, *Appl. Phys. Lett.* **78**, 1979 (2001).
[40] T. Makino, N. T. Tuan, Y. Segawa, C. H. Chia, M. Kawasaki, A. Ohtomo, K. Tamura and H. Koinuma, *Appl. Phys. Lett.* **77**, 1632 (2000).
[41] C. Gourdon and P. Lavallard, *Phys. Status Solidi B* **153**, 641 (1989).
[42] C. Yong-Hoon, B. D. Little, G. H. Gainer, J. J. Song, S. Keller, U. K. Mishra and S. P. DenBaars, *MRS Int. J. Nitride. Semicond. Res.* **4S1**, G2.4 (1999).
[43] D. C. Reynolds, D. C. Look, B. Jogai, C. W. Litton, W. Harsch and G. Cantwell, *Phys. Rev. B* **60**, 2340 (1999).
[44] C. H. Chia, T. Makino, Y. Segawa, M. Kawasaki, A. Ohtomo, K. Tamura and H. Koinuma, *J. Appl. Phys.* **90**, 3650 (2001).
[45] A. Ohtomo, K. Tamura, M. Kawasaki, T. Makino, Y. Segawa, Z. K. Tang, G. Wong, Y. Matsumoto and H. Koinuma, *Appl. Phys. Lett.* **77**, 2204 (2000).
[46] A. Ohtomo, PhD thesis, Tokyo Institute of Technology, 2000.
[47] H. D. Sun, T. Makino, Y. Segawa, M. Kawasaki, A. Ohtomo, K. Tamura and H. Koinuma, *Appl. Phys. Lett.* **78**, 3385 (2001).
[48] C. Klingshirn, *Phys. Status Solidi B* **244**, 3027 (2007).
[49] T. Makino, Y. Segawa, M. Kawasaki and H. Koinuma, *Semicond. Sci. Technol.* **20**, S78 (2005).

Index

2DHG *see* two-dimensional hole gas

a-plane SCVT ZnO substrates 185
AAM *see* anodic alumina membranes
ab initio theory 14–15, 21–2
AB_0 *see* antibonding
absorption spectroscopy 275
acceptor impurities 129–31
acoustic mode deformation potential scattering 64
acoustoelectric properties 306–8
AFM *see* atomic force microscopy
aluminum doping
  GaN/ZnO heteroepitaxy 226–7, 232, 253–4
  hydrothermal growth methods 210
  impurities and native defects 125, 138
  room temperature stimulated emission 332, 333–5, 346–7
  ultraviolet-range devices 302–3
aluminum nitride 222
aluminum/platinum contacts 18–19
aluminum/zinc oxide diodes 103–7
ambient gas effects 289
ammonium doping
  GaN/ZnO heteroepitaxy 227
  hydrothermal growth methods 204
  impurities and native defects 147
amorphous GaN films 223, 232–3
anisotropies
  fundamental properties 11
  impurities and native defects 154, 165–6
annealing temperatures
  electrical properties 69–70
  impurities and native defects 138–40, 146
  optical properties 39–40, 43–5, 55
  Schottky contacts 91
  vapor growth methods 177–8
anodic alumina membranes (AAM) 295–6
antibonding (AB_0) sites 126
antimony doping 73–4, 130, 153
antisites 113, 120–1
arsenic doping
  electrical properties 73–4
  impurities and native defects 130, 135, 145–53
  room temperature stimulated emission 281
atomic force microscopy (AFM)
  GaN/ZnO heteroepitaxy 226–7, 229–30, 235, 237–8
  hydrothermal growth methods 208
  room temperature stimulated emission 334
  Schottky contacts 91, 100–1
  ultraviolet-range devices 296
  vapor growth methods 183–5
autoclaves 194–7

B3LYP hybrid functional 116–17
back-etching 227
band bending 89–90, 109
band gap engineering
  fundamental properties 19–20, 25
  heterovalent heterostructures 22
  homovalent heterostructures 20–2
band gap renormalization
  GaN/ZnO heteroepitaxy 258
  optical properties 51–6
  room temperature stimulated emission 266
band structure 2–5, 7–8
BC *see* bond-center
beryllium doping 281, 286
Bohr radius 51
Boltzmann transport equation 62–4
bond-center (BC) sites 126
boron doping 180–1

Zinc Oxide Materials for Electronic and Optoelectronic Device Applications, First Edition.
Edited by Cole W. Litton, Donald C. Reynolds and Thomas C. Collins.

Bose–Einstein condensation 261, 282
bound excitons 5–6
closely spaced donor–acceptor pairs in ZnO 55–7
free-carrier screening and band gap renormalization 51–6
impurities and native defects 136–45, 154, 158, 165–6
magnetic resonance spectroscopy 154, 158, 165–6
optical properties 38–58
optical spectroscopy 136–45
photoluminescence mechanisms of ZnO and GaN 46–50
polarization effects 40–1
rotator states 44–6
two-electron transitions 42–4
bound magnetic polarons 24
breakdown strength 1–2
Brillouin zone 3
Brook–Herring formula 64

C-V *see* capacitance–voltage
cadmium doping 272–3
cadmium sulfide 39
cadmium telluride 39, 131
cadmium-zinc oxide alloys 20–2
capacitance–voltage (*C–V*) characteristics 90, 96, 257
carbon nanotubes (CNT) 295–6
carrier concentration 211–12
carrier lifetime 320–1
carrier mobility 320
cathodoluminescence (CL)
depth-resolved 92, 98–101, 105, 108, 179
GaN/ZnO heteroepitaxy 239–43, 261
temperature-dependent 151, 178–9
CBE *see* charge-balance equation
CBM *see* conduction-band minimum
CCD *see* charge-coupled devices
charge transport 108–10
charge-balance equation (CBE) 64–6
charge-coupled devices (CCD) 274
chemical potentials 114–15
chemical vapor deposition (CVD)
impurities and native defects 150–1, 163
ultraviolet-range devices 296, 318, 322–3
see also metal organic chemical vapor deposition
chemical vapor transport and condensation (CVTC) 318
chemical vapor transport (CVT) 171–86
circular MSM photoconductors 297, 302
CL *see* cathodoluminescence
CNT *see* carbon nanotubes
co-doping
electrical properties 15–16, 75
native point defects 130–1
combinatorial techniques 345
conductance bands (CB)
structure 2–5, 7–8
ultraviolet-range devices 294–5, 323
conducting surface layers 66
conduction-band minimum (CBM) 117–20, 127
copper doping 129
crevice flaws 205
crystal-field parameters 33–4
crystal growth methods 194–9
crystallinity
GaN/ZnO heteroepitaxy 233
hydrothermal growth methods 190, 200–5, 212–13, 217
vapor growth methods 175–6
Curie temperature 2
Curie–Weiss law 23
current–voltage (*I–V*) characteristics
fundamental properties 17–19, 22
GaN/ZnO heteroepitaxy 253–5, 257
Schottky contacts 90–1, 93–7, 105, 108
ultraviolet-range devices 290, 296, 299–300, 302–3, 316, 322
CVD *see* chemical vapor deposition
CVT *see* chemical vapor transport
CVTC *see* chemical vapor transport and condensation

DAP *see* donor–acceptor pair
DBR *see* distributed Bragg reflectors
Debye length 51
deep level (DL) emission intensity 92–3
deep level transient spectroscopy (DLTS) 102–3, 291
defect localized state (DLS) 293
defect pair spectra
bound excitons 38–46
free electrons 34–5
defects
extended imperfections 205–8
GaN/ZnO heteroepitaxy 221–2, 241–3

hydrothermal growth methods 190, 205–8, 211
migration 121–5
optical properties 49–50
Schottky contacts 92–3, 97–101, 110
ultraviolet-range devices 293–4, 300
see also native defects; native point defects
density functional theory (DFT)
electrical properties 68, 69
native point defects 114, 115–18
optical properties 54
depth-resolved cathodoluminescence (DRCL) 92, 98–101, 105, 108, 179
DFT *see* density functional theory
diamond films 61
differential insertion losses 311–12
differential reflectivity spectra 245–6
dilute magnetic semiconductors (DMS) 22–4
diode lasers *see* laser diodes
dipole–dipole interactions 154, 159
dislocations 205
distributed Bragg reflectors (DBR) 259–60
DL *see* deep level
DLS *see* defect localized state
DLTS *see* deep level transient spectroscopy
DMS *see* dilute magnetic semiconductors
donor impurities 125–29
donor–acceptor pair (DAP) transitions
GaN/ZnO heteroepitaxy 240–1, 245–7
impurities and native defects 136–7, 145–8, 151–3, 163–4
optical properties 48–9, 55–7
doping
acceptor states 70–5
donor states 66–69
electrical properties 66–75
fundamental properties 2
hydrothermal growth methods 204
impurities and native defects 113, 125–31, 135, 145–53, 164
optical properties 47
ultraviolet-range devices 291, 300
vapor growth methods 180–1
see also n-type semiconductors; p-type semiconductors
DRCL *see* depth-resolved cathodoluminescence
drift velocity 11, 13

ECR *see* electron cyclotron resonance
EELS *see* electron energy loss spectroscopy
effective mass approach (EMA) 137
EHP *see* electron–hole plasmas
EL *see* electroluminescence
electrical properties 10–19, 61–86
acceptor states and p-type doping 70–5
annealing temperatures 69–70
conducting surface layers 66
donor states and n-type doping 66–69
GaN/ZnO heteroepitaxy 249–2, 253–5, 257
Hall effect analysis 62–6, 69–1, 75–7
historical development of ZnO semiconductors 61–2
hydrogen diffusion 69–70
hydrothermal growth methods 210–13
impurities and native defects 68, 71, 143–5
intrinsic electronic transport 10–11
n-type doping and donor levels 11–13
p-type doping and dopability 13–17, 22
photoconductivity 76–78
Schottky barriers and ohmic contacts 17–19, 25
single-band conduction 62–5
substitutional acceptors 72–5
substitutional donors 69
two-band mixed conduction 65–6
two-layer model 66–7
vapor growth methods 177–8
electroluminescence (EL)
GaN/ZnO heteroepitaxy 253, 255, 258
heterovalent heterostructures 22
electron cyclotron resonance (ECR) 15
electron drift velocity 11, 13
electron energy loss spectroscopy (EELS) 278
electron–hole plasmas (EHP) 265, 267, 269–70, 272, 347
electron nuclear double resonance (ENDOR) 128, 137–38, 154, 158, 164
electron paramagnetic resonance (EPR)
electrical properties 68, 72
hydrothermal growth methods 214
impurities and native defects 117–19, 122–4, 128, 138, 153–7, 162, 164–6
electron-trap centers 291
EMA *see* effective mass approach
emission spectra
bound excitons 40
free electrons 30–2, 36
room temperature stimulated emission 277–9, 281
ENDOR *see* electron nuclear double resonance

EPD *see* etch pit density
EPR *see* electron paramagnetic resonance
etch channels 205
etch pit density (EPD) 208
etching methods 215–17
exciton–electron scattering 265, 266
exciton–exciton scattering 265, 266, 268–70, 272, 281
exciton–phonon interactions 336–7
excitonic localization 337–41
extended imperfections 205–8
extended metal/zinc oxide diodes 108–10
external magnetic field effects 6–8

Fabry–Perot modes 265, 274, 276, 281
Fermi level
  native point defects 115–16, 118, 120, 127
  pinning 107
FESEM *see* field emission scanning electron microscopy
field effect transistors (FET)
  fundamental properties 2
  room temperature stimulated emission 331
  ultraviolet-range devices 286, 314–15
field emission scanning electron microscopy (FESEM) 322–4
fine structure 155–7, 159
fluid dynamics 196–7
fluorine doping 125
formation energy 114
Fourier transform infrared photocurrent (FTIR-PC) spectroscopy 291
free-carrier screening 51–5
free excitons 5–6
  free-carrier screening and band gap renormalization 51–2
  optical properties 29–36, 38, 46–8, 51–2, 57
  photoluminescence mechanisms of ZnO and GaN 47–8
  rotator states 34–5
  strain splitting 35–6
FTIR-PC *see* Fourier transform infrared photocurrent
full width at half maximum (FWHM)
  GaN/ZnO heteroepitaxy 175–6, 226–8, 232, 243–5, 247–8
  hydrothermal growth methods 207
  room temperature stimulated emission 336–7

g-shifts 161–2, 165
gallium arsenide
  impurities and native defects 151
  optical properties 39–45
  ultraviolet-range devices 286–7
gallium doping
  hydrothermal growth methods 210
  native point defects 125
  room temperature stimulated emission 281
gallium nitride
  compositional analysis 230–1
  electrical properties 61, 75, 249–52, 253–5, 257
  free-carrier screening and band gap renormalization 51–6
  fundamental properties 4, 13–14, 25
  growth of GaN/ZnO 222–30
  heteroepitaxy 221–64
  heterojunction LEDs 253–8
  hybrid devices 252–61
  hydrothermal growth methods 214
  microcavities 259–1
  native point defects 130–1
  optical properties 46–55, 58, 237–49, 253
  optical spectroscopy 138, 142, 149–50
  room temperature stimulated emission 273, 282, 332
  structural analysis 232–6
  surface properties 226–31, 235–7
  ultraviolet-range devices 286–7, 294, 306
  vapor growth methods 185
  yellow PL band 47–50
GDMS *see* glow discharge mass spectrometry
generalized gradient approximation (GGA) 115–16
glow discharge mass spectrometry (GDMS) 176, 209
gold/nickel contacts 19
gold/zinc oxide diodes 87–89, 91, 94–5, 97, 100, 102–7, 110

Hall effect analysis 62–6
  conducting surface layers 66
  GaN/ZnO heteroepitaxy 249–52, 258
  hydrogen diffusion 69–71
  hydrothermal growth methods 211–12
  impurities and native defects 143–5
  n-type doping and donor levels 11–13
  photoconductivity 75–7
  single-band conduction 62–5

two-band mixed conduction 65–6
two-layer model 66–7
vapor growth methods 177–8
Hedin–Lundquist parameterization 54
Heisenberg Spin Hamiltonian 22
heteroepitaxy 221–64
band gap engineering 20, 25
compositional analysis 230–1
electrical properties 249–52, 253–5, 257
growth of GaN/ZnO 222–30
heterojunction LEDs 253–58
hybrid devices 252–61
microcavities 259–61
optical properties 237–49, 253
structural analysis 232–6
surface properties 226–30, 235–7
heterojunction LEDs 253–58
heterovalent heterostructures 22
high-resolution transmission electron microscopy (HR-TEM) 298
hole-capture model 288
homoepitaxial ZnO films
a-plane SCVT ZnO substrates 185
impurities and native defects 140–2
substrate preparation 182–3
vapor growth methods 181–5
homovalent heterostructures 20–2
hopping transport 109
HR-TEM *see* high-resolution transmission electron microscopy
HSE hybrid functional 116–17
HVPE *see* hydride vapor phase epitaxy
hybrid devices
GaN/ZnO heteroepitaxy 252–61
heterojunction LEDs 253–8
microcavities 259–61
$ZnO/LiNbO_3$ photodetectors 311–14
hydride vapor phase epitaxy (HVPE)
gallium nitride 47
GaN/ZnO heteroepitaxy 222, 224–5, 228, 239, 250–1, 253, 261
hydrogen diffusion
electrical properties 69–70, 75
impurities and native defects 125–29, 150
optical properties 57
Schottky contacts 88–90, 99
hydrothermal growth methods 189–220
advantages and disadvantages 190
characterization 190, 200–5
crystallinity 190, 200–5, 212–13, 217
electrical properties 210–13
etching and polishing 215–17
extended imperfections and surface studies 205–8
growth rates 200–5, 208
impurities 190, 199, 204–5, 208–10
industrial scale ZnO crystal growth 197–9
kinetics of growth 200–5
microcrystalline ZnO powder 194
optical properties 200–1, 213–15
phase stability 191–3
properties of bulk ZnO 205–17
single ZnO crystal growth 194–7
solubility of ZnO in aqueous media 190–2
surface cleaning 215–17
techniques 194–9
thermodynamics of growth 190–3
ultraviolet-range devices 291–3
hyperfine interactions 153–7, 165–6

I–V *see* current–voltage
ICP *see* inductively-coupled plasma
IDT *see* interdigital
impurities
bound excitons with shallow donors 136–45, 154, 158–9
deep level effects 158–61, 164–5
extrinsic acceptors 161–6
GaN/ZnO heteroepitaxy 241–3
hydrothermal growth methods 190, 199, 204–5, 208–10
intrinsic acceptors 166
magnetic resonance spectroscopy 135, 153–66
morphological effects 204–5
native point defects 114–15, 121, 125–31
nitrogen and arsenic doping 135, 145–53
optical spectroscopy 135–53
Schottky contacts 92–3, 98
ultraviolet-range devices 300
vapor growth methods 176
see also doping
index of refraction 9
indium doping
GaN/ZnO heteroepitaxy 226–7, 232–4, 247–48, 255–7
hydrothermal growth methods 210, 212
native point defects 125
vapor growth methods 180–1
indium phosphide 34, 40, 107, 209

inductively-coupled plasma mass spectrometry (ICP-MS) 209, 227
inductively-coupled plasma reactive ion etching (ICP-RIE) 289
industrial scale crystal growth 197–9
inelastic exciton–exciton scattering 345–6
infrared (IR) spectroscopy 69, 128
infrared (IR) transmission spectroscopy 215–16
infrared quantum cascade lasers 282
integrated surface acoustic wave (SAW) detectors 286, 305–14, 325
  hybrid ZnO/$LiNbO_3$ photodetectors 311–14
  multilayer ZnO photodetectors 308–11
  principles 305–8
interdigital (IDT) MSM photoconductors 297, 302–6, 311
interface charge transfer 89
internal photoemission spectroscopy (IPS) 90
intrinsic electronic transport 10–11
IPS *see* internal photoemission spectroscopy
IR *see* infrared
isoelectronic dopants
  native point defects 131
  ultraviolet-range devices 291
ITO *see* tin-doped indium oxide

Jahn–Teller distortions 166

Kelvin force probe microscopy (KFLM) 100–1
killer defects 15

LAC *see* level anticrossing
laser diodes (LD)
  GaN/ZnO heteroepitaxy 221
  historical development 61–2
  room temperature stimulated emission 280–1, 331
  vapor growth methods 180–1
laser-assisted molecular beam epitaxy (laser-MBE)
  room temperature stimulated emission 267, 271–2, 346–7
  ultraviolet-range devices 293
lasers
  cavity formation 274–5, 277
  microstructural lasers 275–8
  nanowire lasers 279–80, 281
  powder lasers 276–7, 278–9
  room temperature stimulated emission 265–6, 274–82
LCAO *see* linear combination of atomic orbitals
LD *see* laser diodes
LDA *see* local density approximation
LED *see* light-emitting diodes
level anticrossing (LAC) 159–60
light-emitting diodes (LED)
  fundamental properties 22
  GaN/ZnO heteroepitaxy 221–2, 228–9, 252–8
  historical development 61–2
  room temperature stimulated emission 331
  Schottky contacts 87
  ultraviolet-range devices 286
  vapor growth methods 180–2
light field distributions 313–14
linear combination of atomic orbitals (LCAO) 161–2
lithium doping
  electrical properties 74–5
  hydrothermal growth methods 196, 204, 210–11, 215
  impurities and native defects 129, 142, 145, 147, 161–6
lithium niobate 311–14
LO *see* longitudinal optical
local density approximation (LDA) 115–17, 293
local vibrational modes (LVM) 149–50
localization energy 138–41, 143
localization mechanism of excitons 337–41
longitudinal optical (LO) phonons 136–7, 163–4, 336–7
Love wave mode 308
luminescence spectroscopy
  hydrothermal growth methods 214–15
  impurities and native defects 135, 138–40, 146–7, 162–4
  vapor growth methods 178–80
  see also cathodoluminescence; photoluminescence
LVM *see* local vibrational modes

magnesium doping
  band gap engineering 20–2
  electrical properties 75
  gallium nitride 48
  GaN/ZnO heteroepitaxy 253–5

optical properties 48
room temperature stimulated emission 270–2, 282, 334–6, 338–41
ultraviolet-range devices 286, 308, 315–18
magnetic circular polarized emission (MCPE) 159–60
magnetic field effects 6–8
magnetic resonance spectroscopy
bound excitons with shallow donors 154, 158–59, 165–6
deep level effects 158–61, 164–5
extrinsic acceptors 161–6
hyperfine interactions 153–7, 165–6
impurities and native defects 135, 153–66
intrinsic acceptors 166
spin-Hamiltonian formalism 153–7, 166
manganese doping 204
MBE *see* molecular beam epitaxy
MCPE *see* magnetic circular polarized emission
metal–insulator–metal (MIS) structures 286
metal organic chemical vapor deposition (MOCVD)
band gap engineering 21
electrical properties 75
gallium nitride 47
GaN/ZnO heteroepitaxy 227, 255–59
native point defects 126–28
ultraviolet-range devices 297, 301–2, 306, 318
vapor growth methods 174, 182, 184–5
metal organic vapor phase epitaxy (MOVPE) 222–3, 226–7
metal–semiconductor–metal (MSM) photoconductors 289, 297–306, 317, 325
$Mg_xZn_{1-x}O$ photodetectors 286, 308, 315–18
micro-capping 206
microcavities 259–61
microcrystalline ZnO powder 194
microstructural lasers 275–8
MIS *see* metal–insulator–metal
MOCVD *see* metal organic chemical vapor deposition
mode spacing 275–6
molecular beam epitaxy (MBE)
band gap engineering 20–1
gallium nitride 47–8
GaN/ZnO heteroepitaxy 226–28, 249, 253, 259
room temperature stimulated emission 267, 269–2, 274, 333, 346–7
ultraviolet-range devices 293–4, 317, 318
vapor growth methods 182, 184
morphological effects 203, 204–5
Mott density 265
MOVPE *see* metal organic vapor phase epitaxy
MQW *see* multiple quantum wells
MSM *see* metal–semiconductor–metal
multiple quantum wells (MQW)
excitation intensity 342–4, 345
exciton–phonon interactions 336–7
experimental details 333
GaN/ZnO heteroepitaxy 255–58
localization mechanism of excitons 337–41
optical properties 332–3
quantum confinement 333–6
room temperature stimulated emission 271–2, 281, 331–49
time-resolved photoluminescence 341–2

N bands 266, 267, 270
n-type semiconductors
acceptor states 73–4
band gap engineering 22
donor levels 11–13
donor states 66–69
electrical properties 61–2, 64–69
fundamental properties 2, 11–13, 18–19
GaN/ZnO heteroepitaxy 255
historical development 61–2
impurities and native defects 113, 124, 129, 145
ohmic contacts 18–19
photoconductivity 77–78
room temperature stimulated emission 281, 347
Schottky contacts 89, 97
ultraviolet-range devices 291, 294–5, 300, 302–3
vapor growth methods 177, 185
nanocrystals 278–79
nanoscale resonant tunneling devices 331
nanowire (NW) lasers 279–80, 281
nanowire (NW) photodetectors
experimental results 321–3
noise characteristics 323–5
photoconductive gain 318–23
theoretical background 318–21
ultraviolet-range devices 293, 295–6, 318–15

native defects
  bound excitons with shallow
    donors 136–45, 154, 158–9, 165–6
  deep level effects 158–61, 164–5
  extrinsic acceptors 161–6
  fundamental properties 15
  hydrothermal growth methods 211
  intrinsic acceptors 166
  magnetic resonance spectroscopy 135, 153–66
  nitrogen and arsenic doping 135, 145–53
  optical spectroscopy 135–53
native point defects 113–34
  acceptor impurities 129–31
  acceptor states 71
  antisites 113, 120–21
  co-doping 130–1
  defect migration 121–5
  definitions 113–14
  density functional theory 114, 115–18
  donor impurities 125–9
  donor states 68
  doping 113, 125–31
  electrical properties 68, 71
  electron paramagnetic resonance 117–19, 122–4, 128
  green luminescence 124–5
  isoelectronic impurities 131
  optical properties 49–50
  self-interstitials 113, 119–20, 122
  theoretical framework 114–15
  vacancies 113, 117–19, 121
natively p-type ZnO 14
near band edge (NBE) emission intensity
  GaN/ZnO heteroepitaxy 240–41
  hydrothermal growth methods 213–16
  Schottky contacts 92–3, 98, 101–2
  vapor growth methods 179
negative photoconductivity 295–6
NEP *see* noise equivalent power
neutral donor defect pair spectra 40–1
nickel doping 204
nitrogen doping
  electrical properties 72
  impurities and native defects 129–30, 135, 145–53, 161–6
  ultraviolet-range devices 291
noise equivalent power (NEP) 323–5
normal solubility 191–2
NW *see* nanowires
ODEPR *see* optically detected electron paramagnetic resonance
ODMR *see* optically detected magnetic resonance
ohmic contacts 17–19, 25
optical microscopy 316
optical properties 5–10, 29–61
  annealing temperatures 39–40, 43–5, 55
  band gap renormalization 51–5
  bound excitons 38, 39–58
  closely spaced donor–acceptor pairs in ZnO 55–8
  defect pair spectra 33, 38, 40–41
  external magnetic field effects 6–8
  free and bound excitons 5–6
  free-carrier screening 51–5
  free excitons 29–36, 38, 46–8, 51–2, 58
  gallium nitride 46–57, 58
  GaN/ZnO heteroepitaxy 237–49, 253
  green PL band in ZnO 50, 58
  hydrothermal growth methods 200–1, 213–15
  photoluminescence mechanisms of ZnO and GaN 46–50
  polarization effects 41–2
  room temperature stimulated emission 265, 271–2, 275–7, 332–3
  rotator states 34, 44–6
  spatial resonance dispersion 9–10
  strain field 8–9
  strain splitting of free excitons 35–6
  two polar faces of ZnO 36–8
  two-electron transitions 41–3
  vapor growth methods 178–80
  yellow PL band in GaN 48–50, 58
optical spectroscopy
  bound excitons with shallow
    donors 136–45
  impurities and native defects 135–53
  nitrogen and arsenic doping 135, 145–53
optically detected electron paramagnetic resonance (ODEPR) 119, 124
optically detected magnetic resonance (ODMR) 158, 160–3, 166
oxygen plasma treatments 289–90
oxygen vacancies 117–19

P bands 266, 267, 269–70, 344–7
p-i-n heterojunctions
  GaN/ZnO heteroepitaxy 255, 257

ultraviolet-range devices 286
P-MBE *see* plasma-assisted molecular beam epitaxy
p-n junctions
electrical properties 61
fundamental properties 22
GaN/ZnO heteroepitaxy 253–5, 257
native point defects 129–30
ultraviolet-range devices 286, 325
p-type semiconductors
acceptor states 70–5
band gap engineering 19, 22
dopability 13–17, 22
electrical properties 61–2, 64–6
fundamental properties 2, 13–19
GaN/ZnO heteroepitaxy 252, 256–58
historical development 61–2
impurities and native defects 113, 118, 120–21, 129–31, 145, 150–3, 164
photoconductivity 77–8
room temperature stimulated emission 265–6, 281–2, 347
Schottky barriers and ohmic contacts 17–19, 25
Schottky contacts 89
ultraviolet-range devices 286, 325
palladium/zinc oxide diodes 89, 94, 96–8, 102–3, 105, 110
PAS *see* positron annihilation spectroscopy
Paschen–Back limit 8
PC *see* photoconductivity
persistent photoconductivity (PPC) 76–7, 293–5
phosphorus doping
electrical properties 73–4
impurities and native defects 130, 152
ultraviolet-range devices 291
photo-Hall analysis 288
photoconductive detectors 286, 297–301
photoconductive wireless UV detectors (UV-SAW) 286, 305–14, 325
hybrid $ZnO/LiNbO_3$ photodetectors 311–14
multilayer ZnO photodetectors 308–11
principles 305–8
photoconductivity (PC) 76–78
gain in ZnO nanowires 318–3
negative 295–6
persistent 293–5
ultraviolet-range devices 288–96, 318–23
photoluminescence (PL)
bound excitons 39, 44
free-carrier screening and band gap renormalization 53–6
free excitons 29–31, 38
GaN/ZnO heteroepitaxy 242–9, 261–2
heterovalent heterostructures 22
hydrothermal growth methods 211, 213–16
impurities and native defects 136–7, 140–4, 146–8, 151–3, 160–2
mechanisms of ZnO and GaN 46–50
room temperature stimulated emission 275, 334–6, 338–42
two polar faces of ZnO 36–8
ultraviolet-range devices 294
vapor growth methods 178
physical vapor deposition 279–80
physical vapor transport (PVT) 171–4
piezoelectric surfaces 1, 306–9, 325
PL *see* photoluminescence
plasma-assisted molecular beam epitaxy (plasma-MBE)
band gap engineering 20–1
room temperature stimulated emission 269–270, 272, 274
ultraviolet-range devices 317
platinum/zinc oxide diodes 87–89, 94, 105, 110
PLD *see* pulsed laser vapor deposition
polar-optical mode scattering 64
polariton lasers 259, 261, 282
polarity of ZnO 102–3
polarization effects 39–40
polishing methods 215–17
polycrystalline GaN films 223, 232–3
polycrystalline ZnO films
fundamental properties 1, 25
hydrothermal growth methods 191
room temperature stimulated emission 276–7, 278–9
ultraviolet-range devices 288, 290–91, 297, 301
positron annihilation spectroscopy (PAS) 68, 71, 105
powder lasers 276–7, 278–9
PPC *see* persistent photoconductivity
pulsed laser vapor deposition (PLD)
GaN/ZnO heteroepitaxy 222–4, 228–9, 233–4, 250, 255–8
room temperature stimulated emission 277

pulsed laser vapor deposition (PLD) (*Continued*)
  ultraviolet-range devices 291, 316
purity *see* impurities
PVT *see* physical vapor transport

quantum confinement 333–6
quantum wells (QW)
  GaN/ZnO heteroepitaxy 255–8
  room temperature stimulated emission 270, 271–3, 281
  see also multiple quantum wells
Quasi-cubic model 3–5, 33
QW *see* quantum wells

radio frequency plasma-enhanced chemical vapor deposition (RF-PECVD) 227
radio frequency plasma-enhanced molecular beam epitaxy (RF-PEMBE) 253
radio frequency (RF) magnetron sputtering 289, 312, 314
radio frequency (RF) sources 15–16
Raman spectroscopy 148–9, 248–9
Rayleigh wave mode 308–10
RBS *see* Rutherford backscattering spectroscopy
reactive ion etching (RIE) 289
reflection high energy electron diffraction (RHEED) 224, 229, 232–4
reflection spectra 30–3
reflectivity spectra 245–7
relaxation time approximation (RTA) 62–4
remote oxygen plasma (ROP) 91–3, 97–100, 102–3, 108
remote plasma-enhanced laser-induced chemical vapor deposition (RPE-LICVD) 228
remote plasma-enhanced metal organic chemical vapor deposition (RPE-MOCVD) 21
resistivity 212
resonant tunneling devices 331
retrograde solubility 191–2
RF *see* radio frequency
RF-PECVD *see* radio frequency plasma-enhanced chemical vapor deposition
RF-PEMBE *see* radio frequency plasma-enhanced molecular beam epitaxy
RHEED *see* reflection high energy electron diffraction
RIE *see* reactive ion etching
rocking curves 175–6, 183–5, 207, 216–17, 226, 234–5
Γ point *see* rotator states
room temperature stimulated emission 265–84
  bulk ZnO 267
  emission mechanisms 266–7
  epitaxial layers 267–70
  excitation intensity 342–3, 345
  exciton–phonon interactions 336–7
  experimental details 333
  historical development 266–73
  laser diodes 280–1
  localization mechanism of excitons 337–41
  microstructural lasers 275–8
  nanowire lasers 279–80
  optical properties 265, 271–2, 275–7, 332–3
  powder lasers 276–7, 278–9
  quantum confinement 333–6
  quantum wells 270, 271–3, 281, 331–49
  superlattices 270, 343
  time-resolved photoluminescence 341–2
  ZnCdO/ZnO structures 272–3
  ZnMgO/ZnO structures 270–2
  ZnO lasers 265–6, 274–82
ROP *see* remote oxygen plasma
rotator states
  bound excitons 44–6
  free electrons 34–5
  valence and conductor bands 3–4, 5–6
RPE-LICVD *see* remote plasma-enhanced laser-induced chemical vapor deposition
RPE-MOCVD *see* remote plasma-enhanced metal organic chemical vapor deposition
RTA *see* relaxation time approximation
Rutherford backscattering spectroscopy (RBS) 230–1

SAW *see* surface acoustic wave
scandium doping 332, 333–5, 346–7
scanning electron microscopy (SEM) 151
  GaN/ZnO heteroepitaxy 237, 241–2, 260
  room temperature stimulated emission 280
  ultraviolet-range devices 298, 303, 322–4
Schottky contacts 17–19, 25, 87–112
  charge transport 108–10
  chemical influences 103–8
  early Schottky barrier studies 90

extended metal/zinc oxide diodes 108–10
historical development 87–90
impurities and native defects 92–3, 97–101, 110
polarity of ZnO 102–3
recent studies 91–2
surface cleaning 91–3
surface preparations 93–7, 109
ultraviolet-range devices 286, 301–5
ZnO surface effects 88–90
Schottky–Mott model 17–18
SCVT *see* seeded chemical vapor transport
secondary ion mass spectrometry (SIMS)
GaN/ZnO heteroepitaxy 230
hydrothermal growth methods 209–10
impurities and native defects 138, 145, 148–50
seed veils 205
seeded chemical vapor transport (SCVT) 172–86
seeded physical vapor transport (SPVT) 172
seeded vapor phase transport (SVPT) 12–13
self-compensation 113
self-diffusion 121–2
self-interstitials 113, 119–20, 122
self-seeding 172
SEM *see* scanning electron microscopy
semi-insulating behavior 211
Sezawa wave mode 308–12
SFIT *see* slanted finger interdigital transducers
silver doping 303–4
silver/zinc oxide diodes 95–7
SIMS *see* secondary ion mass spectrometry
single crystal growth methods 194–9, 291–3
single-band conduction 62–5
slanted finger interdigital transducers (SFIT) 312–14
sodium doping 145, 161–6
SOG *see* spin-on-glass
solubility of ZnO in aqueous media 190–2
spatial resonance dispersion 9–10
spatially directed emission 265
spin-Hamiltonian formalism 153–7, 166
spin-on-glass (SOG) films 323–7
spin-orbit interactions
free electrons 32
fundamental properties 4–5, 8
impurities and native defects 162
spintronics 22–4
SPVT *see* seeded physical vapor transport
Stark effect 8
stimulated emission *see* room temperature stimulated emission
strain field 8–9
strain splitting of free excitons 35–6
substitutional acceptors 72–5, 129–30
substitutional donors 69, 126–9
superlattices 270, 343
supersaturation 202
surface acoustic wave (SAW) detectors 286, 305–14, 325
hybrid ZnO/$LiNbO_3$ photodetectors 311–14
multilayer ZnO photodetectors 308–11
principles 305–8
surface functionalization 322–3
surface modified carrier lifetime 320–1
surface modified carrier mobility 320
surface passivation 289–90, 322–3
surface preparations
hydrothermal growth methods 215–17
Schottky contacts 91–7, 109
surface properties
GaN/ZnO heteroepitaxy 226–31, 235–7
hydrothermal growth methods 205–8, 215–17
kinetics 202
polarity 291–3
Schottky contacts 88–90
surface trapping mechanism 321
SVPT *see* seeded vapor phase transport
synchrotron white-beam X-ray topography (SWBXT) 206

tantalum/zinc oxide diodes 87–88, 95, 97, 105–6
tellurium doping 291
TEM *see* transmission electron microscopy
temperature-dependent cathodoluminescence 151, 178–9
TES *see* two-electron satellite
thin-film transistors (TFT) 286, 314–15
time-resolved photoluminescence (TRPL) 294, 341–2, 345
tin-doped indium oxide (ITO) 62
titanium/gold contacts 19
TM *see* transition metal
transient photoconductivity 288
transition metal (TM) ions 2, 17, 22–4

transmission electron microscopy (TEM)
  GaN/ZnO heteroepitaxy 229–1, 237, 252
  impurities and native defects 151
  room temperature stimulated emission 280
  ultraviolet-range devices 298
transmission spectroscopy 237–39
transparent thin film transistors (TTFT)
  electrical properties 62
  Schottky contacts 87
trap level effects 290–1, 294
TRPL *see* time-resolved photoluminescence
TTFT *see* transparent thin film transistors
tunneling 99, 109
two-band mixed conduction 65–6
two-dimensional hole gas (2DHG) 16
two-electron satellite (TES) recombinations 137, 140
two-electron transitions 42–3
two-layer model 66–7

ultrahigh vacuum remote plasma-enhanced chemical vapor deposition (UHV-RPECVD) 259–60
ultrahigh vacuum (UHV) 88, 91
ultraviolet (UV)-range devices 285–329
  ambient gas effects 289
  doping 291
  field-effect phototransistors 286
  fundamental properties 13–14, 19, 20–1, 25
  GaN/ZnO heteroepitaxy 253, 255
  historical development 285–6
  hybrid ZnO/$LiNbO_3$ photodetectors 311–14
  hydrothermal growth methods 189–90, 215
  material parameters 286–8
  metal–insulator–metal structures 286
  $Mg_xZn_{1-x}O$ photodetectors 286, 308, 315–18
  nanowire photodetectors 293, 295–6, 318–25
  negative photoconductivity 295–6
  noise characteristics 323–5
  oxygen plasma treatments 289–90
  p-n junctions and p-i-n photodiodes 286
  persistent photoconductivity 293–5
  photoconductive detectors 286, 297–301
  photoconductivity in ZnO 288–97, 318–23
  principles and mechanisms 285, 288–9, 305–8, 318–21
  Schottky barrier photodetectors 286, 301–5
  spectral response and device parameters 304–5, 323–4
  surface acoustic wave detectors 286, 305–14, 325
  surface passivation 289–90, 322–4
  surface polarity 291–2
  thin-film transistors 286, 314–15
  trap level effects 290–1, 294
UV-SAW *see* photoconductive wireless UV detectors

vacancies 113, 117–19, 121
valence band maximum (VBM) 115–18, 126–7, 130
valence bands (VB)
  structure 2–5, 7–8
  ultraviolet-range devices 294–5
vapor cooling condensation method 255–6
vapor growth methods 171–87
  a-plane SCVT ZnO substrates 185
  characterization 175–80
  chemical vapor transport 171–86
  crystallinity 175–6
  electrical properties 177–8
  homoepitaxial ZnO films 181–5
  in situ doping 180–1
  optical properties 178–80
  physical vapor transport 171–4
  purity characterization 176
  substrate preparation 182–3
  transport theory and growth data 172–5
vapor–liquid–solid (VLS) process 279–80
VBM *see* valence band maximum
Vegard's law 315
vertical etch channels 205
VLS *see* vapor–liquid–solid

wide band gap semiconductors
  electrical properties 61
  fundamental properties 1, 14
  GaN/ZnO heteroepitaxy 234
  native point defects 115
  ultraviolet-range devices 285–6
wurtzite
  fundamental properties 1, 3–4, 11, 20
  hydrothermal growth methods 200

optical properties 35, 41
room temperature stimulated emission 331

X-ray diffraction (XRD)
GaN/ZnO heteroepitaxy 225–9, 232–5
impurities and native defects 143
room temperature stimulated emission 333
ultraviolet-range devices 298
X-ray photoelectron spectroscopy (XPS)
electrical properties 72, 74
GaN/ZnO heteroepitaxy 225
Schottky contacts 91, 99
X-ray rocking curves 175–6, 183–5, 207, 216–17, 226, 234–5
X-ray topography 206–8
XPS *see* X-ray photoelectron spectroscopy
XRD *see* X-ray diffraction

yttrium-stabilized zirconia (YSZ) 226

Zeeman interactions 30, 153
zero-phonon line (ZPL) 147, 163–4
zinc antisites 113, 120–1
zinc blende 3–4
zinc-implanted ZnO 56–8
zinc interstitials 113, 119–20, 122
zinc selenide 39, 281–2
zinc vacancies 121
ZPL *see* zero-phonon line

Printed and bound by CPI Group (UK) Ltd, Croydon, CR0 4YY

24/09/2026

14969324-0002